T0198286

Data-Driven Fluid Mechanics

Data-driven methods have become an essential part of the methodological portfolio of fluid dynamicists, motivating students and practitioners to gather practical knowledge from a diverse range of disciplines. These fields include computer science, statistics, optimization, signal processing, pattern recognition, nonlinear dynamics, and control. Fluid mechanics is historically a big data field and offers a fertile ground for developing and applying data-driven methods, while also providing valuable shortcuts, constraints, and interpretations based on its powerful connections to basic physics. Thus, hybrid approaches that leverage both methods based on data as well as fundamental principles are the focus of active and exciting research.

Originating from a one-week lecture series course by the von Karman Institute for Fluid Dynamics, this book presents an overview and a pedagogical treatment of some of the data-driven and machine learning tools that are leading research advancements in model-order reduction, system identification, flow control, and data-driven turbulence closures.

Miguel A. Mendez is Assistant Professor at the von Karman Institute for Fluid Dynamics, Belgium. He has extensively used data-driven methods for post-processing numerical and experimental data in fluid dynamics. He developed a novel multi-resolution extension of POD which has been extensively used in various flow configurations of industrial interest. His current interests include data-driven modeling and reinforcement learning.

Andrea Ianiro is Associate Professor at Universidad Carlos III de Madrid, Spain. He is a well-known expert in the field of experimental thermo-fluids. He has pioneered the use of data-driven modal analysis in heat transfer studies for impinging jets and wall-bounded flows with heat transfer. He extensively applies these techniques in combination with advanced measurement techniques such as 3D PIV and IR thermography.

Bernd R. Noack is National Talent Professor at the Harbin Institute of Technology, China. He has pioneered the automated learning of control laws and reduced-order models for real-world experiments as well as nonlinear model-based control from first principles. He is Fellow of the American Physical Society and Mendeley/Web-of-Science Highly Cited Researcher with about 300 publications including 5 books, 2 US patents and over 120 journal publications.

Steven L. Brunton is Professor at the University of Washington, USA. He has pioneered the use of machine learning to fluid mechanics in areas ranging from system identification to flow control. He has an international reputation for his excellent teaching and communication skills, which have contributed to the dissemination of his research through textbooks and online lectures.

Data-Driven Fluid Mechanics

Combining First Principles and Machine Learning

Based on a von Karman Institute Lecture Series

Edited by

MIGUEL A. MENDEZ
von Karman Institute for Fluid Dynamics, Belgium

ANDREA IANIRO
Universidad Carlos III de Madrid

BERND R. NOACK
Harbin Institute of Technology, China

STEVEN L. BRUNTON
University of Washington

CAMBRIDGE
UNIVERSITY PRESS

Shaftesbury Road, Cambridge CB2 8EA, United Kingdom

One Liberty Plaza, 20th Floor, New York, NY 10006, USA

477 Williamstown Road, Port Melbourne, VIC 3207, Australia

314–321, 3rd Floor, Plot 3, Splendor Forum, Jasola District Centre, New Delhi – 110025, India

103 Penang Road, #05–06/07, Visioncrest Commercial, Singapore 238467

Cambridge University Press is part of Cambridge University Press & Assessment, a department of the University of Cambridge.

We share the University's mission to contribute to society through the pursuit of education, learning and research at the highest international levels of excellence.

www.cambridge.org
Information on this title: www.cambridge.org/9781108842143

DOI: 10.1017/9781108896214

First published 2023

Printed in the United Kingdom by TJ Books Limited, Padstow Cornwall

A catalogue record for this publication is available from the British Library.

ISBN 978-1-108-84214-3 Hardback

Contents

Contributors

Mohammad Abu-Zurayk
Institute of Aerodynamics and Flow Technology, German Aerospace Center (DLR), Germany

Gianmarco Aversano
Université Libre de Bruxelles, Belgium

Philipp Bekemeyer
Institute of Aerodynamics and Flow Technology, German Aerospace Center (DLR), Germany

Steven L. Brunton
University of Washington, USA

Guy Yoslan Cornejo Maceda
CNRS, LISN, Université Paris Saclay, France

Axel Coussement
Université Libre de Bruxelles, Belgium

Giuseppe D'Alessio
Université Libre de Bruxelles, Belgium
Politecnico di Milano, Italy

Scott Dawson
Illinois Institute of Technology, USA

Stefano Discetti
Universidad Carlos III de Madrid, Spain

Arthur Ehlert
Technische Universität, Berlin, Germany
Industrial Analytics, Berlin, Germany

Daniel Fernex
École Polytechnique Fédérale de Lausanne, Switzerland

Thomas Franz
Institute of Aerodynamics and Flow Technology, German Aerospace Center (DLR), Germany

Stefan Görtz
Institute of Aerodynamics and Flow Technology, German Aerospace Center (DLR), Germany

Andrea Ianiro
Universidad Carlos III de Madrid, Spain

Javier Jiménez
Universidad Politécnica Madrid, Spain

Alex Kuhnle
University of Cambridge, United Kingdom

François Lusseyran
CNRS, LISN, Université Paris Saclay, France

Mohammad R. Malik
Université Libre de Bruxelles, Belgium

Miguel A. Mendez
von Karman Institute for Fluid Dynamics, Belgium

Marek Morzyński
Poznań University of Technology, Poland

Christian N. Nayeri,
Technische Universität Berlin, Germany

Bernd R. Noack,
Harbin Institute of Technology, China
Technische Universität Berlin, Germany

Alessandro Parente
Université Libre de Bruxelles, Belgium

Jean Rabault
Norwegian Meteorological Institute, Norway
University of Oslo, Norway

Matteo Ripepi
Institute of Aerodynamics and Flow Technology, German Aerospace Center (DLR), Germany

Peter J. Schmid
Imperial College London, United Kingdom

Richard Semaan
Technische Universität Braunschweig, Germany

James C. Sutherland
University of Utah, USA

Kamila Zdybal
Université Libre de Bruxelles, Belgium

Preface

This book is for scientists and engineers interested in data-driven and machine learning methods for fluid mechanics. Big data and machine learning are driving profound technological progress across nearly every industry, and they are rapidly shaping research into fluid mechanics. This revolution is driven by the ever-increasing amount of high-quality data, provided by rapidly improving experimental and numerical capabilities. Machine learning extracts knowledge from data without the need for first principles and introduces a new paradigm: use data to discover, rather than validate, new hypotheses and models. This revolution brings challenges and opportunities.

Data-driven methods are an essential part of the methodological portfolio of fluid dynamicists, motivating students and practitioners to gather practical knowledge from a diverse range of disciplines. These fields include computer science, statistics, optimization, signal processing, pattern recognition, nonlinear dynamics, and control. Fluid mechanics is historically a *big data* field and offers a fertile ground to develop and apply data-driven methods, while also providing valuable shortcuts, constraints, and interpretations based on its powerful connections to first principles physics. Thus, hybrid approaches that leverage both data-driven methods and first principles approaches are the focus of active and exciting research. This book presents an overview and a pedagogical treatment of some of the data-driven and machine learning tools that are leading research advancements in model-order reduction, system identification, flow control, and data-driven turbulence closures.

About the Book and the VKI Lecture Series

This book originated from a one-week course at the von Karman Institute for Fluid Dynamics, Belgium (VKI; `www.vki.ac.be/`). The course was hosted by the Université Libre de Bruxelles, Belgium, from February 24 to 28, 2020, in the classic VKI lecture series format. These are one-week courses on specialized topics, selected by the VKI faculty and typically organized 8–12 times per year. These courses have gained worldwide recognition and are among the most influential and distinguished European teaching forums, where pioneers in fluid mechanics have been training young talents for many decades.

The lecture series was co-organized by Miguel A. Mendez from the von Karman Institute, Alessandro Parente from the Université Libre de Bruxelles, Andrea Ianiro from the Universidad Carlos III de Madrid (Spain), Bernd R. Noack from the Harbin Institute of Technology, Shenzhen (China), and TU Berlin (Germany), and Steven L. Brunton from the University of Washington (USA).

Online Material

The book is supported by supplementary material, including codes, experimental and numerical data, exercises, and the video lectures recorded from the course. All material is hosted on the course website:

`www.datadrivenfluidmechanics.com/`

The supplementary material covers more exercises, tutorials, and practicalities than could be included in this book while preserving its conciseness. Readers interested in gaining a working knowledge on the subject are encouraged and expected to download this material, study it along with the book, and test it on their own data.

The Audience

The book is intended for anyone interested in the use of data-driven methods for fluid mechanics. We believe that the book provides a unique balance between introductory material, practical hands-on tutorials, and state-of-the-art research. While keeping the approach pedagogical, the reader is exposed to topics at the frontiers of fluid mechanics research. Therefore, the book could be used to complement or support classes on data-driven science, applied mathematics, scientific computing, and fluid mechanics, as well as to serve as a reference for engineers and scientists working in these fields. Basic knowledge of data processing, numerical methods, and fluid mechanics is assumed.

The Book's Road Map

Like the course from which it originates, this book results from the contribution of many authors. The use of machine learning methods in fluid mechanics is in its early days, and a large team of lecturers allowed the course attendees to learn from the expertise and perspectives of leading scientists in different fields.

Here we provide a road map of the book to guide the reader through its structure and link all the chapters into a coherent narrative. The book chapters can be clustered into six interconnected parts, slightly adapted from the VKI lecture series.

Part I: Motivation. This part includes the first three chapters, which introduce the motivation for data-driven techniques from three perspectives.

Chapter 1, by B. R. Noack and coauthors, opens with a tour de force on machine learning tools for dimensionality reduction and flow control. These techniques are introduced to analyze, model, and control the well-known cylinder wake problem, building confidence and intuition about the challenges and opportunities for machine learning in fluid mechanics. Chapter 2, by J. Jiménez, takes a step back and gives both a historical and a data science perspective. Most of the dimensionality reduction techniques presented in this book have been developed to identify patterns in the data, known as *coherent structures* in turbulent flows. But what are coherent structures? This question is addressed by discussing the relationship between data analysis and conceptual modeling and the extent to which artificial intelligence can contribute to these two aspects of the scientific method. Chapter 3, by S. L. Brunton, gives an overview of how machine learning tools are entering fluid mechanics. This chapter provides a short introduction to machine learning, its categories (e.g., supervised versus unsupervised learning), its subfields (regression and classification, dimensionality reduction and clustering), and the problems in fluid mechanics that can be addressed by these methods (e.g., feature extraction, turbulence modeling, and flow control). This chapter contains a broad literature review, highlights the key challenges of the field, and gives perspectives for the future.

Part II: Methods from Signal Processing. This part brings the reader back to classic tools from signal processing, usually covered in curricula crossed by experimental fluid dynamicists, although with a large variety of depth. This part of the book is motivated by two reasons. First, tools from signal processing are, and will likely remain, the first "off-the shelf" solutions for many practical problems. Examples include filtering, time-frequency analysis, and data compression using filter banks or wavelets, or the use of linear system identification and time series analysis via autoregressive methods. The second reason – and this is a central theme of the book – is that much can be gained by combining machine learning tools with methods from classic signal processing, as later discussed in Chapter 8. Therefore, Chapter 4, by M. A. Mendez, reviews the theory of linear time-invariant (LTI) systems along with their properties and the fundamental transforms used in their analysis: the Laplace, Fourier, and Z transforms. This chapter draws several parallels with more advanced techniques. For example, the use of the Laplace transform to reduce ordinary differential equations (ODEs) to algebraic equations parallels the use of Galerkin methods to reduce the Navier–Stokes equation to a system of ODEs. Similarly, there is a link between the classical Z-transform and the modern dynamic mode decomposition (DMD). Chapter 5, by S. Discetti, complements Chapter 4 by focusing on time–frequency analysis. The fundamental Gabor and continuous/discrete wavelet transforms are introduced along with the related Heisenberg uncertainty principle and multiresolution analysis. The methods are illustrated on a time series obtained from hot-wire anemometry in a turbulent boundary layer and from flow fields obtained via numerical simulations.

Part III: Data-Driven Decompositions. This part of the book consists of four chapters dedicated to a cornerstone (and rapidly growing subfield) of fluid mechanics: modal analysis. This part is mostly concerned with methods for linear dimensionality reduction, originally introduced to identify, and "objectively" define, coherent structures in turbulent flows.

Chapter 6, by S. Dawson, is dedicated to the proper orthogonal decomposition (POD), the first and most popular tool introduced in the fluid mechanics community in the 1970s. This chapter reviews the link between POD with the singular value decomposition (SVD), its essential properties (e.g., optimality, relation to eigenvalue decomposition, and generalization to weighted inner products), its practical computation on discrete data sets, and its extension to continuous systems. This chapter closes with illustrative exercises that guide the reader to practical computation. Chapter 7, by P. Schmid, is dedicated to the dynamic mode decomposition (DMD), a powerful alternative to POD introduced by P. Schmid a decade ago. This chapter reviews the derivation of DMD and its roots in dynamical systems and Koopman operator theory. The main DMD algorithm is presented along with its "sparsity promoting" variant, and this chapter is enriched by three applications to experimental and numerical data, as well as a brief outlook at new extensions and generalizations.

Chapter 8, by M. A. Mendez, presents a generalized framework for deriving, computing, and interpreting *any* linear decomposition. Modal decompositions are analyzed in terms of matrix factorization and viewed as a special case of 2D discrete transforms. This framework is used to combine multiresolution analysis via filter banks with the classic POD, and derive the multiscale POD (mPOD). The mPOD is a recent decomposition that generalized the energy-based (POD-like) and the frequency-based (DMD-like) formalism. This chapter includes several exercises and tutorials, allowing the reader to test these decompositions on experimental data. Finally, this part on modal analysis closes with Chapter 9, by A. Ianiro, with an overview of good practices and applications of modal analysis. This chapter addresses essential questions on the statistical convergence of POD, the impact of random noise, and the possibility to extract phase information about the modes even if the data is not time-resolved. Moreover, this chapter presents interesting applications of the extended POD – in which decompositions of different data sets are correlated – to experimental and numerical data.

Part IV: Dynamical Systems. This part of the book consists of four chapters dedicated to various aspects of dynamical systems. Chapter 10, by S. Dawson, gives a brief overview of linear dynamical systems and linear control. This is one of the most developed disciplines in engineering, with applications across robotics, automation, aeronautics, and mechanical systems in general. Linear techniques provide a standard approach for closed-loop control and have been successfully used in fluid flows. This chapter illustrates the main concepts (state-space representation, controllability and observability, and optimal control) and tools (root locus, pole placement, proportional, integral and derivative (PID) controllers) focusing on a specific example from fluid mechanics, namely the stabilization of a wake flow. An overview of additional control techniques and a brief literature review for flow control are also provided.

Chapter 11, by S. L. Brunton, provides an overview of nonlinear dynamical systems. This chapter introduces fundamental concepts such as flow maps, attracting sets, and bifurcations, and gives a modern perspective on the field, with its current goals and open challenges. These include recent advances in the operator-theoretic views that seek to identify a linear representation of nonlinear systems and identify dynamical systems from data, further discussed in Chapter 12. Chapter 12, also by S. L. Brunton, builds on Chapter 11 and Part III of the book to introduce several advanced topics in model reduction and system identification. This chapter opens with a review of balanced model reduction goals for linear systems and builds the required mathematical background and the fundamentals of balanced POD (BPOD). Linear and nonlinear identification tools are introduced. Among the linear identification tools, this chapter presents the eigensystem realization algorithm (ERA) and the observer Kalman filter identification (OKID). Among the nonlinear identification tools, the chapter presents the sparse identification of nonlinear dynamics (SINDy) algorithm, which leverages the LASSO regression from statistics to identify nonlinear systems from data.

This part closes with Chapter 13, by P. Schmid, providing a modern account of stability analysis of fluid flows. This chapter begins with a brief review of the classic definition of stability (e.g., Lyapunov, asymptotic, and exponential stability) and moves toward a modern formulation of stability as an optimization problem: unstable modes are those along with which the growth of disturbances is maximized. This chapter introduces a powerful, adjoint-based, iterative method to solve such an optimization and shows how to recover common stability and receptivity results from the general framework. Finally, an illustrative application to the problem of tonal noise is given.

Part V: Applications. This part of the book is dedicated to the application of data-driven and machine learning methods to fluid mechanics.

Chapter 14, by B. R. Noack and coworkers, is dedicated to reduced-order modeling. This chapter gives an overview of the classic POD Galerkin approach, reviewing the main challenges in closure and stabilization as well as classic applications. It then moves to emerging cluster-based Markov models and their possible generalization. A detailed tutorial is also provided to offer the reader hands-on experience with reduced-order modeling.

Chapter 15, by A. Parente and coworkers, focuses on the use of data-driven models for studying reacting flows. The numerical simulation of these flows is extremely challenging because of the vast range of scales involved. This chapter gives a broad overview of how machine learning techniques can help reduce the computational burden. The key challenges of high dimensionality are discussed along with an overview of dimensionality reduction methods, ranging from classic principal component analysis (PCA) to local PCA, nonnegative matrix factorization (NMF), and artificial neural network (ANN) autoencoders. The application of these tools to reduce dimensionality in the modeling of transport and chemical reactions is illustrated in a challenging test case.

Chapter 16, by S. Görtz and coworkers, is dedicated to the application of reduced-order modeling for multidisciplinary analysis and design optimization in aerodynamics. The design of an aircraft involves thousands of extremely expensive numerical simulations. This chapter shows how linear and nonlinear dimensionality reduction tools can help speed up the process. POD, cluster POD, and Isomaps, combined with nonlinear regression, are discussed and demonstrated in industrially relevant cases such as the aero/structure optimization of an entire aircraft.

Chapter 17, by B. R. Noack and coworkers, presents how machine learning revolutionizes flow control. This chapter gives first an overview of flow control, its purposes, goals, tools, and strategies. Then, two paradigms for flow control are introduced and compared. On the one hand, there are model-based approaches, rooted in first principles and our ability to derive models that predict how a system responds to inputs. On the other hand, there are model-free approaches rooted in powerful optimization strategies that can "learn" the best control laws from data, by simply interacting with the system. Cluster-based control and linear genetic programming are illustrated, and the chapter closes with a tutorial on an illustrative nonlinear benchmark problem.

Chapter 18, by J. Rabault and A. Kuhnle complements Chapter 17 with an overview of deep reinforcement learning (DRL) for active flow control. Reinforcement learning is one of the three paradigms of machine learning. Contrary to the other two (supervised and unsupervised learning), a reinforcement learning algorithm starts with no data and learns through experience, that is, via trial and error. This framework is meant to tackle decision-making processes, such as teaching a computer to play chess or drive a car or, as the authors show, to control a fluid flow. This chapter introduces the main approaches of reinforcement learning (e.g., Q-learning vs. policy gradient methods), the current research directions, and the recent applications to fluid mechanics problems. Guidelines to practically deploy DRL are given, along with a perspective for the future of the field.

Part VI: Perspectives. The book closes with Chapter 19, by J. Jiménez, with a fascinating perspective and important questions for the field. Combined with Chapter 2, this chapter explores how much the progressive synergy between machine learning and fluid dynamics, fostered by ever-increasing computational capabilities, could promote the "automation" of science and ultimately turn machines into colleagues. This, as masterfully illustrated with a simple case study, ultimately depends on whether "blind" randomized trials can be integrated in the process of formulating hypothesis, eventually giving computers the ability to ask questions rather than just providing answers.

A Note on the Notation

The reader will quickly realize that different chapters have (slightly) different notation. Among these, the same symbol is sometimes used for different purposes, and different symbols are sometimes used for the same quantities. This choice is deliberate and motivated by the wide range of intersected disciplines, each of which has well-established notations. For example, the symbol $\boldsymbol{u}$ usually denotes the actuation in control theory and the velocity field in fluid mechanics. In reinforcement learning, the actuation is denoted by a_t and called the "action" while the sensor measurement is denoted by s_t and called the "state" (it is usually denoted by $\mathbf{y}$ in control theory). Resolving these ambiguities would make it difficult for readers to link the material in this book with the literature of each field.

Moreover, each chapter represents the starting point toward more advanced and specialized literature, in which a standard notation has not yet been settled. Keeping the notation as close as possible to the cited literature helps the reader make essential connections. We hope that the reader will approach each chapter with the required flexibility, and we welcome comments, corrections, and suggestions to benefit students for the next reprint.

Part I

Motivation

1 Analysis, Modeling, and Control of the Cylinder Wake

B. R. Noack, A. Ehlert, C. N. Nayeri, and M. Morzynski

We give a tour de force through select powerful methods of machine learning for data analysis, dynamic modeling, model-based control, and model-free control. Focus is placed on a few Swiss army knife methods that have proven capable of solving a large variety of flow problems. Examples are proximity maps, manifold learning, proper orthogonal decomposition, clustering, dynamic modeling, and control theory methods as contrasted with machine learning control (MLC). In Chapters 14 and 17 of this book, the mentioned machine learning approaches are detailed for reduced-order modeling and for turbulence control. All methods are applied to a classical, innocent looking benchmark: the oscillatory two-dimensional incompressible wake behind a circular cylinder at $Re = 100$ without and with actuation. This example has the advantage of being visually accessible to interpretation and foreshadows already key challenges and opportunities with machine learning.

1.1 Introduction

Machine learning, and more generally, artificial intelligence, is increasingly transforming fluid mechanics (Brunton et al. 2020). This change is based on several trends. First, the efforts from first principles to new theoretical insights have diminishing returns after hundreds of years of theoretical research. Second, multiphysics multiscale problems of engineering interest significantly increase in complexity. Third, fluid mechanics creates increasing amounts of high-quality data from experiments, for example, participle image velocimetry, to simulations. Finally, the methods of computer science become increasingly powerful with increasing performance of computers and with continual discoveries of new algorithms and their improvements.

In this chapter, we focus on machine learning methods for three classical fields of fluid mechanics: analysis of snapshot data, dynamic reduced-order modeling, and the control of a given configuration. Following the literature, we take the transient oscillatory cylinder wake as the most simple, yet nontrivial benchmark. The flow physics is phenomenologically easy to visualize and can be conceptualized as a nonlinear oscillator. Yet, an accurate description poses already challenges. Section 1.2 describes the employed configuration and data.

Analysis may start with the search of a few features helping to categorize or explain fluid mechanics. In the case of the cylinder wake, the amplitude and phase are two such features helping to parameterize drag, lift, and even the flow with good accuracy. Section 1.3 presents two tools for this purpose. First, the *proximity map* cartographs all snapshot data in a two-dimensional plane with *classical multidimensional scaling* (CMDS) as the prominent approach. Second, an automated manifold extraction, *local linear embedding* (LLE), is presented. Both methods allow us to extract the amplitude and the phase of vortex shedding. CMDS can be applied to arbitrary even turbulent data. LLE comes with an estimate of the embedding dimension, if the dynamics is simple enough.

The analysis may continue with the search of a low-dimensional representation of the flow data – as described in Section 1.4. Two different approaches are presented. First, the flow data is represented by a data-driven Galerkin expansion with *proper orthogonal decomposition (POD)*. POD minimizes the averaged error of the expansion residual. Second, the flow data may also be represented by a small number of representative state, called *centroids*. K-means++ clustering achieves this goal by minimizing the averaged representation error between the snapshots and their closest centroids.

The dynamics may be understood by *reduced-order models* (ROM) building on such low-dimensional representations. The spectrum of ROM has a bewildering richness with a myriad of enabling auxiliary methods. Hence, an overview is postponed until Chapter 14. We focus on simple dynamical models of the cylinder wake, illustrating the analytical insights that may be gained (see Section 1.5).

The stabilization of the wake is discussed in Section 1.6. This discussion starts with a classical approach employing ROM for the derivation of the control law. A powerful model-free alternative is *MLC*, which learns the control laws in hundreds or thousands of test runs.

Finally, Section 1.7 summarizes some good practices of analysis, modeling, and control. The chapter cannot be an exhaustive state-of-the-art compendium of machine learning approaches. Instead, the presented methods can be seen as the Swiss army knife of machine learning. They are simple yet powerful and can already be applied in experimental projects with no or limited availability to first principle equations.

1.2 The Cylinder Wake: A Classical Benchmark

This section describes a classical, innocent looking benchmark of modeling and control, the two-dimensional oscillatory flow behind a circular cylinder. The first wake models were proposed over 100 years ago (von Kármán 1911), giving the von Kármán vortex street its name. The configuration and direct numerical simulation are described in Section 1.2.1. The transient flow behavior is outlined in Section 1.2.2. A sketch of the dynamics is previewed in Section 1.2.3.

1.2.1 Configuration and Direct Numerical Simulation

The two-dimensional viscous, incompressible wake behind a circular cylinder is computed. This flow is characterized by the Reynolds number $Re = UD/\nu$ where D represents the cylinder diameter, U the oncoming velocity, and ν the kinematic viscosity of the fluid. The reference Reynolds number is set to $Re = 100$, which is significantly above the onset of vortex shedding at $Re = 47$ (Jackson 1987, Zebib 1987) and also far below the onset of three-dimensional instabilities around $Re = 160$ (Zhang et al. 1995, Barkley & Henderson 1996).

In the following, all quantities are assumed to be normalized with the cylinder diameter D, the oncoming velocity U, and the density of the fluid ρ. The two-dimensional cylinder wake is described by a Cartesian coordinate system (x, y) with the origin in the cylinder center, the x-axis pointing in streamwise, and the y-axis in the transverse direction. The incompressibility condition and Navier–Stokes equations read

$$\nabla \cdot \boldsymbol{u} = 0, \tag{1.1a}$$

$$\partial_t \boldsymbol{u} + \boldsymbol{u} \cdot \nabla \boldsymbol{u} = -\nabla p + \frac{1}{Re} \Delta \boldsymbol{u}, \tag{1.1b}$$

where p represents the pressure, "∂_t" partial differentiation with respect to time, "∇" the Nabla operator, "Δ" the Laplace operator, and "$\cdot$" an inner product or contraction in tensor algebra.

The rectangular computational domain $\Omega_{\rm DNS}$ has a length and width of 50 and 20 diameters, respectively. The cylinder center has a distance of 10 diameter to the front and lateral sides. Summarizing,

$$\Omega_{\rm DNS} = \left\{ (x,y) \in \mathcal{R}^2 : x^2 + y^2 \geq 1/4 \cap -10 \leq x \leq 40 \cap |y| \leq 10 \right\}.$$

On the cylinder, the no-slip condition $\boldsymbol{u} = \boldsymbol{0}$ is enforced. At the front $x = -10$ and lateral sides of the domain $y = \pm 10$, a uniform oncoming flow $\boldsymbol{u}_\infty = (1, 0)$ is assumed. A vanishing stress condition is employed at the outflow boundary $x = 40$.

Simulations are performed with a finite-element method on an unstructured grid with implicit time integration. This solver is third-order accurate in time and second-order accurate in space. Details about the Navier–Stokes and stability solvers are described in Morzyński et al. (1999) and Noack et al. (2003). The triangular mesh consists of 59 112 elements.

The employed initial conditions are based on the unstable steady solution $\boldsymbol{u}_s$ and a small disturbance with the most unstable eigenmode $\boldsymbol{f}_1$. The steady solution is computed with a Newton gradient solver. The eigenmode computation is described in our earlier work (Noack et al. 2003). The disturbance is the real part of the product of the eigenmode and unit phase factor $e^{\mathrm{j}\phi}$. Here, "j" denotes the imaginary unit and ϕ the phase. The amplitude ε is chosen to create a perturbation with a fluctuation energy of 10^{-4}. The resulting initial condition reads

$$\boldsymbol{u}(\boldsymbol{x}, t = 0) = \boldsymbol{u}_s(\boldsymbol{x}) + \varepsilon \mathbb{R} \left\{ \boldsymbol{f}_1(\boldsymbol{x})\, e^{\mathrm{j}\phi} \right\}. \tag{1.2}$$

Sixteen initial conditions are considered. These correspond to equidistantly sampled phases $\phi \in [22.5°, 45°, \ldots, 337.5°, 360°]$. Integration is performed from $t = 0$

to $t = 200$, capturing the complete transient and post-transient state. The time step is $\Delta t = 0.1$, corresponding to roughly one 50th of the period.

1.2.2 Unforced Transients

In this section, the transients from the steady solution to periodic vortex shedding are investigated. The flow is analyzed in the observation domain

$$\Omega := \{(x, y) \in \Omega_{\mathrm{DNS}} : 5 \leq x \leq 15 \wedge 5 \leq y \leq 5\}. \tag{1.3}$$

This domain is about twice as long as the vortex bubble of the steady solution. The streamwise extent is large enough to resolve over one wavelength of the initial vortex shedding as characterized by the stability eigenmode. A larger domain is not desirable, because a small increase in wavenumber during the transient will give rise to large phase differences in the outflow region, complicating the comparison between flow states. The domain is consistent with earlier investigations by the authors (Gerhard et al. 2003, Noack et al. 2003) and similar to the domains of other studies (Deane et al. 1991).

The analysis is based on the inner product of the Hilbert space of square-integrable functions over the observation domain Ω. This inner product between two velocity fields $\boldsymbol{v}$ and $\boldsymbol{w}$ is defined by

$$(\boldsymbol{v}, \boldsymbol{w})_\Omega = \int_\Omega d\boldsymbol{x}\, \boldsymbol{v} \cdot \boldsymbol{w}, \tag{1.4}$$

where "·" denotes the Euclidean inner product. The corresponding norm of the velocity field $\boldsymbol{v}$ reads

$$\|\boldsymbol{v}\|_\Omega = \sqrt{(\boldsymbol{v}, \boldsymbol{v})_\Omega}. \tag{1.5}$$

The flow $\boldsymbol{u}$ is decomposed into a slowly varying base flow $\boldsymbol{u}^D$ and an oscillatory fluctuation $\boldsymbol{u}'$,

$$\boldsymbol{u} = \boldsymbol{u}^D + \boldsymbol{u}'. \tag{1.6}$$

The short-term averaged flow is approximated as the projection of the flow on the one-dimensional affine space containing the steady solution $\boldsymbol{u}_s$ and the post-transient mean flow $\boldsymbol{u}_0$. The superscript "D" comes from the term *distorted mean flow* of mean-field theory (Stuart 1958). In other words,

$$\boldsymbol{u}^D(\boldsymbol{x}, t) = \boldsymbol{u}_s(\boldsymbol{x}) + a_\Delta(t)\, \boldsymbol{u}_\Delta(\boldsymbol{x}), \tag{1.7}$$

with the shift-mode $\boldsymbol{u}_\Delta = (\boldsymbol{u}_0 - \boldsymbol{u}_s) / \|\boldsymbol{u}_0 - \boldsymbol{u}_s\|_\Omega$ and amplitude $a_\Delta = (\boldsymbol{u} - \boldsymbol{u}_s, \boldsymbol{u}_\Delta)_\Omega$. This definition approximates a short-term averaged flow and generalizes the notion in the stability literature where the steady solution is identified with the base flow.

The shift-mode amplitude a_Δ characterizes the mean-flow distortion while the fluctuation energy

$$\mathcal{K} := \|\boldsymbol{u}'\|_\Omega^2 / 2 \tag{1.8}$$

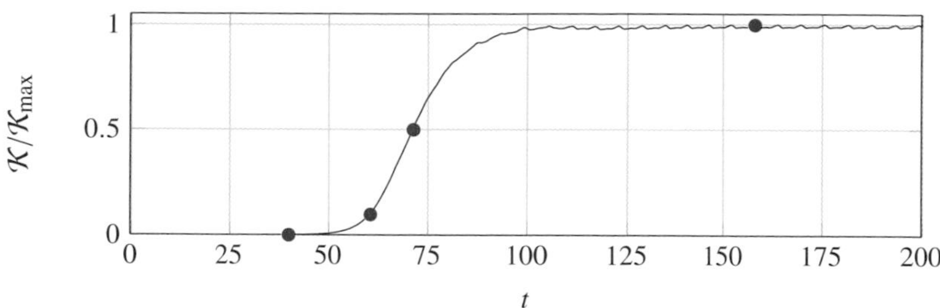

Figure 1.1 Evolution of the turbulent kinetic energy $\mathcal{K}$ with time t associated with an initial condition for $\phi = 22.5°$. The values are normalized with the maximum value $\mathcal{K}_{max}$. Red points indicate normalized fluctuation levels of 0%, 10%, 50%, and 100%. For details, see Ehlert et al. (2020).

parameterizes the fluctuation level. We also refer to $\mathcal{K}$ as *turbulent kinetic energy* (TKE) following the mathematical definition of statistical fluid mechanics, realizing that the flow is laminar, not turbulent.

Figure 1.1 displays the TKE evolution with time. The maximum TKE value $\mathcal{K}_{max}$ is used for normalization. Three dynamic phases can be distinguished. Within the first 30 convective time units D/U, the flow exhibits *linear dynamics* or exponential growth in the neighborhood of the steady solution. This exponential growth can clearly be seen in a logarithmic plot (Noack et al. 2003). In the second, *nonlinear transient phase* for $50 < t < 100$ the flow transitions from the steady solution to the limit cycle with decreasing growth rate. In the *post-transient phase* for $t > 150$, a periodic vortex shedding or, equivalently, limit-cycle dynamics is observed. The figure marks four times for TKE levels near 0%, 10% , 50%, and 100%, corresponding to the linear dynamics phase, the beginning and middle of the nonlinear transient phase and the limit cycle.

In Figure 1.2, the vorticity for the four selected time instants is shown. Positive (negative) values of vorticity are shown in red (blue) bounded by solid (dashed) lines. The three dynamic phases can be distinguished based on the closeness of vortex shedding to the cylinder and on the formation of pronounced individual vortices.

This discussion provides a basis for the time interval $[t_{min}, t_{max}]$ for snapshot selection. A lower bound $t_{min} = 40$ is chosen. This bound guarantees a TKE below 0.01% or, equivalently 10^{-4} of the asymptotic maximum value. The upper bound $t_{max} = 110$ includes few periods on the limit cycle.

1.2.3 Wake Dynamics

Figure 1.3 foreshadows a state-space picture of the transient dynamics. Near the steady solution (bottom of the paraboloid), the flow spirals outward on a plane spanned by the real and imaginary part of the unstable stability mode (bottom right). With increasing fluctuation amplitude, the Reynolds stress deforms the short-term averaged flow in the direction of the shift mode depicted middle left. This deformation leads

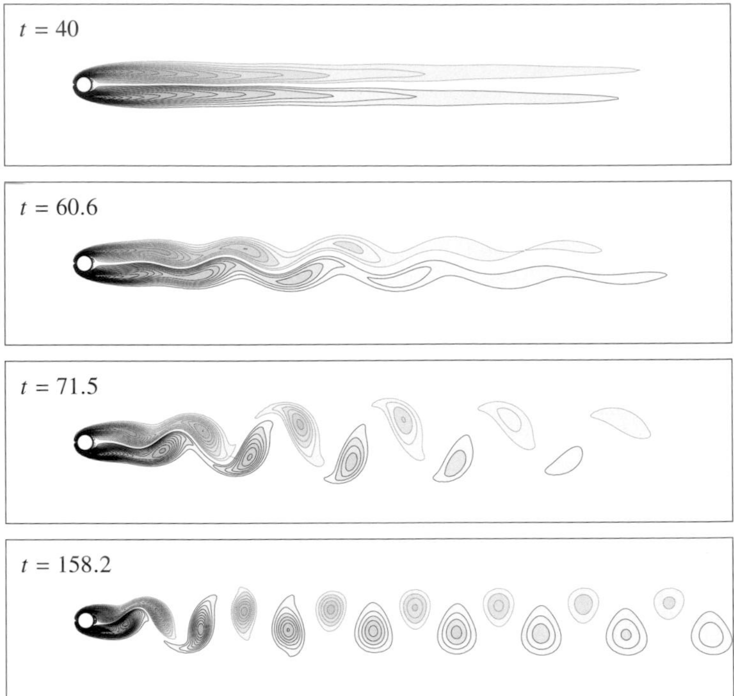

Figure 1.2 Vorticity snapshots corresponding to 0%, 10%, 50%, and 100% fluctuation level for the simulation displayed in Figure 1.1. The flow is visualized by iso-contours of vorticity with positive (negative) values marked by solid (dashed) lines and red (blue) background. The iso-contour levels and color scales are the same for all snapshots. For details, see Ehlert et al. (2020).

to a shorting of the vortex bubble and an upward motion of the fluctuation energy. The state moves outward and upward on a paraboloid until it converges against a limit cycle. The center of this limit cycle is characterized by the mean flow while the fluctuations are well approximated by the first two POD modes. In Chapters 14 and 17, this dynamics will be distilled from the data, dynamically modeled, and reversed by stabilizing control.

1.3 Cartographing the Data with Features

In this section, feature extraction is discussed. For the sake of concreteness, features are considered for an ensemble of M velocity snapshots $\boldsymbol{u}^m(\boldsymbol{x}) = \boldsymbol{u}(\boldsymbol{x}, t^m)$, $m = 1, \ldots, M$. For an oscillatory flow, amplitude and frequencies are important features that can completely or, in case of slow drifts, partially characterize the state. For a turbulent flow, features are far more challenging to design. In case of skin friction reduction of a turbulent boundary layer, features might be the position and amplitude of sweeps and ejections. In the following, two feature extraction methods are presented. First (Section 1.3.1), proximity maps via *classical multidimensional*

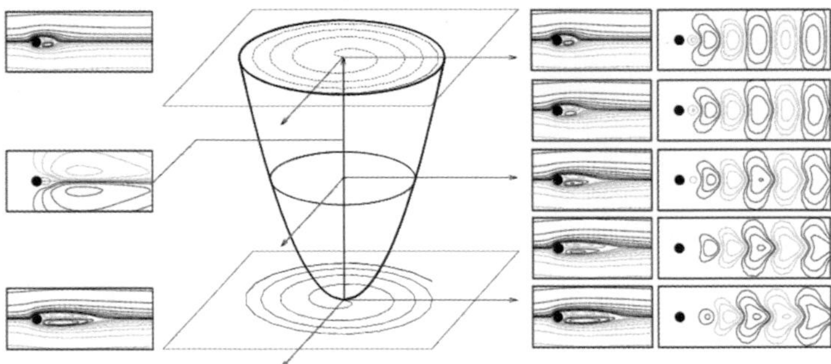

Figure 1.3 Principal sketch of the wake dynamics. The left side displays the mean flow (top), shift-mode (middle) and steady solution (bottom). The right side illustrates interpolated vortex streets on the mean-field paraboloid (middle column). The short-term averaged flows are depicted also as the streamline plots. Adapted from Morzyński et al. (2007).

scaling (CMDS) can be employed for any data. Second (Section 1.3.2), manifold extraction with *local linear embedding* (LLE) is particularly powerful for low-dimensional dynamics.

1.3.1 Proximity Maps with Multidimensional Scaling

The goal of a proximity map is to cartograph high-dimensional snapshots in a visually accessible, often two-dimensional, feature space such that neighborhood relations are preserved as much as possible. Let $\boldsymbol{\gamma}^m = \left[\gamma_1^m, \gamma_2^m\right]^{\mathrm{T}} \in \mathcal{R}^2$ (the superscript "T" denotes the transpose) with $m = 1, \ldots, M$ the two-dimensional feature vectors corresponding to the snapshots $\boldsymbol{u}^m, m = 1, \ldots, M$. In CMDS (Cox & Cox 2000), these features minimize the accumulative error of the distances between the snapshots

$$E = \sum_{m=1}^{M} \sum_{n=1}^{M} \left[\|\boldsymbol{u}^m - \boldsymbol{u}^n\| - \|\boldsymbol{\gamma}^m - \boldsymbol{\gamma}^n\|\right]^2 . \tag{1.9}$$

The translational degree of freedom is removed by requesting centered features, $\sum_{m=1}^{M} \boldsymbol{\gamma}^m = 0$. The rotational degree of freedom is fixed by requiring the first feature coordinate to have maximum variation. In general, for an N-dimensional feature space, the sum of first I variances is maximized for all $I \in \{1, \ldots, N\}$. For the invariance of the error under mirroring, however, there is no cure, as with the sign indeterminacy of POD modes and amplitudes. In fact, the resulting proximity map yields the first two POD amplitudes a_1, a_2. The resulting metric may be tailored to specific applications, for example, identifying regions with similar cost functions (Kaiser et al. 2017b). Since proximity maps are based on preserving neighborhood information, it is strongly related to LLE.

In this formulation, the features γ_1 and γ_2 coincide exactly with the POD mode amplitudes a_1 and a_2 discussed in Section 1.4.1. Hence, the proximity map is identical with the phase portrait of these POD mode coefficients. Numerous generalizations of

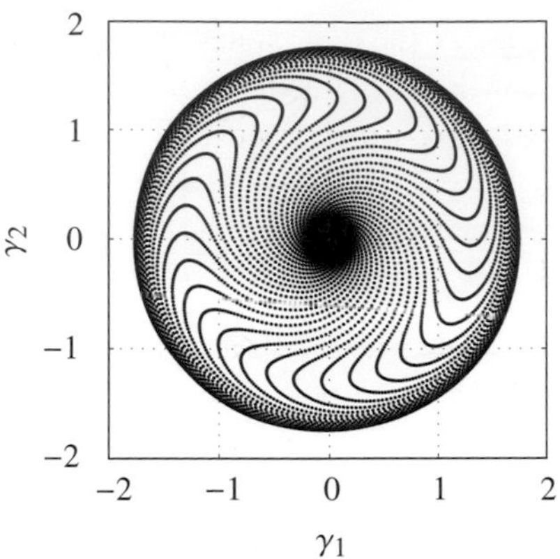

Figure 1.4 LLE of 16 cylinder wake transients. The figures displays the first two embedding coordinates $\boldsymbol{\gamma} = [\gamma_1, \gamma_2]^{\mathrm{T}}$ resulting from $K = 15$ nearest neighbors. For details, see Ehlert et al. (2020).

proximity maps have been proposed. In case of control, the error may include also the cost function to bring similarly performing states closer together (Kaiser et al. 2017).

1.3.2 Identifying the Manifold with Local Linear Embedding

LLE (Roweis & Lawrence 2000) targets to approximate M data points of a typically high-dimensional space by a low-dimensional manifold. In particular, neighboring points in the original data space remain neighbors in the low-dimensional embedding space.

The result is an optimal mapping from the snapshots $\boldsymbol{u}^m$ to N-dimensional features $\boldsymbol{\gamma}^m \in \mathcal{R}^N$. The neighborhood relation is preserved as follows. Let $i_1^m, \ldots, i_K^m$ be the indices of the K closest snapshots to the mth one. Let

$$\boldsymbol{u}^m \approx \sum_{k=1}^{K} w_{mk} \boldsymbol{u}^{i_k^m}$$

be the best approximation of the mth snapshot by its neighbors with optimized nonnegative weights w_{mk} adding up to unity. These constraints on the weights enforce a local interpolation. Then, also the feature vector can be approximated by the same expansion, $\boldsymbol{\gamma}^m \approx \sum_{k=1}^{K} w_{mk} \boldsymbol{\gamma}^{i_k^m}$. Here, K is a design parameter. It must be larger than the dimension of the manifold yet sufficiently small for the assumed locally linear behavior of the manifold. N is increased until convergence of the error is reached. Then, N denotes the dimension of the manifold. For the details, we refer to the original literature (Roweis & Lawrence 2000).

Figure 1.4 displays the LLE feature coordinates of the wake snapshot data. The origin corresponds to the steady solution $\boldsymbol{u}_s$, while the outer circle represents post-transient vortex shedding states. An analysis of the polar representation $\gamma_1 + \mathrm{j}\gamma_2 = r\exp(\mathrm{j}\theta)$ reveals that the distance r correlates well with the fluctuation energy, while θ can be considered the vortex shedding phase (Ehlert et al. 2020).

1.3.3 Other Features

Evidently, many other features can be constructed. An obvious design parameter of CMDS and LLE is the chosen distance between two snapshots, for example, the domain and the considered state variables.

Here, we mention one feature vector which is of large relevance to experiments: time-delay coordinates from sensor signals (Takens 1981). While reduced-order representations of fluid flows reduce the dimension of the state, time-delay coordinates increase the dimension of the measured signal to a level where the trajectories do not cross each other (no false neighbors) and a dynamical system can be identified. Loiseau et al. (2018) discuss the construction of velocity field manifold for the transient cylinder wake from the lift coefficient.

1.4 Low-Dimensional Representations

Low-dimensional state representations are key enablers for understanding, state estimation, dynamic modeling, and control. The starting point of this section is the ensemble of M flow snapshots $\boldsymbol{u}^m(\boldsymbol{x})$, $m = 1, \ldots, M$ (Section 1.2). A low-dimensional data-driven representation is synonymous with an *autoencoder* in machine learning. An autoencoder targets a low-dimensional parameterization of the snapshot data, say in $\mathbb{R}^N$. More precisely, an autoencoder comprises an *encoder* $\boldsymbol{G}$ from the high- or infinite-dimensional state-space to a low-dimensional feature space, for example,

$$\boldsymbol{u}^m \mapsto \boldsymbol{a}^m := \boldsymbol{G}(\boldsymbol{u}^m) \in \mathbb{R}^N, \quad m = 1, \ldots, M \tag{1.10}$$

and a *decoder* or state estimator $\boldsymbol{H}$, for example,

$$\boldsymbol{a}^m \mapsto \hat{\boldsymbol{u}}^m := \boldsymbol{H}(\boldsymbol{a}^m), \quad m = 1, \ldots, M \tag{1.11}$$

for the reconstruction of the state. Ideally, the autoencoder identifies the best possible pair of encoder $\boldsymbol{G}$ and decoder $\boldsymbol{H}$ that minimizes the in-sample error of the estimator/decoder

$$E_{in} := \frac{1}{M} \sum_{m=1}^{M} \|\hat{\boldsymbol{u}}^m - \boldsymbol{u}^m\|_\Omega^2 . \tag{1.12}$$

1.4.1 Optimal Subspaces with Proper Orthogonal Decomposition

POD can be considered as an optimal linear autoencoder onto an N-dimensional affine subspace. Let $\boldsymbol{u}_0$ be the average of the snapshot ensemble, $\boldsymbol{u}_i$, $i = 1, \ldots, N$ be the N

POD modes, and a_i be the corresponding mode coefficients. Then the encoder $\boldsymbol{G}$ of a velocity field $\boldsymbol{u}$ to the mode amplitudes $\boldsymbol{a} = [a_1, \ldots, a_N]^{\mathrm{T}}$ is defined by

$$a_i := (\boldsymbol{u} - \boldsymbol{u}_0, \boldsymbol{u}_i)_\Omega, \quad i = 1, \ldots, N, \tag{1.13}$$

while the decoder $\boldsymbol{H}$ reads

$$\hat{\boldsymbol{u}}(\boldsymbol{x}) = \boldsymbol{u}_0(\boldsymbol{x}) + \sum_{i=1}^{M} a_i \boldsymbol{u}_i(\boldsymbol{x}). \tag{1.14}$$

POD modes are an orthonormal set and sorted by energy content. The optimality condition (e.g., Holmes et al. 2012) implies a minimal in-sample representation error from (1.12). We cannot find another Nth order Galerkin expansion (more precisely, a different N-dimensional subspace) yielding a smaller error.

For the transient wake data, the most energetic POD modes $\boldsymbol{u}_i$ and the amplitude evolution a_i are displayed in Fig. 3 and 4 of Noack et al. (2016). The first two modes correspond to von Kármán vortex shedding. The third mode resolves the change of the mean field. The following modes mix different frequencies and wavelengths.

1.4.2 Coarse-Graining the Data into Bins with Clustering

The key idea of clustering is representing the snapshots by a small number, say K, of centroids $\boldsymbol{c}_k$ with $k = 1, \ldots, K$. Every snapshot $\boldsymbol{u}^m$ can be associated with its closest centroid $\boldsymbol{c}_k$. Thus, the encoder $\boldsymbol{G}$ maps the velocity field $\boldsymbol{u}$ to $k \in \{1, \ldots, K\}$, the index of the closest centroid. In other words, the encoder creates "bins" of similar snapshots. The decoder $\boldsymbol{H}$ approximates the velocity field by the closest centroid $\hat{\boldsymbol{u}} = \boldsymbol{c}_k$. The k-means algorithm aims to minimize the in-sample representation error (Arthur & Vassilvitskii 2007). For generic data, the centroids can be expected to be unique modulo the index numbering. Clustering with K bins cannot yield a lower in-sample error than a POD representation with K modes $\boldsymbol{u}_i$, $i = 0, \ldots, K-1$. Both clustering and POD span a $K-1$-dimensional subspace, but a POD expansion can interpolate states while the centroids are fixed representations of the corresponding snapshot bin.

Figure 1.5 displays the results for 10 centroids visualized with LLE features. The centroids attempt to fill the circular region. One centroid is the steady solution; seven centroids resolve the limit cycle; and a phase opposite pair of centroids represent low-amplitude oscillations.

1.4.3 Comparison and Discussion

POD and clustering are quite different autoencoders. While POD is based on a superposition of modes, centroids are representative states that cannot be superimposed. LLE can be generalized to an autoencoder. The feature vector $\boldsymbol{\gamma}$ is obtained in the encoder step. A decoder can easily be constructed, for instance, with K-nearest neighbor interpolation (Ehlert et al. 2020).

Figure 1.6 displays the reconstruction error of the three methods for the simulation data. All three methods have the largest reconstructing error in the transient phase

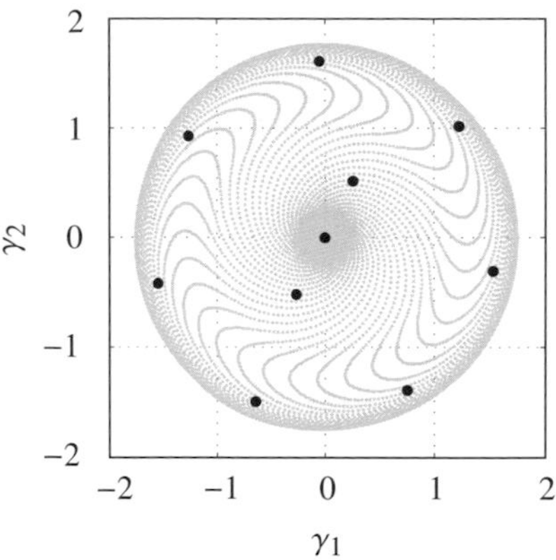

Figure 1.5 Cluster centroids localized in the LLE-based feature space. One centroid represents the steady-state solution; two resolve the opposite transient phases; and the remaining eight centroids are close to the limit cycle. For details, see Ehlert et al. (2020).

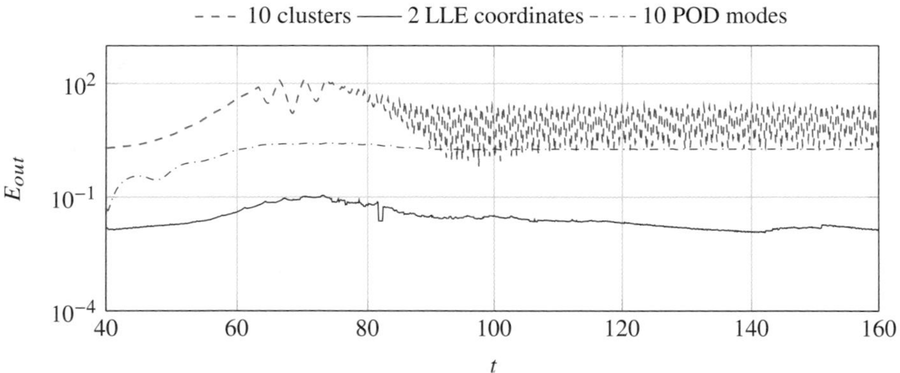

Figure 1.6 Out-of-sample error E_{out} for a new simulation trajectory at $Re = 100$. The solid line corresponds to LLE representations. The red dash-dotted curve and blue dashed curve refer to approximations with 10 centroids and 10 POD modes, respectively. For details, see Ehlert et al. (2020).

between $t = 60$ and $t = 80$. LLE significantly outperforms both POD and clustering by up to three orders of magnitudes, highlighting the two-dimensional manifold of the Navier–Stokes dynamics and a niche application of LLE. As expected, clustering performs worst lacking any intrinsic interpolation. The low error of the LLE-based autoencoder demonstrates that the dynamics is effectively two-dimensional. Yet, about 50 POD modes are necessary for a similar resolution as corroborated by Loiseau et al. (2018) for a similar manifold approximation. Evidently, POD-based representations are not efficient for slowly changing oscillatory coherent structures.

Data-driven Galerkin expansions have been optimized for a myriad of purposes. *Dynamic mode decomposition* (DMD) (Rowley et al. 2009, Schmid 2010) can extract stability modes from initial transients and Fourier modes from converged post-transient data. However, the performance for transient wakes is disappointing while a recursive DMD can keep some advantages of POD and DMD (Noack et al. 2016). Flexible state-dependent modes may significantly improve the accuracy of a low-order representation (Siegel et al. 2008, Tadmor et al. 2011, Babaee & Sapsis 2016).

1.5 Dynamic Models: From Proper Orthogonal Decomposition to Manifolds

In this section, a path to a least-order model for the transient cylinder wake is pursued. The starting point is the POD Galerkin method (Section 1.5.1). Then (Section 1.5.2), mean-field theory is employed to significantly simplify the dynamics. The simplification culminates in a manifold model with the Landau equation as dynamics (Section 1.5.3).

1.5.1 Proper Orthogonal Decomposition Galerkin Method

The traditional Galerkin method (Fletcher 1984) derives the dynamics of the Galerkin approximation (1.14) with orthonormal modes from the Navier–Stokes equations (1.1). Under weak conditions, a constant-linear-quadratic system of ordinary differential equations is obtained

$$\frac{da_i}{dt} = c_i + \sum_{j=1,\ldots,N} l_{ij} a_j + \sum_{j,k=1,\ldots,N} q_{ijk} a_j a_k. \tag{1.15}$$

These conditions may include the incompressibility of the flow, a stationary domain, stationary Dirichlet, periodic, or von Neumann boundary conditions, and smoothness of the flow. The coefficients c_i, l_{ij}, and q_{ijk} are functions of the modes and of the Reynolds number. The coefficients may also be identified from numerical solutions or experimental data (Galletti et al. 2004, Cordier et al. 2013), for instance, if the flow domain is too small or if the turbulent fluctuations are not resolved in (1.14). We refer to exquisite textbooks (Fletcher 1984, Holmes et al. 2012) for details. Schlegel and Noack (2015) have derived necessary and sufficient conditions for bounded solutions, which can be considered a requirement for a physical sound model.

1.5.2 Mean-Field Model

The Galerkin system (1.15) can be significantly simplified exploiting the manifold dynamics depicted in Figure 1.3 and derived in Section 1.5.1. The flow is dominated by a zeroth and first harmonics component (see (1.6)).

The zeroth component is on the line from the steady solution $\boldsymbol{u}_s$ to the mean flow $\overline{\boldsymbol{u}}$. This line approximates the one-period averaged flow $\boldsymbol{u}^D$ and is parameterized by the shift mode $\boldsymbol{u}_3$ and its amplitude a_3 (see (1.7)). The shift mode is effectively a backflow in the wake. The streamlines look like a fly and are depicted in Figure 1.3 (left middle subfigure).

The first harmonics represents von Kármán vortex shedding, which may be resolved by a cosine $\boldsymbol{u}_1$ and sine $\boldsymbol{u}_2$ mode, ignoring the shape deformation for a moment:

$$\boldsymbol{u}'(\boldsymbol{x},t) = a_1(t)\,\boldsymbol{u}_1(\boldsymbol{x}) + a_2(t)\,\boldsymbol{u}_2(\boldsymbol{x}). \tag{1.16}$$

The modes may be inferred from Figure 1.3 in the rightmost column. In the following, $\boldsymbol{u}_1$, $\boldsymbol{u}_2$, and $\boldsymbol{u}_3$ are assumed to be orthonormalized. The first three POD modes of the transient yield are already a good approximation of these modes.

Following mean-field arguments (Noack et al. 2003), the Galerkin system (1.15) simplifies to a self-amplified, amplitude limited oscillator,

$$da_1/dt = \sigma a_1 - \omega a_2, \qquad \sigma = \sigma_1 - \beta a_3, \tag{1.17a}$$

$$da_2/dt = \sigma a_2 + \omega a_1, \qquad \omega = \omega_1 + \gamma a_3, \tag{1.17b}$$

$$da_3/dt = \sigma_3 a_3 + \alpha\left(a_1^2 + a_2^2\right). \tag{1.17c}$$

The oscillator has three parameters σ_1, ω_1, σ_3 for linear dynamics and three parameters for α, β, γ for the weakly nonlinear effects of Reynolds stress. Intriguingly, *sparse identification of nonlinear dynamics* (SINDy) extracts precisely this sparse dynamical system from transient simulation data (Brunton, Proctor & Kutz 2016a).

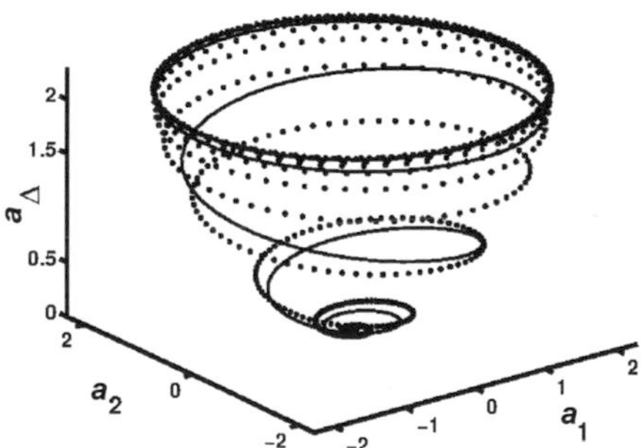

Figure 1.7 Transient dynamics of the cylinder wake from the DNS (solid line) and the mean-field Galerkin system (1.17) (dashed line). Here, $a_\Delta = a_3$. Phase portrait from Tadmor and Noack (2004).

Figure 1.7 shows that the mean-field Galerkin system (1.17) and the direct numerical simulation agree well. A detailed analysis (Tadmor & Noack 2004) quantitatively corroborates this agreement for the manifold and the temporal evolution.

1.5.3 Manifold Model

The mean-field Galerkin system can be further simplified exploiting the slaving of the shift mode to the fluctuation level. In fact, the calibrated $\sigma_3 \approx -6\sigma_1$ shows a much quicker convergence to manifold than the growth of the fluctuation. Hence, (1.17c) is well approximated by the algebraic equation

$$a_3 = \frac{\alpha_3}{|\sigma_3|}\left(a_1^2 + a_2^2\right), \tag{1.18}$$

explaining the mean-field parabola shape in the figure. Equation (1.18) in (1.17a), (1.17b) and the introduction to polar coordinates $a_1 = r\cos\theta$, $a_2 = r\sin\theta$ lead to the famous Landau equation (Landau & Lifshitz 1987) for a supercritical instability,

$$dr/dt = \sigma_1 r - \beta' r^3, \quad d\theta/dt = \omega_1 + \gamma' r^2. \tag{1.19}$$

These equations show an exponential growth by linear instability and a cubic damping by Reynolds stress and mean-field deformation. The frequency changes as well. The nonlinearity parameters β' and γ' can easily be derived from (1.17). Intriguingly, this equation is found to remain accurate even far from the onset of vortex shedding.

The resulting Landau model does not include higher harmonics. Even worse, we have assumed fixed modes $\boldsymbol{u}_1$, $\boldsymbol{u}_2$. Yet, the prominent stability mode near the steady solution is distinctly different from the POD modes characterizing the limit cycle. a_3-dependent modes can cure this shortcoming (Morzyński et al. 2006), but the model is still blind to higher harmonics. A more accurate model is based on the LLE feature coordinates $\gamma_1 = r\cos\theta$ and $\gamma_2 = r\sin\theta$ and an identified Landau equation for the dynamics of r and θ. The LLE autoencoder incorporates the mode deformation and higher harmonics (Ehlert et al. 2020). Loiseau et al. (2018) has derived such a model based on similar premises. The mean-field Galerkin model can be generalized for two (and more) frequencies (Luchtenburg et al. 2009).

1.6 Control: Model-Based and Model-Free Approaches

This section previews two flow control approaches. Section 1.6.1 follows the classical paradigm of model-based control design, while Section 1.6.2 outlines a model-free machine-learned control optimization.

1.6.1 Model-Based Control

As a control benchmark (Hinze & Kunisch 2000), we aim to stabilize the cylinder wake with a transversal volume force in the near wake (see Figure 1.8). This admittedly academic volume force actuator is surprisingly often used in the computational flow control literature and significantly simplifies the discussion. As added complexity, experimental conditions are emulated by the hot-wire measurement of the streamwise velocity component s. The goal is stabilizing sensor-based control law (see, again, Figure 1.8).

The volume force can be shown to lead to an extra term gb in the mean-field system (Gerhard et al. 2003),

$$da_1/dt = \sigma a_1 - \omega a_2 + gb, \qquad \sigma = \sigma_1 - \beta a_3, \tag{1.20a}$$

$$da_2/dt = \sigma a_2 + \omega a_1, \qquad \omega = \omega_1 + \gamma a_3, \tag{1.20b}$$

$$a_3 = \alpha' \left(a_1^2 + a_2^2\right), \qquad \alpha' = \alpha/|\sigma_3|. \tag{1.20c}$$

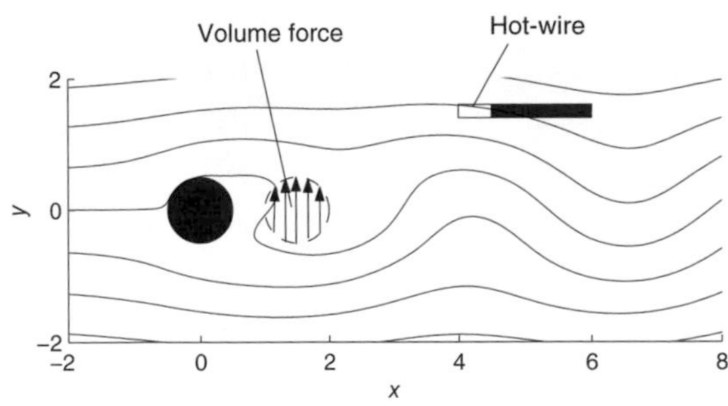

Figure 1.8 Cylinder wake configuration with volume force actuation and hot-wire sensor.

In this differential algebraic system, a_3 is slaved to the fluctuation energy (1.18). Here, b is the actuation command, for example, the induced acceleration in the circular region, while the positive gain g quantifies the effect on the dynamics and depends, for instance, on the size and location of the volume force support. Without loss of generality, the actuation term only effects a_1 by suitable rotation of the modes $\boldsymbol{u}_1$, $\boldsymbol{u}_2$.

The forced growth rate of the fluctuation energy $K = r^2/2 = (a_1^2 + a_2^2)/2$ reads

$$\frac{dK}{dt} = a_1 \frac{da_1}{dt} + a_2 \frac{da_2}{dt} = \sigma r^2 + g\, b\, a_1. \tag{1.21}$$

The fluctuation energy can be mitigated with negative actuation power $g\, b\, a_1$, that is, b has to have the opposite sign of a_1. For simplicity, a linear control law is assumed,

$$b = -k a_1, \qquad k > 0. \tag{1.22}$$

The control gain k shall ensure a forced exponential decay rate $\sigma_c < 0$ of the amplitude r. This implies with (1.22) and (1.21),

$$\frac{dK}{dt} = \sigma_c r^2 = \sigma r^2 + g\, k\, a_1^2. \tag{1.23}$$

Averaging over one period and exploiting $\overline{a_1^2} = r^2/2$ yields the control gain k and thus the control law

$$b = 2 \frac{\sigma_c - \sigma}{g} a_1. \tag{1.24}$$

The implications of this law are plausible: the higher the unforced growth rate and the higher the desired damping, the larger the volume force amplitude must become. Contrary, the larger the gain g of actuation on the dynamics, the smaller the volume force needs to be. It should be born in mind that σ is r dependent.

For the sensor-based control, the state $\boldsymbol{a} = [a_1, a_2]^{\mathrm{T}}$ is estimated from the sensor signal with a dynamic observer. The resulting stabilization effect is shown in Figure 1.9. While the model allows complete stabilization, the fluctuation energy of the DNS has been reduced by 30% in the observation region $x < 6$. The model is only partially accurate as it ignores shedding mode changes due to actuation and convection

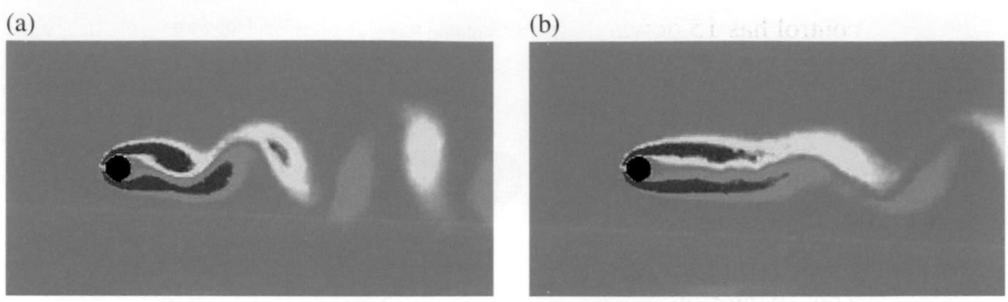

Figure 1.9 Model-based cylinder wake stabilization. (a) unforced and (b) controlled flow. Here, the vorticity field is shown: red and blue mark negative and positive vorticity, while green corresponds to potential flow.

effects. The maximum reduction of the fluctuation level is 60%, leading to a complete stabilization of the near wake and a residual fluctuation in the far wake.

1.6.2 Model-Free Machine Learning Control

In Section 1.6.1, the inaccuracy of the model has led to a reduced control performance. In model-free control, such model-based errors are avoided. Instead, a regression problem is solved: find a control law $b = K(s)$ that minimizes the cost function J, for example, the fluctuation energy. Any regression solver requires repeated tests of control laws in the full plant for the optimization problem. A linear ansatz for the control law allows for a gradient-based approach, for example, the downhill simplex method, for parameter optimization. If no structure of the control law shall be assumed, for example, in case of strongly nonlinear dynamics, MLC has proven to be very powerful. MLC is based on genetic programming as regression solver and has optimized the nonlinear control law in dozens of experiments and simulations (Noack 2019). The observed testing requires hundreds to one thousand control laws from simple single-input single-output (SISO) to complex nonlinear multiple-input multiple-output (MIMO) control.

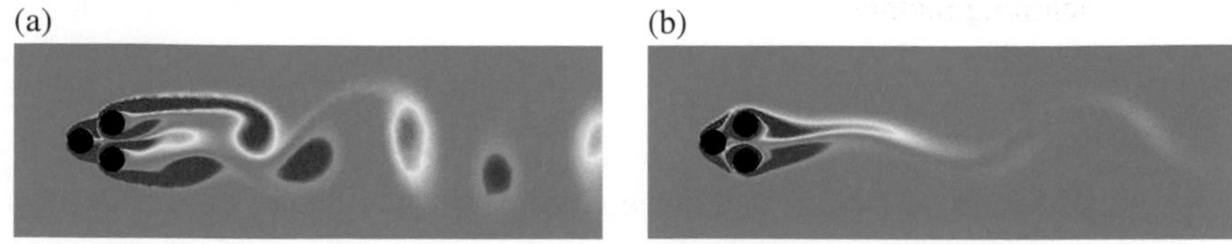

Figure 1.10 Machine learning control of the fluidic pinball with 3 cylinder rotations responding to 15 undisplayed downstream sensors. Contour levels of vorticity for the unforced flow (a) and controlled flow (b) by the control law. Solid lines and dashed lines represent respectively positive and negative vorticity. For details, see Cornejo Maceda et al. (2019).

Figure 1.10 provides a synopsis for MLC applied to the fluidic pinball configuration (Cornejo Maceda et al., (2019), (2021)). Forty-two percent reduction of the effective drag power is achieved accounting for actuation energy. The enabling MIMO feedback control has 15 downstream sensors (not displayed) and commands the rotation of the three cylinders. The mechanism is a combination of open-loop Coanda forcing and closed-loop phasor control (as in the cylinder wake example).

1.7 Conclusions

This chapter provides examples of machine learning applications in fluid mechanics. The examples are from analysis, modeling, and control of the oscillatory cylinder wake. All applications are based on regression problems, that is, finding a function that minimizes a cost. Feature extraction leads to a mapping from the flow to few coordinates optimizing some neighborhood/distance criteria. An autoencoder comprises an encoder and decoder for a low-dimensional flow representation that minimizes the representation error for the snapshot data. Dynamic modeling relies on a mapping from the state to the change of state. State estimation is a function from the measured sensor signals to the flow field. Control design leads to a function from the state or sensor signals to the actuation command minimizing a given cost. Closures can be seen as control terms that minimize the tracking error between computed and observed states. More examples can be added (Brunton & Noack 2015).

Machine learning provides powerful tools for regression solvers based on existing data ("curve fitting") or based on in situ optimization ("variational problems"). The discussed data-based regressions – classical multidimensional scaling, local linear embedding, proper orthogonal decomposition, clustering – can be used for a wide range of problems. Genetic programming is a powerful tool for solving variational problem, control design or closures, without required data but with in situ testing.

The oscillatory cylinder wake and its stabilization looks like an innocently simple benchmark. Yet, it shows already that machine learning methods need to be carefully chosen. For instance, local linear embedding does a remarkable job of compressing the data to a two-dimensional manifold. In contrast, 50 POD modes are required for a similar representation error making understanding, state estimation, dynamic modeling, and model-based control next to impossible. We will elaborate on machine learning methods and more advanced applications for reduced-order modeling (Chapter 14) and for control (Chapter 17). We highly recommend three introductory books for machine learning: Abu-Mostafa et al. (2012) as a deep introduction to how one can learn from data, Wahde (2008) for optimization solvers, and Burkov (2019) for an inspiring overview of machine learning methods.

2 Coherent Structures in Turbulence: A Data Science Perspective

J. Jiménez*

This chapter and Chapter 19 explore how far the scientific discovery process can be automated. After discussing the scientific method and its possible relation with artificial intelligence techniques, this first chapter deals with how to extract correlations and models from data, using examples of how computers have made possible the identification of structures in shear flows. It is concluded that, besides the data-processing techniques required for the extraction of useful correlations, the most important part of the discovery process remains the generation of rules and models.

2.1 Introduction

This chapter is intended to be used together with Chapter 19 of this book. Both deal with the problem of how much the process of scientific discovery can be automated and, in this sense, address the core question of the use of artificial intelligence (AI) in fluid mechanics, or in science in general. The two chapters are, however, very different. The present one discusses techniques and results that have been current in fluid mechanics for the past few decades. Specifically in turbulence research. They owe a lot to the growth of computers, and have changed the field to an extent that would have been hard to predict 40 years ago. But, during this period, computers were basically used as research tools, and the way research was done was not very different from the way it had always been.

Chapter 19 deals with one way in which the role of computers in research could develop in the near future, and will argue that the scientific method is about to undergo changes by which computers may evolve from being used as research tools to being considered minor research "participants." Since that chapter discusses nascent technology, closer to AI than what has been used up to now in fluid mechanics, the scientific results in it will necessarily be more limited than those in the present chapter. But the methods will be more interesting for being new. In this sense, the present chapter deals with the past, and Chapter 19 deals with a possible future.

We begin by specifying what science and the scientific method means for us in these two chapters. There are two meanings to the word "science." The first one is a

* This work was supported by the European Research Council under the Coturb grant ERC-2014.AdG-669505.

systematic method for discovering how things work, and the second is the resulting body of knowledge about how they work. Depending on who you ask, one aspect is considered to be more important than the other. Empiricists tend to focus on the scientific method (Poincaré 1920), and practitioners tend to be more interested in the results (Kuhn 1970). These two chapters deal mostly with methods while illustrating them with examples from real applications to fluid mechanics.

We take the view that the basic goal of the scientific method is to search for rules that can be used to make predictions. But the second important goal is to make those rules "beautiful" for the researcher (Jho 2018). This is not an arbitrary requirement because, at least for now, science depends on the work of scientists, and people work better when they like what they are working with. The logarithmic function can be defined by a computer algorithm to evaluate it, but most mathematicians would be more likely to use logarithms if they are defined as

$$d\log(t)/dt = 1/t.$$

The well-known conundrum about why mathematics is useful to the physical sciences, even when it is primarily developed by mathematicians for its aesthetic value (Wigner 1960), may have more to do with the motivation of mathematicians and physicists than with mathematics.

The classical point of view is that a prerequisite for science is the determination of causes (*Does* A cause B?), preferably complemented by mechanisms (*How* does A cause B?). Causality without mechanisms is unlikely to lead to quantitative predictions, and mechanisms without causality have a high probability of being wrong. Both causes and mechanisms have to be testable, preferably on data sets different from the ones used to train them. The ancient Greeks knew about the alternation of night and day, and imagined causes and mechanisms involving gods and chariots. They explained the facts known to them, but the Greeks did not have data from other planets, or even from other latitudes, and we know today that their mechanisms are hard to generalize. One possible characterization of these two chapters is that they explore how to determine causality in physical systems, and whether recent developments provide us with new tools for doing so.

There are several distinctions that need to be made before we begin our argument. We have mentioned the importance of causes, but research is not always geared toward them. The classical description of the scientific method is summarized in Figure 2.1, which emphasized its iterative character. Causality is encapsulated in the modeling and testing steps, S2 and S3, and the emphasis is often put on these two steps rather than on the data-gathering one, S1. In fact, the loop in Figure 2.1 often starts with step S2, and only later are hypotheses tested against observations, as in S2–S1–S3. The implied relation is not always causal. Even in data-driven research, where observations precede hypotheses, the argument is often that "if A precedes B," A is likely to be the cause of B, although it is generally understood that correlation and causation are not equivalent. A classical example is the observation of night and day mentioned earlier. The correlation between earlier days and later nights is perfect, but it does not imply causation (Mackie 1974).

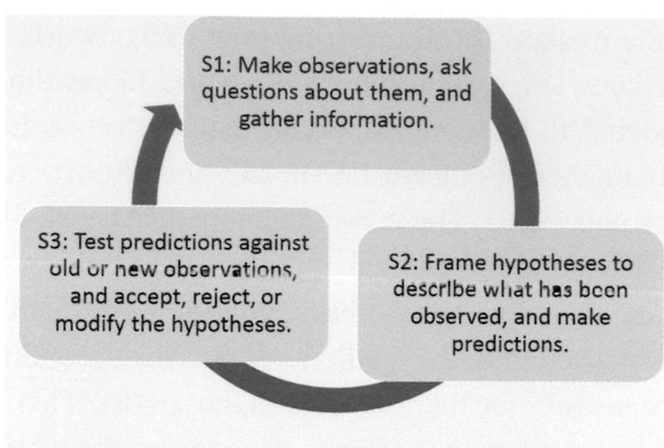

Figure 2.1 The scientific method.

Moreover, the concept of causality is not without problems, and it has been argued that it is indistinguishable from initial conditions in systems described by differential equations, such as fluid mechanics (Russell 1912). While this is true, and we could formulate science as a quest to classify initial conditions in terms of their outcome, the result may not be very informative, especially in turbulence and other chaotic systems, which lose memory of their previous evolution after a while. We will restrict ourselves here to a prescribed-time definition of cause, of the type of: "The falling tree causes the crashing noise," even if we know that the fall of the tree has its own reasons for happening, and that intermediate causes depend on the time horizon that we impose on them. Such investigations require something beyond correlations, and imply an active intervention of the observer in the generation of data. This is particularly true when we are interested in the possibility of controlling our system, instead of simply describing it. Consider the difference between formulating sub-grid models for large-eddy simulations, and devising strategies to decrease drag in a pipe.

Simplifying a lot, the difference between the two points of view, which mostly affects how and when step S1 in Figure 2.1 is undertaken, can be summarized as follows:

- *Observational science.* Searches for correlations between different observations, and generalizes them into models. It can be described as the science of prediction.
- *Interventional science*. Relies on the results of experiments in which some condition is changed by the observer, generally in the hope of uncovering "causal" (i.e., if this, then that), or "counterfactual" relations (if not this, then not that). Interventional science is the science of control.

The present chapter is oriented toward predictions, and therefore toward establishing correlations among observations, without being especially interested in causality. Chapter 19 is oriented toward causality and control, and looks very different from the present one. In the aforementioned example, Chapter 19 will want to know about the falling tree because it might be interested in doing something to prevent it from falling. The present chapter will center (although not completely) on establishing correlations between tree health and noise levels.

There is a second distinction that, although related to the previous one, is independent from it. It has do with how AI, or intelligence in general, is supposed to work (Nilsson 1998), and is traditionally divided into *symbolic* (Newell & Simon 1976) and *sub-symbolic* (Brooks 1990). Symbolic AI is the classical kind, which manipulates symbols representing real-world variables according to rules set by the programmer, typically embodied into an "expert system." For example, assuming that we agree that "vortices are localized pressure minima" (or any favorite personal definition), symbolic AI encapsulates this knowledge into a set of rules to identify and isolate the vortices. On the contrary, sub-symbolic AI is not interested in rules, but in algorithms to do things. For example, given enough snapshots of a flow in which vortices have been identified (maybe manually by the researcher), we can train a neural network to distinguish them from vortex sheets or from other flow features. The interesting result of sub-symbolic AI is not the rule, but the algorithm, and it does not imply that a rule exists. After training our system, we may not know (or care) what a vortex is, or what distinguishes it from a vorticity sheet, but we may have a faster way of distinguishing one from the other than would have been possible using only preordained physics-based rules. Copernicus and the Greeks had symbolic representations of the day-night cycle, although with very different ideas of the rules involved. Most other living beings, which can distinguish night from day and usually predict quite accurately dawn and dusk, are (probably) sub-symbolic. The classical scientific method, including causality, is firmly symbolic: we are not only interested in the result, but also in the rule. But there is a small but growing body of scientists and engineers who feel that data and a properly trained algorithm are all the information required about Nature, and that no further rule is necessary (Coveney et al. 2016, Succi & Coveney 2018). Observational science is sub-symbolic at heart, although we will argue that this point of view is incomplete even for it. Interventional science at least aims for symbolism.

In fluid mechanics, the availability of enough data to even consider sub-symbolic science is a recent development, made possible by the numerical simulations of the 1990s, which gave us for the first time the feeling that "we knew everything," and that any question that could be posed to a computer would eventually be answered (Brenner et al. 2019, Jiménez 2020a). The present chapter is partly a review of this work. Of course, even without considering practical problems of cost, this did not mean that all questions were answered, because they first had to be put to the computer by a researcher. But this discussion has to be postponed to Chapter 19.

Some readers could be excused for wondering whether these two chapters, and especially the present one, deal with AI at all. They may be right. Artificial intelligence, as the term is mostly used at the moment, is a collection of techniques for the analysis of large quantities of data. In terms of the scientific method, this corresponds to the generation of empirical knowledge, usually from observations, but this is only a small part of the overall method (distributed among steps S1 and S2 in Figure 2.1). What interests us here is the full process of data exploitation, from their generation to their final incorporation into hypotheses. It will become clear that AI is only occasionally useful in the process, even if computers have become indispensable at most stages of it.

The rest of this chapter is structured as follows. Section 2.2 discusses the concept of coherent structure, including, in Section 2.2.1, examples from free-shear flows and, in Section 2.2.2 and Section 2.2.3, examples from wall-bounded flows. Section 2.3 summarizes the results and discusses the relation between data analysis and conceptual modeling.

2.2 Coherent Structures

Although turbulence is often treated as a random process in which questions are posed in terms of statistics, a competing point of view that describes it in terms of structures and eddies has had its followers from the beginning. Thus, while Reynolds (1894) centered on the statistics of the fluctuations, Richardson (1920) described them in terms of "little and big whorls." And while Kolmogorov (1941) framed his cascade theory as a statistical relation between fluctuation intensity and scale, Obukhov (1941) interpreted it in terms of interactions among eddies. There is probably a reason for the different emphases. Reynolds was an engineer and Kolmogorov a mathematician. Both could disregard individual fluctuations, which they saw from the outside. Richardson and Obukhov were primarily atmospheric scientists, and atmospheric turbulent structures cannot be ignored, because they are large enough for us to live *within* them.

These notes take the view that randomness is an admission of ignorance that should be avoided whenever possible (Voltaire 1764), and that turbulence is a deterministic dynamical system satisfying the Navier–Stokes equations. Specifically, we will be interested in whether the description of the flow can be simplified by decomposing it into "coherent" structures that can be extracted by observation or predicted from theoretical considerations, and whether the dynamics of these structures can be used to model the evolution of the flow, or even to guide us into ways of controlling it.

Up to fairly recently, the statistical point of view was all but inevitable, because it is difficult to extract structural information from single-point measurements. It was not until shear-flow visualizations in the 1970s revealed structures whose lives were long enough to visually influence scalar tracers (Kline et al. 1967, Corino & Brodkey 1969, Brown & Roshko 1974) that the structural point of view began to gain acceptance. And it was only after probe rakes, numerical simulations, and PIV experiments routinely provided multidimensional flow fields that that point of view became fully established. In the same way, the more recent introduction of time-resolved three-dimensional flow information, mostly from simulations (Perlman et al. 2007, Lozano-Durán & Jiménez 2014, Cardesa et al. 2017) has opened a window into the dynamics of those structures, not only allowing us to conceive structures as predictable, but forcing us to try to predict them.

However, any attempt to simplify complexity should be treated with caution, because it implies neglecting something that may be important. There are several ways of approaching these simplifications, some of which are described in detail in other parts of this course. Many are based on projecting the equations of motion onto

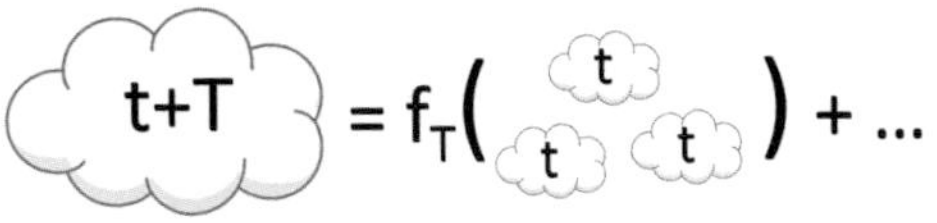

Figure 2.2 The evolution of the coherent structures is expected to depend mostly on a few other structures, plus perhaps nonessential residues.

a smaller set of variables, typically a linear subspace or a few Fourier modes, but this approach becomes less justifiable as the Reynolds number increases, or as the system becomes more extended. If we consider a pipe at even moderate Reynolds number, whose length is several hundred times its diameter, we can expect to find several thousands eddies of any definition, with widely varying scales, far enough from each other to be essentially independent. Projection methods become less useful in those cases because they do not respect locality, and treat together unrelated quantities.

Moreover, the Navier–Stokes equations are partial differential equations in physical space (except for the pressure in incompressible flows), but not in Fourier space, and our approach will be to treat the evolution of the flow as largely local, and to look for solutions that evolve relatively independently from other similar solutions far away. This view of turbulence is based on the "hope" that at least part of its dynamics can be described in terms of a relatively small number of elementary objects that, at least for some time, depend predominantly on a "few" similar objects and on themselves, with only minor contributions from an 'unstructured' background. We will refer to them as "coherent structures." The underlying model is sketched in Figure 2.2, but it is important to understand that this simple dependence, and therefore the existence of coherent structures, is mostly a hope that requires testing at every step of the process.

"Self" dependence suggests properties that the structures should possess to be relevant and practical, such as that they should be strong enough with respect to their surroundings to have some dynamics of their own, and that they should be observable, predictable, or computable. These requirements generally imply that coherent structures are either "engines" that extract energy from some relevant forcing, sinks that dissipate it, or "repositories" that hold energy long enough to be important for the general budget of the flow.

The emphasis on structure does not exclude statistics. The descriptions of Reynolds (1894) and Kolmogorov (1941) have been more useful in applications than most structural approaches, and it cannot be forgotten that turbulence is as important for engineering as it is for science. Moreover, statistics are required even when analysing structure. Turbulent flows are chaotic and high-dimensional (Tennekes & Lumley 1972, Keefe et al. 1992) and, even if dissipation probably restricts them to a finite-dimensional attractor in phase space, they explore that space widely. In essence, anything that is not strictly inconsistent with the equations of motion is bound to occasionally happen in a turbulent flow, and it is important to approach structure identification statistically, and to make sure that any observed phenomenon, even if intellectually appealing, occurs often enough to influence the overall dynamics.

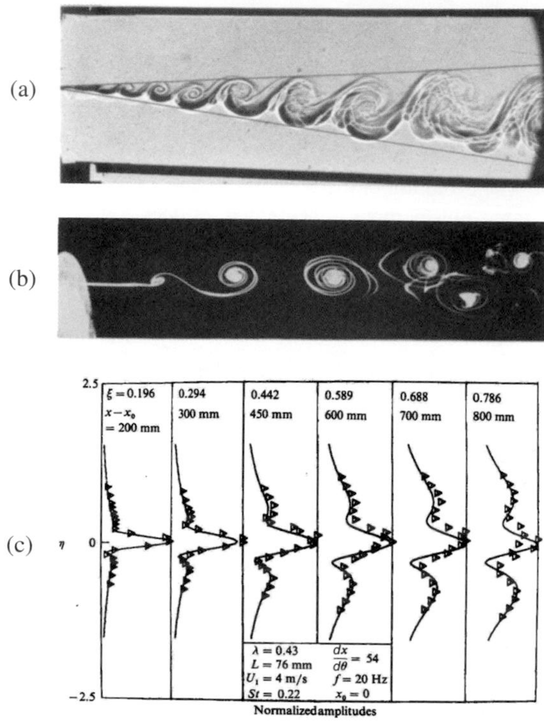

Figure 2.3 (a) Shadowgraph of a turbulent-free mixing layer, from Brown and Roshko (1974). Flow is from left to right. (b) Transition in a laminar mixing layer, from Freymuth (1966). (c) Fluctuation amplitude in a forced turbulent mixing layer. Lines are linear stability theory, and symbols are laboratory measurements. From Gaster et al. (1985). All reproduced with permission.

On the other hand, rare events can be important if they can be exploited for control purposes (Kawahara 2005), or if they are harmful enough to justify investing in avoiding them, for example, tornadoes. Turbulence, as an example of natural macroscopic chaos, may be one of the first systems in which we may hope for the engineering implementation of a "Maxwell daemon," in the sense of extracting useful work from apparently random fluctuations. In a way, soaring albatrosses, or glider pilots riding thermals, have been doing this for a long time.

2.2.1 Free-Shear Flows

Probably the most influential early visualizations of coherent structures in a fully turbulent flow were the free-shear layers in Brown and Roshko (1974) (Figure 2.3(a)). They were not the earliest; streaks and ejections had been observed before in wall-bounded flows by Kline et al. (1967) and by Corino and Brodkey (1969), but the mechanism for the structures in Brown and Roshko (1974) was immediately obvious, because similar structures were known in transitional flows at lower Reynolds numbers, and had been theoretically analyzed in terms of linear stability

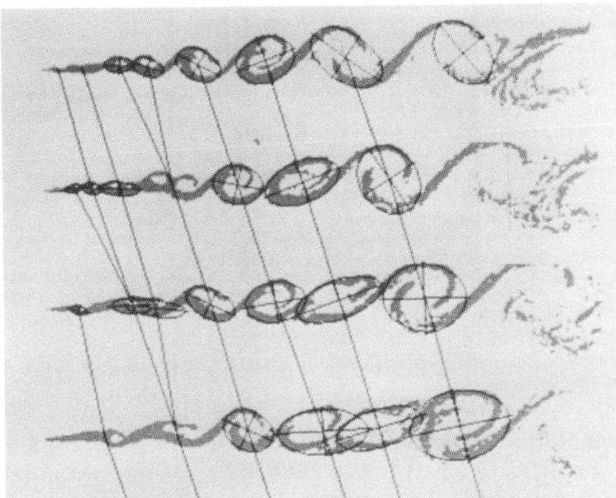

Figure 2.4 Automatic tracking of structures in a free-shear layer, from Hernán and Jiménez (1982), reproduced with permission.

theory by Freymuth (1966) (Figure 2.3(b)). The mean velocity profile of free-shear flows, such as shear layers, wakes and jets, are modally unstable, and the associated Kelvin–Helmholtz instability is well understood. In fact, even if the stability analysis is linear, while turbulence is not, it was soon confirmed that the linearized results match the large-scale dynamics of forced shear layers well (Gaster et al. 1985), within well-understood limits (Figure 2.3(c)). We noted in the introduction that it is important to have mechanisms to supplement and support raw observations. This was the case for the structures of free-shear flows, and it was probably the main reason for their early scientific acceptance. The instability of shear layers is used industrially today to enhance mixing and to reduce noise.

Of interest for the present course is that these experiments were some of the first computer-analyzed temporal series of images in any experimental flow. A film of the experiments in Bernal (1981) was processed by Hernán and Jiménez (1982) to catalogue the properties of the structures and of their interactions (Figure 2.4). It may add historical perspective to the present course that that paper, which included automatic structure identification, classification and temporal analysis, refers in several places to "syntactic programming," "formal grammars," and similar techniques of current AI flavor. The paper did not even claim to have invented those techniques, but cited existing textbooks on image processing.

2.2.2 Wall-Bounded Flows

Structures in wall-bounded flows were observed even earlier than in free-shear layers (Kline et al. 1967, Corino & Brodkey 1969), but the lack of a theoretical model comparable to the linear instability of the latter leads to continued disagreements about the meaning of the wall-bounded structures. It had been found early that the mean velocity profile of a turbulent channel is modally stable (Reynolds & Tiederman 1967), and it was not until it was understood that modally stable perturbations can transiently

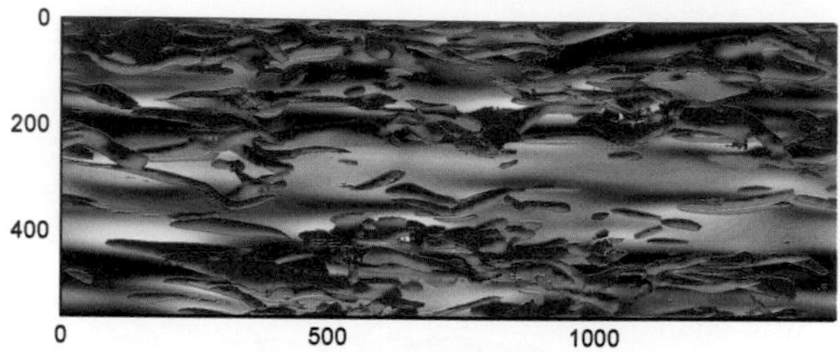

Figure 2.5 Near-wall structure of a turbulent channel. The flow is from left to right and the figure only includes the region below $y^+ \approx 60$, looking toward the wall. The gray background is the shear at the wall and represents the streamwise-velocity streaks, and the colored shorter objects are vortices with either positive or negative streamwise vorticity. The axes labels are in viscous units. From Jiménez (2002), reproduced with permission of the Licensor through PLSclear.

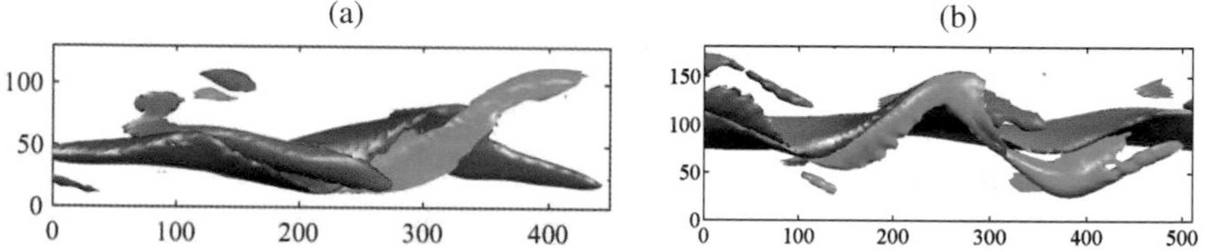

Figure 2.6 Near-wall structure of a turbulent channel. The flow is from left to right, and the figure represents the region below $y^+ \approx 80$, looking toward the wall. The gray objects are the streamwise-velocity streak, and the colored shorter objects are the vortices responsible for the sweeps and ejections, with either positive or negative streamwise vorticity. The axes labels are in viscous units. (a) Minimal channel at $Re_\tau = 180$. (b) A weakly oscillatory wave in a minimal autonomous channel, from Jiménez and Simens (2001), reproduced with permission.

grow by considerable amounts (Butler & Farrell 1993) that the question of wall-bounded turbulent structures could be treated with some rigor.

Visualizations show that the region near the wall of boundary layers, pipes, and channels is dominated by long "streaks" (Kline et al. 1967), eventually shown to be jets of high or low streamwise velocity, sprinkled with shorter fluid eruptions, or "bursts" (Kim et al. 1971). It was initially hypothesized that bursts were due to the intermittent breakup of the near-wall streaks, but even the original authors acknowledged that their visualizations could be consistent with more permanent objects being advected past the observation window. The term "burst" eventually became associated with the "sweeps" and "ejections" observed by stationary velocity probes, which respectively move fluid toward and away from the wall (Lu & Willmarth 1973). After the careful visual analysis by Robinson (1991) of a temporally resolved film of one of the first numerical simulations of a turbulent boundary layer (Spalart 1988), it became clear that, at least in the viscous layer near the wall, the sweeps and ejections known from single-point measurements reflected the passing of shorter quasi-streamwise vortices, intermittent in space but not necessarily in time (see Figure 2.5). It was also soon understood that ejections and sweeps create the streaks by deforming the mean velocity profile, but the origin and dynamics of the bursts remained unclear.

The next step was taken by Jiménez and Moin (1991), who shrank the dimensions of a channel simulation to the minimum value that could accommodate a single sweep–ejection pair and a short streak segment, thus allowing the study of their interactions "in isolation." These simulations showed that wall turbulence is largely independent of the chaotic interaction among structures in the flow, thus satisfying one of the conditions for coherence in Figure 2.2, and that this minimal unit was unambiguously intermittent in time. These and other authors eventually converged to the description of a "cogeneration" cycle in which the vortices sustain the streak, and an undetermined instability of the streak creates the vortices (Jiménez & Moin 1991, Hamilton et al. 1995, Jiménez & Pinelli 1999, Schoppa & Hussain 2002).

A parallel, and largely independent, development was led by Nagata (1990), who studied wall turbulence from the point of view of dynamical systems, and was able to compute fully nonlinear steady-state solution of the Navier–Stokes equations in Couette flow. These solutions are typically unstable, and therefore unlikely to be found in a real flow, but the idea was that the system evolves relatively slowly in their phase-space neighborhood, or in that of related oscillatory solutions, and that the relatively large fraction of time spent near them would "anchor" the flow statistics. In fact, these simple solutions look extraordinarily similar to the bursting solutions in minimal channels (Figure 2.6), and predict fairly well the amplitude and dimensions of the observations in the viscous layer of real flows (Jiménez et al. 2005). They remain to this day one of the main reasons to believe that coherent structures are important components in the dynamics of wall-bounded turbulence. A review of their general significance is Kawahara et al. (2012).

The reason why Figures 2.5 and 2.6 can be compared visually is that they are restricted to the viscous layer of the flow, where the internal Reynolds number of the structures, $\Delta_y^+ = u_\tau \Delta_y / \nu$, is low, where u_τ is the friction velocity, ν is the kinematic viscosity, and Δ_y is the height of the structure. The resulting structures are relatively smooth and can be recognized as "objects." As the Reynolds number of the flow increases and some of the structures become larger, their internal Reynolds number grows, and their shape can best be described as a fractal (see Figure 2.7(a)). Such objects should be treated statistically, and their study had to wait until the Reynolds number of the simulations grew and methods of analysis for the large data sets involved were developed.

The first such studies examined the structure of vorticity in channels. Individual intense vortices had been studied in isotropic turbulence (Vincent & Meneguzzi 1991, Jiménez et al. 1993), where they are associated with dissipation. They were known to be predominantly concentrated in low-velocity regions in channels (Tanahashi et al. 2004), but del Álamo et al. (2006) showed that the relevant objects in wall-bounded turbulence were not individual vortices but vortex tangles, or "clusters" (Figure 2.7(b)). They are approximately collocated with the ejections, which explains their association with low-velocity regions, and form a self-similar family of clusters (not of individual vortices) across the logarithmic layer. The minimal-channel technique was extended by Flores and Jiménez (2010) to the logarithmic layer, where

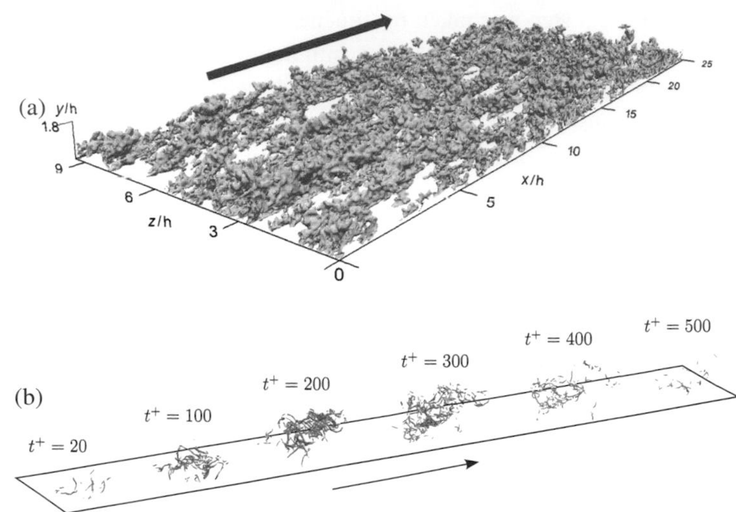

Figure 2.7 (a) Reynolds-stress structures in the outer layer of a turbulent channel, $Re_\tau = 2000$ (Hoyas & Jiménez 2006). The flow is from left to right. The gray objects are the ejections, and approximately correspond to low-velocity streaks. The yellow ones are sweeps. Image courtesy of A. Lozano-Durán. (b) Temporal evolution of a vortex cluster along its lifetime. Channel at $Re_\tau = 4200$. The maximum height of the cluster is $u_\tau \Delta_y / \nu \approx 300$ at which time the cluster is attached to the wall. From Lozano-Durán and Jiménez (2014), reproduced with permission.

they used it to show that the vortex clusters also burst intermittently, and that this process is associated with the deformation of the large-scale streaks found in that region. Probably related to these clusters are the self-similar "hairpin packets" that have been intensively studied by experimentalists (Adrian 2007), although there is little information about their temporal behavior.

The next step was to analyze whether a similar organization exists for sweeps and ejections, which had been studied for a long time as being responsible for carrying the tangential Reynolds stress. This was done by Lozano-Durán et al. (2012), using three-dimensional data from simulations. The structures were defined by thresholding the Reynolds stress, and proved to also form self-similar families of side-by-side pairs of a sweep and an ejection, which are the logarithmic-layer equivalent of the quasi-streamwise vortices of the buffer layer. A typical flow field in which these Reynolds-stress structures have been isolated by thresholding their intensity is given in Figure 2.7(a), confirming that they are no longer smooth objects, and that it is difficult to analyze them individually rather than statistically. The study of large-scale structures of either velocity or Reynolds stress has since been extended to boundary layers (Sillero 2014), homogeneous shear flows (Dong et al. 2017), and pipes (Lee & Sung 2013), with broadly similar results. The elementary unit is a sweep–ejection pair with a vortex cluster in between, located in the lateral boundary between a high- and a low-velocity streak. We will use this structure interchangeably with the term "burst."

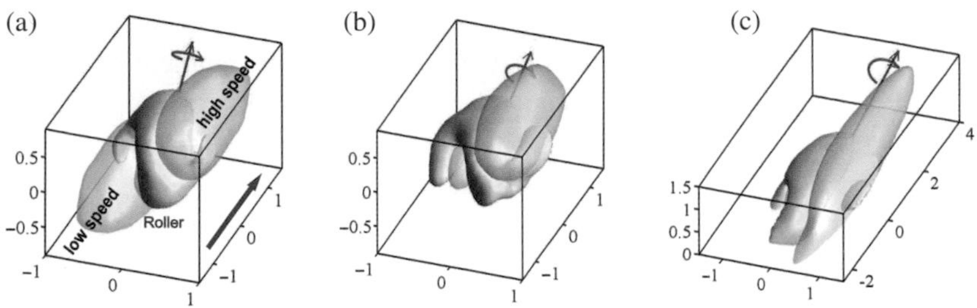

Figure 2.8 Effect of the inhomogeneity of the mean velocity profile on the flow field conditioned to sweep–ejection pairs. The central opaque S-shaped object is an isosurface of the magnitude of the conditional perturbation vorticity, at 25% of its maximum. The two translucent objects are isosurfaces of the conditional perturbation streamwise velocity, $u^+ = \pm 0.6$. The axes are scaled with the average of the diagonal sizes of the sweep and the ejection. (a) Homogeneous shear turbulence. (b) Detached eddy in a channel, $Re_\tau = 950$. (c) As in (b), for an attached eddy. Adapted from Dong et al. (2017), reproduced with permission.

As temporally resolved data sets became available, individual bursts were tracked by Lozano-Durán and Jiménez (2014), finally answering the question of whether ejections are permanent passing objects or transient ones. They have a definite lifetime, shown in Figure 2.7(b) for a vortex cluster, but this lifetime is longer than the passing time across a stationary probe, so that the signal that the latter sees is due to advection. Individual bursts are born and disappear at an approximately constant distance from the wall but, as they grow toward the middle of their lifetime, some of them become large enough for their root to become attached to the wall. On the general principle that the largest eddies carry the Reynolds stresses (Tennekes & Lumley 1972), these attached eddies are responsible for most of the momentum transfer in the flow (Townsend 1976), where the definition of "large" is linked to the Corrsin (1958) scale, which separates the eddies that feel the effect of the shear from the smaller ones that do not (Jiménez 2018a). The temporal behavior of the large eddies is also self-similar. The lifetime of an eddy whose maximum height is Δ_y is approximately an eddy turnover, $u_\tau T/\Delta_y \approx 1$.

The confirmation that bursts are transient reopened the question of their origin. There is persuasive evidence that they are essentially linearizable Orr (1907b) solutions, which transiently extract energy from the mean flow as they are tilted forward by the shear (Jiménez 2013, Encinar & Jiménez 2020). In retrospect, this was to be expected, because the only energy source in the flow is the mean shear, but it leaves open the question of how the bursts are seeded. Orr bursts are temporary structures of the wall-normal velocity; they are born weak, leaning backward with respect to the shear; they intensify as the shear tilts them toward the normal to the wall, and eventually disappear when the forward tilt becomes too pronounced. In the process, some of their energy is transferred to the streaks and remains in the flow. This is a plausible mechanism for turbulent-energy production, but there is no obvious process to restart the eddies as backward-leaning structures. There is overwhelming "circumstantial" evidence that the picture sketched earlier corresponds to the kinematics of the flow, but using the same argument as in free-shear flows, its definitive acceptance will only come when a complete model is available. Note that the model presented here refers only to the shear, and does not reserve any special role for the wall. In fact, the structures studied by Dong et al. (2017) in a homogeneous shear flow, which has no walls, behave very similarly to those in channels, and there is a smooth transition from the symmetric sweep–ejection pairs of the homogeneous shear to the very asymmetric wall-attached ones (Figure 2.8). An extended recent review of these and other aspects of the structures of wall-bounded turbulent flows is Jiménez (2018a).

2.2.3 Conceptual Experiments

Although most of the results mentioned earlier exploit correlations among data that had not been compiled in response to specific questions, settling controversies often requires direct experimentation of the kind discussed in the introduction as "interventional science." Conceptual, "hypothesis-driven," experiments have been a fixture of physics for a long time. In simple dynamical situations, such as point masses or moving clocks, the outcome of such exercises can often be used to settle the soundness of a particular model with little more than pencil, paper, and a good intellect, but in complex cases, such as turbulence, the guessing almost always has to be substituted by actual experimentation on the modified system.

The first experiment of this type in structural turbulence was probably Cimbala et al. (1988), and predates simulations. There was some controversy at the time about whether the structures in turbulent wakes were remnants of vortex shedding by the obstacle that creates the wake, or local instabilities of the wake profile. Their experiment showed that the latter was the case by artificially damping the structures present in the wake at some distance downstream, and observing how they reappear afterward.

After computers started to be powerful enough to simulate turbulence, conceptual experiments became mostly computational, because simulations permit many "unphysical" manipulations that are not easily implemented otherwise. In wall-bounded flows, we have already mentioned the minimal experiments by Jiménez and Moin (1991). Later, to settle a controversy about whether turbulence near the wall was autonomous or forced by outer motions, Jiménez and Pinelli (1999) performed simulations in which the flow equations were artificially damped away from the wall, and complemented these experiments by selective modifications of the equations of motion and of the boundary conditions, intended to falsify different versions of the regeneration cycle. Mizuno and Jiménez (2013) did the opposite, damping only the near-wall region. Both experiments showed that the undamped part of the flow survives by itself, and that wall turbulence at a given distance from the wall can be studied relatively independently of other wall distances. The homogeneous-shear simulations by Dong et al. (2017), mentioned earlier, were intended to complement these experiments, removing both the near-wall region and the outer flow.

2.3 Discussion

The previous sections have reviewed some of what is known about structures in shear flows. Except for the earliest visualizations of free-shear layers, all the data mentioned here have been "big" in relation to the dates in which they were generated and analyzed: from tens of Mbytes for the first channel simulations in the 1980s, to hundreds of Tbytes for the current time-resolved data sets. The field has been driven as much by storage capacity as by computational power or experimental methods, and the analysis of the data has always involved new techniques that were not available, or required, before the new data came online. It would appear that the description of these methods should have been the main thrust of a set of notes on the application of AI to fluid mechanics, but a rereading of the previous sections shows that this has not been the case. Most of them are dedicated to the discussion of physical models. We have noted several times that the quest for structures is driven by the hope that they simplify the description and control of the flow, and that their acceptance by the community depends on the acceptance of these simplifications.

Coherent structures are not an obvious fit for fluid mechanics, which is best described by continuous vector fields. Structures are patterns that we hope are useful for us, but we should be weary of the evidence that humans are notoriously prone to finding spurious patterns. Consider constellations in the night sky, or divination from tea leaves, and note that even those spurious patterns keep being considered relevant because there is a community that finds them useful.

The Kelvin–Helmholtz rollers of free-shear layers were considered useful from the start because their mechanism is clear and offers a distinct avenue for the description and control of the flow. The bursts and streaks of wall-bounded flows are still considered less definitive because the description of their regeneration cycle remains incomplete.

If any conclusion can be drawn from the present discussion is that the scientific cycle in Figure 2.1 should be considered as a whole, that data analysis should include a parallel development of physical models, and that no single isolated step should be taken as final. Apparently convincing flow descriptions in terms of a reduced set of structures or modes should be taken with suspicion unless they are complemented by plausible mechanisms. In the nomenclature introduced in Section 2.1, sub-symbolic AI only becomes scientifically conclusive once it becomes symbolic.

Appealing mechanisms should also be carefully examined for evidence that they are really important for the flow. No single description is likely to span the full dynamics of turbulence, and we are probably bound to always "make do" with partial results. But it is important to consider the phenomenon as a whole, and to recognize which part of it is described by our models. One of the reasons to appreciate the bursting model in Section 2.2.2 is that it accounts for 60% of the skin friction using only 10% of the flow volume (Jiménez 2018a). This is appealing, but it is fair to wonder what happens to the other 40%. One of the most dangerous sentences in scientific research is: "It is a start."

3 Machine Learning in Fluids: Pairing Methods with Problems

S. Brunton*

The modeling, optimization, and control of fluid flows remain a grand challenge of the modern era, with potentially transformative scientific, technological, and industrial impact. An improved understanding of complex flow physics may enable drag reduction, lift increase, mixing enhancement, and noise reduction in domains as diverse as transportation, energy, security, and medicine. Fluid dynamics is challenged by strong nonlinearity, high-dimensionality, and multiscale physics; both modeling and control may be thought of as non-convex optimization tasks. Recent advances in machine learning and data-driven optimization are revolutionizing how we approach these traditionally intractable problems. Indeed, machine learning may be considered as a growing set of techniques in data-driven optimization and applied regression that are tailored for high-dimensional and nonlinear problems. Thus, the modeling, optimization, and control of complex fluid systems via machine learning are complementing mature numerical and experimental techniques that are generating increasingly large volumes of data.

3.1 Overview

The large range of spatial and temporal scales in a complex fluid flow requires exceedingly high-dimensional measurements and computational discretization to resolve all relevant features, resulting in vast data sets and time-intensive computations. There is no indication that the increasing volumes of data generated and analyzed in fluid mechanics will slow any time soon, and we are decades away from fully resolving the most complex engineering flows (aircraft, submarines, etc.) fast enough for iterative optimization and in-time control. Despite the large number of degrees of freedom required to describe fluid systems, there are often dominant patterns that emerge, characterized by energetically or dynamically important coherent structures (see Figure 3.1 and Chapter 2). These coherent structures provide a compact representation of these flows, and capturing their evolution dynamics has been the focus of intense research in reduced-order modeling for decades, as described in Chapter 1. Advances in machine learning has promised a renaissance in the analysis and understanding of such complex data, extracting patterns in high-dimensional, multimodal data that are beyond the ability of humans to grasp. As more complete measurements

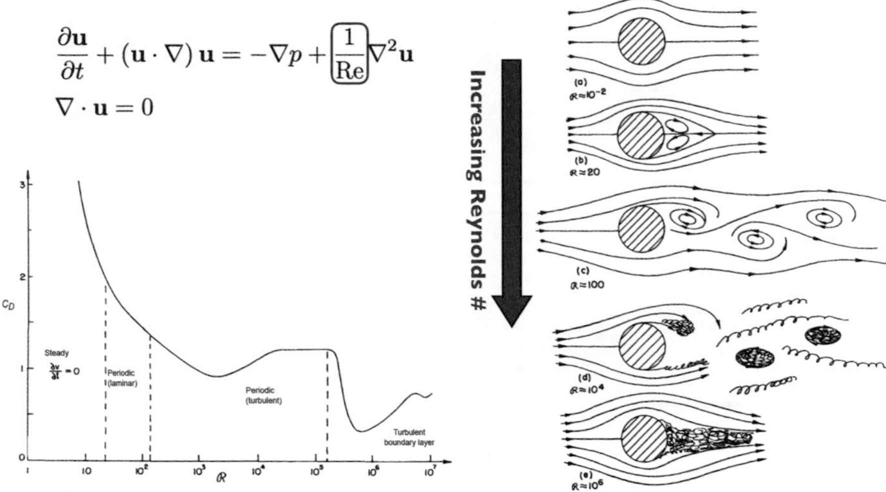

Figure 3.1 Fluid dynamics are often characterized by complex, multiscale dynamics in space and time. For example, by increasing the Reynolds number, the flow past a circular cylinder becomes unsteady and turbulent. Figure adapted from Feynman et al. (2013), reprinted by permission of Basic Books, an imprint of Hachette Book Group, Inc.

become available for more complicated fluid systems, the data associated with these investigations will only become increasingly large and unwieldy. Data-driven methods are central to managing and analyzing this data, and generally, we consider data methods to include techniques that apply equally well to data from numerical or experimental systems. In fact, data-driven methods in machine learning provide a common framework for both experimental and numerical data.

In this chapter, we will provide an overview of the established and emerging applications of machine learning in fluid dynamics, emphasizing the potentially transformative impact in modeling, optimization, and control; for more details and references, see Brenner et al. (2019) and Brunton et al. (2020). We take the perspective that machine learning is a collection of applied optimization techniques that build models based on available data, which is abundant in fluid systems. Because of the data-intensive nature of fluid dynamics, machine learning complements efforts in both experiments and simulations, which generate incredible amounts of data. In machine learning, it is just as important to understand when methods fail as when they succeed, and we will attempt to convey the limitations of current methods, exploring what is easy and difficult in machine learning. We will also emphasize the need to incorporate physics into machine learning to promote interpretable and generalizable models. Finally, we cannot overstate the importance of cross-validation to prevent overfitting.

Fluid flow fields may be naturally viewed as images, as in Figure 3.2. Thus, there is considerable low-hanging fruit in applying standard techniques from machine learning for image analysis to flow fields, for example for occlusion inference (Scherl et al. 2020) and super-resolution (Fukami et al. 2018, Erichson et al. 2020). However, fluid dynamics departs from many of the typical application domains of machine learning, such as image recognition, in that it is fundamentally governed by *physics*. Many of the

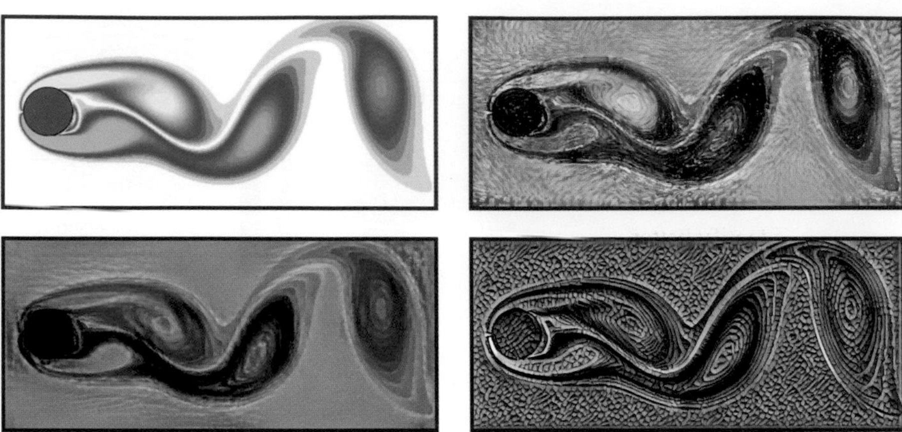

Figure 3.2 Illustration of how simple it is to add *texture* to an image using modern neural networks. Is it possible to add the texture of turbulence to a laminar flow?

goals of fluid dynamics involve actively or passively manipulating these dynamics for an engineering objective through intervention, which inherently changes the nature of the data collected; see Chapters 17 through 19. Thus, the dynamic and physical nature of fluids provides both a challenge and an opportunity for machine learning, potentially providing deeper theoretical insights into algorithms by applying them to systems with known physics.

Although fluid data is vast in some dimensions (i.e., fully resolved velocity fields may easily contain millions or billions of degrees of freedom), it may be relatively sparse in others (e.g., it may be expensive to sample many geometries and perform parametric studies). This heterogeneity of the data can limit the available techniques in machine learning, and it is important to carefully consider whether or not the model will be used for interpolation within a parameter regime or for extrapolation, which is generally much more challenging. In addition, many fluid systems are naturally nonstationary, and even for stationary flows it may be prohibitively expensive to obtain statistically converged results when behavior is broadband; see Chapter 9. Typically, engineering goals that involve a working fluid are generally not simple classification tasks, but often involve more subtle multi-objective optimization, for example to maximize lift while minimizing drag. Finally, because fluid dynamic systems are central to transportation, health, and defense systems, it is essential that machine learning solutions are interpretable, explainable, and generalizable, and it is often necessary to provide guarantees on performance.

Fluid dynamics challenges often amount to large, high-dimensional, non-convex optimization problems for which machine learning techniques are becoming increasingly well adapted. Although the opportunities of machine learning in fluid mechanics are nearly limitless, we roughly categorize efforts into (1) modeling, (2) closed-loop control, and (3) engineering optimization. These categories necessarily share some overlap. The traditional wisdom in fluid dynamics is that modeling is essential to support the downstream goals of optimization and control. However, effective optimization and control may bypass modeling. In fact, by manipulating the flow and exciting transients, we may indeed generate valuable data to enrich our modeling efforts.

Flow feature extraction
– There is a tremendous opportunity to improve efforts in dimensionality reduction to identify better coordinates for compact descriptions. This topic is related to coherent structures from Chapter 2 and will be discussed in Section 3.3.

Modeling flow dynamics
– Machine learning is providing new computational techniques to develop models from data, which may improve our understanding of key dynamic mechanisms for efficient and accurate prediction, estimation, and control. This will be discussed in Section 3.4 and Chapter 14.

Control and optimization
– Flow control is a long-held goal of fluid dynamics. Sensor-based feedback control of a fluid flow may be viewed as a non-convex optimization based on measurement data. Machine learning is poised to transform the field of turbulence control, with potential impact across the industrial and technological landscape. It may also be used for design optimization in fluids, for example to optimize geometry and parameters for multi-objective performance goals. This will be discussed in Section 3.5 and Chapter 17.

3.1.1 Kinematic and Dynamics Modeling of Fluid Flows

Modeling fluid flows has been the focus of efforts in mathematical physics and applied engineering for centuries. Flow modeling is particularly useful for understanding essential flow physics, design and optimization, and active control. Fluid dynamics is a notoriously challenging field of classical physics, especially the characterization of turbulence, which is nonlinear, high-dimensional, and multiscale. Although the Navier–Stokes equations provide a precise mathematical model, it is difficult to use this representation for design, optimization, and control. In fact, much of the observed richness observed in fluids is often obscured by the simplicity of the governing equations, as elegantly observed by Feynman:

> The test of science is its ability to predict. Had you never visited the earth, could you predict the thunderstorms, the volcanos, the ocean waves, the auroras, and the colorful sunset?

Instead of analyzing the governing equations directly, we typically simulate high-dimensional discretized equations or perform laboratory experiments to explore the phenomenology associated with a specific relevant configuration. However, simulations and experiments are expensive, both in financial and human resources, and they are often not suitable for iterative optimization or real-time control. Thus, considerable effort has gone into obtaining accurate and efficient reduced-order models that capture the essential flow mechanisms at a fraction of the cost of a fully resolved flow solver.

This section will distinguish between kinematic and dynamic flow modeling, which will be important for later sections.

There are several important aspects of flow modeling that we will investigate here. Modeling may be roughly categorized into two complementary efforts: dimensionality reduction and reduced-order modeling. Dimensionality reduction involves extracting dominant patterns that may be used as reduced coordinates where the fluid is compactly and efficiently described. Reduced-order modeling describes the evolution of the flow in time as a parameterized dynamical system, although it may also involve developing a statistical map from parameters to averaged quantities, such as drag. Models describe how the flow varies, either in time, with a parameter (e.g., for optimization), or with actuation (e.g., for control). Identifying and extracting relevant flow features is essential to describe and model the flow. In fact, much of the progress in mathematical physics has revolved around identifying coordinate transformations that simplify dynamics and capture essential physics, such as the Fourier transform and modern data-driven modal decompositions such as the proper orthogonal decomposition (POD); see Chapter 6.

Reduced-order modeling encompasses model reduction, which begins with the governing equations, and system identification, which begins with data. Model reduction, such as Galerkin projection of the Navier–Stokes equations onto an orthogonal basis of POD modes, benefits from a close connection to the governing equations; however, it is intrusive, requiring human expertise to develop models from a working simulation. System identification provides a flexible and data-driven alternative, although often resulting in black box models that lack a deep connection to the physics; see Chapters 7 and 8 for more details on system identification. Machine learning constitutes a rapidly growing body of algorithms that may be used for advanced system identification. A central goal of modeling is to balance efficiency and accuracy, which are often dueling objectives. When modeling physical systems, interpretability and generalizability are also critical considerations. Other unique aspects of data-driven modeling of physical systems include partial prior knowledge of the governing equations, constraints, and symmetries. Finally, the volume, variety, and fidelity of data will impact the viability of various machine learning methods.

We will separate the discussion into the use of machine learning for (1) flow feature extraction to model flow *kinematics*, and (2) modeling *dynamics*. Of course, there is some overlap in these topics. For example, the dynamic mode decomposition (DMD) and related Koopman operator approaches are simultaneously concerned with obtaining effective coordinates to represent coherent structures (*kinematics*) and a model for how these coherent structures evolve (*dynamics*). Similarly, in the discussion of closure models for turbulence, there are several approaches, ranging from using super-resolution to fill in low-resolution flow images (*kinematics*), to directly modeling the closure terms as a dynamical system (*dynamics*). Mathematically, the discussion of kinematics amounts to obtaining a change of coordinates from some variables or measurements $\mathbf{x}$ to a new set of variables $\mathbf{a}$ through a possibly nonlinear function $\boldsymbol{\varphi}$:

$$\mathbf{a} = \boldsymbol{\varphi}(\mathbf{x}). \tag{3.1}$$

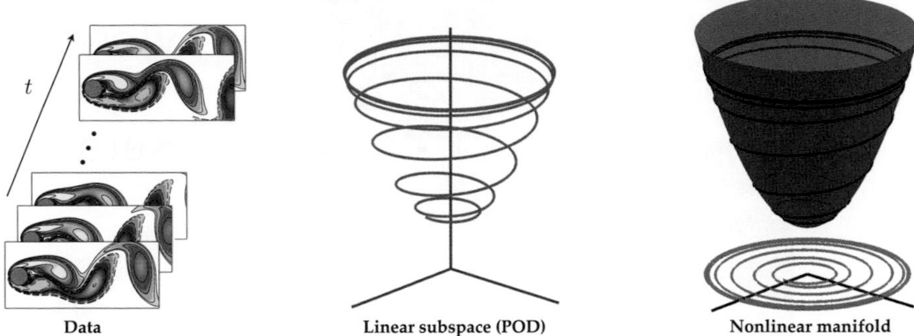

Figure 3.3 Illustration of linear versus nonlinear embeddings of fluid flow data for flow past a circular cylinder. It is possible to describe the dominant evolution in a three-dimensional linear subspace obtained via POD (Noack et al. 2003). However, in this subspace, the flow actually evolves on a two-dimensional submanifold, given by a parabolic inertial manifold. In general, linear subspaces are simpler to obtain, although nonlinear embeddings may yield enhanced reduction.

The original variables $\mathbf{x}$ may be high-resolution flow fields from a simulation or particle image velocimetry (PIV) experiment, and $\mathbf{a}$ may represent the amplitudes of a modal expansion of the flow in terms of dominant coherent structures (Taira et al. 2017), resulting in *dimensionality reduction*. In contrast, the variables $\mathbf{x}$ may contain under-resolved measurements and $\mathbf{a}$ may be the desired high-resolution flow field, which amounts to a *super-resolution* problem. Similarly, the discussion on dynamics amounts to obtaining a differential equation model of the form

$$\frac{d}{dt}\mathbf{a} = \mathbf{f}(\mathbf{a}, \mathbf{u}, t; \boldsymbol{\beta}), \tag{3.2}$$

in terms of these variables $\mathbf{a}$, some control inputs $\mathbf{u}$, time t, and a set of parameters $\boldsymbol{\beta}$.

3.2 Machine Learning Basics

Machine learning is one of the great technological developments of our generation, and it is undoubtedly changing the art of what is possible across the scientific and industrial landscapes, especially in the field of fluid mechanics (Duraisamy et al. 2019, Brenner et al. 2019, Brunton, Noack & Koumoutsakos 2020, Brunton, Hemati & Taira 2020). Simply put, machine learning refers to the ability of a computer system to learn and improve from experiential data, rather than being explicitly or deterministically programmed. The rise of machine learning is driven by the confluence of (1) vast and increasing volumes of data; (2) tremendous advances in computational hardware and orders of magnitude less expensive data storage and transfer; (3) advanced optimization algorithms in applied math and statistics; (4) increasingly expressive architectures (e.g., neural networks); and (5) significant investment by industry, leading to open source development tools (see Figure 3.4). For example, the ImageNet database of labeled images (Deng et al. 2009) and increasingly powerful GPU technology were critical in the development and demonstration of

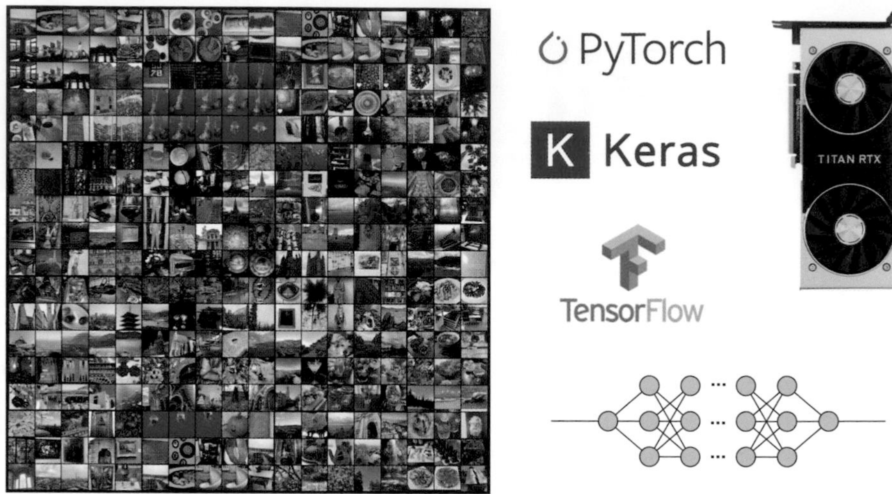

Figure 3.4 The success of machine learning in the past decade can largely be attributed to (1) large-scale labeled training data, such as image databases; (2) advanced computational hardware, such as GPUs; (3) improved neural network architectures and optimization algorithms to train them; and (4) industry investment, leading to open source tools. GPU image used courtesy of NVIDIA, www.nvidia.com. PyTorch, the PyTorch logo, and any related marks are trademarks of Facebook, Inc.

modern deep learning architectures. There is tremendous potential for data science and machine learning to revolutionize nearly every aspect of our modern industrial world, and few fields stand to benefit as clearly or as significantly as the field of fluid mechanics. Indeed, fluid dynamics is one of the original big data fields, and many high-performance computing architectures, experimental measurement techniques, and advanced data processing and visualization algorithms were driven by the decades of research on fluid systems. However, it is important to recognize that machine learning will not *replace* experimental and numerical efforts in fluid mechanics, but will instead complement them (see Figure 3.5), much as the rise of computational fluid dynamics (CFD) complemented experimental efforts (Brunton, Hemati & Taira 2020).

Roughly speaking, there are five major stages in machine learning: (1) asking a question, defining a hypothesis to be tested, or identifying a function to be modeled; (2) preparing a data pipeline, either through collecting and curating the training data (e.g., for supervised and unsupervised learning), or setting up an interactive environment (e.g., for reinforcement learning); (3) identifying the model architecture and parameterization; (4) defining a loss function to be minimized; and (5) choosing an optimization strategy to optimize the parameters of the model to fit the data. Human intelligence is critical in each of these stages. Although considerable attention is typically given to the learning architecture, it is often the data collection and optimization stages that require the most time and resources. It is also important to note that known physics (e.g., invariances, symmetries, conservation laws, constraints, etc.) may be incorporated in each of these stages (Battaglia et al. 2018, Loiseau

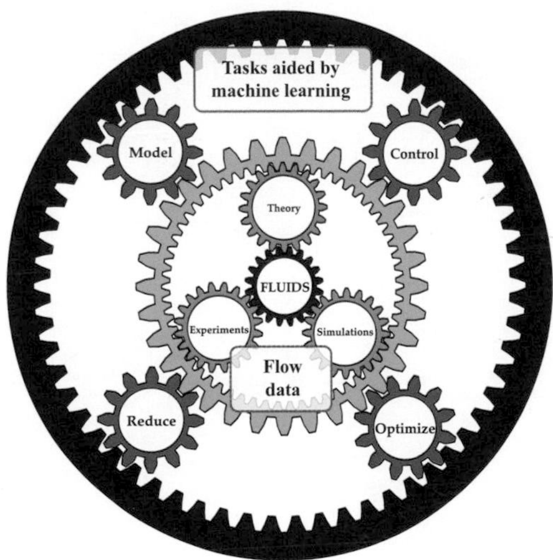

Figure 3.5 Schematic overview of the use of machine learning in fluid dynamics.

et al. 2018, Lusch et al. 2018, Champion et al. 2019, Cranmer et al. 2019, Zheng et al. 2019, Greydanus et al. 2019, Noé et al. 2019, Champion et al. 2020, Zheng et al. 2019, Cranmer et al. 2020, Finzi et al. 2020, Raissi et al. 2020, Zhong & Leonard 2020). For example, rotational invariance is often incorporated by enriching the training data with rotated copies, and translational invariance is often captured using convolutional neural network (CNN) architectures. Additional physics and prior knowledge may also be incorporated as extra loss functions or constraints in the optimization problem (Loiseau et al. 2018, Lusch et al. 2018, Champion et al. 2019, Zheng et al. 2019, Champion et al. 2020). There is also considerable work in developing physics-informed neural network architectures (Lu et al. 2019, Raissi et al., 2019, 2020), networks that capture Hamiltonian or Lagrangian dynamics (Cranmer et al. 2019, 2020, Greydanus et al. 2019, Finzi et al. 2020, Zhong & Leonard 2020), and that uncover coordinate transformations where nonlinear systems become approximately linear (Takeishi et al. 2017, Yeung et al. 2017, Lusch et al. 2018, Mardt et al. 2018, Wehmeyer & Noé 2018, Otto & Rowley 2019).

3.2.1 Machine Learning Categorizations

Machine learning is a growing set of optimization and regression techniques to build models from data. There are a number of important dichotomies with which we may organize the variety of machine learning techniques. Here, we will group these into *supervised*, *unsupervised*, and *semi-supervised* techniques, based on the extent to which the training data is labeled. Reinforcement learning is another major branch of machine learning research, which will be described in detail in Chapter 18. A rough organization is shown in Figure 3.6. In many of these approaches, a function mapping inputs to outputs must be learned. There are many approaches, including linear and nonlinear regression, genetic programming (i.e., symbolic regres-

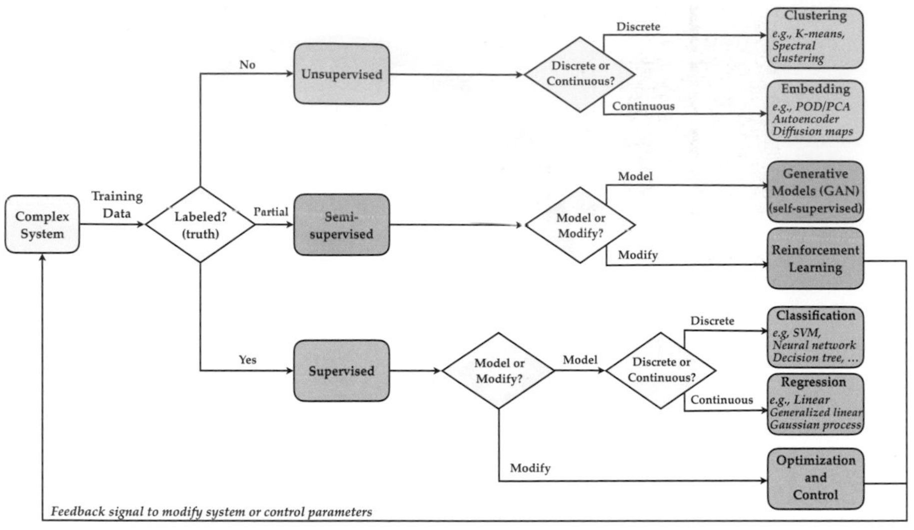

Figure 3.6 Schematic overview of various machine learning techniques.

sion) (Holland 1975, Koza 1992), and neural networks (Goodfellow et al. 2016). In addition, it is increasingly common to add *regularization* terms to the optimization problem to promote certain beneficial problems in the resulting model. For example, promoting sparsity often results in models that are more interpretable (i.e., fewer parameters to interpret), and less prone to overfitting (i.e., fewer parameters to overfit) (Hastie et al. 2009, Brunton, Proctor & Kutz 2016a, Brunton & Kutz 2019). Including priors and regularizers to embed known physics is a particularly exciting area of development.

Supervised Learning

Supervised learning assumes the availability of labeled training data, and it is concerned with learning a function mapping the training data to the label. Many of the most powerful approaches in machine learning are supervised, as there are many data sets where we have empirical knowledge of the labels, but where it is challenging to develop a deterministic algorithm to describe the function between the data and the labels. Supervised learning solves this *fuzzy* problem by leveraging data rather than hand-coded algorithms.

If the labels are discrete, such as categories to classify an image (e.g., dog vs. cat), then the supervised learning task is a *classification*. If the labels are continuous, such as the value of the lift coefficient for a particular airfoil shape, then the task is a *regression*. The labels may also correspond to an objective that should be minimized or maximized, resulting in an optimization or control problem.

Before the rise of deep learning, support vector machines (SVMs) (Schölkopf & Smola 2002) and random forests (Breiman 2001) dominated classification tasks, and were industry standard methods. Continuous regression methods, such as linear regression, logistic regression, and Gaussian process regression are still widespread.

Figure 3.7 Generative art based on Arcimbaldo's *La Primavera*. Reproduced from `https://commons.wikimedia.org/wiki/File:Deepdream-arcimboldo.gif`.

Unsupervised Learning

Unsupervised learning, also known as data mining or pattern extraction, determines the underlying structure of a data set without labels. Again, if the data is to be grouped into distinct categories, then the task is clustering, while if the data has a continuous distribution, the task is an embedding.

There are several popular approaches to clustering. The most common and simple clustering algorithm is the k-means algorithm, which partitions data instances into k clusters; an observation will belong to the cluster with the nearest centroid, resulting in a partition of the data space into Voronoi cells. Spectral clustering (Ng et al. 2002) is also quite common. Likewise, there are many approaches to learn embeddings. The singular value decomposition (SVD) and related principal component analysis (PCA) and proper orthogonal decomposition (POD) are perhaps the most widely used linear embedding techniques, resulting in an orthogonal low-dimensional basis that captures most of the variance in a data set. These embeddings have recently been generalized in the neural network *autoencoder*, discussed in Section 3.2.2.

Semi-Supervised and Reinforcement Learning

Semi-supervised learning, where only partially labeled data is available, is becoming increasingly popular. In many situations, it is prohibitively expensive to generate a labeled data set that is large enough to train a modern deep learning architecture. Similarly, in many cases, full state information is not available, for example in control scenarios.

Generative adversarial networks (GANs) (Goodfellow et al. 2014) are an important and growing avenue of semi-supervised learning. GANs result in a generative model that creates new data to mimic the distribution of the training data. A GAN consists of two networks, the generative network, and a critic network that decide if the generated data is fictitious or not. These two networks are trained in an adversarial fashion so that with even a small amount of labeled training data, this architecture is able to begin training itself. Many of the most impressive recent trends in neural networks, such as deep fakes, are based on GANs. Figure 3.7 shows generative art based on Arcimbaldo's *La Primavera*. Interestingly, the original art may be considered as a predecessor of the current *deep dream* trend in generative art.

Reinforcement learning (RL) (Sutton & Barto 2018) is another important class of machine learning algorithms in which an actor is learning to interact with an environment. RL will be the subject of Chapter 18. Often, the reward information about whether or not a given action was beneficial will not occur regularly; for example, in the game of chess, it is not always obvious exactly which moves led to a player winning or losing. Thus, only partial information is available to guide the control strategy, making it closely related to semi-supervised learning, although it is typically considered a distinct branch of machine learning. Reinforcement learning has been widely used to learn game strategies, for example in Alpha Go (Mnih et al. 2015). Increasingly, these approaches are being applied to scientific and robotic applications (Mnih et al. 2015, Silver et al. 2016).

3.2.2 Neural Networks

Neural networks are particularly powerful and expressive architectures (Goodfellow et al. 2016). Many of the most impressive advances in machine learning over the past decade have involved deep neural network architectures that are trained on powerful graphics processing units. Here, we will review some of the most prominent architectures, although new developments are taking place every day in the field.

Before beginning, it is worth noting that neural networks have a long and rich history spanning decades (Hopfield 1982, Hinton & Sejnowski 1986, Gardner 1988), beginning with the perceptron (Rosenblatt 1958). However, early neural networks did not reach their full potential, lacking sufficiently large data sets, computational hardware, and optimization algorithms. There are several reasons for the recent success of neural networks. First, there are vast and increasing volumes of data in nearly every field, providing the data necessary to train increasingly large networks. Second, dramatic improvements to computational hardware make it possible to train these large multi-layer networks. Third, there have been considerable advances in non-convex optimization, with back-propagation being a particularly critical development (Rumelhart et al. 1986). The universal approximation theorem has also been important to establishing the fact that a sufficiently deep neural network is capable of arbitrary function approximation (Hornik et al. 1989). The modern era of *deep learning* (Goodfellow et al. 2016) began with the incredible success of neural networks on the ImageNet image classification task (Krizhevsky et al. 2012), beating out all other algorithms. Since then, neural networks have dominated nearly every task in image processing and natural language processing.

Artificial neural networks (ANNs) seek to mimic the learning capabilities of animals, and they generally consist of a set of neurons connected in a directed graph to process input data and produce an output. Modern *deep learning* refers to ANNs that have many distinct *layers* of neurons, much as the human visual system has several distinct processing layers; each layer learns a different level of abstraction. Each individual neuron performs an operation on the input data, known as an *activation function*. Traditionally, sigmoidal activation functions were used to mimic the firing thresholds of biological neurons, although now there are many activation functions, with the rectified linear unit (ReLU) being a popular choice. Because of the layered structure of many ANNs, mathematically, these may be described by a composition of functions

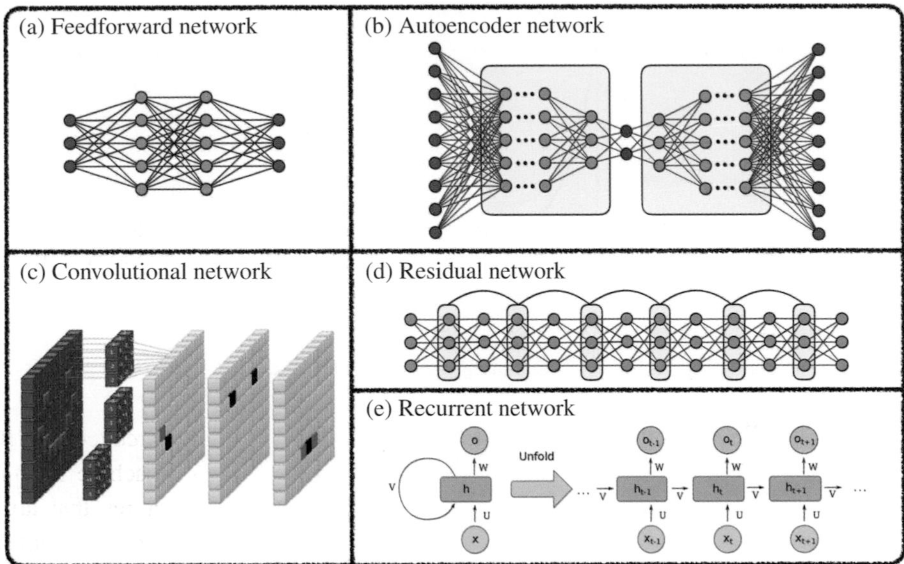

Figure 3.8 (a) Simple feedforward neural network. (b) Deep autoencoder network. (c) Convolutional neural network. Reproduced from `https://commons.wikimedia.org/wiki/File:3_filters_in_a_Convolutional_Neural_Network.gif`. (d) Deep residual network. (e) Recurrent neural network. Reproduced from `https://en.wikipedia.org/wiki/File:Recurrent_neural_network_unfold.svg`.

$$\mathbf{y} = f_1(f_2(\cdots(f_n(\mathbf{x};\boldsymbol{\theta}_n);\cdots);\boldsymbol{\theta}_2);\boldsymbol{\theta}_1), \tag{3.3}$$

where $\boldsymbol{\theta}_j$ describe the network weights for each layer (i.e., the weights of the graph) that must be learned.

There are several challenges in training these large networks, as there may be millions or billions of parameters in a modern ANN. Backpropagation essentially uses the chain rule for the composition of functions to compute the gradient of the objective function with respect to the weights. These gradient computations make it possible to optimize parameters with gradient descent; stochastic gradient descent algorithms, such as ADAM, are typically used because of the large scale of the optimization problem (Kingma & Ba 2014). Other challenges include vanishing gradients and the need for good network initializations. These considerations all depend on having an effective network architecture and loss function, which are meta-challenges. More details on network training can be found in Goodfellow et al. (2016) and Brunton and Kutz (2019).

We will now review a number of commonly used architectures. The simple feedforward architecture, shown in Figure 3.8(a), is useful for approximating input–output functions. Figure 3.8(b) shows a deep autoencoder network. Autoencoders generalize the principal component analysis (PCA) to nonlinear coordinate embeddings; the connection between PCA and linear autoencoders was first established by Baldi

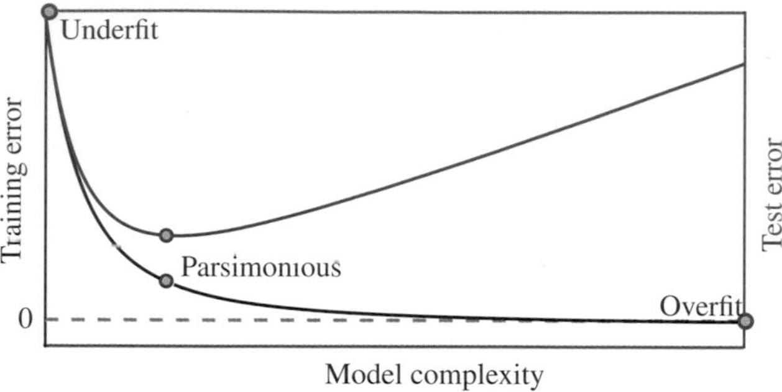

Figure 3.9 Schematic illustrating the importance of cross-validation to prevent overfitting.

and Hornik (1989), and the first use of deep nonlinear autoencoders in fluids was in Milano and Koumoutsakos (2002). Recently, deep residual networks (Szegedy et al. 2017), shown in Figure 3.8(d), have enabled better optimization performance by including jump connections in exceedingly deep networks. Figure 3.8(e) depicts a recurrent neural network (RNN), which, unlike feedforward networks, has feedback connections from later layers back to earlier layers. The long-short term memory (LSTM) network (Hochreiter & Schmidhuber 1997) is a particular variant of an RNN that is useful for dynamical systems and other time-series data, such as audio signals. Finally, Figure 3.8(c) depicts a CNN, which is a useful architecture for processing data with spatial correlations, such as images.

3.2.3 Cross-Validation

When discussing machine learning, it is critical to introduce the notion of cross-validation. Cross-validation plays the same role in machine learning that grid refinement studies play in CFD. Because many modern machine learning architectures have millions of degrees of freedom, it is easy to overfit the model to the training data. Thus, it is essential to split the data into training and testing data; the model is trained using the training data and evaluated on the testing data. Depending on the volume of available data, it is common to randomly split the data into training and testing data, and then repeat this process many times to get statistically averaged results. Figure 3.9 shows the idea behind cross-validation, where a model may overfit the training data unless it is cross-validated against test data.

3.3 Flow Feature Extraction

Pattern extraction is a core strength of machine learning from which fluid dynamics stands to benefit directly. A tremendous amount of machinery has been developed for image and audio processing, and it is fortunate that fluid flow measurements, both spatial and temporal, may readily capitalize on these techniques. Here, we begin by discussing linear and nonlinear dimensionality reduction techniques, also known as

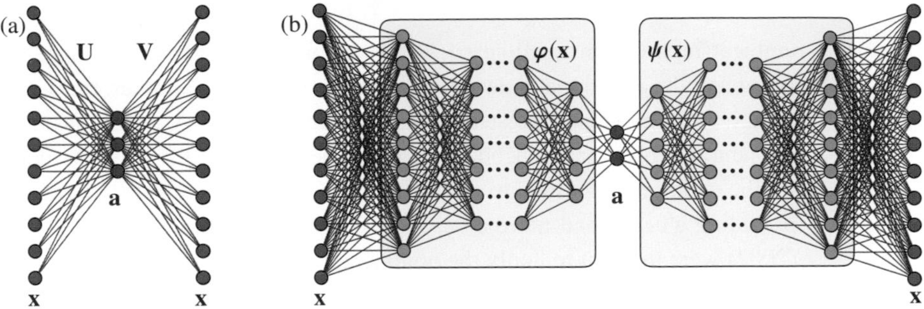

Figure 3.10 Illustration of shallow, linear autoencoder (a), versus deep nonlinear autoencoder (b). In the linear autoencoder, the node activation functions are linear, so that the encoder matrix $\mathbf{U}$ and decoder matrix $\mathbf{V}$ are chosen to minimize the loss function $\|\mathbf{x} - \mathbf{VUx}\|$. In the nonlinear autoencoder, the node activation functions may be nonlinear, and the encoder φ and decoder ψ are chosen to minimize the loss function $\|\mathbf{x} - \psi\varphi(\mathbf{x})\|$. In each case, the latent variable $\mathbf{a}$ describes the low-dimensional subspace or manifold on which the data evolves. Figure modified from Brunton, Noack and Koumoutsakos (2020).

embeddings of the high-dimensional data into either a linear subspace or a nonlinear manifold, as depicted in Figure 3.3. These techniques are related to the classical POD approach (see Chapter 6), but may be extended for nonlinear mappings with machine learning. Next, we discuss the use of clustering and classification, a well-developed field of machine learning, for decomposing flow behaviors.

3.3.1 Dimensionality Reduction

It is common in fluid dynamics to define an orthogonal linear coordinate transformation from physical coordinates into a *modal* basis. For example, the Fourier transform is one such change of basis that has advantages for computational speed and accuracy. However, the Fourier transform is rather restrictive on allowable boundary conditions and geometries. The proper orthogonal decomposition (Lumley 1970) provides a generalization of the Fourier transform, resulting in an orthogonal basis for complex geometries based on empirical measurements. Sirovich (1987) developed the snapshot POD, reducing the computation to a singular value decomposition of a matrix of flow field snapshots. In the same year, Sirovich applied the same POD algorithm to extract a low-dimensional feature space for human faces, which originated much of the modern computer vision literature (Sirovich & Kirby 1987).

POD is closely related to principal component analysis (PCA), which is a cornerstone method of applied statistics that is used to mine dominant correlations in high-dimensional data. POD/PCA is also known under other names, including the Hotelling transformation, Karhunen–Loève decomposition, and empirical orthogonal functions. It is possible to build a simple neural network, called an autoencoder, to compress high-dimensional data for a compact representation, shown in Figure 3.10. This network comprises an encoder network that embeds high-dimensional data into a latent space, followed by a decoder network that lifts from the latent space back to the ambient high-dimensional space. When the neural network activation functions are linear and the encoder and decoder are transposes of one another, the autoencoder can be shown to be closely related to the standard POD/PCA decomposition. However, the neural network autoencoder is more general, and with nonlinear activation units for the nodes, it is possible to represent *nonlinear* analogues of POD/PCA, potentially providing a better and more compact coordinate system. Milano and Koumoutsakos (2002) were the first to apply the nonlinear autoencoder for dimensionality reduction in fluid systems, building nonlinear embeddings to represent near wall turbulence.

Because of the universal approximation theorem (Hornik et al. 1989), which states that a sufficiently large neural network can represent an arbitrarily complex input–output function, *deep* neural networks will continue be leveraged to obtain more effective nonlinear coordinates for increasingly complex flows. However, it is important to note that deep learning requires extremely large volumes of training data, and the resulting models are typically only good for interpolation and cannot be trusted for extrapolation beyond the training data. In many modern machine learning applications, such as image classification, the training data is becoming so vast that it may be expected that most future classification tasks will fall under interpolation of the training data. For example, the ImageNet data set in 2012 (Krizhevsky et al. 2012) contained over 15 million labeled images, which sparked the current movement in *deep learning* (LeCun et al. 2015, Goodfellow et al. 2016). We are still far away from this paradigm in fluid mechanics, with neural network solutions providing targeted and focused models that do not readily generalize. However, it may be possible in the coming years and decades to curate large and complete enough fluid databases that such deep interpolation may be more universally useful.

3.3.2 Clustering and Classification

Clustering and classification are core machine learning techniques, with dozens of mature algorithms to choose from, depending on various factors, such as the size of the data and the number of desired categories. Clustering is an unsupervised data mining technique that identifies similarity-based groupings in the data. The most common and simple clustering algorithm is the k-means algorithm, which partitions data instances into k clusters; an observation will belong to the cluster with the nearest centroid, resulting in a partition of the data space into Voronoi cells. Kaiser et al. (2014) used k-means clustering to dramatically reduce the effective dimension of a high-dimensional fluid mixing layer, resulting in an accurate and efficient data-driven model. This cluster-based representation scaled with the number of clusters, not the ambient measurement dimension of the fluid, enabling tractable Markov transition models to describe the evolution of the flow from one state to another. Their modeling approach is also interpretable, in the sense that each cluster centroid may be associated with a physical flow field. In Amsallem et al. (2012) the k-means clustering approach was used to partition phase space into distinct regions where separate local reduced-order bases were constructed, resulting in improved stability and robustness to parameter variations.

Classification is a related supervised learning approach where the algorithm learns to group data into several labeled categories. Colvert et al. (2018) developed a neural network classifier to characterize bluff-body wake topology, such as 2S, 2P+2S, 2P+4S, and so on, in the wake of a pitching airfoil from local measurements of the vorticity. Wake detection is believed to be a biologically relevant behavior, for example to detect or avoid certain species of fish. Related work investigated the classification performance with different types of sensors (Alsalman et al. 2018) and other classifiers, such as the k-nearest neighbors (KNN) algorithm (Wang & Hemati 2017). Neural network classifiers may also be combined with dynamical systems models, for example to detect flow disturbances and estimate their parameters (Hou et al. 2019). Community detection (Meena et al. 2018) and other network-based analyses (Nair & Taira 2015) have been applied with great success to model and understand flows. Sparse representation for classification (Wright et al. 2009) has also been applied to classify wakes by Bright et al. (2013) in one of the earliest examples of classification in fluid dynamics. Extensions to DMD (Kramer et al. 2015) and sparse sensor optimization (Brunton, Brunton, Proctor & Kutz 2016a) have since been developed.

3.3.3 Randomized Linear Algebra and Sparse Sampling

Despite the high ambient dimension of fluid measurements and simulations, there are often low-dimensional coherent structures that adequately describe the flow. Randomized linear algebra (Mahoney 2011, Halko, Martinsson & Tropp 2011, Liberty et al. 2007) is a growing field that leverages this low-dimensional structure to accelerate computations, such as POD based on the SVD (Sarlos 2006, Rokhlin et al. 2009, Halko, Martinsson, Shkolnisky & Tygert 2011). Randomized algorithms rely on the fact that if a large data matrix has structure, then this structure is often preserved after multiplying with a smaller random matrix to project onto a random lower-dimensional subspace where computations are more efficient. Randomized DMD has also been developed recently for fluid flows (Bistrian & Navon 2016, Erichson et al. 2019). In addition to facilitating fast numerical linear algebra, low-dimensional structure also facilitates sparse and efficient measurements (Manohar et al. 2018). If there are dominant patterns in the flow that are known, it is unnecessary to measure the flow in every location. Instead, it is possible to measure at a few key locations and infer the amplitude of these patterns, and therefore estimate the flow. Sparse measurements are desirable for control applications, where rapid computations and decisions reduce latency and improve robust performance. Classification and detection of fluid events based on measurements are a crucial capability for intelligent flow control systems, and it is desirable to perform such categorical tasks with limited or incomplete information (Bright et al. 2013, Kramer et al. 2015, Brunton, Brunton, Proctor & Kutz 2016a, Callaham et al. 2019).

3.3.4 Super Resolution and Experimental Flow Processing

Billions of dollars of industrial machine learning research has focused on image science, providing robust techniques that may be directly leveraged for flow field analysis, as flow fields closely resemble images. Super resolution is a particularly intriguing field of research, where it is possible to statistically infer a high-resolution image from low-resolution measurements, given sufficiently rich high-resolution training data. At first, super-resolution sounds too good to be true, but in reality it is simply a statistical inference based on past experience. If one is filling in an image of a mountain landscape, and there is a low-resolution patch of green, it is much more likely that this region is a tree or a bush, and not a green elephant; thus, super resolution seeks to find the microstructure that is most consistent with the neighboring low-resolution pixels and a library of high-resolution examples. Several approaches have been developed for super resolution, for example based on a library of examples (Freeman et al. 2002), neighbor embedding (Chang et al. 2004), sparse representation in a library (Yang et al. 2010), and most recently based on CNNs (Dong et al. 2014). Both fluid simulations and experiments stand to benefit from super-resolution techniques, as improved resolution would improve our ability to resolve relevant flow structures and dynamic mechanisms. For example, particle image velocimetry (PIV) (Adrian 1991, Willert & Gharib 1991) would directly benefit from improved resolution, potentially improving hardware around the world with what essentially amounts to a software update. Large eddy simulations (LES) (Germano et al. 1991, Meneveau & Katz 2000) suffer from similar resolution limitations, and the ability to infer sub-grid-scale models with advanced machine learning techniques is an area of active research. Fukami et al. (2018) recently developed a super-resolution algorithm based on a CNN and demonstrated that it is possible to reconstruct turbulent flows from coarse data, preserving the energy spectrum.

Machine learning may also be used for flow cleansing and imaging in experiments. There are several learning approaches that are used for speeding up PIV (Knaak et al. 1997) and for particle tracking (Labonté 1999), with impressive demonstrations for three-dimensional Lagrangian particle tracking (Ouellette et al. 2006). Deep CNNs are now being used to directly construct velocity fields from PIV image pairs, bypassing traditional cross-correlation techniques (Lee et al. 2017). Robust principal component analysis (RPCA) (Wright et al. 2009) has recently been applied to fluid flow filtering (Scherl et al. 2020), making it possible to remove outliers and spurious measurements.

3.4 Modeling Flow Dynamics

3.4.1 Linear Models: Dynamic Mode Decomposition and Koopman Analysis

Many of the classical techniques in system identification may be considered as machine learning, as they are data-driven models that hopefully generalize beyond the training data. A majority of classical system identification methods result in linear models (Juang & Pappa 1985, Juang 1994, Ljung 1999), which are simple to analyze and use for control design (Dullerud & Paganini 2000, Skogestad & Postlethwaite 2005). Despite the powerful tools for linear model reduction and control, the assumption of linearity is often overly restrictive for real-world flows. Turbulent fluctuations are inherently nonlinear, and often our goal is not to stabilize an unstable laminar solution, but instead to modify the nature of the turbulent statistics. However, linear modeling and control can be quite useful for stabilizing unstable fixed points, for example to maintain a laminar boundary layer profile or suppress oscillations in an open cavity (Kim & Bewley 2007, Brunton & Noack 2015). Refer to Chapter 12 for an overview of system identification and to Chapter 10 for an overview of linear dynamical systems.

The DMD (Schmid 2010, Tu, Rowley, Luchtenburg, Brunton & Kutz 2014, Kutz, Brunton, Brunton & Proctor 2016) is a recent technique introduced by Schmid (2010) in the fluid dynamics community to extract spatiotemporal coherent structures from high-dimensional time-series data, resulting in a low-dimensional linear model for the evolution of these dominant coherent structures. DMD is an entirely data-driven regression technique, and is equally valid for time-resolved experimental and numerical data. Shortly after being introduced, it was shown that DMD is closely related to the Koopman operator (Rowley et al. 2009, Mezić 2013), which is an infinite-dimensional linear operator that describes how *all* measurement functions of the fluid state will evolve in time. There have also been connections between the linear resolvent analysis and Koopman (Sharma et al. 2016). Because the original DMD algorithm is based on linear measurements of the flow field (i.e., direct measurements of the fluid velocity or vorticity field), the resulting models are generally not able to capture nonlinear transients, but are well suited to capture periodic phenomena. See Chapter 7 for more details on DMD and Koopman operator theory.

To improve the performance of DMD, researchers are using machine learning to uncover nonlinear coordinate systems where the dynamics appear linear (Brunton & Kutz 2019). The extended DMD (eDMD) (Williams, Rowley & Kevrekidis 2015) and variational approach of conformation dynamics (VAC)) (Noé & Nuske 2013, Nüske et al. 2016) enrich DMD models with nonlinear measurements, leveraging kernel methods (Williams, Rowley & Kevrekidis 2015) and dictionary learning approaches (Li et al. 2017) from machine learning. Although the resulting models are simple and linear, obtaining nonlinear coordinate systems may be arbitrarily complex, as these special nonlinear measurement functions may not have simple, or even closed-form solutions. Fortunately, this type of arbitrary function representation problem is ideally suited for neural networks, especially the emerging deep learning approaches with large multilayer networks. Deep learning is also being used to identify these nonlinear coordinate systems, related to eigenfunctions of the Koopman operator (Takeishi et al. 2017, Yeung et al. 2017, Lusch et al. 2018, Mardt et al. 2018, Wehmeyer & Noé 2018, Otto & Rowley 2019). The VAMPnet architecture (Mardt et al. 2018, Wehmeyer & Noé 2018) uses a time-lagged auto-encoder and a custom variational score to identify Koopman coordinates on an impressive protein folding example. The field of fluid dynamics may benefit from leveraging these techniques from neighboring fields, such as molecular dynamics, which have similar modeling issues, including stochasticity, coarse-grained dynamics, and massive separation of timescales.

3.4.2 Modeling Dynamics with Neural Networks

Neural networks are among the most powerful and expressive machine learning architectures, and they are well suited to represent the evolution of complex dynamical systems. Even for simple continuous-time dynamics, the discrete-time flow map that advances the system state forward by some amount of time may be arbitrarily complex, and generally the map becomes more complex for longer advances in time. Fortunately, deep learning is ideal for this type of arbitrary function approximation. The history of neural networks in dynamics goes back decades, with early examples modeling ordinary and partial differential equations (Gonzalez-Garcia et al. 1998, Lagaris et al. 1998). More recently, neural networks are used for nonlinear system identification to model fluids (Glaz et al. 2010, Zhang et al. 2012, Semeraro et al. 2017). LSTM networks (Hochreiter & Schmidhuber 1997) are a type of RNNs (Elman 1990) that are increasingly being used to model dynamical systems and extreme events (Wan et al. 2018, Vlachas et al. 2018). More generally, deep learning is providing a new approach to model systems in physics (Raissi & Karniadakis 2018, Cranmer et al. 2019, 2020, Noé et al. 2019, Raissi et al. 2019, Greydanus et al. 2019, Finzi et al. 2020, Zhong & Leonard 2020).

Although deep learning is a promising technique to model dynamical systems in fluid mechanics, there are several challenges that must be addressed. Deep learning relies on tremendous volumes of training data that are sufficiently rich so as to represent any example that will likely be seen in the future. Further, deep learning is inherently interpolative, simply providing a model based on the training examples. In many computer vision and speech recognition examples, the training data sets are so vast that nearly all future tasks may be viewed as an interpolation on the training data, although this scale of training has not been achieved to date in fluid mechanics. Therefore, neural network models should be used with extreme care for extrapolation tasks. Neural networks are also prone to overfitting, and the resulting models are often opaque and uninterpretable. Finally, the structure and scale of neural networks make it difficult to incorporate partially known physics, such as symmetries, constraints, and conserved quantities, although this is an area of active research.

3.4.3 Parsimonious Nonlinear Models

Parsimony is a common theme in the natural sciences, where the simplest model describing observed data is often presumed to be the correct model. Moreover, simplicity is a powerful technique to promote generalizability in models. Although neural networks are a powerful branch of machine learning, it is important not to forget that there are many other techniques that identify parsimonious models, balancing predictive accuracy with model complexity. These minimalistic models tend to prevent overfitting and promote both interpretability and generalizability. One of the seminal approaches to parsimonious modeling used genetic programming to discover conservation laws and parsimonious governing equations (Bongard & Lipson 2007, Schmidt & Lipson 2009). Sparse regression has also been used to identify ordinary (Brunton, Proctor & Kutz 2016a) and partial differential equations (Rudy et al. 2017, Schaeffer 2017).

Sparse model identification has been successfully adapted to obtain reduced-order models of several fluid systems, including fluid flow past a cylinder and the shear-driven cavity flow (Brunton, Proctor & Kutz 2016a, Loiseau & Brunton 2018). Because the SINDy algorithm is based on an iteratively thresholded least squares regression, it is straightforward to incorporate known constraints via the Karush–Kuhn–Tucker (KTT) equations, as used by Loiseau and Brunton to enforce energy conserving constraints on the quadratic nonlinearities (Loiseau & Brunton 2018). Later, it was shown that SINDy models can be obtained from sparse, physical measurements of a system, such as time-series measurements of lift and drag forces (Loiseau et al. 2018), producing models that may be more robust to mode deformation, moving geometry, and varying operating conditions, compared with Galerkin projection models. More recently, there have been several applications of this approach in fluid dynamics for reduced-order modeling (Murata et al. 2019), for turbulence closure modeling (Beetham & Capecelatro 2020, Schmelzer et al. 2020), and in the related fields of electroconvection and plasma physics (Guan et al. 2020, Kaptanoglu et al. 2020).

3.4.4 Closure Models with Machine Learning

One of the most promising applications of machine learning in fluid dynamics is to develop advanced turbulence closure models (Duraisamy et al. 2019). The extreme separation of scales, in both space and time, for the turbulent flows over engineering relevant configurations (e.g., aircraft wings, submarines, etc.), makes it exceedingly costly to resolve all scales in simulation, as shown in Figure 3.11. Even with Moore's law, we are decades away from resolving all scales in complex turbulent flows. Because large, energetic coherent structures are often the most relevant for design, it is common to truncate small scales and model their effect on the large scales with a *turbulence closure model*, shown in Figure 3.12. There are several common approaches, including Reynolds averaged Navier–Stokes (RANS) and LES models, which make it possible to apply simulations to approximate the flow over engineering relevant configurations. However, the underlying physics and accuracy of these turbulence closures is often dubious, and models often require careful tuning to match fully resolved simulations or experiments.

There are several recent efforts to incorporate data-driven methods from machine learning to improve turbulence modeling, as nicely summarized in Kutz (2017) and Duraisamy et al. (2019). In one prominent line of research, researchers have used various machine learning techniques to identify and model discrepancies in the Reynolds stress tensor between RANS model and high-fidelity simulations (Ling &

Production Range
Inertial Range
Dissipation Range
-5/3
Energy
Log E(k)
Wavenumber, Log k

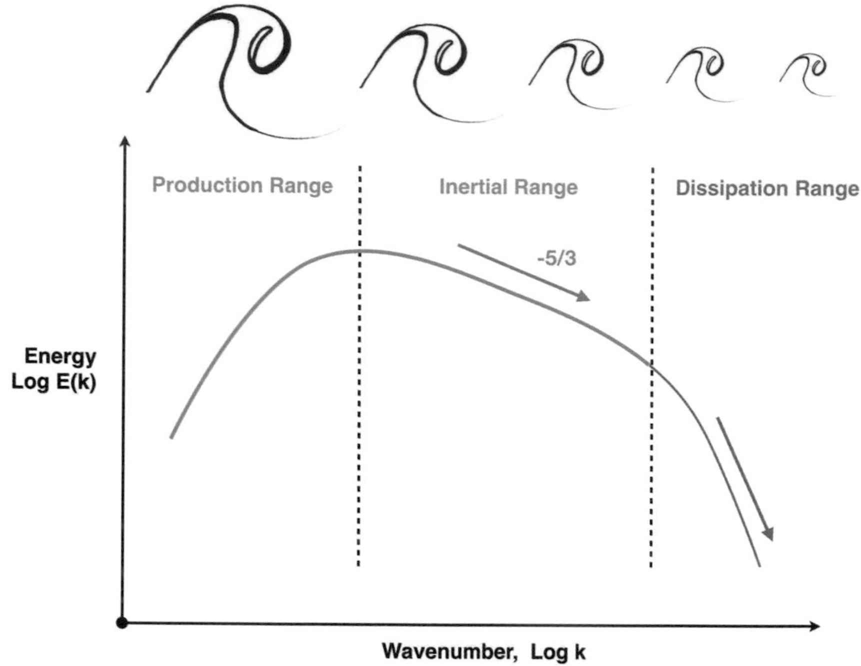

Figure 3.11 Turbulence is characterized by a large range of spatial and temporal scales, as illustrated in this famous turbulence energy cascade. Resolving all scales for industrially relevant flows has been notoriously challenging. See Pope (2000) for details.

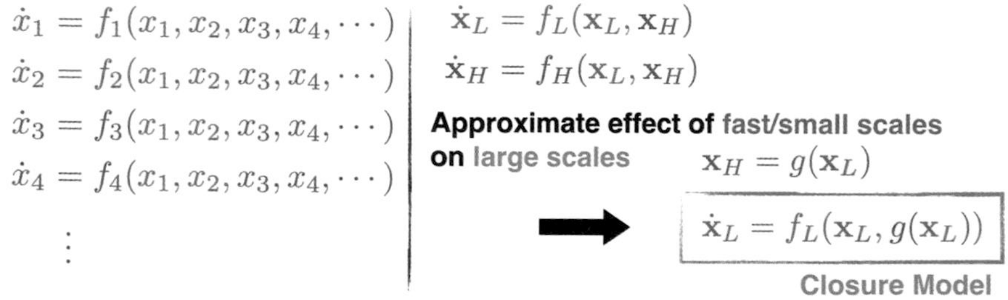

$\dot{x}_1 = f_1(x_1, x_2, x_3, x_4, \cdots)$
$\dot{x}_2 = f_2(x_1, x_2, x_3, x_4, \cdots)$
$\dot{x}_3 = f_3(x_1, x_2, x_3, x_4, \cdots)$
$\dot{x}_4 = f_4(x_1, x_2, x_3, x_4, \cdots)$
$\vdots$
$\dot{\mathbf{x}}_L = f_L(\mathbf{x}_L, \mathbf{x}_H)$
$\dot{\mathbf{x}}_H = f_H(\mathbf{x}_L, \mathbf{x}_H)$
Approximate effect of fast/small scales on large scales
$\mathbf{x}_H = g(\mathbf{x}_L)$
$\dot{\mathbf{x}}_L = f_L(\mathbf{x}_L, g(\mathbf{x}_L))$
Closure Model

Figure 3.12 The turbulence closure problem may be formulated as modeling the effect of high frequency structures on the energy containing low frequency structures.

Templeton 2015, Ling, Kurzawski & Templeton 2016, Parish & Duraisamy 2016, Xiao et al. 2016, Singh et al. 2017, Wang, Wu & Xiao 2017). Ling and Templeton (2015) compare support vector machines, Adaboost decision trees, and random forests to classify and predict regions of high uncertainty in the Reynolds stress tensor. Wang, Wu and Xiao (2017) went on to use random forests to build a supervised model for the discrepancy in the Reynolds stress tensor. Xiao et al. (2016) leveraged sparse online velocity measurements in a Bayesian framework to infer these discrepancies. In a related line of work, Parish and Duraisamy (2016) develop the *field inversion and machine learning* modeling framework that builds corrective models based on inverse modeling. This framework was later used by Singh et al. (2017) to develop a neural network enhanced correction to the Spalart–Allmaras RANS model, with excellent

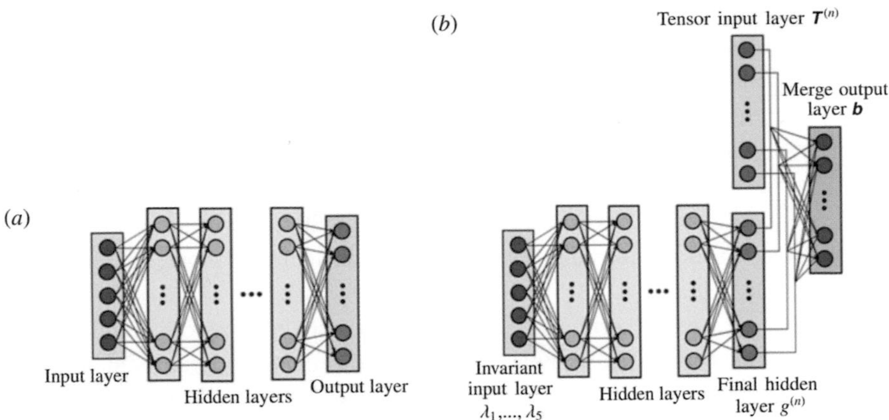

Figure 3.13 Comparison of standard neural network architecture (a) with modified neural network architecture for identifying Galilean invariant Reynold stress models (b), reproduced with permission from Ling, Kurzawski and Templeton (2016).

performance. One particular interesting application of machine learning for turbulence modeling was developed by Ling, Kurzawski and Templeton (2016), which involved the first *deep* neural network architecture (i.e., with 8–10 hidden layers) trained to model the anisotropic Reynolds stress tensor, as shown in Figure 3.13. What was so interesting about this work is that they developed a novel architecture that incorporates a multiplicative layer to embed Galilean invariance into the tensor predictions. This provides an innovative and simple approach to embed known physical symmetries and invariances into the learning architecture (Ling, Jones & Templeton 2016). For LES closures, Maulik et al. (2019) employ artificial neural networks to predict the turbulence source term from coarsely resolved quantities.

3.5 Control and Optimization

Fluid flow control is one of the grand challenge problems of the modern era, with nearly limitless potential to enable advanced technologies. Indeed, working fluids are central to many trillion dollar industries (transportation, health, energy, defense), and even modest improvements to flow control could have transformative impact through lift increase, drag reduction, mixing enhancement, and noise reduction. However, flow control is generally a highly non-convex and high-dimensional optimization problem, which has remained challenging despite many concerted efforts (Brunton & Noack 2015). Fortunately, machine learning may be considered as a growing body of data-driven optimization procedures that are well suited to highly nonlinear and non-convex problems. Thus, there are renewed efforts to solve traditionally intractable flow control problems with emerging techniques in machine learning.

There is a considerable effort applying reinforcement learning (Sutton & Barto 2018) to fluid flow control problems (Beintema et al. 2020, Rabault & Kuhnle 2020), as will be discussed more in Chapter 18. In particular, controlling the motion of fish (Gazzola et al. 2014, Gazzola et al. 2016, Novati et al. 2017, Verma et al. 2018) and of robotic gliders (Reddy et al. 2018) has experienced great strides with more powerful reinforcement learning architectures. These techniques are also being used to optimize the flight of uninhabited aerial vehicles (Kim et al. 2004, Tedrake et al. 2009) and for path planning applications (Colabrese et al. 2017). It is believed that advances will continue, as reinforcement learning with deep neural networks is an active area of machine learning research, with considerable progress, for example in Alpha Go (Mnih et al. 2015).

Genetic algorithms (Holland 1975) and genetic programming (Koza 1992) have also been widely applied to fluid flow control (Dracopoulos 1997, Fleming & Purshouse 2002, Duriez et al. 2017, Noack 2019). Neural networks have also been used for flow control, with decades of rich history (Phan et al. 1995, Lee et al. 1997). Recently, deep learning has been used to improve model predictive control (MPC) efforts (Kaiser et al. 2018, Bieker et al. 2019) with impressive performance gains. Finally, machine learning approaches have also been used for aerodynamic and hydrodynamic shape and motion optimization (Hansen et al. 2009, Strom et al. 2017).

In another important vein of research, machine learning and sparse optimization (Brunton & Kutz 2019) are being leveraged for sensor and actuator placement (Manohar et al. 2018). Sensors and actuators are the workhorses of active flow control. Developments in sensing and actuation hardware will continue to drive advances in flow control (Cattafesta III & Sheplak 2011), including smaller, higher-bandwidth, cheaper, and more reliable devices that may be integrated directly into existing hardware, such as wings or flight decks. Many competing factors impact control design, and a chief consideration is the latency in making a control decision, with larger latency imposing limitations on robust performance (Skogestad & Postlethwaite 2005). Thus, as flow speeds increase and flow structures become more complex, it is increasingly important to make fast control decisions based on efficient low-order models, with sensors and actuators placed strategically to gather information and exploit flow sensitivities. Optimal sensor and actuator placement are one of the most challenging unsolved problems in flow control (Giannetti & Luchini 2007, Chen & Rowley 2011). Nearly every downstream control decision is affected by these sensor or actuator locations, although determining optimal locations amounts to an intractable brute force search among the combinatorial possibilities. Therefore, the placement of sensors and actuators is typically chosen according to heuristics and intuition. Figure 3.14 shows schematically how sensors and actuators are used for flow control. Recent work by Manohar et al. (2018) has demonstrated efficient, near-optimal sensor placement based on sparse optimization. This is a promising area of development for flow control.

3.6 Challenges for Machine Learning in Fluid Dynamics

Applying machine learning methods to model dynamical systems from physics, such as fluid dynamics, poses a number of unique challenges and opportunities. In physical

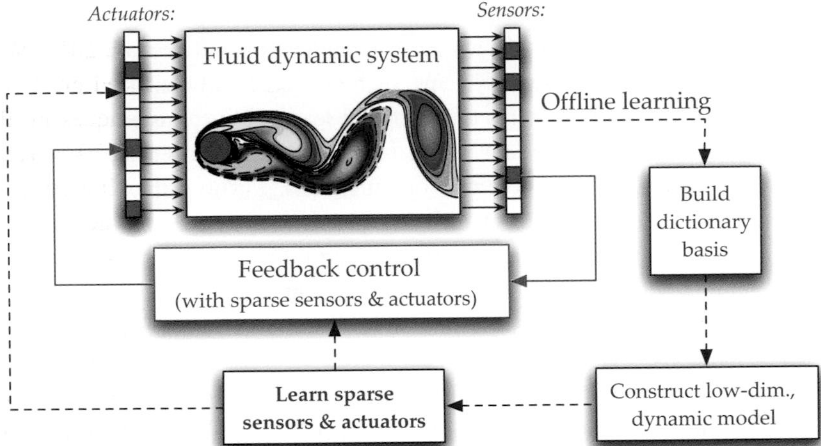

Figure 3.14 Schematic illustrating optimal sensor and actuator design for closed-loop feedback fluid flow control. Learning sparse sensor and actuator locations is now possible using convex optimization, and coherent structures in the fluid facilitate sparse sensing and sparse actuation. The offline learning represents a slow optimization, enabling online feedback (red).

systems, interpretability and generalizability are essential, as obtaining a simple model that explains a range of phenomena is a cornerstone of physics. In fact, a well-crafted model will yield hypotheses for new phenomena that have not been observed before. This principle is clearly exhibited in the parsimonious formulation of classical mechanics in Newton's second law, and later by the Hamiltonian and Lagrangian formulations, which provide an extremely general model framework (Abraham et al. 1978, Marsden & Ratiu 1999).

Unsteady fluid mechanical systems may be viewed as high-dimensional dynamical systems, with the associated challenges of varying parameters, multiscale dynamics, noise and disturbances, latent and hidden variables, and transients, all of which require care when using machine learning. In machine learning for dynamics, there are often two dueling tasks of discovering unknown physics and improving models by incorporating partially known physics. In many learning architectures, such as neural networks, it is unclear how to incorporate partially known physics, which often come in the form of symmetries, constraints, and conservation laws. For example, in fluid mechanics, energy-preserving constraints are known to improve the long-time stability and performance of nonlinear models, while standard Galerkin projection methods often suffer from stability issues (Carlberg et al. 2017). However, a major drawback of regression-based methods is the possible loss of existing symmetries in the governing equations, which may otherwise be included in the physics-based Galerkin projection methods described previously (Noack et al. 2011, Balajewicz et al. 2013, Carlberg et al. 2015, Schlegel & Noack 2015). A notable exception is the physics-constrained multilevel quadratic regression used to identify models in climate and turbulence (Majda & Harlim 2012).

It is believed that the widespread application and development of machine learning techniques to physical systems, such as fluids, will benefit both fields. Indeed, one of the remarkable features of biological learning systems is their ability to model and interact with a *physical* world, learning representations of the rules that are useful for complex multi-objective optimizations. Pushing the limits of machine learning by applying them to challenging physical systems may well yield transformative breakthroughs in how we structure and train models. However, machine learning is unlikely to displace established experimental and numerical methods, and will rather complement these approaches, much as CFD has complemented earlier experimental methods.

It is important to be clear about where *not* to employ machine learning. For example, it is highly unlikely that machine learning will be more efficient than the fast Fourier transform (FFT) in most applications. Similarly, when optimal or near-optimal solutions are known, it may be best to use these solutions. However, it is rare to have such optimal solutions in the real-world, and so machine learning will likely be a useful alternative in these cases, where heuristics and simplified models are currently employed.

Part II

Methods from Signal Processing

4 Continuous and Discrete Linear Time-Invariant Systems

M. A. Mendez

This chapter reviews the fundamentals of continuous and discrete linear time-invariant (LTI) systems with single-input single-output (SISO). We start from the general notions of signals and systems, the signal representation problem, and the related orthogonal bases in discrete and continuous forms. We then move to the key properties of LTI systems and discuss their eigenfunctions, the input–output relations in the time and frequency domains, the conformal mapping linking the continuous and the discrete formulations, and the modeling via differential and difference equations. Finally, we close with two important applications: (linear) models for time-series analysis and forecasting and (linear) digital filters for multi-resolution analysis. This chapter contains seven exercises, the solution of which is provided in the book's webpage.[1]

4.1 On Signals, Systems, Data, and Modeling of Fluid Flows

A *signal* is any function that conveys information about a specific variable (e.g., velocity or pressure) of interest in our analysis. A signal can be a function of one or multiple variables; it can be continuous or discrete; and can have infinite or finite duration or extension (that is be non-null only within a range of its domain). Signals are produced by *systems*, usually in response to other signals or due to the interaction between different interconnected *subsystems*. In the most general setting, a system is any entity that is capable of manipulating an *input* signal and producing an *output* signal.

At such a level of abstraction, a vast range of problems in applied science fall within the framework of this chapter. For a fluid dynamicist, the flow past an airfoil is a system in which the inputs are the flow parameters (e.g., free-stream velocity and turbulence) and control parameters (e.g., the angle of attack), and the outputs are the drag and lift components of the aerodynamic force exchanged with the flow.

Any measurement chain is a system in which the input is the quantity *to be measured*, and the output is the quantity that *is measured*. A hot-wire anemometer, for example, is a complex system that measures the velocity of flow by measuring the heat loss by a wire that is heated by an electrical current. Any signal processing

[1] www.datadrivenfluidmechanics.com/download/book/chapter4.zip

technique for denoising, smoothing, and filtering can be seen as a system that takes in input the raw signal and outputs its processed version (e.g., with enhanced details or reduced noise).

Regardless of the number of subsystems composing a system, and whether the system is a physical system, a digital replica of it, or an algorithm in a computer program, the relations between input and output are governed by a *mathematical model*. The derivation of such models is instrumental for applications encompassing simulation, prediction, or control. Models can be phrased with various degrees of sophistication, depending on the purposes for which these are developed. They usually take the form of partial differential equations (PDEs), ordinary differential equations (ODEs) or difference equations (DEs), or simple algebraic relation. Different models might have different ranges of *validity* (hence different degrees of generalization) and might involve different levels of complexity in their *validation*.

Models can be derived from two different routes, hinging on *data* and *experimentation* in various ways. The first route is that of *fundamental principles*, based on the division of a system into subsystems for which empirical observations have allowed to derive well established and validated relations. In the example of the hot-wire anemometer, the subsystems operate according to Newton's cooling law for forced convection, the resistive heating governed by Joule's law, the thermoelectric laws relating the wire resistance to its temperature, and the Ohm's and Kirchhoff's laws governing the electric circuits that are designed to indirectly measure the heat losses. Arguably, none of these laws were formulated with a hot-wire anemometer in mind. Yet, their range of validity is wide enough to accommodate *also* such an application: these laws generalize well.

In the example of the flow past an airfoil, the system is governed by Navier–Stokes equations, which incorporates other laws such as constitutive relations for the shear stresses (e.g., Newton's law for a Newtonian fluid) and the heat fluxes (e.g., Fourier's Law for conduction). These closure relations led to the notion of fluid properties such as dynamic viscosity and thermal conductivity and were also derived in simple experiments that did not target any specific application.

Because of their remarkable level of generalization, we tend to see these laws as simple mathematical representations of the "laws of nature." Whether nature is susceptible to mathematical treatment is a question with deep philosophical aspects. As engineers, we accept the pragmatic view of relying on models and laws if these are *validated* and useful. This is the foundation of our scientific and technical background.

The second route is that of *system identification* or *data-driven approach*, based on the inference of a suitable model from a (usually large) set of input–output data. The model might be constrained to a certain parametric form (e.g., with a given order in the differential equations) or can be completely inferred from data. In the first case, we face a regression problem of identifying the parameters such that the model *fits* the data. In the second case, an algorithm proposes possible models (e.g., in Genetic Programming, see Chapter 14) or uses such a complex parametrization that an analytic form is not particularly interesting (e.g., in Artificial Neural Networks, see Chapter 3).

This paradigm is certainly not new. An excellent example of system identification is the method proposed by the Swedish physicist Angström to measure thermal conductivity of a material in 1861 (Sundqvist 1991). This method consists in using a long metal rod in which a harmonic heat wave is produced by periodically varying the temperature at one end. The spatial attenuation of the heat wave is modeled by the unsteady one-dimensional heat equation, leaving the thermal conductivity as an unknown. A simple formula can be derived for the harmonic response[2] of the system, and the thermal conductivity is then identified by fitting the model to data.

Although the data-driven paradigm has a long history (Ljung 2008), its capabilities have been significantly augmented and empowered by our increasing ability to generate enormous amounts of data and by the powerful tools popularized by the ongoing machine learning revolution. Some of these are described in Chapters 1 and 3. The big challenge (and the big opportunity) in big data is to combine physical modeling with the data-driven modeling. It is the author's opinion that the combination of machine learning tools and the general framework of signals and systems can accommodate the formulation of many problems of applied science and, with the required caution, problems in fluid mechanics.

Caution is certainly required as the Navier–Stokes equations governing fluid flows feature the entire spectra of complexities and challenges that a system analyst could think of. In most configurations of interest, fluid flows are *nonlinear*: a linear combination of input does not produce a predictable linear combination of outputs. It is thus not possible to predict the response of the system from a dictionary of known outputs. Fluid flows are often *high dimensional* and involve many scales: the amount of information required to identify the state of a flow system is tremendous. Fluid flows are often chaotic systems and are thus *unpredictable*: an infinitesimal change in initial or boundary conditions quickly yields very different instantaneous states of the system. Decades of fluid mechanics research aimed at models that are of *reduced order*, which include engineering *closure laws* (e.g., turbulence modeling) or treat the flows in terms of statistical quantities that are usually predictable.

While the Navier–Stokes equations are on the top of the ladder of complexity in the signal-system framework, this chapter treats systems that are at the bottom of the ladder: systems that are *linear*, *single dimensional*, *deterministic*, and *time invariant*. These are a subclass of linear time (or Translation) invariant systems (LTI). The interest in such a review is justified by three reasons. The first is that LTI systems are still often encountered in practice, both in physical systems[3] and in most signal processing operations. The second is that many nonlinear systems can be treated reasonably well as linear systems if the range of operations is close enough to a fixed state; this is what led to the remarkable success of many linear control methods described in Chapter 10. The third reason is that this chapter provides the background material to understand time-frequency analysis in Chapter 5 and the multi-resolution analysis underpinning the data decompositions in Chapter 8, the state-space models and the linear control theory in Chapter 10, and the system identification tools in Chapter 12.

[2] The harmonic response of a system is described in Section 4.7.

[3] See the previous example of Angström's system.

Most of the material presented in this chapter can be found in classic textbooks on signal and systems (Ljung & Glad 1994, Oppenheim et al. 1996, Hsu 2013), signal processing (Williamson 1999, Hayes 2011, Ingle & Proakis 2011), orthogonal transforms (Wang 2009), system identification (Ljung 1999, Oppenheim 2015), or control theory (Sigurd Skogestad 2005, Ogata 2009). We begin this chapter by introducing the relevant notation.

4.2 A Note about Notation and Style

In their most general form, continuous and discrete LTI systems admit multiple inputs and respond with multiple outputs (MIMO systems) or have a single input and respond with a single output (SISO systems)[4]. This classification and the relevant notation are further illustrated in the block diagram in Figure 4.1.

In a continuous SISO system, inputs and outputs are denoted, respectively, as continuous functions $u(t), y(t) \in \mathbb{R}$ with $t \in \mathbb{R}$. Following the signal processing literature, for discrete systems, these are denoted using an index notation, as $u[k], y[k] \in \mathbb{R}$ with $k \in \mathbb{Z}$. In this chapter, discrete and continuous signals are assumed to be linked by a sampling process, hence the time domain is discretized as $t \longrightarrow t_k = k\Delta t$ with an index $k \in \mathbb{Z}$, sampling period $\Delta t = 1/f_s$, and (constant) sampling frequency f_s. Therefore, the notation $u(t_k)$ is equivalent to the index notation $u[k]$, but the second makes no link to the sampling process nor the time axis.

In a MIMO system, both the inputs and the outputs are vectors $\mathbf{u}(t), \mathbf{u}[k] \in \mathbb{R}^{n_I}$ and $\mathbf{y}(t), \mathbf{y}[k] \in \mathbb{R}^{n_O}$, with n_I and n_O the number of inputs and outputs. MIMO systems are better treated in state-space representation presented in Chapter 10, hence this chapter only focuses on SISO systems.

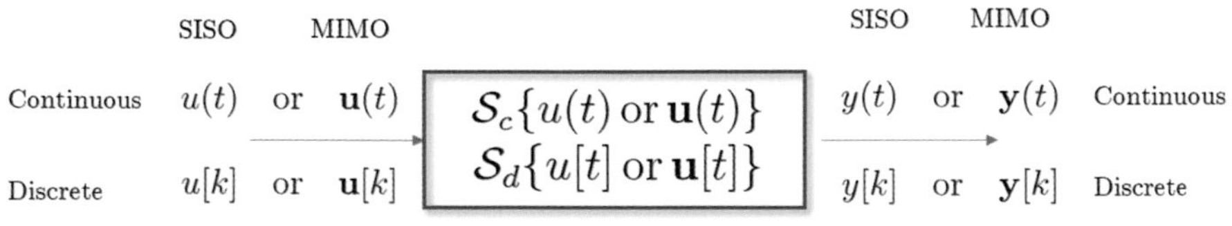

	SISO		MIMO		SISO		MIMO	
Continuous	$u(t)$	or	$\mathbf{u}(t)$	$\mathcal{S}_c\{u(t) \text{ or } \mathbf{u}(t)\}$	$y(t)$	or	$\mathbf{y}(t)$	Continuous
Discrete	$u[k]$	or	$\mathbf{u}[k]$	$\mathcal{S}_d\{u[t] \text{ or } \mathbf{u}[t]\}$	$y[k]$	or	$\mathbf{y}[k]$	Discrete

Figure 4.1 Block diagram representation of systems, classified as single-input single-output (SISO) or multiple-input multiple-output (MIMO) and continuous or discrete.

Signals and systems can contain a deterministic and a stochastic part. While the focus is mostly on the deterministic part, the treatment of stochastic signals is briefly recalled in Section 4.8.

[4] Or can be a combination of the two as in MISO/SIMO systems.

4.3 Signal and Orthogonal Bases

For reasons that will become clear in Section 4.4, it is convenient to represent a signal in a way that allows decoupling the contribution of every time instance. In other words, we seek to define a signal with respect to a very localized *basis* that allows for *sampling* the signal at a given time. With such a (unitary) basis, the sampling process can be done by direct comparison: for example, we say that a continuous signal $u(t)$ has $u(3) = 2$ because at time $t = 3$ this signal equals two times the element of the basis that is unitary (in a sense to be defined) at $t = 3$ and zero elsewhere. Mathematically, this "comparison" process is a *correlation*, the signal processing equivalent of the *inner product*. This notion is more easily introduced for discrete signals, considered in Section 4.3.2. Continuous signals are treated in 4.3.1.

4.3.1 Discrete Signals

Consider a discrete signal of finite duration $u[k] = [1,3,0,4,5]^T$, that is,[5] $k \in [0,\dots,4]$, which can be arranged as a column vector $\mathbf{u} \in \mathbb{R}^{5\times 1}$. In many applications, a finite duration signal is assumed to be a special case of an infinite duration signal, which is zero outside the available points. This is a common practice referred to as *zero padding* that will be further discussed in Chapter 8. This signal can be written as a linear combination of shifted unitary impulses:

$$\mathbf{u} = 1 \underbrace{\begin{bmatrix}1\\0\\0\\0\\0\end{bmatrix}}_{\delta[l-0]} + 3 \underbrace{\begin{bmatrix}0\\1\\0\\0\\0\end{bmatrix}}_{\delta[l-1]} + 0 \underbrace{\begin{bmatrix}0\\0\\1\\0\\0\end{bmatrix}}_{\delta[l-2]} + 4 \underbrace{\begin{bmatrix}0\\0\\0\\1\\0\end{bmatrix}}_{\delta[l-3]} + 5 \underbrace{\begin{bmatrix}0\\0\\0\\0\\1\end{bmatrix}}_{\delta[l-4]} . \tag{4.1}$$

The elementary basis to describe such a signal is thus the set:

$$\boldsymbol{\delta}_k = \delta[l-k] = \begin{cases} 0 & \text{if } l \neq k \\ 1 & \text{if } l = k \end{cases} \quad \text{and} \quad \sum_{l=-\infty}^{l=\infty} \delta[l-k] = 1 . \tag{4.2}$$

Note that two notations are introduced. $\boldsymbol{\delta}_k$ is a *vector* of the same size of $\mathbf{u}$, which is zero except at $l = k$, where it is equal to one; $\delta[l-k]$ is a *sequence* of numbers collecting the same information. In this sequence, k is the index spanning the position of the impulse, while the index l spans the time (shift) domain. Infinite duration signals are vectors of infinite length. For two such signals $a[k], b[k]$ or vectors $\mathbf{a}, \mathbf{b}$, the inner product for the "comparison procedure" is

$$\langle \mathbf{a}, \mathbf{b} \rangle = \boldsymbol{b}^\dagger \boldsymbol{a} = \sum_{k=0}^{n_t-1} a[k]\,\overline{b}[k] , \tag{4.3}$$

[5] Note that we will here use a "Python-like" indexing, that is, starting from $k = 0$ rather than $k = 1$.

where $b^\dagger = \overline{b}^T$ is the Hermitian transpose, with the superscript ^{T} denoting transposition and the overline denoting conjugation. With such a basis, the value of the signal at a specific location $u[k] = \mathbf{u}_k$ can be written as

$$\boxed{\mathbf{u}_k = u[k] = \langle \mathbf{u}, \boldsymbol{\delta}_k \rangle_l \quad \text{or} \quad u[k] = \sum_{l=-\infty}^{\infty} u[l]\delta[k-l]\,.} \tag{4.4}$$

The operation on the left is a *correlation*: for a given location of the impulse k, the inner product is performed over the index l spanning the entire length of the signal and the result is a scalar – the signal's value at the index k. The operation on the right is a *discrete convolution* and the result is a signal: the entire set of shifts will be spanned. We shall return to the algebra of this operation in Chapter 8.

Note the flipping of the indices $\delta[l-k]$ in (4.2) to $\delta[k-l]$ in (4.4). In the first case, the location of the impulse k is *fixed* and l spans the vector entries; in the second case, within the summation, the time domain k is fixed and l spans the possible locations of the delta functions[6].

The inner product in a vector space is the fundamental operation that allows for a rigorous definition of intuitive geometrical notions such as the length of a vector and the angle between two vectors. The length (l^2 norm) of a vector $||\mathbf{a}||$ and the cosine of the angle β between two vectors $\mathbf{a}$ and $\mathbf{b}$ of equal size are defined, respectively, as

$$||\mathbf{a}|| = \sqrt{\langle \mathbf{a}, \mathbf{b} \rangle} = \sqrt{\mathbf{b}^\dagger \mathbf{a}} \quad \text{and} \quad \cos(\beta) = \frac{\langle \mathbf{a}, \mathbf{b} \rangle}{||\mathbf{a}||\,||\mathbf{b}||}. \tag{4.5}$$

In signal processing, the first quantity is the root of the signal's *energy*[7], defined as $\mathcal{E}\{y\} = ||\mathbf{y}||^2$ while the second is the *normalized correlation* between two signals.

Signals with finite energy are *square-summable*. If $\cos\beta = 0$, two vectors are *orthogonal* and two signals are *uncorrelated*; if $\cos\beta = 1$, two vectors are *aligned* and two signals are *perfectly correlated*.

Notice that the projection of a vector $\mathbf{a}$ onto a vector $\mathbf{b}$ (see Figure 4.2) is

$$\mathbf{a}_b = ||\mathbf{a}|| \cos(\beta) = \frac{\langle \mathbf{a}, \mathbf{b} \rangle}{||\mathbf{b}||}. \tag{4.6}$$

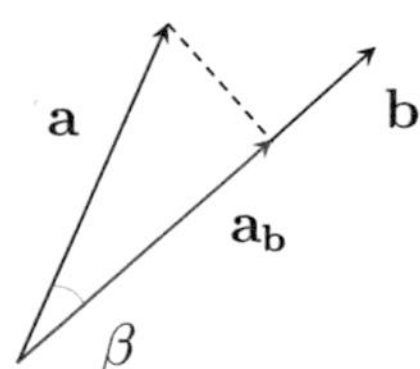

Figure 4.2 Projections and inner products.

Hence if $\mathbf{b}$ is a basis vector of unitary length, the inner product equals the projection. When this occurs, as in most of the cases presented in what follows, the notion of *inner product* and *projection* are used interchangeably.

The basis of shifted impulses has a very special property: it is *orthonormal*. This means that the inner product of two basis elements (in this case the shifted delta functions) gives zero unless the same basis element is considered in which case we

[6] This distinction is irrelevant for a symmetric function such as $\boldsymbol{\delta}$, but it is essential in the general case: if $\delta[k-l]$ is replaced by $\delta[l+k]$ in (4.4), the operation is called cross-correlation.

[7] Note that the notion of energy is used in signal processing for the square of a signal independently of whether this is actually linked to physical energy.

recover its unitary norm (energy). We return to this property in Chapter 8. Before moving to continuous signals, it is worth introducing another important signal that is linked to delta functions, namely the Heaviside step function. This is defined as

$$u_S[k-l] = \begin{cases} 1 & \text{if } k \geq l \\ 0 & \text{if } l < k \end{cases} \quad \text{that is} \longrightarrow u_S[k-l] = \sum_{r=-\infty}^{k} \delta[r-l]. \tag{4.7}$$

The difference between two shifted step functions generates a box function $u_{\mathcal{B}a,b}$, which is unitary in the range $k = [a, b-1]$ and is zero outside. The difference between two step functions shifted by a single step is a delta function, that is, $\delta[k-l] = u_S[k-l] - u_S[k-l-1]$. In the discrete setting, this is equivalent to a differentiation. Hence, *the delta function is the derivative of the step function* and the summation in (4.7) shows that *the step function is an integral of the delta functions*. The reader should close this subsection with an exercise on some distinctive features of discrete signals.

Exercise 1: Continuous versus Discrete

Consider the continuous signal $y(t) = e^{-0.2t}\cos(2\pi t)$ where t is in seconds. Assuming that the signal is sampled at $f = 100$Hz, and that $n_t = 320$ points are collected from $t = 0$, what is the resulting discrete signal? Plot this signal and comment on its link with the continuous one.

4.3.2 Continuous Signals

The extension of the previously introduced notion to continuous signals brings several complications, a detailed resolution of which is out of the scope of this chapter. Interested readers are referred to Kaiser (2010) for a gentle introduction. The main difficulty arises from the need to define an inner product space which can generalize (4.3) for functions while allowing for a basis of impulses, that is, functions that are zero at all but one point. This generalization is provided by the Hilbert space within which the inner product of two complex-valued functions $a(t), b(t) \in \mathbb{C}$ is

$$\langle a, b \rangle = \int_{-\infty}^{\infty} a(t)\,\overline{b}(t)\,dt. \tag{4.8}$$

The notion of energy (norm), correlation, and projection in (4.5)–(4.6) extends to continuous signals using the inner product[8] in (4.8). The generalization of (4.4) requires the definition of a continuous delta function $\delta(t)$ with the same properties as $\delta[k]$: it is nonzero only at a given time and it is absolutely integrable. This can be constructed as the limit of a suitably chosen function having unit area over an infinitesimal time interval. An example is a normalized Gaussian $G(t, \mu, \sigma)$, which has unit integral regardless of its standard deviation σ:

$$\int_{-\infty}^{\infty} G(t, \mu = 0, \sigma)dt = \frac{1}{\sqrt{2\pi\sigma^2}} \int_{-\infty}^{\infty} e^{\frac{-t^2}{2\sigma^2}}\,dt = 1\,. \tag{4.9}$$

[8] Note that the upper case is used for the norms of a continuous function, that is, $||a(t)||_2 = \sqrt{\langle a, a \rangle}$, with the inner product in (4.8), is the L^2 norm of the function $a(t)$.

This function gets narrower and taller, as $\sigma \longrightarrow 0$, to the point at which it becomes infinite at $t = 0$ and null everywhere else while still having unit area. This is the definition of a continuous Dirac delta function, which in its shifted form is

$$\delta(t-\tau) = \begin{cases} 0 & \text{if } t \neq \tau, \\ \infty & \text{if } t = \tau \end{cases} \quad \text{and} \quad \int_{-\infty}^{+\infty} \delta(t-\tau)dt = 1. \tag{4.10}$$

This function is not an ordinary one, as its integration poses several technical difficulties. Without entering into details of measure and distribution theory (Richards & Youn 1990), we shall accept this as a *generalized function* that serves well our purpose of sampling a continuous signal. From the definition in (4.10), it is easy to derive the sifting (sampling property):

$$\begin{aligned} \boxed{y(t) = \langle y(\tau), \delta(t-\tau)\rangle = \int_{-\infty}^{\infty} y(\tau)\delta(t-\tau)d\tau} &= \int_{-\infty}^{\infty} y(t)\delta(t-\tau)d\tau \\ &= y(t)\int_{-\infty}^{\infty} \delta(t-\tau)d\tau = y(t). \end{aligned} \tag{4.11}$$

The equivalence $y(\tau) = y(t)$ in the integral results from the product $y(\tau)\delta(t-\tau)$ being null everywhere but at t; then, in the last step it is possible to move $y(t)$ outside the integral as this is independent from the integration domain τ.

As for the discrete case, it is interesting to introduce the unitary step function and its link with the delta function as

$$u_S(t-t_0) = \begin{cases} 1 & \text{if } t > t_0, \\ 0 & \text{if } t < t_0 \end{cases} \quad \text{that is,} \longrightarrow u_S(t-t_0) = \int_{-\infty}^{t} \delta(\tau - t_0)d\tau. \tag{4.12}$$

Notice that the step function is not defined at $t = t_0$. This definition is the continuous analogue of (4.7). To show that the delta function is the derivative of the step function, we must introduce the notion of *generalized derivative*. For a continuous signal $u(t)$, denoting as u' and $u^{(n)}$ its first and nth derivative, integration by part using an appropriate test function $\xi(t)$ gives

$$\int_{-\infty}^{\infty} \xi(t)u^{(n)}(t)dt = (-1)^n \int_{-\infty}^{\infty} \xi^{(n)}u(t)dt, \tag{4.13}$$

where the test function $\xi(t)$ is assumed to be continuous and differentiable at least n times and is such that $\xi(t) \longrightarrow 0$ for $t \longrightarrow \pm\infty$. The first derivative of u_S is

$$\begin{aligned} \int_{-\infty}^{\infty} \xi(t)\boxed{u_S'(t-t_0)}dt &= -\int_{-\infty}^{\infty} \xi'(t)u_S(t-t_0)dt = -\int_{t_0}^{\infty} \xi'(t) \\ &= \xi(t_0) - \xi(\infty) = \xi(t_0) = \int_{-\infty}^{\infty} \xi(t)\boxed{\delta(t-t_0)}dt. \end{aligned} \tag{4.14}$$

The first equality results from direct application of (4.13); the second from $u_S(t-t_0)$ being zero in $t < t_0$. After integration, the sifting property of the delta function (4.11) is used, and the result is obtained by equivalence of last step with the first, holding on the fact that $\xi(t)$ is arbitrary.

4.4 Convolutions and Eigenfunctions

In a *linear system*, the input–output relation satisfies *homogeneity* and *superposition* conditions. For continuous ($\mathcal{S}_c\{u(t)\} = y(t)$) and discrete ($\mathcal{S}_d\{u[k]\} = y[k]$) systems, these are:

Homegeneity:

$$\begin{aligned} \mathcal{S}_c\{a\, u(t)\} &= a\mathcal{S}_c\{\, u(t)\} = ay(t) \quad \text{for all } a \in \mathbb{C}, \\ \mathcal{S}_d\{a\, u[k]\} &= a\mathcal{S}_d\{\, u[k]\} = ay[k] \quad \text{for all } a \in \mathbb{C}. \end{aligned} \tag{4.15}$$

Superposition:

$$\begin{aligned} \mathcal{S}_c\left\{\int_{-\infty}^{\infty} u(t)\right\}dt &= \int_{-\infty}^{\infty} \mathcal{S}_d\{u(t)\}dt = \int_{-\infty}^{\infty} y(t)dt, \\ \mathcal{S}_d\left\{\sum_{n=1}^{N} u_n[k]\right\} &= \sum_{n=1}^{N} \mathcal{S}_d\{u_n[k]\} = \sum_{n=1}^{N} y_n[k], \end{aligned} \tag{4.16}$$

where we considered a finite summation of N inputs for the discrete case and an infinite summation of infinitesimally close inputs for the continuous one. Combining these properties, we see that a linear combination of inputs results in the same linear combination of outputs:

$$\begin{aligned} \mathcal{S}_c\left\{\int_{-\infty}^{\infty} a(\tau)u(t,\tau)\right\}d\tau &= \int_{-\infty}^{\infty} a(\tau)\mathcal{S}_d\{u(t,\tau)\}d\tau = \int_{-\infty}^{\infty} a(\tau)y(t,\tau)d\tau, \\ \mathcal{S}_d\left\{\sum_{n=1}^{N} a_n u_n[k]\right\} &= \sum_{n=1}^{N} a_n\mathcal{S}_d\{u_n[k]\} = \sum_{n=1}^{N} a_n y_n[k]. \end{aligned} \tag{4.17}$$

A system is *time-invariant* (or *translation-invariant*, if time is replaced by space) if the response to the input does not change over time (or space), that is,

$$\begin{aligned} &\text{if } \mathcal{S}_c\{u(t)\} = y(t) \quad \text{then} \quad \mathcal{S}_c\{u(t-t_0)\} = y(t-t_0) \text{ for all } t_0 \in \mathbb{R}, \\ &\text{if } \mathcal{S}_d\{u[k]\} = y[k] \quad \text{then} \quad \mathcal{S}_d\{u[k-k_0]\} = y[k-k_0] \text{ for all } k_0 \in \mathbb{Z}. \end{aligned} \tag{4.18}$$

A system is linear time/translation invariant (LTI) if it is both lincar *and* time/translation invariant. These two properties, combined with the signal representations in (4.10)–(4.11), make the analysis of LTI systems particularly simple: the response to *any* input can be fully characterized from the response to a *single impulse.*

Defining as $h(t) = \mathcal{S}_c\{\delta(t)\}$ the impulse response of a continuous system, the response to any input is

$$
\begin{aligned}
y(t) = \mathcal{S}_c\{u(t)\} = \mathcal{S}_c\left\{\int_{-\infty}^{\infty} u(t)\delta(t-\tau)d\tau\right\} = \int_{-\infty}^{\infty} u(t)\mathcal{S}_c\{\delta(t-\tau)\}d\tau \\
\rightarrow \boxed{y(t) = \int_{-\infty}^{\infty} u(\tau)\, h(t-\tau)d\tau = \int_{-\infty}^{\infty} h(\tau)\, u(t-\tau)d\tau}.
\end{aligned}
\tag{4.19}
$$

The last two integrals are equivalent forms of the convolution integral, hinging on its commutative property. Similarly, defining $h[k] = \mathcal{S}_d\{\delta[k]\}$ the impulse responses of a discrete system, the response to any input is

$$
\begin{aligned}
y[k] = \mathcal{S}_d\{u[k]\} = \mathcal{S}_d\left\{\sum_{l=-\infty}^{\infty} u[k]\delta[k-l]\right\} = \sum_{l=-\infty}^{\infty} u[k]\mathcal{S}_d\{\delta[k-l]\} \\
\rightarrow \boxed{y[k] = \sum_{l=-\infty}^{\infty} u[l]\, h[k-l] = \sum_{l=-\infty}^{\infty} h[l]\, u[k-l]},
\end{aligned}
\tag{4.20}
$$

having introduced the discrete convolution and its commutative property.

Note that the system response obtained via (4.19) or (4.20) is independent from the initial state of a system. Such a response is thus a *particular* solution, that is, after the transitory from the initial condition vanishes.

A special case is produced when the input is an exponential of the form $u(t) = e^{st}$ for continuous systems and $u[k] = z^k$ for the discrete ones, with $s, z \in \mathbb{C}$. In what follows, we write s in a Cartesian form as $s = \sigma + \mathrm{j}\omega$ and z in a polar form as $z = \rho e^{\mathrm{j}\theta}$; the convenience in this is evident in Section 4.7. For the moment, note that these two variables are the continuous and discrete *complex frequencies*.

In continuous systems, from (4.19), the output is

$$
y(t) = \mathcal{S}_c\{e^{st}\} = \int_{-\infty}^{\infty} h(\tau)\, e^{(t-\tau)s}d\tau = \left(\int_{-\infty}^{\infty} h(\tau)\, e^{-\tau s}d\tau\right) e^{st} = \lambda_c(s)e^{st}. \tag{4.21}
$$

In discrete systems, from in (4.20), the output is

$$
y[k] = \mathcal{S}_d\{z^k\} = \sum_{l=-\infty}^{\infty} h[l]\, z^{k-l} = \left(\sum_{l=-\infty}^{\infty} h[l]\, z^{-l}\right) z^k = \lambda_d(z)z^k. \tag{4.22}
$$

In both cases, this result shows that LTI system responds to a complex exponential with *the same input* multiplied by a complex number (λ_d or λ_c). This number solely depends on the complex frequencies $z \in \mathbb{C}$ and $s \in \mathbb{C}$. Therefore, these special input functions are *eigenfunctions* of the LTI operators and their complex eigenvalues are

$$
\lambda_c(s) = H(s) = \int_{-\infty}^{\infty} h(\tau)\, e^{-\tau s}d\tau, \tag{4.23a}
$$

$$
\lambda_d(z) = H(z) = \sum_{l=-\infty}^{\infty} h[l]\, z^{-l}. \tag{4.23b}
$$

These are, respectively, the Laplace transform and the Z-transform of the impulse response. These are the *transfer functions* of the LTI systems and link input and output in the *complex* frequency domain. For time varying or nonlinear systems, the notion of transfer function is not useful. Finally, note that in discrete systems, the transfer function is a continuous function of z.

Exercise 2: Convolution with Impulse Responses

To identify a SISO system, a step function test is carried out. A regression analysis reveals that the response of the system to a step u_S is well described by the function

$$y_S = \frac{1}{5} - \frac{1}{5}e^{-t}\left(\cos(2\,t) + \frac{1}{2}\sin(2\,t)\right). \tag{4.24}$$

Compute the response of the system to an exponential input $u(t) = e^{-t}$ in the continuous and the discrete domains, assuming that the system can be sampled at $f_s = 10$ Hz or at $f_s = 3$ Hz for $T = 5s$.

4.5 Causal and Stable Systems

In most applications of interest, signals are assumed to be null at time $t < 0$. This is important for the impulse response of systems that are *causal* in which the impulse response is $h(t) = 0$ for $t < 0$ and $h[k] = 0$ for $k < 0$. This means that no output can be produced *before* the input, and hence the system is not anticipatory. Models of physical systems and *online* data processing must be causal. On the other hand, many data processing schemes operating *off-line* are *not causal* (e.g., zero-phase filters described in Section 4.8).

As anticipated in the previous exercise, the convolution integral and summations in (4.19)–(4.20) for causal systems become

$$y(t) = \mathcal{S}_c\{u(t)\} = \int_0^t h(\tau)\,u(t-\tau)d\tau = \int_0^t u(\tau)\,h(t-\tau)d\tau, \tag{4.25a}$$

$$y[k] = \mathcal{S}_d\{u[k]\} = \sum_{l=0}^{k} h[l]\,u[k-l] = \sum_{l=0}^{k} u[l]\,h[k-l]\,. \tag{4.25b}$$

The upper limit is replaced by t or k since signals and impulse responses are null for $\tau > t$ or $k > l$, and the lower one is replaced by 0 since both are null for $t < 0$ or $l < 0$. The continuous and discrete transfer functions in (4.23) become

$$\lambda_c(s) = H(s) = \int_{0^-}^{\infty} h(\tau)\,e^{-\tau s}d\tau \quad \text{and} \quad \lambda_d(z) = H(z) = \sum_{n=0}^{\infty} h[n]\,z^{-n}. \tag{4.26}$$

In the continuous case, $t = 0$ *must* be included in the integration; hence the lower bound is tuned to accommodate for any peculiarity occurring at $t = 0$ (notably an impulse). Nevertheless, to avoid the extra notational burden, the minus subscript in the lower bound is dropped in what follows.

Finally, another important class of interest is that of *stable* systems. The stability analysis of complex systems is a broad topic (see Chapters 10 and 13). Here, we limit the focus to bounded-input/bounded-output (BIBO) stability. A system is BIBO stable if its response to *any* bounded input is a bounded output. This requires that the impulse response of continuous and discrete signals satisfies

$$\int_0^\infty |h(t)|dt < \infty \quad \text{and} \quad \sum_{k=0}^{\infty} |h[k]| < \infty\,. \tag{4.27}$$

A different notion is that of *asymptotic* stability, which is related to the *internal* stability of a system. A system is *asymptotically stable* if every initial state, in the absence of inputs, produces a bounded response that converges to zero. In an LTI system, asymptotic stability implies BIBO stability, but the reverse is not true.

4.6 Linear Time-Invariant Systems in Their Eigenspace

In Section 4.5, we have seen that complex exponentials are eigenfunctions of LTI systems. Great insights on a system's behavior can be obtained by projecting their input–output relation onto the system's eigenfunctions. The projection of signals into complex exponentials leads to the Laplace transform in the continuous domain and the Z transform in the discrete domain. This section is divided into four subsections. We start with some definitions.

4.6.1 Laplace and Z Transforms

Laplace Transforms.
Given a continuous signal and causal signal $u(t)$, the Laplace transforms are

$$U(s) = \mathcal{L}\{u(t)\} = \int_{-\infty}^{\infty} u(t)e^{-st}\,dt = \int_0^\infty u(t)e^{-st}\,dt\,. \tag{4.28}$$

The first integral is the *bilateral* transform; the second is the *unilateral* transform. As we here focus on causal signals (i.e., $u(t < 0) = 0$), these are identical. Nevertheless, these are different tools required for different purposes: the first is suitable for infinite duration signals for which it can be linked to the Fourier Transform (Section 4.7); the second is developed for solving initial value problems, as it naturally handles initial conditions.

These integrals converge, and hence the Laplace transforms exist, if $u(t)e^{-st} \to 0$ for $t \to \pm\infty$ (only $t \to +\infty$ for the *unilateral*). This requires that the signal is of *exponential order*, that is, grows more slowly than a multiple of some exponential: $|u(t)| \leq Me^{\alpha t}$. If this is the case, the range of values $\mathbb{R}\{s\} > \alpha$ is the *region of convergence* (ROC) of the transform.

The reader is referred to Beerends et al. (2003), Wang (2009), Hsu (2013) for a review of all the properties of the Laplace transform; we here focus on the key operations enabled by this powerful tool and we omit formulation of the inverse Laplace transform, as it requires notions of complex variables theory that are out of the scope of this chapter. The Python script EX3.PY for solving Exercise 3 provides the commands to compute both the transform and its inverse using the Python library SYMPY[9].

The key property of interest in this chapter is that of time derivation, which can be easily demonstrated using integration by parts. The bilateral ($\mathcal{L}^b$) and unilateral ($\mathcal{L}^u$) transforms of a time derivative are

$$\mathcal{L}^b\{u'(t)\} = sU(s) \quad \text{and} \quad \mathcal{L}^u\{u'(t)\} = sU(s) - u(0)\,. \tag{4.29}$$

That is *differentiation in the time domain corresponds to multiplication by s in the frequency domain*; similarly, one can show that *integration in the time domain corresponds to division by s*. Notice that no distinction between $\mathcal{L}^b$ and $\mathcal{L}^a$ is needed for a system initially at rest and the *bilateral transform cannot handle initial conditions*.

Finally, compare the inner product in (4.28) with (4.8), taking $a(t) = u(t)$. Note that the Laplace transform is a projection of the signal $u(t)$ onto an exponential basis $b_{\mathcal{L}}(t,s) = e^{-\sigma + j\omega}$. It is left as an exercise to show that the Laplace basis is not orthogonal, unless $\sigma = 0$. This is the basis of the continuous Fourier transform.

Z Transform.
Given a discrete and causal signal $u[k]$, the Z transforms are

$$U(z) = \mathcal{Z}\{u[k]\} = \sum_{k=-\infty}^{+\infty} u[k] z^{-k} = \sum_{k=0}^{\infty} u[k] z^{-k}\,. \tag{4.30}$$

The same distinction on *bilateral* or *unilateral* transforms, as well as their equivalence for causal signals, applies to the discrete transform. Observe that the Z transform is a *continuous function* of z. These summations converge, and hence the transforms exist, if $|u[k]z^{-k}| < 1$. The domain within which this occurs is the ROC of the transforms, and this is typically within a domain such that $|z| > \alpha$. As before, we avoid a review of all the properties of this transform, and the formulation of its inverse (see Wang 2009, Hayes 2011).

The Z transform equivalent of the time differentiation and integration properties of the Laplace transform are the time-shifting properties

$$\mathcal{Z}\{u[k-1]\} = z^{-1}\,U(z) \quad \text{and} \quad \mathcal{Z}\{u[k+1]\} = z\,U(z)\,. \tag{4.31}$$

Finally, compare (4.30) with (4.3) taking $a[k] = u[k]$ to see that the Z transform is a projection of the signal onto a basis of powers $b_{\mathcal{Z}}[k] = \rho^{-k} e^{j\theta k}$. This basis is not orthogonal unless $\rho = 1$. This is the basis of the discrete Fourier transform.

The Z transform is not suited for compression purposes, as the dimensionality of the problem in the transformed domain is *increased*: the original signal has a (finite) dimension t_k, but the same information in the frequency domain is mapped onto a *continuous complex plane z*.

[9] see `www.sympy.org/en/index.html`.

4.6.2 Discrete and Continuous Frequencies

If a discrete signal is obtained by sampling a continuous one, the link between these two transforms reveals the important impact of the sampling process on the frequency domain. The discretization $t_k = k\Delta t$ creates a point-wise equivalence such that any signal (causal or not) can be equivalently written as

$$u[k] = \sum_{l=-\infty}^{\infty} u[l]\delta[k-l] \longleftrightarrow u(t_l) = \sum_{l=-\infty}^{\infty} u(t_l)\delta(t_l - l\Delta t)\,. \tag{4.32}$$

With $e^{-s\,t_k} = e^{-s\,k\Delta t} = z^k$, the Laplace transform of this discrete signal is

$$\begin{aligned}\mathcal{L}\{u[k]\} &= \int_{-\infty}^{\infty} \sum_{l=-\infty}^{\infty} u[k]\delta(t-l\Delta t)e^{-st}\,dt \\ &= \int_{-\infty}^{\infty} \sum_{l=-\infty}^{\infty} u[k]\delta(t-l\Delta t)z^{-k}\,dt \\ &= \sum_{l=-\infty}^{\infty} u[k]z^{-k} \int_{-\infty}^{\infty} \delta(t-l\Delta t)dt = \mathcal{Z}\{u[k]\}.\end{aligned} \tag{4.33}$$

The equivalence of these transforms relies on the change of variables $z = e^{s\Delta t}$. This maps s onto the z while preserving angles in the two domains: such mapping is a *conformal mapping*. Introducing $s = \sigma + \mathrm{j}\omega$ and $z = \rho e^{\mathrm{j}\theta}$ shows that $\rho = e^{\sigma\Delta t}$ and $\theta = \omega\Delta t$.

Figure 4.3 shows several important features of this mapping. Observe that the imaginary axis ω in the continuous domain is mapped onto an angular coordinate $\theta \in [-\pi, \pi]$ or $\theta \in [0, 2\pi]$. This offers yet another way of introducing the notion of *aliasing* and the Nyquist–Shannon sampling theorem (encountered in Exercise 1), arising from the fact that the frequency domain of a digital signal is periodic. Another important observation is that the left side of the complex domain $s = \sigma < 0$ is mapped *inside* the unit circle $\rho = 1$, while the axis $s = \sigma = 0$ is mapped *on* the unit circle.

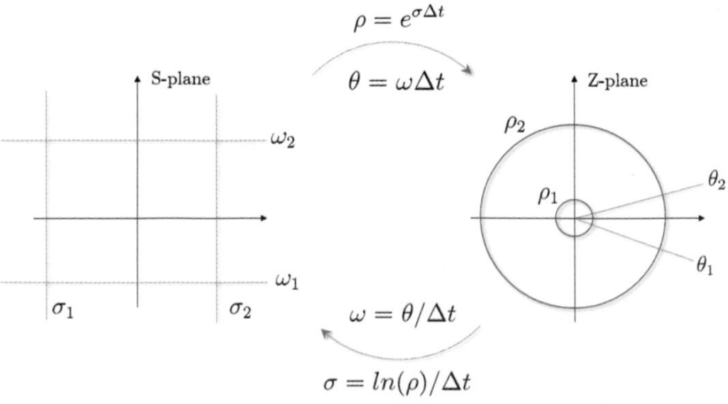

Figure 4.3 The conformal mapping linking the Laplace and the Z transforms.

4.6.3 The Convolution Theorem

While the convolution integrals and summations in (4.19) and (4.20) provide the input–output relation in the time domain using the impulse response, further insights can be obtained by analyzing this relation in the frequency domain. We here give, without proofs[10], one of the most important results of signal processing, known as the convolution theorem: *a convolution in the time domain is a multiplication in the frequency domain.* For continuous and discrete systems, this means

$$Y(s) := \mathcal{L}\{y(t)\} = \mathcal{L}\left\{\int_0^t h(\tau)\, u(t-\tau) d\tau\right\} = H(s)U(s), \tag{4.34a}$$

$$Y(k) := \mathcal{Z}\{y[k]\} = \mathcal{Z}\left\{\sum_{l=0}^{k} h[l]\, u[k-l]\right\} = H(z)U(z)\,. \tag{4.34b}$$

4.6.4 Differential and Difference Equations

Continuous LTI systems can be described in terms of (linear) differential equations, while discrete LTI systems can be described in terms of (linear) difference equations. The Laplace and the Z transforms are powerful tools to solve these equations because of the properties in (4.29) and (4.31). In both cases, the equations have constant coefficients and are often acronymized as LCCDE.

Differential Equations.
The general form of the LCCDE of a continuous SISO LTI system with input $u(t)$ and output $y(t)$ reads

$$\sum_{n=0}^{N_b} a_n\, y^{(n)}(t) = \sum_{n=0}^{N_f} b_n\, u^{(n)}(t), \tag{4.35}$$

where $N_f \geq N_b$ is the order of the system[11]. The coefficients a_n are called *feedback coefficients*; the coefficients b_n are *feedforward coefficients.*

The LCCDE provides an *implicit* representation of a system since the input–output relation can be revealed only by solving the equation. Introducing the Laplace transform in a LCCDE is an operation similar to the Galerkin projection underpinning

[10] See Wang (2009) for more details.

[11] The condition $N_f \geq N_b$ is necessary to ensure that the system is *realizable*, that is, both stable and causal. More about this is provided in footnote 12.

reduced-order modeling (ROM, see Chapters 1 and 14). Recalling that the Laplace transform is a projection onto the basis $b_{\mathcal{L}}(t)$, (4.35) leads to:

$$\begin{aligned}
\Big\langle \sum_{n=0}^{N_b} a_n y^{(n)}(t), b_{\mathcal{L}}(t) \Big\rangle &= \Big\langle \sum_{n=0}^{N_f} b_n u^{(n)}(t), b_{\mathcal{L}}(t) \Big\rangle \\
\to \sum_{n=0}^{N_b} a_n \mathcal{L}\Big\{ y^{(n)}(t) \Big\} &= \sum_{n=0}^{N_f} b_n \mathcal{L}\Big\{ u^{(n)}(t) \Big\} \\
\to Y(s)\Big(\sum_{n=0}^{N_b} a_n s^{(n)} \Big) &= U(s)\Big(\sum_{n=0}^{N_f} b_n s^{(n)} \Big), \\
H(s) = \frac{Y(s)}{U(s)} = \frac{\sum_{n=0}^{N_f} b_n s^{(n)}}{\sum_{n=0}^{N_b} a_n s^{(n)}} &= \frac{b_{N_f}}{a_{N_b}} \frac{\prod_{n=0}^{N_f}(s - z_n)}{\prod_{n=0}^{N_b}(s - p_n)}.
\end{aligned} \tag{4.36}$$

The transfer function of LTI systems is a polynomial rational function of s, with the coefficients of the polynomials being the coefficients of the LCCDE. In the factorized form, z_n and p_n are, respectively, the *zeros* and the *poles* of the system[12]. Note that since the coefficients a_n, b_n are real, these can either be purely real or appear in complex conjugate pairs. These coefficients have a straightforward connection with the LCCDE, which can immediately be recovered from the transfer function. The zeros z_n are associated to inputs $e^{z_n t}$ in which the transfer function is *null* and thus leads to no output; the poles p_n corresponds to resonances, inputs $e^{p_n t}$ in which the transfer function is infinite and leads to the blowup of the system.

The poles are eigenvalues of the matrix $\mathbf{A}$, advancing a linear system in its state-space representation (see Chapters 10 and 12). In a stable system, poles are located in regions of the s-plane that are "not accessible" by any input, that is, outside the ROC of the transfer function. Defining the ROC of $H(s)$ as $\mathbb{R}\{s\} > \alpha$, and observing that the poles are by definition outside the ROC, stability is guaranteed if $\alpha = 0$, that is, if *the ROC includes the imaginary axis*. This is equivalent to imposing that all the poles are located in the left side of the s-plane, that is, $\mathbb{R}\{p_n\} < 0 \; \forall n \in [0, N_f]$. This result can also be derived from the BIBO stability condition in (4.27).

Difference Equations.

In the discrete case, the general form of LCCDE associated to SISO LTI systems with input $u[t]$ and output $y[t]$ reads

$$\sum_{n=0}^{N_b} a_n y[k-n] = \sum_{n=0}^{N_f} b_n u[k-n], \text{ that is, } y[k] = \sum_{n=0}^{N_b} b_n^* u[k-n] - \sum_{n=1}^{N_f} a_n^* y[k-n]. \tag{4.37}$$

[12] A transfer function that has more zeros than poles (i.e., $N_f > N_b$) is said to be *improper*. In this case, $\lim_{s\to\infty} |H(s)| = +\infty$, which violates stability: this implies that at large frequencies, a finite input can produce an infinite output. Moreover, after the polynomial division, the transfer function brings polynomial terms in s. The inverse Laplace transform of these are (generalized) derivatives of the delta functions; hence the corresponding impulse response $h(t) = \mathcal{L}^{-1}\{H(s)\}$ violates causality.

The order of the system[13] is $\max(N_b, N_f)$. The form on the right plays a fundamental role in filter implementation, time-series analysis, and system identification and is known as *recursive form* of the differnce equation. Note that the *feedback* and *feedforward* coefficients in the recursive form are simply $a_n^* = a_n/a_0$ and $b_n^* = b_n/a_0$, respectively; hence the coefficient b_0^* is the *static gain* of the system.

As for the continuous case, projecting (4.37) onto the Z basis $b_Z(t)$ via the Z transform and using (4.31) yields the transfer function of a discrete system:

$$\begin{aligned}\Big\langle\sum_{n=0}^{N_b} a_n y[k-n], b_Z[k]\Big\rangle &= \Big\langle\sum_{n=0}^{N_f} b_n u[k-n], b_Z[k]\Big\rangle \\ \rightarrow \sum_{n=0}^{N_b} a_n \mathcal{Z}\Big\{y[k-n]\Big\} &= \sum_{n=0}^{N_f} b_n \mathcal{Z}\Big\{u[k-n]\Big\} \\ \rightarrow Y(z)\Big(\sum_{n=0}^{N_b} a_n z^{-n}\Big) &= U(z)\Big(\sum_{n=0}^{N_f} b_n z^{-n}\Big), \\ H(z) = \frac{Y(z)}{U(z)} = \frac{\sum_{n=0}^{N_f-1} b_n z^{-n}}{\sum_{n=0}^{N_b-1} a_n z^{-n}} &= \frac{b_0}{a_0} z^{N_b-N_f} \frac{\prod_{n=0}^{N_f-1}(1-\zeta_n z^{-1})}{\prod_{n=0}^{N_b-1}(1-\pi_n z^{-1})},\end{aligned} \tag{4.38}$$

where ζ_n and π_n are, respectively, the zeros and poles of the discrete transfer function. Observe that the factored form of the discrete transfer function is usually given in terms of polynomials of z^{-1} rather than z.

The link between zero and poles in continuous and discrete domains is given by the conformal mapping in Figure 4.3. In the absence of inputs, the poles control the evolution of a linear system from its initial condition (i.e., the homogeneous solution of the LCCDE). The Dynamic Mode Decomposition (DMD) introduced in Chapter 7 is a powerful tool to identify the poles π_n of a system from data, and to build linear reduced-order models by projecting the data onto the basis of eigenfunctions $z^{\pi_n k}$.

Finally, in analogy with the continuous case, a discrete system is stable if its poles are outside the ROC of the transfer function. Defining the ROC of $H(z)$ as $|z| \geq \alpha$, one sees that this occurs if $\alpha = 1$: the ROC includes the unit circle and hence all the poles have $|\pi_n| < 1$. This can be derived from the BIBO stability condition in (4.30).

Exercise 3: Transfer Function Analysis

Consider the system in Exercise 2. Compute the transfer function and the system output from the frequency domain, then identify the LCCDE governing the input–output relation. Then, assuming that a discrete system is obtained sampling the continuous domain, derive a recursive formula that mimics the input–output link of the continuous system. Test your result for a sampling frequency of $f_s = 3$ Hz and $f_s = 10$ Hz.

[13] Note that in (4.37) the restriction $N_b \geq N_f$ is not needed to enforce causality: by construction, the output $y[k]$ only depends on past information.

4.7 Application I: Harmonic Analysis and Filters

BIBO stability guarantees that the output produced by a stationary input is also stationary. It is thus interesting to consider only the portion of the complex planes s and z associated to infinite duration signals, that is, $s = \mathrm{j}\omega$ and $z = e^{\mathrm{j}\theta}$. These correspond to harmonic eigenvalues of the LTI system, hence lead to a *harmonic response*. From Laplace and Z transforms, we move to continuous and discrete Fourier transforms in Section 4.7.1. A system that manipulates the harmonic content of a signal is a filter; these are introduced in Section 4.7.2 along with their fundamental role in multi-resolution decompositions.

4.7.1 From Laplace to Fourier

Consider the bilateral Laplace and Z transform of a signal along the imaginary axis $s = \mathrm{j}\omega$ and the unitary circle $z = e^{\mathrm{j}\theta}$, respectively,

$$U(\mathrm{j}\omega) = \mathcal{F}_C\{u(t)\} = \int_{-\infty}^{\infty} u(t)e^{-\mathrm{j}\omega t}\,dt, \tag{4.39a}$$

$$U(\mathrm{j}\theta) = \mathcal{F}_D\{u[k]\} = \sum_{k=-\infty}^{+\infty} u[k]e^{-\mathrm{j}\theta k}. \tag{4.39b}$$

These are the continuous (CT) and the discrete (time) Fourier Transforms (DTFT). Both are continuous functions, with the second being periodic of period 2π because of the conformal mapping introduced in Figure 4.3. Comparing these to (4.28) and (4.30) shows that the bilateral Laplace and Z transforms are the Fourier transforms of $u(t)e^{-\sigma t}$ and $u[k]\rho^{-k}$. Without these exponentially decaying modulations, the conditions for convergence are more stringent: signals must be absolutely integrable and absolutely summable[14].

The main consequence is that *infinite duration stationary signals do not generally admit a Fourier transform.* This explains why the manipulations of these signals by an LTI system are better investigated in terms of some of their statistical properties, such as *autocorrelation* or *autocovariance*, as illustrated in Section 4.8. A special exception are periodic signals for which (4.39a) and (4.39b) lead to Fourier *series*, and the problem of convergence becomes less stringent.

In stable continuous and discrete LTI systems, satisfying (4.30), the impulse response always admits Fourier transform: these can be obtained by replacing $s = \mathrm{j}\omega$ and $z = e^{\mathrm{j}\theta}$ in the transfer function. This leads to *frequency transfer functions*, which are complex functions of real numbers[15] (ω or θ), customarily represented by plotting $\log(|H(x)|)$ and $\arg(H(x))$ versus $\log(x)$ in a *Bode plot*, with $x = \omega$ or $x = \theta$.

[14] This condition is sufficient but not necessary: some non-square integral functions do admit a Fourier transform. Important examples are the constant function $u(t) = 1$ or the step function $u_S(t)$. Moreover, note that the Fourier transform can be obtained from the Laplace and Z transform, only for signals that are absolutely integrable or summable. For instance, the Laplace transform of $e^{\alpha t}$ with $\alpha > 0$ has ROC $s > \alpha$, while the Fourier transform does not exist.

[15] These are often called *real* frequencies as opposed to the *complex frequencies* s and z.

The modulus of the frequency transfer function is the *amplitude response*; its argument is the *phase response*.

If the Fourier transform (or series) exists for both inputs and outputs, the properties of the Laplace and Z transform apply: the harmonic contents of the output is $Y(\mathrm{j}\omega) = H(\mathrm{j}\omega)U(\mathrm{j}\omega)$ in the continuous domain; $Y(\mathrm{j}\theta) = H(\mathrm{j}\theta)U(\mathrm{j}\theta)$ in the discrete one.

Discrete signals of finite duration $u(t_k) = \mathbf{u} \in \mathbb{R}^{n_t \times 1} = u[k]$, with $k \in [0, n_t - 1]$, are usually extended to infinite duration signals assuming periodic boundary conditions. The frequency domain is thus discretized into bins $\theta_n = n\Delta\theta$ with $n \in [0, n_f - 1]$ and $\Delta\theta = 2\pi/n_F$. The mapping to the continuous frequency domain, from Figure 4.3, gives $f_n = 2\pi\omega_n = n f_s/n_f$, with $f_s = 1/\Delta t$ the sampling frequency. With both time and frequency domain discretized, the Fourier pair are usually written as

$$U[n] = \frac{1}{\sqrt{n_t}} \sum_{k=0}^{n_t-1} u[k] e^{-2\pi \mathrm{j} \frac{n\,k}{n_f}} \iff u[k] = \frac{1}{\sqrt{n_t}} \sum_{n=0}^{n_f-1} U[n] e^{2\pi \mathrm{j} \frac{n\,k}{n_f}} . \tag{4.40}$$

The equations in (4.49) are, respectively, the discrete Fourier transform (DFT) and its inverse. Note that the normalization $1/\sqrt{n_t}$ is used for later convenience: we see in Chapter 8 that (4.49) can be written as matrix multiplications with the columns of the matrix being orthonormal vectors. Finally, if $n_f = n_t$ and n_t is a power of 2, this multiplication can be performed using the famous FFT (fast Fourier transform) algorithm (see Loan 1992), reducing the computational cost from n_t^2 to $n_t \log_2(n_t)$. An excellent review of the DFT is provided by Smith (2007b).

Exercise 4: Frequency Response Analysis

Consider the discrete system derived in Exercise 3, but now assume that the static gain is unitary. Compute the frequency transfer function of this system and show that this can be seen as a *low-pass filter*. Study how the frequency response changes if the coefficients a_1 or a_2 are set to zero. Finally, derive the system that should have the *complementary* transfer function and show its amplitude response. Is this response also complementary?

4.7.2 Multiresolution Analysis and Digital Filters

Filters are at the center of most signal processing applications, and the theory behind their design is a vast subject (see Smith 2007a). Among the essential applications discussed in this book are feedback control design (Chapter 10), multiresolution analysis (MRA) and wavelet decomposition (Chapter 5), and multiscale proper orthogonal decomposition (Chapter 8).

In feedback control, a controller manipulates the feedback coefficients of an LCCDE by introducing a control input, which is a function of the output. Therefore, filter design methods can be used to design the actuation such that the transfer function of the controlled system rejects certain disturbances (see Bode design methods (Distefano 2013)).

In MRA, filters are used to decompose signals. While the DFT represents a signal as a linear combination of harmonics, MRA represents it as a combination of frequency bands called *scales*. A packet of similar frequencies can be assembled into bases called *wavelets*, hence the connection to Chapter 5.

The MRA partitions the spectra of signal into n_M scales, each taking a portion of the signal's content in bands $[0, f_1], [f_1, f_2], \ldots, [f_{M-1}, f_s/2]$. In Chapter 8, these will be identified by a *frequency splitting vector* $F_V = [f_1, f_2, f_3, \ldots, f_{M-1}]$.

The MRA of a discrete signal can be written as

$$u[k] = \sum_{m=1}^{M} s_m[k] = \mathcal{F}^{-1}\left\{\sum_{m=1}^{M} H_m(f_n)U(f_n)\right\} \text{ with } \sum_{m=1}^{M} |H_m(f_n)| = 1\,, \tag{4.41}$$

where s_m is the portion of the signal in the scale m, within the frequency range $f_n \in [f_{m-1}, f_m]$ and $H_m(f_n)$ is the transfer function of the filter that isolates that portion. Therefore, $|H_m(f_n)| \approx 1$ for $f_n \in [f_{m-1}, f_m]$ and $|H_m(f_n)| \approx 0$ otherwise. The assumption on the right enables a lossless decomposition.

The MRA requires the definition of one low-pass filter for the range $[0, f_1]$, one high-pass filter for the range $[f_{M-1}, f_s/2]$, and $M - 2$ band-pass filters. Because these are complementary, all these filters can be obtained from a set of low-pass filters, as described at the end of this section. Therefore, to learn MRA, one should first learn how to construct a low-pass filter with a given cutoff frequency f_c.

We now focus on the two main families of filters and the most common design methods. Let us consider a specific example, with $f_s = 2k$ Hz and $f_c = 200$ Hz. In the digital frequency domain, we map the sampling frequency to $\theta_s = 2\pi$ and the cutoff to $\theta_c = \pi/5$.

The transfer function of the ideal low-pass filter is

$$H_{id}(\mathrm{j}\theta) = \begin{cases} e^{-\mathrm{j}\alpha_d\theta} & \text{if } |\theta| \le \theta_c, \\ 0 & \text{if } \theta_c < |\theta| \le \pi. \end{cases} \tag{4.42}$$

This leads to the impulse response

$$h_{id}[k] = \frac{1}{2\pi}\int_{-\pi}^{\pi} H(\mathrm{j}\theta)e^{-\mathrm{j}k\theta}\,d\theta = \frac{\sin[k - \alpha_d]\theta_c}{\pi(k - \alpha_d)}\,. \tag{4.43}$$

The need for a delay α is evident after introducing FIR filters. Notice that having a linear phase $\alpha\theta$ delays the input without distorting its waveform. The modulus of this frequency response and a portion of its transfer function are shown in Figure 4.4(a) with continuous black curves.

Such an ideal filter is not realizable in the time domain because its impulse response is not causal ($h[k] \neq 0$ for $k < 0$) and is not absolutely summable. The ideal constraints must be relaxed. The most popular categories are IIR and FIR filters.

1) Infinite Impulse Response Filters (IIR)

These filters are based on a continuous function that mimics the ideal low-pass filter. The most common are Butterworth, Chebyshev, and elliptic filters (Hayes 2011,

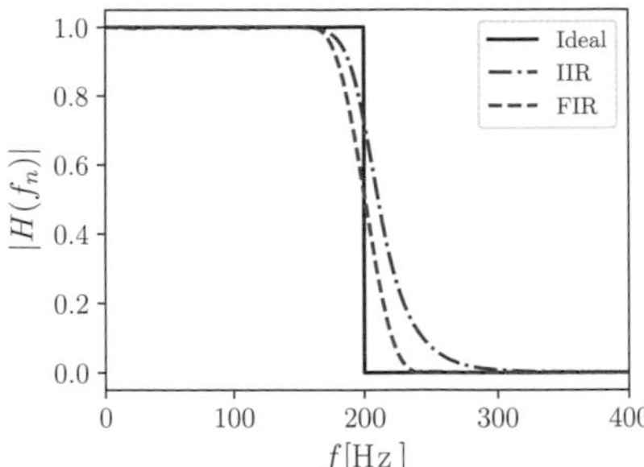

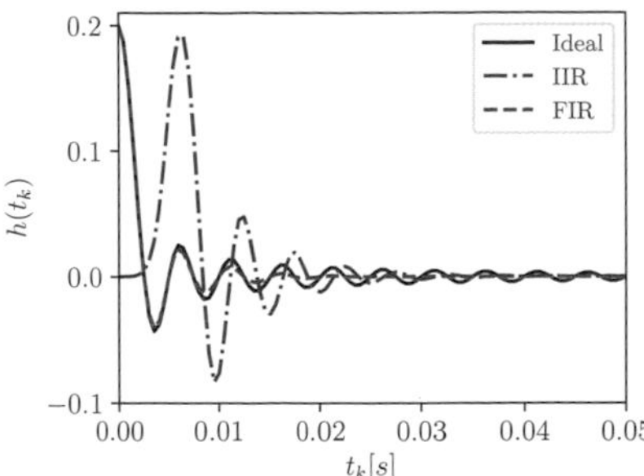

Figure 4.4 (a): Absolute value of the frequency response of an ideal (black continuous), an IIR (red dash-dotted) and an FIR (blue dashed) low-pass filter. (b): impulse responses associated to the frequency responses on the left. In the case of the FIR, the response is shifted backward by α_d.

Oppenheim & Schafer 2009). These filters have no zeros and N poles, with N the filter order, equally spaced around the unit circle.

Once these poles are computed, the continuous frequency response function can be readily obtained in its factor form and the last step consists in identifying the associated recursive formula as in Exercise 4. However, note that mapping from s to z is usually performed using the bilinear transform[16] rather than the standard mapping $z = e^{\Delta t s}$ that is used in Exercise 4, since this has the advantage of mapping $f_s/2$ to π and thus prevents aliasing. The nonlinearity in the bilinear transform results in a *wrapping* of higher frequencies so the correct cutoff frequency should first be *pre-warped* to account for the distortion in the frequency calculation[17].

Software packages such as SciPy in Python or Matlab offer the functions BUTTERWORTH to design a Butterworth filter with given order and cutoff frequency (see Python script Ex5.py).

The red curves in Figure 4.4(a) show the amplitude response and the impulse response of a Butterworth filter of order 11. The main advantage of these filters is their capability of well approximating the ideal filter using a limited order, which requires storing few coefficients in their recursive formulation. On the other hand, these filters tend to become *unstable* as the order increases (and the poles approach the unit circle). Moreover, their phase delay is generally not constant, and this potentially introduces phase distortion. Finally, note that since the impulse response of these filters is infinite, these cannot be implemented in the time domain via simple convolution, but via the recursive solution of the filter's LCCDE.

[16] which reads

$$s = \frac{2}{\Delta t}\left(\frac{1 - z^{-1}}{1 + z^{-1}}\right) \longleftrightarrow z = \frac{1 + \Delta t/2\, s}{1 - \Delta t/2\, s}.$$

[17] The pre-warp can be achieved using $f_c' = f_s/\pi \tan(\pi f_c/f_s)$. Therefore, if the desired cutoff frequency is $f_c = 200$ Hz with a sampling frequency $f_s = 1000$ Hz, the filter should target a cutoff frequency of $f_c' = 231.26$ Hz to compensate for the warping due to the bilinear transform.

2) Finite Impulse Response Filters (FIR)

These filters are constructed in the discrete domain and have no poles (no feedback coefficients in their recursive formulation). This leads to a finite impulse response. The classic design method is the *windowing* technique, which consists in multiplying the impulse response of the ideal filter in (4.42) by a window $w[k]$, which is zero outside the interval $0 \leq k \leq N$, with N the filter order.

Taking N as an odd number, these windows are symmetric about the midpoint, that is, $w[k] = w[N-k]$; this results in the lag $\alpha = (N-1)/2$ in the output with respect to the input. The need for a *lag* in the ideal filter in (4.42)–(4.43) is now clear: if $\alpha = 0$, the windowed impulse response is centered in 0, and the filter is noncausal. If the filtering is performed "off-line," it is possible to obtain a zero-phase filter by centering the windowed impulse response in $k = 0$.

Common functions are the *Hanning*, *Hamming*, or *Blackman* and *Kaiser* windows. The windowing in the time domain corresponds to a convolution in the frequency domain between the ideal filter and the Fourier transform of the window function. This *smooths* the transition from the band-pass to the band-stop region. Software packages as SciPy in Python or Matlab offer the functions firwin and fir1 to design FIR filters with a given order, cutoff frequency and window function.

In addition to the linear phase response, these filters are also always stable because of the lack of poles. Moreover, the finite length of their impulse response enables their implementation via convolution. Note, however, that FIR filters require much larger order to achieve performances comparable with IIR filters. Figure 4.4(b) shows, in the dashed blue line, the amplitude function and the corresponding impulse response (shifted by $(N-1)/2$) of an FIR filter designed using a Hamming window of order $N = 111$. The higher the filter order, the larger the window multiplying the ideal impulse response, the more this filter approach is the ideal one. On the other hand, increasing the filter order increases the sensitivity of the filter to the finite duration of the signal.

An FIR formulation makes the calculation of complementary filters particularly simple thanks to the constant phase response, solely linked to the filter order. To illustrate this, consider a signal $u[k]$, and its DFT $U[n]$. Let us filter this signal with a low-pass filter to obtain $u_{\mathcal{L}}[k]$ using an FIR filter with frequency transfer function $H_{\mathcal{L}} = |H_{\mathcal{L}}|e^{-\mathrm{i}\alpha k}$. The filter operation in the frequency domain reads $U_{\mathcal{L}} = |H_{\mathcal{L}}|e^{-\mathrm{i}\alpha\theta}U$. Let $U_{\mathcal{H}} = |H_{\mathcal{H}}|e^{-\mathrm{i}\alpha\theta}U$ denote the high-pass filter that gives the signal $u_{\mathcal{H}}[k]$ having complementary spectra (i.e., $|H_{\mathcal{L}}| + |H_{\mathcal{H}}| = 1$) and *same order* and thus *same* phase delay. Using the shifting properties of the Fourier transforms, the link between the low-pass and the high-pass counter parts in the time and frequency domain sets is

$$
\begin{aligned}
k\,\text{Domain} &: u_{\mathcal{H}}[k] = u[k+\alpha] - u_{\mathcal{L}}[k], \\
\theta\,\text{Domain} &: H_{\mathcal{H}}(\mathrm{i}\theta)U(\mathrm{i}\theta) = U(\mathrm{i}\theta)e^{\mathrm{i}\alpha\theta} - H_{\mathcal{L}}(\mathrm{i}\theta)U(\mathrm{i}\theta) \\
&\to H_{\mathcal{H}}(\mathrm{i}\theta) = e^{\mathrm{i}\alpha\theta} - H_{\mathcal{L}}(\mathrm{i}\theta)\,.
\end{aligned}
\tag{4.44}
$$

Note that the backward shifting $u[k+\alpha]$ in the time domain cancels the phase delay produced by the low-pass filter before performing the subtraction. It is easy to show that because of the linearity of the convolution, the impulse responses of complementary high-pass ($h_{\mathcal{H}}$) and low-pass ($h_{\mathcal{L}}$) filters are linked by[18] $\delta[n] = h_{\mathcal{L}}[n] + h_{\mathcal{H}}[n]$.

Finally, we close with the practical implementation of MRA in "off-line" conditions for which it is possible to release the constraints of causality and use zero-phase filters. These are usually implemented by operating on the signal twice (first on $u[k]$ and then on $u[-k]$), to artificially cancel the phase delay of the operation. In SCIPY and in MATLAB this is performed using the function FILTFILT.

If the phase delay is canceled, complementary filters can be computed by taking differences of the frequency transfer functions (which become real functions). Therefore, if the first scale with band-pass $[0, f_1]$ is identified by the frequency transfer function $H_1(f) = H_{\mathcal{L}}(f, f_1)$, the second scale with band-pass $[f_1, f_2]$ is identified by a filter with transfer function $H_2(f) = H_{\mathcal{L}}(f, f_2) - H_{\mathcal{L}}(f, f_1)$. The transfer function of the general band-pass filter is $H_m(f) = H_{\mathcal{L}}(f, f_m) - H_{\mathcal{L}}(f, f_{m-1})$, while the last scale is identified by the high-pass filter with $H_M(f) = 1 - H_{\mathcal{L}}(f, f_M)$.

This set of cascaded filters is known as a *filter bank* and is at the heart of the pyramid algorithm for computing the discrete wavelet transform (Strang 1996, Mallat 2009), where it is combined with sub-sampling at each scale. The general architecture of this decomposition is summarized in Figure 4.4. Observe that at the limit at which all the frequency bands become unitary, the MRA becomes a DFT.

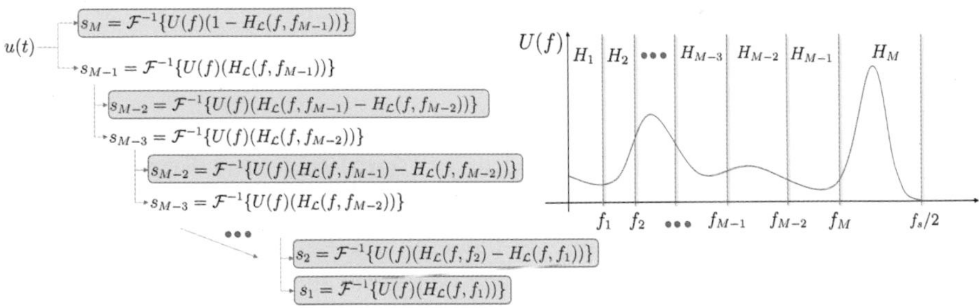

Figure 4.5 Pyramid-like algorithm to compute the MRA of a signal. Each band-pass scale is computed as the difference of two low-pass filters and the terms in blue are preserved to form the summation in (4.41). The graph on the right shows a pictorial representation of the partitioning of the signals spectra.

[18] Note that this is not the only method of obtaining a high-pass filter from a low-pass filter: another approach is to reverse the frequency response $H_{\mathcal{L}}$, flipping it from left to right about the frequency $f_s/4$ for $f > 0$ and from right to left about $-f_s/4$ for $f < 0$ (see Smith (1997)). The impulse response of the resulting high-pass filter is $h_{\mathcal{H}}[n] = (-1)^n h_{\mathcal{L}}[n]$. The two methods are equivalent if the cutoff frequency separating the transition bands is $f_s/4$. This is the case encountered when performing MRA via dyadic wavelets as discussed in Chapter 5.

Exercise 5: Multi-Resolution Analysis

Consider the synthetic signal

$$u(t_k) = a_1 \sin(2\pi f_1 t_k) e^{\frac{-(t_k-\tau_1)^2}{b_1}} + a_2 \sin(2\pi f_2 t_k) e^{\frac{-(t_k-\tau_2)^2}{b_2}} \\ + a_3 \sin(2\pi f_3 t_k) + \mathcal{N}(0, 0.2), \tag{4.45}$$

with $a_1 = 2$, $a_2 = a_3 = 1$, $f_{1,2,3} = 1, 20, 90$ (in Hz), $\tau_1 = 1$, $\tau_2 = 2.2$ (in s) and $b_1 = b_2 = 0.05$. $\mathcal{N}(0, 0.2)$ denotes Gaussian noise with zero mean and standard deviation 0.2. The signal is sampled over $n_t = 4\,096$ points at a sampling frequency of $f_s = 1k$ Hz.

(1) Compute "by hand" the impulse response h_L of a low-pass FIR filter of order $N_O = 5$, with cutoff frequency $f_c = 10$ Hz using a Hamming window. Check your answers using Python's function FIRWIN.

(2) Prepare a filter h_L of order 511 using FIRWIN. Apply this to the signal using four methods: (a) direct convolution, (b) FFT-based convolution *with* SCIPY's function FFTCONVOLVE, (c) FFT-based convolution using FFT and IFFT, and (d) a recursive implementation solving the filter's LCCDE.

(3) Construct a filter bank with frequency splitting vector $F_V = [10, 70, 100, 300]$ and show the portions of the signal within the identified scales. Use filters with Hamming window and an order of $N_O = 511$.

4.8 Application II: Time-Series Analysis and Forecasting

LTI systems are the simplest model in time-series analysis and forecasting. In these applications, treating signals and systems as fully deterministic is too optimistic, and it is thus essential to consider stochastic signals: predictions have a certain probability range (see Guidorzi 2003, Brockwell & Davis 2010). This section briefly reviews the main features of stochastic signals and systems in Section 4.8.1. Section 4.8.2 reviews the basic tools for forecasting, using classic linear regression. Only the discrete domain is considered. More advanced techniques are discussed in Chapter 12.

4.8.1 Stochastic Linear Time-Invariant Systems

A stochastic signal (or the stochastic portion of a signal) is a member of an ensemble of signals characterized by a set of probability density functions. For a comprehensive review of stochastic signals, the reader is referred to classic textbooks (Ljung 1999, Hsu 2013, Oppenheim 2015). Here, we briefly recall how LTI systems manipulate stochastic signals.

The notion and the role of the impulse response remain the same as for deterministic signals: given an input (stochastic) discrete signal, the response of the system is governed by the convolution sum in (4.21). On the other hand, the notion of frequency spectra requires some adaptation, as stochastic signals do not generally admit a Fourier transform and focus must be placed on properties that are deterministic *also* in a stochastic signal. These are the statistical properties.

Accordingly, the time-invariance in LTI systems is extended in terms of invariance of the statistical properties. This is linked to the notion of *stationarity*. Stationarity can be weak or strong. A stochastic signal $v[k]$ is stationary in a strict sense (strong stationarity) if its distributions remain invariant over time. Weak stationarity (or stationarity in a wide-sense) requires that only its *time average* μ_v and *autocorrelation* $r_{vv}[m]$ of a signal v are time-invariant. These are defined as

$$\mu_v = \mathbb{E}\{v[k]\} \quad \text{and} \quad r_{vv}[m] = \mathbb{E}\{v[k]v[k+m]\}\,, \tag{4.46}$$

with $\mathbb{E}$ the expectation operator.

In the analysis of the LTI system's response to stochastic signals, the link between a specific input and the corresponding output is not particularly interesting. Instead, we focus on the link between *the statistical properties* of the input and the output. In particular, we consider how the properties in (4.46) are manipulated. Let $y_v[k]$ be the response of the system to the stochastic signal $u_v[k]$. The *expected* (time average of the) output μ_y is

$$\mu_y = \mathbb{E}\left\{\sum_{l=-\infty}^{\infty} h[l]u_v[k-l]\right\} = \sum_{l=-\infty}^{\infty} h[l]\mathbb{E}\{u_v[k]\} = \mu_v \sum_{l=-\infty}^{\infty} h[l]\,. \tag{4.47}$$

This is a direct application of the homogeneity (4.16) and superposition (4.17). We thus see that in a BIBO stable system (satisfying (4.27)) the output average is finite if the input average is finite[19]. The input–output relation for the autocorrelation function has a more involved derivation, here omitted (see Oppenheim et al. 1996). Given r_{uu} and r_{yy} the input and the output autocorrelations, one retrieves

$$r_{yy}[m] = \sum_{l=-\infty}^{\infty} r_{uu}[m-l]r_{hh}[l] \text{ where } r_{hh}[k] = \sum_{l=-\infty}^{\infty} h[k]h[l+k]\,. \tag{4.48}$$

The sequence $r_{hh}[k]$ is the autocorrelation of the impulse response and the operation on the left is a convolution. In words: *the autocorrelation of the output is the convolution of the autocorrelation of the input with the autocorrelation of the impulse response*. This equation extends the convolution link in (4.21) to the autocorrelation functions. These functions admit Fourier transform, so the convolution theorem can be used to see the link in the frequency domain:

$$R_{yy}(\mathrm{j}\omega) = C_{hh}(\mathrm{j}\omega)R_{uu}(\mathrm{j}\omega)\,, \tag{4.49}$$

where $R_{yy}(\mathrm{j}\omega)$, $C_{hh}(\mathrm{j}\omega)$ and $R_{uu}(\mathrm{j}\omega)$ are the Fourier transform of $r_{yy}[k]$, $r_{hh}[k]$, and $r_{uu}[k]$, respectively. These are the *power-spectral densities* of y, h and u. Hence we see that an LTI system acts on the frequency content of the autocorrelation function of a stochastic signal.

[19] We also see why a high-pass filter has $\sum_k h[k] = 0$ while a low-pass filter has $\sum_k h[k] = 1$.

4.8.2 Time-Series Forecasting via Linear Time-Invariant Systems

Consider the explicit form of the LCCDE of an LTI in (4.37) and assume that the input signal (u) has both a deterministic (u_d) and a stochastic (u_s) part (i.e., $u = u_d + u_s$). Because of homogeneity, the output of the LTI system also has a deterministic (y_d) and a stochastic part (y_s). We could split these as follows:

$$y[k] = y_d[k] + y_s[k] = \sum_{n=0}^{N_b} b_n^* u_d[k-n] - \sum_{n=1}^{N_f} a_n^* y_d[k-n] + y_s\,. \tag{4.50}$$

Many models can be obtained depending on the assumptions on u_s, and hence y_s (see Guidorzi 2003, Brockwell & Davis 2010, Nielsen 2019). For example, the stochastic part y_s can be taken as white noise with zero average or as the output of a moving average filtering[20] of white noise. Any other filter can be used to allow controlling and/or modeling the frequency content of the stochastic contribution using (4.48) and (4.49).

To illustrate the main steps of time-series forecasting, let us consider the simplest approach of y_s being white noise. If the system is known (i.e., the coefficients a_n^* and b_n^* are known), the recursive equation (4.50) can be written as a matrix multiplication[21]. Assume that we have collected n_t samples of the input u. Let $\mathbf{u}_d^{(-l)}, \mathbf{y}_d^{(-l)} \in \mathbb{R}^{n_t\times 1}$ be the vectors collecting the deterministic inputs and outputs shifted backward by a lag l. Then, the matrix form of (4.50) is

$$\begin{bmatrix} | \\ \mathbf{y}^{(0)} \\ | \end{bmatrix} = \begin{bmatrix} | & | & & | & | & & | \\ \mathbf{u}^{(0)} & \mathbf{u}^{(-1)} & \cdots & \mathbf{u}^{(-N_f)} & \mathbf{y}^{(-1)} & \cdots & \mathbf{y}^{(-N_b)} \\ | & | & & | & | & & | \end{bmatrix} \begin{bmatrix} b_0 \\ \vdots \\ b_{N_b} \\ a_1 \\ \vdots \\ a_{N_F} \end{bmatrix} + \mathbf{y}_s\,. \tag{4.51}$$

We define $\mathbf{H} := [\mathbf{u}^{(0)}, \ldots, \mathbf{u}^{(-N_f)}, \mathbf{y}^{(-1)}, \ldots, \mathbf{y}^{(N_b)}]$ the Hankel matrix of the LTI system and $\mathbf{w} = [b_0, \ldots, b_{N_b}, a_1, \ldots, a_{N_f}]^T$ the vector of coefficients. The output of the LTI system is

$$\mathbf{y}^{(0)} = \mathbf{H}\mathbf{w} + \mathbf{y}_s\,. \tag{4.52}$$

Time-series forecasting via LTI systems begins with system identification. An excellent tutorial on the topic is provided by Semeraro and Mathelin (2016). The first goal is to identify the set of coefficients $\mathbf{w}$ from the input/output vectors. The stochastic part is considered as noise and the determinist part is our *expectation*. We thus seek to solve the system $\mathbf{y}^{(0)} \approx \mathbf{H}\mathbf{w}$. Like most regression problems, this problem is ill-posed: $\mathbf{H}$ is rectangular and there is no guarantee that a unique solution exists. Like all *linear* regression problems, the solution is found by minimizing a regularized cost function of the form

[20] A "moving average" filter is a filter with constant impulse response.

[21] This needs to be evaluated from the first to the last entry of $\mathbf{y}^{(0)}$.

$$J(\mathbf{w}) = ||\mathbf{y}^{(0)} - \mathbf{H}\mathbf{w}||_2^2 + \alpha R(\mathbf{w}), \tag{4.53}$$

with $\alpha \in \mathbb{R}$ acting as a smoothing parameter and $R(\mathbf{w})$ a regularizing function.[22] Classic choices are $R(\mathbf{w}) = ||\mathbf{w}||_2$ (l^2 penalty), or $R(\mathbf{w}) = ||\mathbf{w}||_1$ (l^1 penalty), or a combination of the two. The first is known as Tikhonov regularization, the second as LASSO regularization, and the third as Elastic Net. These classic tools from machine learning (see Vladimir Cherkassky 2008, Murphy 2012, Bishop 2016) are also employed in Chapter 12.

The reader should notice that the regression method can be generalized easily: one could replace the predictive equation (4.52) by a more complex model (e.g., an artificial neural network or template of polynomial nonlinearities as in Chapter 12) and minimize a cost function like (4.53) using an arsenal of optimization strategy.

Exercise 6: Time-Series Forecasting via LTI Systems

Generate a synthetic stochastic system with coefficients $(a_1, a_2) = (-1/2, 0)$ and $(b_0, b_1, b_2) = (2, 1.8, -1.2)$. Consider a deterministic input signal $u[k] = u(t_k) = \sin(8\pi t_k) + 3\exp(-(t_k - 3)^2/2)$ with $t_k = k\Delta t$, $\Delta t = 0.02$ and $k \in [0, 1999]$. The stochastic contribution $y_s[k]$ is white noise in $[0, 1]$.
Use the set of 2 000 points in input and output to identify the set of coefficients from the data, by solving (4.53) with a Tikhonov regularization. Consider $\alpha = 1$, and recall that the minimization of $J(\mathbf{w})$ in this kind of regression (known also as Ridge regression) has the simple solution

$$\mathbf{w} = (\mathbf{H}^T\mathbf{H} + \alpha\mathbf{I})^{-1}\mathbf{H}\,\mathbf{y}^{(0)},$$

with $\mathbf{I}$ the identity matrix. Is the expected result recovered? Is the identified LTI system capable of predicting what happens if $t_k > 5$? What happens for $\alpha = 0$ (i.e., ordinary least square)?

4.9 What's Next?

This chapter reviewed the fundamentals of signals and systems and presented LTI systems in case of SISO. We have seen that the input–output relation can be derived from knowledge of the impulse response of a system and via the convolution integral. It was shown that complex exponentials are eigenfunctions of these systems and that important transforms can be derived by projecting input and output signals onto these eigenfunctions. In the eigenspace of the LTI systems, convolutions become multiplications.

[22] readers familiar with Lagrangian multipliers should recognize in (4.53) an augmented cost function with α the Lagrangian multiplier.

Chapter 10 presents LTI systems, the so-called *state-space* representation, which is more common in dynamical system theory and which allows for straightforward generalization to MIMO systems. Chapter 11 reviews the analysis of nonlinear systems, while Chapter 12 describes the system identification more broadly and also considers nonlinear systems.

This chapter reviewed the link between the continuous and the discrete world and the impact of the discretization on the eigenfunctions of an LTI system. Special values of the complex frequencies, called poles, yield infinite response of a system and are linked to the notion of stability, reviewed in Chapter 13. The reader should recognize that the identification of these poles from large data sets is the essence of the DMD described in Chapter 7.

Finally, this chapter also introduced the fundamentals of MRA, which complements wavelet theory in Chapters 5 and 8 on the multiscale proper orthogonal decomposition.

5 Time–Frequency Analysis and Wavelets

S. Discetti

Spectral analysis is the cornerstone of some of the most celebrated turbulent flow theories. When focusing on the identification and modeling of coherent flow patterns, on the other hand, the need of localizing the spectral content in time or space often arises. Time–frequency analysis encompasses an arsenal of techniques aimed to localize the spectral content of a signal in time (although for velocity fields we could clearly refer to space and spatial frequencies). This chapter gives an overview of techniques for time–frequency analysis. We start the journey with the windowed Fourier transform (WFT) in continuous and discrete form, and introduce the concepts of shifting and stretching to localize a variety of scales. The bounds imposed by the uncertainty principle will clearly arise. Then, we *adapt* to those bounds introducing the continuous wavelet transform (CWT) and its link to multi-resolution analysis (MRA). The fundamentals of the discrete wavelet transform (DWT) and its intimate relation with filter banks will be outlined. Finally, two applications are presented: a time–frequency analysis of hot-wire data in a turbulent pipe flow using the WFT and the CWT; filtering and compression of velocity fields using a filter based on the DWT. MATLAB codes to practice with the provided examples are provided on the book's webpage.[1]

5.1 Introduction

The Fourier transform (FT) of a function $f(t)$ is an integral transform representing its projection into harmonic functions. From the inner product introduced in Chapter 4, the FT is defined as follows:

$$\mathcal{F}(\omega) = \int_{-\infty}^{+\infty} f(t) e^{-\mathrm{j}\omega t} \, dt. \tag{5.1}$$

The FT is a powerful and efficient technique to solve a wide variety of problems, ranging from data analysis and filtering, image compression, communications, and so on (Smith 2007b). Nonetheless, the FT has severe limitations when applied to non-stationary signals. Indeed, the FT describes the signal's frequency content, but does not provide time localization, i.e., it does not determine the time in which a given frequency occurs in the signal. This information is lost because of the infinite integration bounds in (5.1) and the infinite support of the harmonic basis $e^{-\mathrm{j}\omega t}$.

[1] www.datadrivenfluidmechanics.com/download/book/chapter5.zip

Consider, for example, the signal $x(t)$ represented in Figure 5.1, with its (discrete) FT normalized with respect to its maximum and represented as a function of the frequency $\nu = \omega/2\pi$. The signal exhibits two different frequencies in well-localized time intervals. Even though a clear signature of the two main frequencies is present in the spectrum, no information is retained on the time localization of those frequencies.

Although this example might appear extreme at first glance, there is a wide range of applications where the localization in time is relevant, for example, timescale modification, sinusoidal modeling of musical compositions, cross-synthesis, and so on. In fluid mechanics, the localization of scales in space-time has strong potential implications in the identification and modeling of the flow dynamics (Farge 1992, Schneider & Vasilyev 2010). For such cases, different implementations are sought, possibly maintaining the appeal of computationally efficient algorithms such as the fast Fourier transform (FFT, Cooley & Tukey 1965).

Chapter 4 introduced the limits of the FT and the MRA. This chapter focuses on techniques for time–frequency analysis of signals, including a discussion of the capabilities and limits of the WFT, and the wavelet-based multi-resolution methods. The relation between the DWT and filter banks is also briefly discussed. The objective is to introduce the fundamental concepts of time–frequency analysis, making no pretence of being exhaustive. For a more comprehensive mathematical background, the reader is referred to several excellent literature contributions on the topic (Strang 1989, Daubechies 1990, Daubechies 1992, Cohen 1995, Strang & Nguyen 1996, Torrence & Compo 1998, Van den Berg 2004, Mallat 2009, Kaiser 2010, Chui 1992).

5.2 Windowed Fourier Transform

5.2.1 Windowed Fourier Transform and Gabor Transform

The most intuitive workaround for the time localization limits of the FT is to split the signal into shorter time segments and compute the Fourier spectrum on each of them. This allows localizing in time the frequency content of the signal, with time and frequency resolution depending on the segment size. Recall that the smallest resolved frequency and the frequency resolution of the spectrum are inversely proportional to the signal's time duration (in this case, the segment). Similarly, shorter or larger segments enable more or less precise time localization. The delicate compromise between time and frequency resolution is further discussed in this section.

The signal-splitting method outlined earlier is equivalent to pre-multiplying the signal by a top-hat window, equal to 1 in the selected time segment and 0 elsewhere. The window is then shifted along the signal to obtain a time–frequency description. This is the underlying idea of the Gabor transform (Gabor 1946) in which the kernel of the FT is multiplied by a sliding window function. This transform is defined as

$$\mathcal{G}(\tau,\omega) = \int_{-\infty}^{\infty} f(t)\overline{g}(t-\tau)e^{-j\omega t}\,dt, \tag{5.2}$$

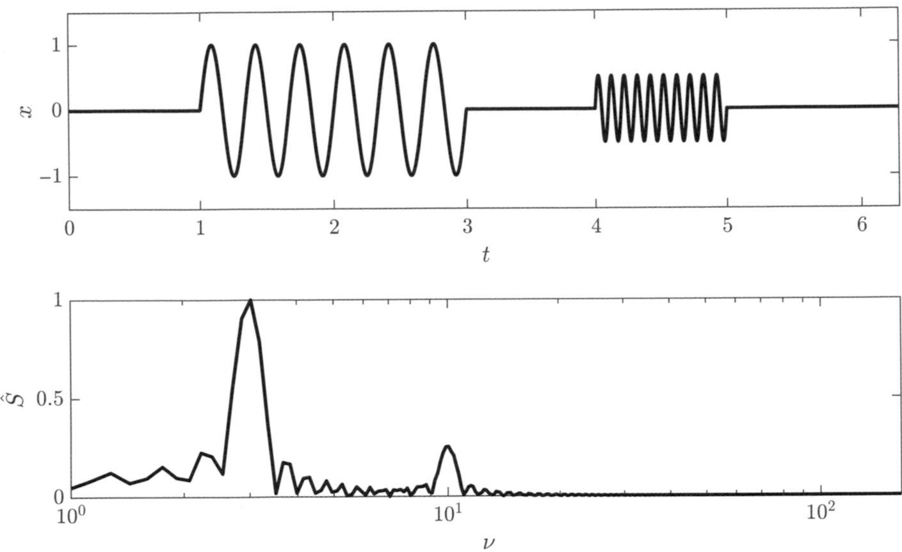

Figure 5.1 Example of non-stationary signal (top) and corresponding normalized Fourier spectrum (bottom).

where the overbar indicates the complex conjugate. The window function $g(t)$ is often selected to be a Gaussian function:

$$g(t) = \frac{1}{\alpha\sqrt{2\pi}} e^{-\frac{1}{2}\left(\frac{t}{\alpha}\right)^2}. \tag{5.3}$$

The parameter α can be used to stretch the window along the time axis. The window can be translated over the time sequence with the shifting parameter τ to obtain the temporal description.

The Gabor transform shares the same linearity properties of the FT and can be inverted with the relation

$$f(t) = \frac{1}{2\pi} \int_{-\infty}^{+\infty} \int_{-\infty}^{+\infty} \mathcal{G}(\tau,\omega) g(t-\tau) e^{\mathrm{j}\omega\tau} d\omega d\tau \tag{5.4}$$

if the L^2 norm of g(t) is equal to 1.

Figure 5.2 shows the Gabor transform of the signal described by the following equations:

$$x(t) = \begin{cases} sin(2\pi\nu_1 t) & \text{if } 1 \leq t \leq 3, \\ 0.5sin(2\pi\nu_2 t) & \text{if } 4 \leq t \leq 5, \\ 0 & \text{otherwise,} \end{cases} \tag{5.5}$$

with $\nu_1 = 3$ and $\nu_2 = 10$ being the frequencies of the two portions of the signal.

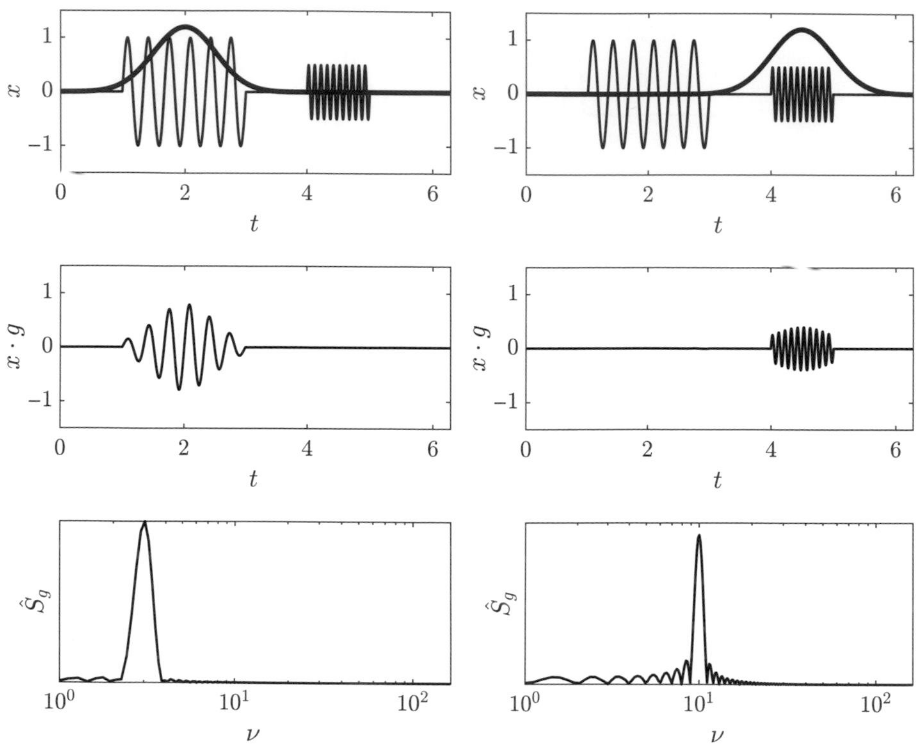

Figure 5.2 Example of application of the Gabor transform on the signal in (5.5). In both cases $\alpha = 0.5$, while $\tau = 2.0, 4.5$ for the left and the right column, respectively. From top to bottom: raw signal with superposed sliding window of the Gabor transform in thick black line; premultiplied signal; Fourier spectrum of the signal premultiplied by the Gaussian kernel at $\tau = 2.0, 4.5$.

This is the same signal analyzed with FT in Figure 5.1. A Gaussian kernel with $\alpha = 0.5$ is here chosen. The selected kernel is expected to have sufficient width to capture the low-frequency oscillations and to be reasonably well localized to identify the high-frequency oscillations occurring on a shorter time segment. The Gabor kernel is then centered at $\tau = 2.0$ and $\tau = 4.5$, i.e., at the center of the regions interested by the two sinusoidal oscillations. Observing the spectrum of the Gabor transform coefficients, the two frequencies are well localized both in the frequency and time domain, with a single sharp peak in the spectrum at $\nu = 3$ and $\nu = 10$ at the two selected time instants, respectively.

5.2.2 Discrete Gabor Transform

In real applications the signals are discretized, thus needing the definition of a discrete Gabor transform. In addition to the discretization of time and frequencies introduced in Chapter 4, we have to include also a discretization of the time shift $\tau = n\tau_0$, with

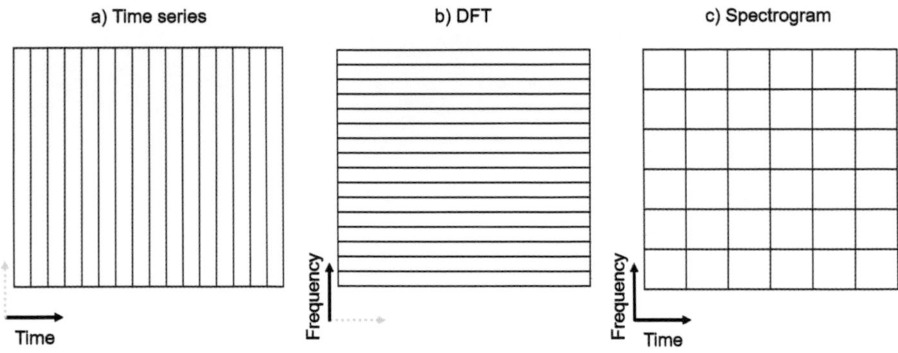

Figure 5.3 Graphical explanation of the different description of the signal in the time domain, frequency domain, and with the spectrogram obtained by the Windowed Fourier Transform.

$n \in \mathbb{N}$, with τ_0 being a fundamental time unit. With the sampling frequency being ν_s, the time grid is $t = k\Delta t$, with $\Delta t = 1/\nu_s$ and $k = 0, \ldots, N-1$. The corresponding discrete Gabor transform is thus given by

$$\mathcal{G}_{m,n} = \sum_{k=0}^{N-1} f_k g(k\Delta t - n\tau_0) e^{-j2\pi m\nu_0}, \tag{5.6}$$

with f_k being the discrete values of $f(t)$ sampled at $t = k\Delta t$.

A careful choice of the fundamental time τ_0 and frequency ν_0 for discretization must be taken. A proper sampling is achieved with $\tau_0 = \Delta t$ and $\nu_0 = \nu_s/N$. In the time domain, boundary treatment (e.g., zero-padding, see Chapter 4) might be beneficial, while in the frequency domain the same equally spaced bins of the discrete FT are generally adopted (i.e., $m = 0, \ldots, N-1$).

Once the lattice of time and frequency is selected, (5.6) can be used to compute a spectrogram. A conceptual sketch to exemplify the differences between the representation in the time–frequency domain and the spectrogram is shown in Figure 5.3. A signal can be described purely in the time domain (i.e., the raw signal), with no frequency resolution. On the other hand, a description in the frequency domain can be obtained via the FT at the expense of withdrawing the time resolution in the integration process. The spectrogram obtained with a time-shifting windowing FT is able to achieve a temporal description of the frequency content by partly giving up frequency resolution. While in Figure 5.3 an equal share between time and frequency resolution is shown for simplicity of illustration, the real balance is tuned according to the properties of the kernel used for the windowing.

This concept is made clear by inspection of different spectrograms computed on the signal defined by (5.5), reported in Figure 5.4. For this computation, $t_0 = \Delta t$ and $\nu_0 = \nu_s/N$ are selected. The parameter α, i.e., the "time-width" of the windowing function $g(t)$, has a significant impact on the transform. According to (5.3), a small value of α corresponds to a relatively narrow window, thus having high

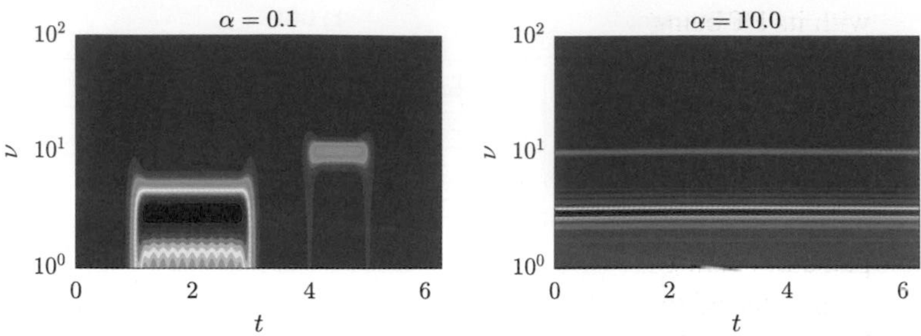

Figure 5.4 Spectrogram of the signal described by (5.5) for two different values of the scaling parameter $\alpha = 0.1$ (left) and $\alpha = 10.0$ (right).

temporal-localization capability (i.e., good temporal resolution) but poor frequency resolution. Intuitively, a narrow window isolates short sequences during the time-shifting process, thus hindering the identification of the low-frequency component of the signal and lacking in spectral resolution. On the other hand, a large value of α delivers a broader window, which in turn has poor temporal localization, but better frequency resolution and capability of identifying the low-frequency component of the signal.

Clearly, if $\alpha \to 0$, the spectrogram tends to perfect temporal localization with no frequency resolution (i.e., a pure time-domain description). On the other hand, for $\alpha \to \infty$, the kernel of the Gabor transform tends to the kernel of the FT, thus giving up completely the time resolution in favor of full frequency resolution.

The backbone of this limit of the Gabor transform is the Heisenberg uncertainty principle. In quantum mechanics, according to the uncertainty principle, position and momentum of a particle cannot be simultaneously ascertained with arbitrary precision. In more general terms, it is not possible to arbitrarily localize both a function and its FT. If a signal is sharply localized (for instance a delta function), its FT is broadband; vice versa, a Fourier spectrum with only one non-null component corresponds to a periodic continuous sinusoidal function.

In strict mathematical terms, the uncertainty principle in the framework of time–frequency analysis can be formulated as follows:

$$\left(\frac{1}{\|f\|^2}\int_{-\infty}^{+\infty} t^2|f(t)|^2 dt\right)\left(\frac{1}{2\pi\,\|f\|^2}\int_{-\infty}^{+\infty} \omega^2|\hat{f}(\omega)|^2 d\omega\right) \geq \frac{1}{4}. \tag{5.7}$$

The integrals in parentheses are also known as *dispersions*.

The choice of the Gaussian kernel in the Gabor transform is not accidental. It can be shown that Gaussian functions have an optimal dispersion product, thus minimizing the combined uncertainty in time–frequency localization. The reader can indeed easily verify that, if

$$f(t) = \frac{1}{\alpha\sqrt{2\pi}} e^{-\frac{t^2}{2\alpha^2}}, \tag{5.8}$$

with its FT being

$$\mathcal{F}(\omega) = e^{-\frac{1}{2}\alpha^2\omega^2}, \tag{5.9}$$

then the left-hand side of (5.7) results in $1/4$.

Furthermore, the integrals on the left-hand side have an interesting interpretation in terms of the so-called *Heisenberg's boxes*, i.e., ideal rectangles in the time–frequency plane indicating the uncertainty in time–frequency definition. These "rectangles" are often referred to as *time–frequency atoms*, already indicated in the conceptual sketch highlighted in Figure 5.3. Interestingly enough, the dispersions for a Gaussian function in (5.8) are proportional to α^2 and $1/\alpha^2$, respectively; for increasing α, the Gaussian function becomes less localized in space but more localized in frequency, and vice versa.

Example 1: DFT versus WFT

Consider the signal $x(t), t \in [0, 2\pi]$ described by (5.5) and sampled with $N = 2\,048$ points.

1. Compute the spectrum of the signal using the DFT.
2. Compute the spectrogram of the signal using the WFT with Gaussian kernel (as in (5.3)). Explore the effect of the standard deviation α of the kernel on the time and frequency resolution.

Solution

The solution is provided in the form of MATLAB code in the supplementary material. The lattice for time and frequency is set as $\tau_0 = \Delta t$, $\nu_0 = \nu_s/N$. The reader can modify the script accordingly to generate Figures 5.1 and 5.4.

5.3 Wavelet Transform

5.3.1 Fundamentals

As discussed in Section 5.2, the main shortcoming of the WFT is the compromise to be sought between time and frequency resolution, i.e., the capability of localizing features in the time and frequency domains. The resolution can be maximized by choosing windows with the minimal area of the corresponding Heisenberg box in the time–frequency plane. However, if our analysis is restricted to one single window choice, then only one scale is selected (within the uncertainty of time–frequency localization from the Heisenberg principle). To achieve a well-resolved description in a set of scales, a process that uses a library of windows spanning the time–frequency domain should be conceived. "Broad" windows, with excellent frequency localization but poor temporal resolution, can be used to describe the low-frequency part of the signal. For such scales, the time localization is

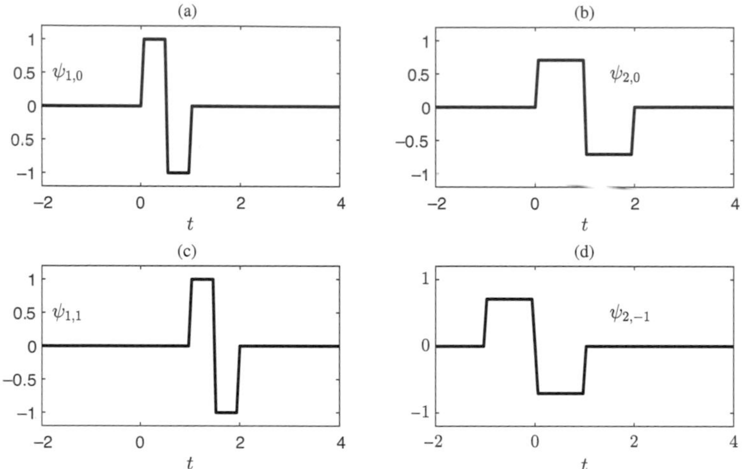

Figure 5.5 Haar mother wavelet (a), with examples of scaled (b), shifted (c), and scaled–shifted (d) versions.

less relevant since low-frequency signals are by definition less localized in time. "Narrow" windows, on the other hand, can deliver excellent time localization for the high-frequency scales. This multi-resolution approach is the cornerstone of *wavelet* theory.

The main ideas behind the wavelet concept are the *scaling* and *shifting* processes, already introduced in the WFT. In wavelet theory, the starting point is a function referred to as "mother wavelet" $\psi(t)$, which is shifted along the signal to provide time localization, and scaled to capture scales of different size. The family of wavelets is thus generated as follows:

$$\psi_{a,b}(t) = \frac{1}{\sqrt{a}}\psi\left(\frac{t-b}{a}\right). \tag{5.10}$$

The variables a and b are, respectively, the scaling and shifting parameters. The term $1/\sqrt{a}$ is introduced to obtain functions with unitary norm. The wavelet is then shifted across the time domain and progressively scaled to create a collection of time–frequency descriptions of the signal, i.e., a multi-resolution analysis.

An illustrative example of this process is shown in Figure 5.5. The earliest example of a wavelet was the Haar wavelet, introduced in 1911 by Alfred Haar (1911) as the Haar sequence (the term wavelet appeared only several decades later).

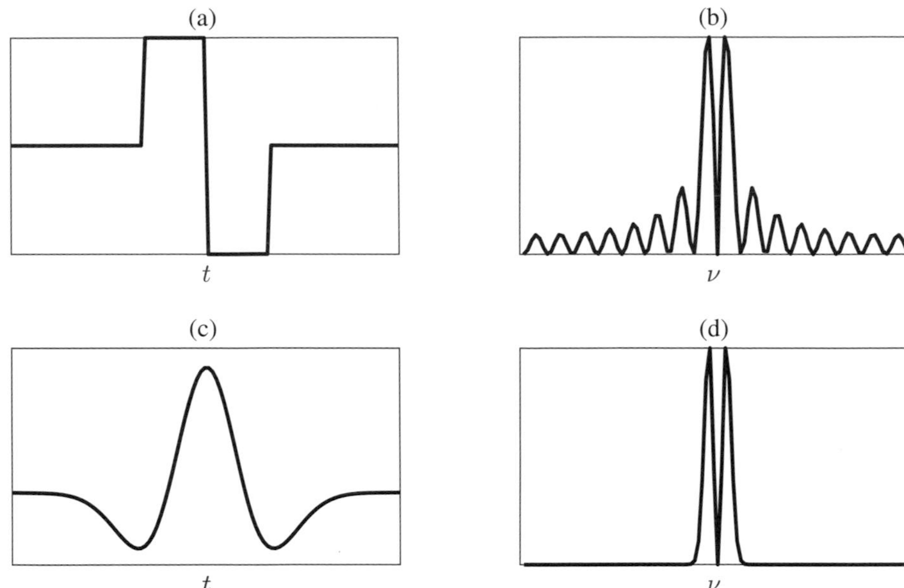

Figure 5.6 Haar (top row) and Mexican-hat (bottom row) wavelets: time representation (left column) and corresponding frequency spectra (right column).

It is defined as follows:

$$\psi(t) = \begin{cases} 1 & \text{if } 0 \leq t < 1/2, \\ -1 & \text{if } 1/2 \leq t < 1, \\ 0 & \text{otherwise.} \end{cases} \tag{5.11}$$

Figure 5.5 includes the mother wavelet $\psi_{1,0}$, a scaled Haar wavelet $\psi_{2,0}$ (i.e., with scaling parameter $a = 2$, thus enlarging the support of the wavelet and, consequently, the corresponding scale), a shifted Haar wavelet $\psi_{1,1}$ (i.e., centered in $t = 3/2$, being $b = 1$) and a scaled-shifted Haar wavelet $\psi_{2,-1}$.

The Haar wavelet has the disadvantage of being noncontinuous (although this is not an issue for the discrete formulation); nonetheless, this turns out to be an advantage for the description of signals with sharp changes. This feature is particularly appreciated, for instance, for edge detection in image processing.

The Haar wavelet is highly localized, thus giving good time localization but poor frequency resolution. This is a direct consequence of the uncertainty principle outlined previously: a compact support in the time-domain results in a broadband frequency spectrum (see Figure 5.6(b)).

Depending on the application and on the desired properties of time–frequency resolution, there is a vast variety of mother wavelets already available in the literature and in software toolboxes. A classic example is the *Mexican-hat* wavelet:

$$\psi_{a,b}(t) = \frac{2}{\sqrt{3a}\pi^{1/4}} \left[1 - \left(\frac{t-b}{a} \right)^2 \right] e^{-\frac{(t-b)^2}{a^2}}. \tag{5.12}$$

The Mexican-hat wavelet and its FT are shown in Figure 5.6(c,d). The Gaussian kernel, which has optimal time–frequency bandwidth product in (5.7), enables the best compromise for the localization in the time and the frequency domains. From the comparison, it is clear that while the spectrum of the Haar wavelet decays with ν^{-1}, the frequency spectrum of the Mexican-hat wavelet is much sharper.

5.3.2 The Continuous Wavelet Transforms

The CWT (Grossmann & Morlet 1984) is defined as the following integral transform:

$$\mathcal{W}_\psi[f(t)](a,b) = \int_{-\infty}^{+\infty} f(t)\overline{\psi}_{a,b}(t)dt. \tag{5.13}$$

Notice that, similarly to the Gabor transform (5.2), the CWT depends on two parameters, the scale a and the shift b. Additionally, as in the WFT, a wide variety of CWTs can be defined by selecting the proper wavelet ψ for the desired application. The main constraint for the selection of the wavelet is the admissibility condition, i.e.,

$$C_\psi = \int_{-\infty}^{+\infty} \frac{|\hat{\psi}(\omega)|^2}{|\omega|} d\omega < \infty, \tag{5.14}$$

with $\hat{\psi}(\omega)$ being the FT of the wavelet function.

The interpretation of the CWT is simple. The signal is processed by a wavelet translated with the translation parameter b, and the result is stored; then the process is repeated after scaling the wavelet with the scaling parameter a to extract a description of the signal at a different scale. This is then carried out in a range of scales, thus covering both the time and the frequency domain.

The CWT is linear (i.e., $\mathcal{W}_\psi[\alpha f(t) + \beta g(t)](a,b) = \alpha\mathcal{W}_\psi[f(t)] + \beta\mathcal{W}_\psi[g(t)]$) and invertible with the relation

$$f(t) = \frac{1}{C_\psi} \int_0^{+\infty} \int_{-\infty}^{+\infty} \frac{1}{a^2} \mathcal{W}_\psi[f](a,b)\psi_{a,b}(t)db\, da. \tag{5.15}$$

The existence of the inverse CWT clearly relies on the admissibility of the wavelet, i.e., $C_\psi < \infty$.

It is important to underline here that we shifted from the time–frequency representation of the Gabor transform to the timescale representation of the CWT. Here two important remarks are needed. First, the relation between scales and frequency might not be immediate, considering that often wavelets have irregular shape (see for instance the Daubechies wavelets, Daubechies 1992). A simple method to extract such relation has been proposed by Meyers et al. (1993). It is based on convolution of the wavelet with a cosine wave, and searching for the scale a that maximizes the correlation. This is a useful exercise for wavelets where a dominant frequency

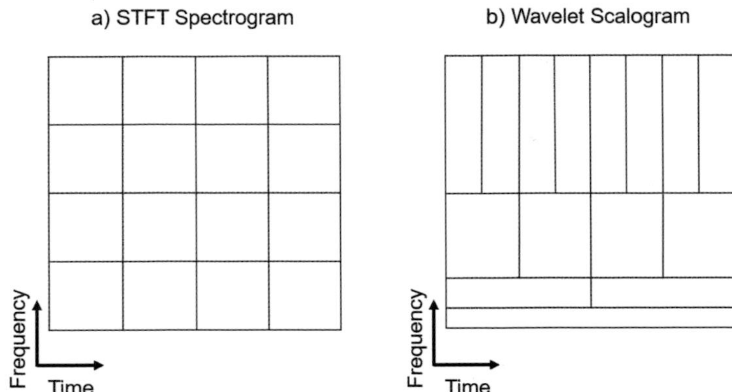

Figure 5.7 Graphical explanation of the difference between the time/frequency description of the windowed Fourier transform (a) and the timescale description of the wavelet analysis (b).

correlation is expected, such as the Morlet wavelet, which is a harmonic function multiplied by a Gaussian. In this case, a linear relation between scale and wavelength is observed (Meyers et al. 1993). Second, the shape of the time–frequency atoms depends on the scaling parameter a. A conceptual sketch of the timescale description obtained in this framework is illustrated in Figure 5.7. Interestingly, it can be easily shown (see chapter 4 of Mallat (2009)) that the sides of the time–frequency atoms scale respectively with a along the time axis, and $1/a$ along the frequency axis, respectively. The area of the time–frequency atom remains constant, and only depends on the wavelet function. It is thus clear that increasing the scaling parameter a "broadens" the atom in time domain and "narrows" it in the frequency domain.

A comparison of the performances of the Gabor transform and the CWT is carried out in Section 5.4.1 using hot-wire data from a pipe-flow experiment in the facility CICLoPE (Centre for International Cooperation in Long Pipe Experiments).

5.3.3 Wavelet Series: the Path to Discrete Wavelet Transform

In practical applications, the data are normally available on a discrete grid, in time (for instance the sampling instants of a hot-wire probe) and/or in space (e.g., the vector grid of Particle Image Velocimetry, or of numerical simulations). It is thus natural to define the wavelet transform in a discrete environment.

Furthermore, it should be remarked that, similarly to the Gabor transform, the CWT description is fundamentally "redundant", i.e., depending on the choice of the shifting and scaling parameters the timescale space can be oversampled. We often search instead for a different representation of our data (i.e., a different basis), with the aim of highlighting certain features (see Part III). Thus, if we have N samples, we search for a set of N numbers in a different space, which allow us to extract patterns or features otherwise hidden in the time (or space) representation.

The wavelet family can be written as

$$\psi_{m,n} = \frac{1}{\sqrt{a_0^m}}\psi(a_0^{-m}x - nb_0), \tag{5.16}$$

with a_0 and b_0 being positive parameters and m, n being integer numbers. Typically $a_0 = 2$ and $b_0 = 1$, thus leading to

$$\psi_{m,n} = \frac{1}{\sqrt{2^m}}\psi(2^{-m}(x - n2^m)). \tag{5.17}$$

This results in progressive binary dilations, also called *voices* or *octaves*, and translations of $n2^m$. As will be made clear in Section 5.3.5, one of the main advantages of this choice is the computational efficiency of its implementation.

A prominent role is played by *orthogonal* wavelets, i.e., fulfilling the condition

$$\langle \psi_{m,n}\psi_{p,q}\rangle = \int_{-\infty}^{+\infty} \psi_{m,n}(t)\psi_{p,q}(t)dx = \delta_{m,n}\delta_{p,q}, \tag{5.18}$$

where $\langle \cdot, \cdot \rangle$ is the inner product in L^2 and $\delta_{i,j}$ is the Kronecker delta

$$\delta_{i,j} = \begin{cases} 1 & i = j, \\ 0 & \text{otherwise.} \end{cases} \tag{5.19}$$

In case of decomposing with orthogonal wavelets, a given function $f(t)$ can be uniquely defined by its Discrete Wavelet Transform, i.e.,

$$\mathcal{W}_{m,n} = (f(t), \psi_{m,n}(t)), \tag{5.20}$$

with the inverse DWT being

$$f(t) = \sum_{m=-\infty}^{+\infty} \sum_{n=-\infty}^{+\infty} \mathcal{W}_{m,n}\psi_{m,n}(t). \tag{5.21}$$

This important property opens the question of how to generate a family of orthogonal wavelets. Section 5.3.4 highlights the link between a general procedure to generate wavelets and the multi-resolution analysis introduced in Chapter 4.

5.3.4 Multi-Resolution Analysis

The concept of multi-resolution approximation (or MRA) forms the fundamental cornerstone for the design of discrete orthogonal wavelets. The concept is here briefly introduced; for a more extensive treatment, the reader is referred to Mallat (2009).

The MRA computes the approximation of a generic function f as a set of orthogonal projections on a sequence of spaces $\{\mathcal{V}_m\}_{m\in\mathbb{Z}}$.

Consider the sequence of closed subspaces $\{\mathcal{V}_m\}_{m\in\mathbb{Z}}$ in $\mathcal{L}^2(\mathbb{R})$. This can be used for multi-resolution approximations if the following properties are satisfied:

- $\mathcal{V}_m \subset \mathcal{V}_{m+1}, \forall m \in \mathbb{Z}$
- $\cup_{m=-\infty}^{+\infty}\mathcal{V}_m$ spans $\mathcal{L}^2(\mathbb{R})$

- $\cap_{m=-\infty}^{+\infty} \mathcal{V}_m = \{0\}$, i.e., the only intersection between the subspaces is the null function
- if $f(x) \in \mathcal{V}_m$, then $f(2x) \in \mathcal{V}_{m+1}, \forall m \in \mathbb{Z}$
- there exists a function ϕ in $\mathcal{V}_0$, called a *scaling function* or *father wavelet*, such that $\phi(x-n)$ with $n \in \mathbb{Z}$ is an orthonormal basis of $\mathcal{V}_0$ with respect to the inner product in $\mathcal{L}^2(\mathbb{R})$.

A simplified view of a 3-level MRA in the form of a Venn diagram is shown in Figure 5.8. It can be demonstrated that an orthogonal basis for each subspace $\{\mathcal{V}_m\}_{m\in\mathbb{Z}}$ can be constructed from the basis of V_0 as follows:

$$\{\phi_{m,n}(x) = \sqrt{2^m}\phi(2^m x - n), n \in \mathbb{Z}\}, \tag{5.22}$$

since $\phi \in \mathcal{V}_0 \subset \mathcal{V}_1 \subset \mathcal{V}_2 \subset \ldots \mathcal{V}_m$. For example, an orthonormal basis for $\mathcal{V}_1$ can be written from the basis of $\mathcal{V}_0$ as $\{\phi_{1,n}(x) = \sqrt{2}\phi(2x-n), n \in \mathbb{Z}\}$. Thus the scaling function ϕ can be expressed as a linear combination of $\{\phi_{1,n}(x)\}$, i.e.,

$$\phi(x) = \sum_{n\in\mathbb{Z}} c_n \phi_{1,n}(x) = \sqrt{2}\sum_{n\in\mathbb{Z}} c_n \phi(2x-n), \tag{5.23}$$

referred to as the *scaling equation* or *refinement equation.*

We need now to determine the set of orthonormal wavelets corresponding to the defined MRA. Consider, for this purpose, the orthogonal complement $\mathcal{W}_0$ of $\mathcal{V}_0$ in $\mathcal{V}_1$, i.e., the set of functions in $\mathcal{V}_1$ that are orthogonal to any function in $\mathcal{V}_0$:

$$\mathcal{W}_0 = \{f \in \mathcal{V}_1 : \langle f, g\rangle = 0 \quad \forall g \in \mathcal{V}_0\}. \tag{5.24}$$

Our candidate to build a set of orthogonal wavelets is a function $\psi \in \mathcal{W}_0$ that is analogous to the scaling function, i.e., it generates a basis of $\mathcal{W}_0$ from integer translation $\psi(x-n)$. Since $\psi \in \mathcal{W}_0 \subset \mathcal{V}_1$ and $\phi_{1,n}$ is a basis for $\mathcal{V}_1$, then

$$\psi(x) = \sum_{n\in\mathbb{Z}} a_n \phi_{1,n}. \tag{5.25}$$

Recall that $\psi \in \mathcal{W}_0$ and $\phi \in \mathcal{V}_0$, then $\langle\psi,\phi\rangle = 0$ from which we obtain that

$$\langle\psi,\phi\rangle = \sum_{n\in\mathbb{Z}} a_n c_n = 0. \tag{5.26}$$

One simple solution for the previous equation is to set $a_n = (-1)^n c_{1-n}$. We can now define the corresponding *mother wavelet*

$$\psi(x) = \sum_{n\in\mathbb{Z}} (-1)^n c_{1-n} \phi(2x-n), \tag{5.27}$$

which generates the family of orthogonal wavelets corresponding to the MRA defined earlier. Interestingly enough, if we interpret the scaling function as a low-pass filter, multiplying by $(-1)^n$ its impulsive response leads us to define its complementary high-pass filter.

As discussed in Chapter 4, this is true because the dyadic construction in (5.22) gives to these complementary filters equal portions of each scale's spectra.

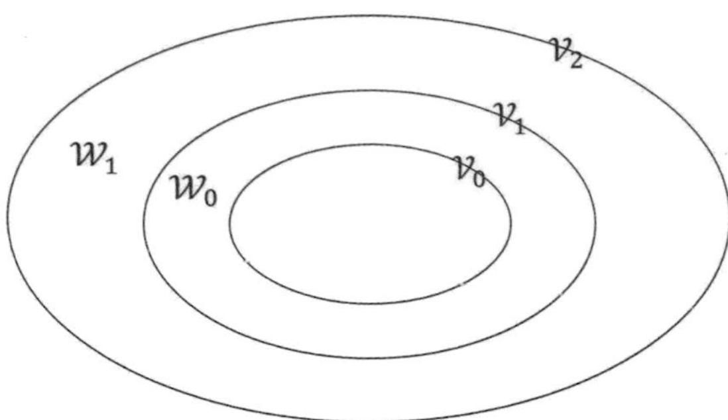

Figure 5.8 Conceptual sketch with Venn diagram of a three-level MRA. The spaces $\mathcal{V}_0, \mathcal{V}_1$, and $\mathcal{V}_2$ of the MRA are indicated with the corresponding symbol located on their boundary. The respective orthogonal complements $\mathcal{W}_0$ and $\mathcal{W}_1$ are indicated with the symbol located within the corresponding domain.

Example: the Haar wavelet

The simplest example of construction of orthogonal wavelets via MRA is the Haar wavelet. Consider a generic function $f \in \mathcal{L}^2$, and suppose we want to approximate it with a piecewise constant approximation at different levels of resolution f_n of length 2^{n-1}. For example, f_0 is a piecewise constant approximation over segments of length 1, f_1 on length $1/2$, and so on. This is equivalent to selecting a scaling function in $\mathcal{V}_0$,

$$\phi_{0,n}(x) = I_{[n,n+1)}(x), \tag{5.28}$$

with I_A being the indicator function, equal to 1 for all elements belonging to A and 0 elsewhere, and $n \in \mathbb{Z}$.

It can be easily shown that the functions $\phi_{0,n}$ are orthogonal, and that they span the entire $\mathcal{V}_0$ with a set of integer translation of $\phi_{0,n} = \phi(x-n)$. This function is the *Haar scaling function*.

If we consider now the piecewise approximation at level $\mathcal{V}_1$, this is carried out by functions that are piecewise constant over length $1/2$. In this space, an orthonormal basis is constituted by the functions $\phi_{1,n}(x) = \sqrt{2} I_{[n/2,(n+1)/2)}(x)$ (with the factor $\sqrt{2}$ to account now for the shorter support of the functions). Intuitively, these functions can be obtained by scaling and shifting the function $\phi_{0,n}$, thus obtaining

$$\phi_{1,n}(x) = \sqrt{2}\phi(2x-n). \tag{5.29}$$

This is the same result of (5.23) outlined earlier. From (5.23), for the Haar system we obtain

$$c_n = \begin{cases} 1/\sqrt{2} & \text{if } n = 0, 1, \\ 0 & \text{otherwise.} \end{cases} \tag{5.30}$$

From (5.27), we can simply derive the corresponding mother wavelet, which is the discrete form of the Haar wavelet ψ reported in (5.11). It is straightforward to show that ϕ and ψ are orthogonal, as was expectable since $\phi \in \mathcal{V}_0$ and $\psi \in \mathcal{W}_0$, which is the orthogonal complement of $\mathcal{V}_0$.

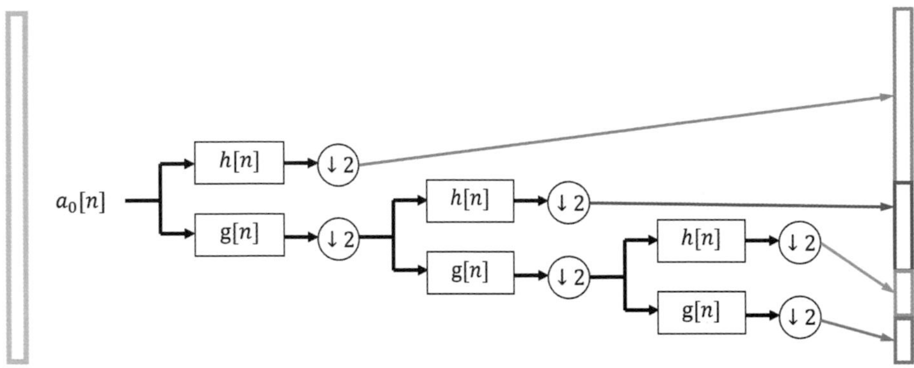

Figure 5.9 Sketch of the computation process of the fast wavelet transform using filter banks.

Now, if we consider the piecewise approximation of a generic function f at the level f_0 and f_1, and compute the difference between them, we discover that it resembles a set of scaled and translated versions of the Haar wavelet ψ. This simple result is just a consequence of the fact that $\mathcal{W}_0$ is the orthogonal complement of $\mathcal{V}_0$, i.e., the part of f that is "discarded" in the projection on $\mathcal{V}_0$ belongs to $\mathcal{W}_0$ and, as such, can be expressed as a linear combination of the wavelet function ψ.

5.3.5 The Discrete Wavelet Transform and Filter Banks

Consider the generic function $f_1(t) \in \mathcal{V}_1$. It can be written as a linear combination of $\phi_{1,n} = \sqrt{2}\phi(2x - n)$, or alternatively it can be expressed as the sum of its projections onto $\mathcal{V}_0$ and its orthogonal complement $\mathcal{W}_0$, i.e.,

$$f_1(t) = \sum_{n\in\mathbb{Z}} \alpha_{1,n}\phi_{1,n}(t) = \sum_{n\in\mathbb{Z}} \alpha_{0,n}\phi_{0,n}(t) + \sum_{n\in\mathbb{Z}} \beta_{0,n}\psi_{0,n}(t). \tag{5.31}$$

This operation is essentially a change of basis. The first term on the RHS is referred to as **approximation** (it is indeed the best approximation of $f(t)$ at the resolution level of $\mathcal{V}_0$), and the second term is the **detail**, i.e., the part of f that is missing to obtain f_1.

As a general result, extending this procedure to the subsequent levels,

$$f(t) = \sum_{n\in\mathbb{Z}} \alpha_{0,n}\phi_{0,n}(t) + \sum_{m\in\mathbb{N}}\sum_{n\in\mathbb{Z}} \beta_{m,n}\psi_{m,n}(t), \tag{5.32}$$

i.e., each function $f(t)$ can be expressed as the sum of the approximation at the resolution level $\mathcal{V}_0$ and a sequence of details at the levels $m \geq 0$.

This process is normally implemented as a two-channel multi-rate filter bank (Mallat 1989), which convolves the signal with a low-pass filter $g[n]$ and a high-pass filter $h[n]$. The high-pass filter retains the details at the scale m, while the low-pass filter contains the approximation. Both descriptions are subsampled by a factor of 2; then the low-pass filtered signal is again passed through the high- and low-pass filters to obtain the approximation and details at the subsequent scale. At the end of the process, the DWT delivers an approximation at the level 0 and a set of details at different scales. A sketch of this process is shown in Figure 5.9.

In order to establish the connection between DWT and filter banks, consider a signal $x[n]$ of N samples, and the simplest two-channel filter bank, composed by a low-pass filter that performs a moving average of the values of the signal, and a high-pass filter that computes the differences. The low-pass filter has coefficients $g[0] = g[1] = 1/2$, while the high-pass filter has coefficients $h[0] = -h[1] = 1/2$. The two filters separate the frequency components of the signal x into two bands, low and high frequencies. In matrix form:

$$Gx = \frac{1}{2}\begin{bmatrix} 1 & 1 & & & \\ & 1 & 1 & & \\ & & 1 & 1 & \\ & & & \ddots & \end{bmatrix}\begin{bmatrix} x_0 \\ x_1 \\ x_2 \\ \vdots \end{bmatrix}, \tag{5.33}$$

$$Hx = \frac{1}{2}\begin{bmatrix} 1 & -1 & & & \\ & 1 & -1 & & \\ & & 1 & -1 & \\ & & & & \cdots \end{bmatrix}\begin{bmatrix} x_0 \\ x_1 \\ x_2 \\ \vdots \end{bmatrix}. \tag{5.34}$$

The convolution of x with each of the filter produce $2N$ components. Recall that we are searching for a new representation, i.e., a new basis, thus we aim to obtain N components. The filtering operation is thus followed by downsampling by a factor of two (indicated with the symbol $\downarrow 2$), which removes the odd components of the filtered signals. This can be expressed simply in matrix form as an identity matrix with missing odd rows, and multiplying by $\sqrt{2}$ for energy normalization (since downsampling is dropping half of the components). Combining the operations of filtering and downsampling in matrix form, we obtain:

$$(\downarrow 2)G = \frac{1}{\sqrt{2}}\begin{bmatrix} 1 & 1 & & & & \\ & & 1 & 1 & & \\ & & & & \cdots & \end{bmatrix}, \tag{5.35}$$

$$(\downarrow 2)H = \frac{1}{\sqrt{2}}\begin{bmatrix} 1 & -1 & & & & \\ & & 1 & -1 & & \\ & & & & \cdots & \end{bmatrix}. \tag{5.36}$$

This operation can be applied recursively, as outlined in Figure 5.9, i.e., the low-pass filtered part is passed again through the low- and high-pass filter, and downsampled, and so on. But how does this relate to wavelets and MRA? The answer is simple: the MRA is performing exactly the same operation, filtering (through ϕ) and downsampling. The reader should now start to glimpse the relation with the

Haar wavelet and the simple two-channel filter bank proposed here. From (5.23), considering the coefficients in (5.30),

$$\phi(t) = \phi(2t) + \phi(2t - 1), \tag{5.37}$$

and, in general, for discrete signals and for the levels j and $j - 1$,

$$\phi_{j-1,n} = \frac{1}{\sqrt{2}}[\phi_{j,2n} + \phi_{j,2n+1}]. \tag{5.38}$$

For the signal x of N samples, this would result in $N/2$ coefficients, with the same output of applying the process of filtering and downsampling (it can be checked easily that the operation can be written in matrix form exactly as in (5.35)).

Similarly, from (5.27),

$$\psi_{j-1,n} = \frac{1}{\sqrt{2}}[\phi_{j,2n} - \phi_{j,2n+1}], \tag{5.39}$$

which is equivalent to (5.36).

This is the celebrated Mallat's pyramid algorithm, which is based on decomposing a function at a certain resolution level as the sum of approximation and detail at a coarser resolution. One of the main advantages of this algorithm is the efficiency of its implementation. The fast wavelet transform requires a number of operations that scales linearly with the number of samples of the signal, thus overcoming in efficiency the FFT (whose complexity scales with $N \log_2(N)$).

The inverse process (referred as *synthesis* or reconstruction) is based on upsampling the signal filling the gaps with zeros, and on convolution with the same filters.

Example 2: Calculate the DWT of a signal

Consider the discrete signal:

$$x = [4, 6, 9, 5, 1, 3, 4, 4].$$

1. Compute the 1-level DWT with Haar wavelet and express it in the form of a matrix transformation.
2. Compute the 3-level DWT with Haar wavelet and express it in the form of a matrix transformation.
3. Compute the inverse DWT of the wavelet decomposition of the previous point.

Solution

1. The father wavelet is given by (5.28), with the corresponding mother wavelet being given by (5.27). After application of the filters and downsampling according to the matrix form of (5.35) and (5.36), we obtain

$$\alpha^{(1)} = [5\sqrt{2}, 7\sqrt{2}, 2\sqrt{2}, 4\sqrt{2}]; \ \beta^{(1)} = [-\sqrt{2}, 2\sqrt{2}, -\sqrt{2}, 0]. \tag{5.40}$$

In matrix form, this is equivalent to

$$x = W^{(1)} x_W^{(1)}, \tag{5.41}$$

with $x_W^{(1)} = [\alpha^{(1)}, \beta^{(1)}]^T$ and

$$W^{(1)} = \frac{1}{\sqrt{2}} \begin{bmatrix} 1 & 0 & 0 & 0 & 1 & 0 & 0 & 0 \\ 1 & 0 & 0 & 0 & -1 & 0 & 0 & 0 \\ 0 & 1 & 0 & 0 & 0 & 1 & 0 & 0 \\ 0 & 1 & 0 & 0 & 0 & -1 & 0 & 0 \\ 0 & 0 & 1 & 0 & 0 & 0 & 1 & 0 \\ 0 & 0 & 1 & 0 & 0 & 0 & -1 & 0 \\ 0 & 0 & 0 & 1 & 0 & 0 & 0 & 1 \\ 0 & 0 & 0 & 1 & 0 & 0 & 0 & -1 \end{bmatrix}. \quad (5.42)$$

Notice that the columns of $W^{(1)}$ are orthonormal, and that by construction, $W^{(1)} = [(\downarrow 2)G^T (\downarrow 2)H^T]$.
It is also worth evaluating the energy of the signal and the share of it between the coefficients:

$$E = \sum_{n=1}^{8} x_i^2 = 200,$$
$$E_{\alpha^{(1)}} = 188,$$
$$E_{\beta^{(1)}} = 12,$$

i.e., 94% of the energy is contained in the approximation coefficients. This is also a consequence of the signal having a nonzero mean. This means that we can preserve 94% of the information of the signal by retaining only half of the wavelet coefficients. This is the cornerstone of signal compression using DWT.

2. To achieve a 3-level DWT, we have to apply recursively the same procedure of the previous point to the approximation at each level. This leads to

$$\alpha^{(2)} = [12, 6]; \quad \beta^{(2)} = [-2, -2], \quad (5.43)$$
$$\alpha^{(3)} = [9\sqrt{2}]; \quad \beta^{(3)} = [3\sqrt{2}]. \quad (5.44)$$

The set of coefficients and the corresponding DWT matrix are as follows:

$$x_W^{(3)} = [\alpha^{(3)}, \beta^{(3)}, \beta^{(2)}, \beta^{(1)}] = [9\sqrt{2}, 3\sqrt{2}, -2, -2, -\sqrt{2}, 2\sqrt{2}, -\sqrt{2}, 0], \quad (5.45)$$

$$W^{(3)} = \begin{bmatrix} \frac{1}{\sqrt{8}} & \frac{1}{\sqrt{8}} & \frac{1}{2} & 0 & \frac{1}{\sqrt{2}} & 0 & 0 & 0 \\ \frac{1}{\sqrt{8}} & \frac{1}{\sqrt{8}} & \frac{1}{2} & 0 & -\frac{1}{\sqrt{2}} & 0 & 0 & 0 \\ \frac{1}{\sqrt{8}} & \frac{1}{\sqrt{8}} & -\frac{1}{2} & 0 & 0 & \frac{1}{\sqrt{2}} & 0 & 0 \\ \frac{1}{\sqrt{8}} & \frac{1}{\sqrt{8}} & -\frac{1}{2} & 0 & 0 & -\frac{1}{\sqrt{2}} & 0 & 0 \\ \frac{1}{\sqrt{8}} & -\frac{1}{\sqrt{8}} & 0 & \frac{1}{2} & 0 & 0 & \frac{1}{\sqrt{2}} & 0 \\ \frac{1}{\sqrt{8}} & -\frac{1}{\sqrt{8}} & 0 & \frac{1}{2} & 0 & 0 & -\frac{1}{\sqrt{2}} & 0 \\ \frac{1}{\sqrt{8}} & -\frac{1}{\sqrt{8}} & 0 & -\frac{1}{2} & 0 & 0 & 0 & \frac{1}{\sqrt{2}} \\ \frac{1}{\sqrt{8}} & \frac{1}{\sqrt{8}} & 0 & -\frac{1}{2} & 0 & 0 & 0 & -\frac{1}{\sqrt{2}} \end{bmatrix}. \quad (5.46)$$

Again, we notice that 81% of the energy is in the approximation and 90% in the level 3 (i.e., in only 2 coefficients). Similarly, we retain 94% of the energy by retaining only the 3 coefficients with magnitude larger than or equal to $2\sqrt{2}$.

3. The inverse transform can be carried out simply by inverting the previous matrices. Nonetheless, it is instructive to carry out the process according to the filter bank implementation, i.e., recursive upsampling and convolution. For the 3-level decomposition, we first upsample including zeros, and then use the filters.

$$\alpha^{(3)}_{\uparrow 2} = [9\sqrt{2}, 0] \;\; \beta^{(3)}_{\uparrow 2} = [3\sqrt{2}, 0], \tag{5.47}$$

$$\hat{\alpha}^{(3)}_{\uparrow 2} = [9, 9] \;\; \hat{\beta}^{(3)}_{\uparrow 2} = [3 - 3], \tag{5.48}$$

$$\alpha^{(2)} = \hat{\alpha}^{(3)}_{\uparrow 2} + \hat{\beta}^{(3)}_{\uparrow 2} = [12, 6]. \tag{5.49}$$

At the next level,

$$\alpha^{(2)}_{\uparrow 2} = [12, 0, 6, 0] \;\; \beta^{(2)}_{\uparrow 2} = [-2, 0, -2, 0], \tag{5.50}$$

$$\hat{\alpha}^{(2)}_{\uparrow 2} = [6\sqrt{2}, 6\sqrt{2}, 3\sqrt{2}, 3\sqrt{2}] \;\; \hat{\beta}^{(2)}_{\uparrow 2} = [-\sqrt{2}, \sqrt{2}, -\sqrt{2}, \sqrt{2}], \tag{5.51}$$

$$\alpha^{(1)} = \hat{\alpha}^{(2)}_{\uparrow 2} + \hat{\beta}^{(2)}_{\uparrow 2} = [5\sqrt{2}, 7\sqrt{2}, 2\sqrt{2}, 4\sqrt{2}]. \tag{5.52}$$

Finally, for the last level,

$$\alpha^{(1)}_{\uparrow 2} = [5\sqrt{2}, 0, 7\sqrt{2}, 0, 2\sqrt{2}, 0, 4\sqrt{2}, 0], \qquad \beta^{(1)}_{\uparrow 2} = [-\sqrt{2}, 0, 2\sqrt{2}, 0, -\sqrt{2}, 0, 0, 0], \tag{5.53}$$

$$\hat{\alpha}^{(1)}_{\uparrow 2} = [5, 5, 7, 7, 2, 2, 4, 4], \qquad \hat{\beta}^{(1)}_{\uparrow 2} = [-1, 1, 2, -2, -1, 1, 0, 0], \tag{5.54}$$

$$x = \hat{\alpha}^{(1)}_{\uparrow 2} + \hat{\beta}^{(1)}_{\uparrow 2} = [4, 6, 9, 5, 1, 3, 4, 4]. \tag{5.55}$$

The interesting exercise of carrying out the DWT after thresholding the wavelet coefficients is left to the reader.

5.4 Tutorials and Examples

Time–frequency analysis, and particularly MRA with wavelets, has found numerous applications in fluid mechanics, especially in turbulence modeling and simulation (see e.g., Meneveau 1991, Farge 1992, Schneider & Vasilyev 2010). In this section, simple examples of application of the concepts outlined earlier are proposed, with the target aimed toward direct application on data analysis, rather than modeling.

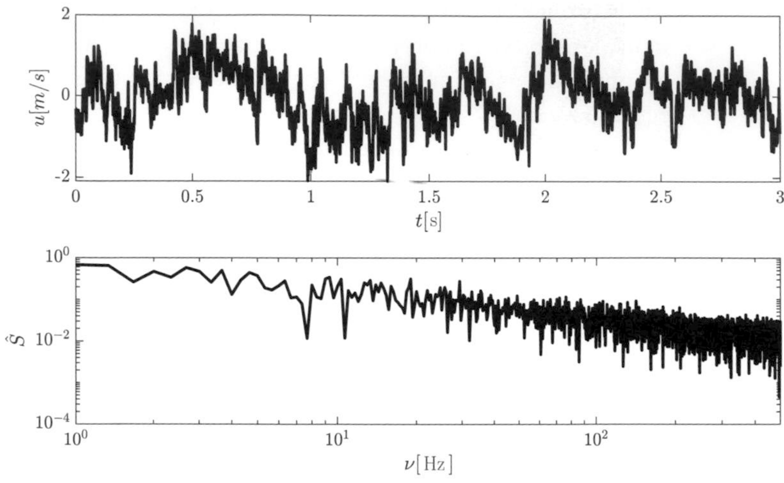

Figure 5.10 Velocity fluctations for Example 3 (top) and corresponding Fourier spectrum (bottom), normalized by the maximum value.

5.4.1 Time–Frequency Analysis of Hot-Wire Data: Continuous Wavelet Transform versus Windowed Fourier Transform

In this example, a signal is analyzed using the WFT with Gabor kernel and the CWT to obtain a time–frequency description. The signal is a hot-wire data set captured in the CICLoPE facility at the Universitá Alma Mater Studiorum di Bologna (Talamelli et al. 2009). The full data set comprises synchronized flow-field measurements with Particle Image Velocimetry (PIV) at low repetition rate, and hot-wire measurements with good temporal resolution. A full description of the experiment is reported in Discetti et al. (2019). The data are available in the Turbase repository (`https://turbase.cineca.it/init/routes/#/logging/view_dataset/81/tabmeta`).

In this section, we consider the data corresponding to Case 1 of the data set. The file contains the variable `ut`, which corresponds to a 20-minute recording at 10 kHz on 6 channels. The first 5 channels correspond to the simultaneous recording of a rake of 5 hot-wires at the same streamwise and azimuthal position but increasing wall distance from row 1 to 5. The 6th channel stores the recording time instant of the PIV system.

For the purpose of this analysis, we consider only data from the first row, i.e., from the closest probe to the wall, within the time segment $1s < t \leq 3s$, subsampled at 1 kHz. The velocity fluctuations (i.e., after removing the time-average velocity) and the corresponding Fourier spectrum are reported in Figure 5.10 for reference.

By visual inspection of the raw signal, we can notice the signature of relatively low-frequency fluctuations, superposed with a range of high-frequency fluctuations of lower intensity. We aim to localize these events through a time–frequency analysis.

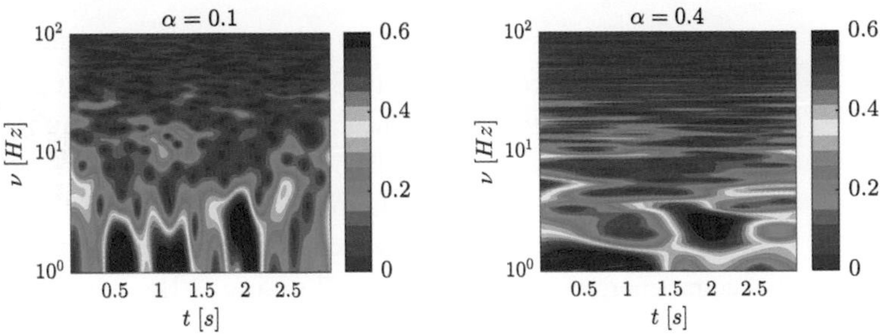

Figure 5.11 Spectrogram for Example 3, with α = 0.1 s (left) and α = 0.4 s (right). The spectrograms have been normalized with their corresponding maximum value.

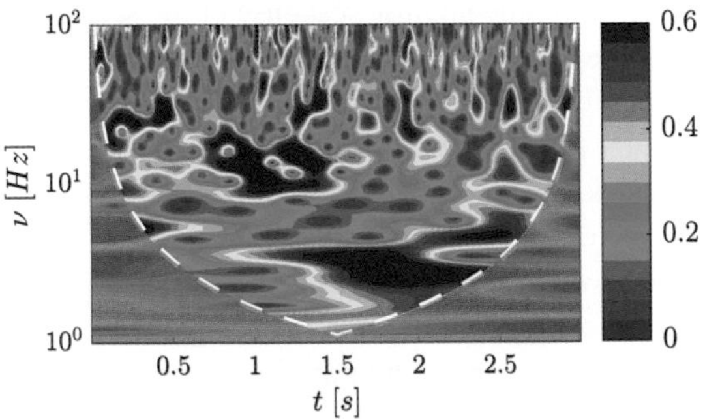

Figure 5.12 Continuous wavelet transform for Example 3. The wavelet coefficients are reported in absolute value and normalized with respect to the maximum.

The quantitative analysis, with computation of the spectrogram and the scalogram, is presented in the following example.

Example 3: Computing the Gabor spectrogram and the CWT scalogram of hot-wire data

Consider the hot-wire experimental data presented before. Extract the data from the closest probe to the wall, within the time segment 1 s $< t \leq$ 3 s, and subsample the sequence to 1 kHz.

1. Compute the Gabor spectrogram of the signal with different kernel size (for instance α = 0.1 s and α = 0.4 s).
2. Compute the CWT of the same signal and discuss the differences between the scalogram and the spectrogram.

Solution

1. The spectrogram, computed with Gabor kernel with $\alpha = 0.1$ s and $\alpha = 0.4$ s is illustrated in Figure 5.11. Once again, it is evident that in the case of the narrow window ($\alpha = 0.1$ s) a good time localization is achieved, at the expense of the frequency resolution. On the other side, with $\alpha = 0.4$ s, it is possible to identify some low-frequency features (for example, a peak of spectral energy is observed at $t \approx 2$ s $-\nu \approx 2$ Hz, which is corresponding to an intense fluctuation that is clearly identifiable in the raw signal and with period of approximately 0.5 s). This feature was evidently distorted for $\alpha = 0.1$ s, which is not capable of identifying low-frequency features due to the poor frequency resolution. Indeed the width of the kernel is 0.2 s if measured at $\pm\alpha$, thus all frequencies below ≈ 5 Hz are not detectable. This qualitative consideration should be of course quantitatively supported by computing the impulse response of the selected Gabor kernel.
2. The corresponding scalogram from the CWT is shown in Figure 5.12. This can be easily computed with the command `cwt` in MATLAB and is generally implemented in libraries for the vast majority of programming languages. In MATLAB the Morse wavelet is set by default. At low frequencies, it can be observed that there is a good frequency resolution but relatively poor time localization. For low frequencies, indeed, wavelets with large support are used. Flowing along the spectrum of scales, we clearly see that the temporal localization progressively improves, at the expense of frequency resolution. This is in line with the conceptual sketch of the wavelet transform reported in Figure 5.7. Since the signal is of finite length, a cone of influence bounds the region where edge effects become relevant, i.e., a region where the scaled wavelet partially extends outside of the time domain of the raw signal.
   Interestingly enough, a visual comparison of Figures 5.11 and 5.12 exemplifies the multi-resolution capabilities of the CWT. In the low-frequency range, the CWT scalogram and the spectrogram with $\alpha = 0.4$ s share significant similarities; for frequencies in the range $5 - 20$ Hz, there is more agreement with the spectrogram with $\alpha = 0.1$ s. This comparison highlights the potential of CWT over the WFT: using a scaled version of the kernel function at different frequencies, it is possible to carry out a tunable MRA of the signal, depending on the scale to be observed.

The solution is provided in the supplementary material in the form of MATLAB code.

5.4.2 Filtering and Compression

A great amount of literature on wavelets is devoted to denoising and/or compressing data. For the case of additive noise, the noise contribution spreads uniformly over

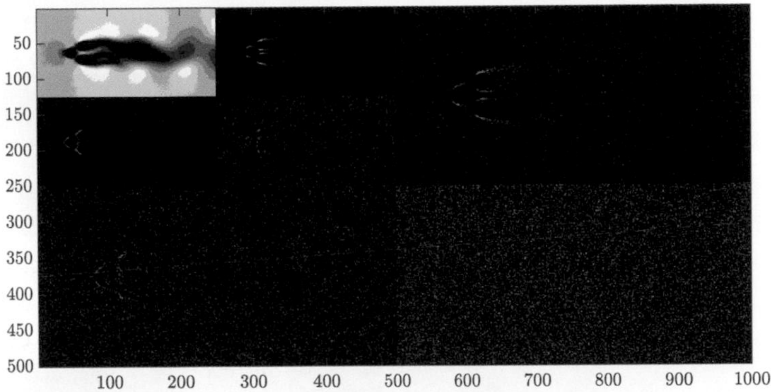

Figure 5.13 Representation of the two-level DWT coefficients with the Haar wavelet. The coefficients are arranged in a 500×1000 matrix, i.e., with the same format of the data in input.

the set of wavelet coefficients; nonetheless, the coefficients of the signal are typically rapidly decreasing in magnitude along the spectrum of scale, thus paving to way to data compression and filtering by retaining only the most important ones. One of the advantages of using wavelets for this purpose is their capability to preserve edges, which is one of the most desired features in image compression.

In this example, DWT is used to analyze a flow field and synthesize it after a thresholding procedure to remove the noise.

The test case is the flow in the wake of three cylinders with diameter D, arranged with the axis on the vertices of an equilateral triangle with side length equal to $3D/2$. The downstream side of the triangle is orthogonal to the freestream flow, and it is located at $x = 0$, centered on the y-axis. This configuration is known as *fluidic pinball* (Deng et al. 2020), and is a very interesting test case for flow control applications (Raibaudo et al. 2020).

An instantaneous flow field from DNS data at $Re = 130$ (referred to as a *chaotic regime* (Deng et al. 2020)) is considered as a reference signal for the following example. For simplicity, only the streamwise component of the velocity field is analyzed.

Example 4: Filtering and compression of velocity fields

Consider the instantaneous flow field provided in the file *pinball_000001.mat*, available in the supplementary material. The file contains the matrices U and V, being respectively the streamwise and crosswise velocity fields on a 500×1000 points grid (obtained after interpolation of the data from the original DNS grid). The grid is stored in the file *GridStruc.mat* in the form of matrices X and Y.

1. Superpose to the streamwise velocity field an additive Gaussian noise, with standard deviation equal to $0.05U_\infty$ (with U_∞ being the freestream

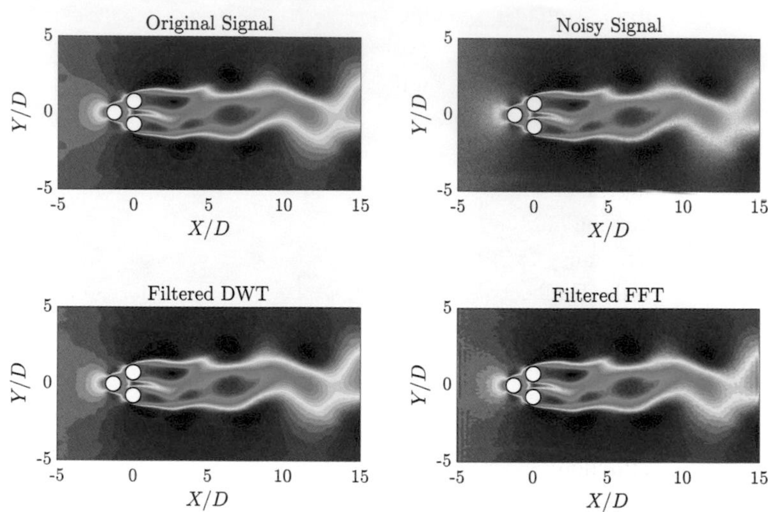

Figure 5.14 Comparison of the reference field of a streamwise velocity component, a field contaminated with noise, and fields filtered via DWT and FFT.

velocity). Compute a 2-level decomposition of the perturbed streamwise velocity field with the Haar wavelet, and visualize the approximation and detail coefficients of the two levels.

2. Compute a 4-level DWT with a Daubechies wavelet of the 4th order, and threshold the wavelet coefficients to retain only the most significant 2%. Compare the results with an FFT-based filtering of your choice.

Solution

1. The wavelet coefficients can be easily computed using standard packages for wavelet calculations, available for a wide variety of programming languages. The solution presented here uses the MATLAB functions *wavedec2*, *appcoef2*, and *detcoef2*, which compute, respectively, the full set of coefficients, the approximation, and the detail coefficients. The approximation and detail coefficients have been scaled to 8-bit level representation, and arranged in the same image for visualization purposes. The coefficients of the 2-level decomposition of the perturbed streamwise velocity field are reported in Figure 5.13. The image is divided into quadrants: the top-right and bottom-left are the coefficients representative of the relevant horizontal and vertical edges, respectively, at the first level; the bottom-right includes the coefficients of the relevant scales on the diagonal direction at level 1; the

top-left quadrant includes details and approximation of level 2, following the same scheme as level 1.
It is clear that the detail coefficients of level 1 are mostly including noise, and in part the regions of strong gradients close to the cylinders. At the subsequent level, the details start including also some features of the shear layers in the wake. The approximation in level 2 is a low-pass filtered version of the original field (included in Figure 5.14 for reference).

2. The noisy field has been filtered using a 4-level approximation with a 4th-order Daubechies wavelet, retaining only 2% of the coefficients with a soft-thresholding procedure (Donoho & Johnstone 1994, Donoho 1995). An FFT-based filtering procedure with a similar amount of retained coefficients is carried out and included in Figure 5.14. The wavelet-based filter provides slightly better results, with the relatively small margin possibly due to the simplicity of this low-Reynolds number flow field.

The full solution is provided in the supplementary material in the form of MATLAB code.

5.5 Conclusions

In this chapter, the fundamentals of time–frequency analysis have been reviewed. The simple, yet powerful, concepts of *scaling* and *shifting* of windows to localize in time the frequency content of the signal have been introduced in the framework of the WFT. The consequences of Heisenberg's uncertainty principle on the localization precision of features in the time and frequency domains have been outlined.

The MRA with wavelets allows storing approximations of the signal at different scales, thus configuring tunable time–frequency resolution. In this chapter, an overview of the basic concepts behind the CWT and the DWT has been outlined, together with the description of a procedure to determine a suitable family of orthogonal wavelets. Some hints for the implementation of the DWT using two-channel multirate filter banks have also been included.

The multi-resolution capabilities of wavelets are exploited in data compression (e.g., the JPEG2000 algorithm is based on wavelets), denoising, speech recognition, finance, and several areas of physics (Van den Berg 2004). In general, all signals arising from multiscale processes are prone to effective use of wavelets. Turbulence is clearly part of this paradigm. In the first age of wavelet use in turbulence, the main applications were mostly (but not exclusively) targeted to turbulence modeling, efficient methods for computational fluid mechanics, and coherent structures segmentation. In the coming years, it is foreseeable that the interest in multiscale time–frequency methods will be further fostered by the current trend toward modal decompositions blending energy optimality and spectral analysis (see e.g., Towne et al. 2018, Mendez, Balabane & Buchlin 2019, Floryan & Graham 2021).

Part III

Data-Driven Decompositions

6 The Proper Orthogonal Decomposition

S. Dawson

The proper orthogonal decomposition (POD) is one of the most ubiquitous data analysis and modeling techniques in fluid mechanics. Since many of the properties of the POD are inherited from the singular value decomposition (SVD), we start with a discussion of the SVD and describe those of its properties that are particularly useful for understanding the POD. Our discussion of the POD starts by characterizing the POD as a decomposition that is specific to a given data set before discussing how the same concept arises when considering dynamical systems that are continuous in space and time. We will describe several variants of the POD that have emerged over the last half a century and how they are related, such as spectral and space-only POD. As well as giving a broad overview, we will discuss some of the technical details often omitted and/or taken for granted when POD is applied in practice, such as how to incorporate nonstandard inner product weights. We finish by using a simple example to demonstrate properties and methods of implementation for the POD.

6.1 Introduction

The POD is one of the most widely used techniques for data analysis in fluid mechanics. Over the past 50 years, its usage in the community has grown and evolved alongside developments in experimental measurement techniques, the introduction and rapid development of computational methods for simulating fluid flows, theoretical developments in dynamical systems, and the ability to store and process increasing quantities of data feasibly. We now discuss the broad motivation for POD and related techniques. Suppose we have data that is a function of both space and time $\boldsymbol{y}(x,t)$. A logical first step toward understanding the salient features of the data, and perhaps the dynamics of the underlying system responsible for its generation, is to perform a separation of variables:

$$\boldsymbol{y}(x,t) = \sum_{j=1}^{m} \phi_j(x)\boldsymbol{a}_j(t). \tag{6.1}$$

That is, we decompose the data into a sum of spatial modes $\phi_j(x)$, and their time-varying coefficients (or equivalently, temporal modes) $\boldsymbol{a}_j(t)$. There are many choices for such a decomposition. For example, one might choose to perform a Fourier transform in space or time, thus obtaining a Fourier basis. Rather than using a predefined basis, POD chooses a decomposition based on the data itself, though we will see later that in certain cases such data-driven modes can actually converge to Fourier modes. Note that we will also see that this decomposition can be further generalized to allow for time-dependency in the spatial structures.

Since many of the properties of POD are inherited from the SVD, we start with a discussion of the SVD and its properties more generally, before discussing how it can be utilized to define and compute POD of a given data set.

6.2 The Singular Value Decomposition

In short, POD can be viewed as the result of taking the SVD of a suitably arranged data matrix. Because of this, and as many of the properties of POD are inherited directly from those of the SVD, we start with a discussion of the SVD and its properties more generally, before discussing how it can be utilized to define and compute POD upon a given data set. This section might seem like a lot of rather dry mathematics, but it will all end up being useful later in this chapter.

6.2.1 The Singular Value Decomposition: Definition

Suppose we have a matrix $\boldsymbol{X}$ that has n rows and m columns. Most of the time, if $\boldsymbol{X}$ is a data matrix then its entries are real, though here for generality we assume that they may be complex. Suppose that $\boldsymbol{X}$ has rank $q \leq \min(m,n)$. The (reduced) SVD of $\boldsymbol{X}$ gives

$$\boldsymbol{X} = \boldsymbol{\Phi}\boldsymbol{\Sigma}\boldsymbol{\Psi}^*, \tag{6.2}$$

where $\boldsymbol{\Phi} \in \mathbb{C}^{n\times q}$, $\boldsymbol{\Psi} \in \mathbb{C}^{m\times q}$, and $\boldsymbol{\Sigma} \in \mathbb{R}^{q\times q}$ is a diagonal matrix with diagonal entries consisting of the singular values $\sigma_1 \geq \sigma_2 \geq \cdots \geq \sigma_q > 0$ (the $\sigma_q > 0$ comes from the fact that we have truncated the SVD to the rank of $\boldsymbol{X}$). Additionally, the columns of $\boldsymbol{\Phi}$ and $\boldsymbol{\Psi}$ are orthonormal, so $\boldsymbol{\Phi}^*\boldsymbol{\Phi} = \boldsymbol{\Psi}^*\boldsymbol{\Psi} = \boldsymbol{I}_q$, where $\boldsymbol{I}_q$ is a $q \times q$ identity matrix.[1] Note in particular that the SVD exists for all matrices. This is in contrast with, for example, an eigendecomposition, which only exists for diagonalizable square matrices.

Thus far, we have not explained the meaning of $\cdot^*$, which denotes the adjoint of a matrix. Most typically, this denotes the conjugate transpose (or just the transpose in the case of a real matrix), but for reasons that will become apparent later in the chapter, we want to keep its definition more general. To define the adjoint, we can first think about the matrix $\boldsymbol{X}$ as mapping from a vector in $\mathbb{C}^m$ to a vector in $\mathbb{C}^n$ through matrix-vector multiplication. Throughout this chapter, all vector quantities can be thought of as column vectors, unless otherwise specified. That is, $\boldsymbol{z} \in \mathbb{C}^m$ maps to $\boldsymbol{y} \in \mathbb{C}^n$ via

[1] Note that the definition of a reduced SVD given here differs from the full SVD, which defines $\boldsymbol{\Phi}$ and $\boldsymbol{\Psi}$ as being square unitary matrices, with $\boldsymbol{\Sigma}$ appropriately padded with zeros.

$y = Xz$. If we assume that we have inner products defined on $\mathbb{C}^m$ and $\mathbb{C}^n$ by $\langle z_1, z_2 \rangle_m$ and $\langle y_1, y_2 \rangle_n$, respectively, then the adjoint of a matrix A is defined as the matrix that satisfies

$$\langle A z_1, y_2 \rangle_n = \langle z_1, A^* y_2 \rangle_m . \tag{6.3}$$

That is, the adjoint is what you get if you move an operator to the other side of an inner product. As an aside, note that (6.3) can be very useful in its own right: if one of m or n is much smaller than the other, then it can be substantially easier to evaluate this inner product in the lower-dimensional space. Note also that this definition of an adjoint reduces to the conjugate transpose matrix in the case where the inner products are the "standard", that is,

$$\langle z_1, z_2 \rangle_m = z_2^\dagger z_1, \quad \langle y_1, y_2 \rangle_n = y_2^\dagger y_1, \tag{6.4}$$

where $\cdot^\dagger$ denotes the conjugate transpose.

6.2.2 The Singular Value Decomposition: Properties

This section lists various properties of the SVD that will be utilized in this chapter. We do not give proofs for all of these properties, and interested readers are encouraged to consult standard references in linear algebra (e.g., Van Loan & Golub 1983, Strang 1988, Horn & Johnson 2012) for further details.

Optimality

The best rank-r approximation to a matrix X (in a least-squares sense) is obtained by taking the first r components of the SVD of X. That is, if X_r is a rank-r approximation of X, we have

$$\underset{X_r}{\operatorname{argmin}} \|X - X_r\|_F = \Phi_r \Sigma_r \Psi_r^*, \tag{6.5}$$

where Φ_r and Ψ_r are the first r columns of Φ and Ψ in the SVD of X, and Σ_r is a diagonal matrix consisting of the first r singular values of X. Here the subscript F denotes the Frobenius norm, given by

$$\|X\|_F = \sqrt{\sum_{j=1}^{n} \sum_{k=1}^{m} |X_{jk}|^2} = \sqrt{\sum_{j=1}^{\min(n,m)} \sigma_j^2}, \tag{6.6}$$

which is the same as the Euclidean norm if we were to squeeze the entries of the matrix X into a single long vector. As will be seen later, this property is responsible for the "optimality" of the POD. We also observe from the right inequality in (6.6) a direct connection between the Frobenius norm and the singular values of a matrix. Intuitively, this connection exists because the singular values contain all of the information about the size of the components in the matrix X, with the Φ and Ψ matrices containing orthonormal vectors. Looking again at the quantity that is minimized on the left-hand side of (6.5), this connection between the Frobenius norm and the SVD also leads to

$$\min_{X_r} \|X - X_r\|_F = \sqrt{\sum_{j=r+1}^{\min(n,m)} \sigma_j^2}. \tag{6.7}$$

As an additional aside, note that we could also consider the L_2 operator norm of a matrix, defined by

$$\|X\|_2 = \max_{\|\psi\|_2=1} \|X\psi\|_2 = \sigma_1, \tag{6.8}$$

with the second equality arising because the unit vector ψ that maximizes $\|X\psi\|_2$ is the first right singular vector of X, ψ_1. Moreover, note that (6.5) also holds when the Frobenius norm is replaced by the operator L_2 norm. In this case, the equivalent of (6.7) is

$$\min_{X_r} \|X - X_r\|_2 = \sigma_{r+1}. \tag{6.9}$$

Relationship between left and right singular vectors

Let ϕ_j and ψ_j denote the jth left and right singular vectors (i.e., the jth columns of $\mathbf{\Phi}$ and $\mathbf{\Psi}$) of a matrix X, respectively, with corresponding singular value σ_j. These are related by

$$\sigma_j \phi_j = X\psi_j, \tag{6.10}$$

$$\sigma_j \psi_j = X^* \phi_j. \tag{6.11}$$

These relationships follow from the fact that $X\mathbf{\Psi} = \mathbf{\Phi\Sigma\Psi}^*\mathbf{\Psi} = \mathbf{\Phi\Sigma}$, and similarly $X^*\mathbf{\Phi} = \mathbf{\Psi\Sigma\Phi}^*\mathbf{\Phi} = \mathbf{\Psi\Sigma}$.

Relationship to eigendecompositions

The left and right singular vectors are eigenvectors of the matrices XX^* and X^*X respectively, with

$$XX^*\phi_j = \sigma_j^2 \phi_j, \tag{6.12}$$

$$X^*X\psi_j = \sigma_j^2 \psi_j. \tag{6.13}$$

These relationships readily follow from (6.10) and (6.11). Note also that the correlation matrices XX^* and X^*X are both Hermitian,[2] since $(XX^*)^* = XX^*$ and $(X^*X)^* = X^*X$.

Permutation of rows and columns

Suppose we want to rearrange the rows of X. We can represent this rearrangement via left multiplication of X by a permutation matrix P (i.e., P is a matrix with exactly one in each row and column, and zeros elsewhere). An SVD of PX is given by

$$PX = (P\mathbf{\Phi})\mathbf{\Sigma\Psi}^*. \tag{6.14}$$

[2] Recall that a Hermitian (equivalently, self-adjoint) matrix A satisfies $A = A^*$.

Similarly, if Q is a permutation matrix that rearranges the columns of $\boldsymbol{X}$ via right multiplication, then $\boldsymbol{X}Q$ has the SVD

$$\boldsymbol{X}Q = \boldsymbol{\Phi\Sigma}(\boldsymbol{\Psi}Q)^*. \tag{6.15}$$

Dyadic expansion

The SVD can be expanded as a sum of rank-1 matrices via

$$\boldsymbol{X} = \boldsymbol{\Phi\Sigma\Psi}^* = \sum_{j=1}^{q} \sigma_j \phi_j \psi_j^*. \tag{6.16}$$

Computing the SVD with a weighted inner product

In practice, the numerical computation of the SVD is most typically done using inbuilt routines (e.g., numpy.linalg.svd in Python). Such routines implicitly assume that the SVD is to be computed with standard inner products for both the columns and rows of the matrix, as in (6.4). This section details how such standard routines may be utilized to perform the SVD with a nonstandard inner product. In particular, suppose that we replace the standard inner products by

$$\langle z_1, z_2 \rangle_m = z_2^\dagger W_m z_1, \quad \langle \boldsymbol{y}_1, \boldsymbol{y}_2 \rangle_n = \boldsymbol{y}_2^\dagger W_n \boldsymbol{y}_1, \tag{6.17}$$

where as before z_j and $\boldsymbol{y}_j$ are rows and columns of the matrix, respectively. In order for these to satisfy the definition of an inner product, the weight matrices $\boldsymbol{W}_m$ and $\boldsymbol{W}_n$ must be positive definite. Note that the adjoint of $\boldsymbol{X}$ with these inner products is given by

$$\boldsymbol{X}^* = \boldsymbol{W}_m^{-1} \boldsymbol{X}^\dagger \boldsymbol{W}_n, \tag{6.18}$$

which comes from the fact that

$$\begin{aligned}
\langle \boldsymbol{y}, \boldsymbol{X}\boldsymbol{z}, \rangle_n &= \boldsymbol{z}^\dagger \boldsymbol{X}^\dagger \boldsymbol{W}_n \boldsymbol{y} \\
&= \boldsymbol{z}^\dagger \boldsymbol{W}_m \boldsymbol{W}_m^{-1} \boldsymbol{X}^\dagger \boldsymbol{W}_n \boldsymbol{y} \\
&= \left\langle (\boldsymbol{W}_m^{-1} \boldsymbol{X}^\dagger \boldsymbol{W}_n) \boldsymbol{y}, \boldsymbol{z} \right\rangle_m \\
&:= \langle \boldsymbol{X}^* \boldsymbol{y}, \boldsymbol{z} \rangle_m.
\end{aligned}$$

$\boldsymbol{W}_n$ and $\boldsymbol{W}_m$ are often positive diagonal matrices, where the diagonal entries contain integration weights. Henceforth, we assume that this is the case, which makes the definition of their square roots $\boldsymbol{W}_m^{1/2}$ and $\boldsymbol{W}_n^{1/2}$ unambiguous. The SVD of $\boldsymbol{X}$ with these inner products can be computed as follows:

1. Form the weighted matrix $\boldsymbol{X}_w = \boldsymbol{W}_n^{1/2} \boldsymbol{X} \boldsymbol{W}_m^{-1/2}$.
2. Compute the SVD of the weighted matrix $\boldsymbol{X}_w = \boldsymbol{\Phi}_w \boldsymbol{\Sigma} \boldsymbol{\Psi}_w^\dagger$ using standard routines.
3. The singular vectors (which are orthonormal with respect to their respective norms) are given by columns ϕ_j and ψ_j of $\boldsymbol{\Phi} = \boldsymbol{W}_n^{-1/2} \boldsymbol{\Phi}_w$ and $\boldsymbol{\Psi} = \boldsymbol{W}_m^{-1/2} \boldsymbol{\Psi}_w$, respectively, with corresponding singular values as given in $\boldsymbol{\Sigma}$.

With this computation, it is easy to verify that the identified singular vectors are orthonormal with respect to their respective inner products ($\mathbf{\Phi}^{\dagger} W_n \mathbf{\Phi} = I$, $\mathbf{\Psi}^{\dagger} W_m \mathbf{\Psi} = I$), and that they satisfy the eigenvalue problems discussed in Section 6.2.2 for the general inner products and adjoints:

$$
\begin{aligned}
\boldsymbol{X}_w \boldsymbol{X}_w^{\dagger}(\phi_w)_j &= \sigma_j^2(\phi_w)_j, \\
(\boldsymbol{W}_n^{1/2}\boldsymbol{X}\boldsymbol{W}_m^{-1/2})(\boldsymbol{W}_m^{-1/2}\boldsymbol{X}^{\dagger}\boldsymbol{W}_n^{1/2})(\boldsymbol{W}_n^{1/2}\phi_j) &= \sigma_j \boldsymbol{W}_n^{1/2}\phi_j, \\
\boldsymbol{X}(\boldsymbol{W}_m^{-1}\boldsymbol{X}^{\dagger}\boldsymbol{W}_n)\phi_j &= \sigma_j^2\phi_j, \\
\boldsymbol{X}\boldsymbol{X}^{*}\phi_j &= \sigma_j^2\phi_j,
\end{aligned}
$$

and similarly

$$
\begin{aligned}
\boldsymbol{X}_w^{\dagger}\boldsymbol{X}_w(\psi_w)_j &= \sigma_j^2(\psi_w)_j, \\
(\boldsymbol{W}_m^{-1/2}\boldsymbol{X}^{\dagger}\boldsymbol{W}_n^{1/2})(\boldsymbol{W}_n^{1/2}\boldsymbol{X}\boldsymbol{W}_m^{-1/2})(\boldsymbol{W}_m^{1/2}\psi_j) &= \sigma_j \boldsymbol{W}_m^{1/2}\psi_j, \\
(\boldsymbol{W}_m^{-1}\boldsymbol{X}^{\dagger}\boldsymbol{W}_n)\boldsymbol{X}\psi_j &= \sigma_j^2\psi_j, \\
\boldsymbol{X}^{*}\boldsymbol{X}\psi_j &= \sigma_j^2\psi_j,
\end{aligned}
$$

where here $(\phi_w)_j$ and $(\psi_w)_j$ are the jth columns of $\mathbf{\Phi}_w$ and $\mathbf{\Psi}_w$, respectively.

6.3 The Proper Orthogonal Decomposition of Discrete Data

Suppose that we measure data that is a function of both space and time, and assemble all data into a matrix, such that each column of the matrix represents a "snapshot" of all data that is measured at a given instance in time, while each row consists of all of the measurements of a given quantity across all times. For example, if we measure two components of the velocity of a fluid (u and v) at spatial locations $x_1, x_2, \ldots, x_n$, and at times $t_1, t_2, \ldots, t_m$, then the data matrix becomes the $n \times m$ matrix

$$
\boldsymbol{X} = \begin{bmatrix}
u(x_1,t_1) & u(x_1,t_2) & \cdots & u(x_1,t_m) \\
u(x_2,t_1) & u(x_2,t_2) & \cdots & u(x_2,t_m) \\
\vdots & \vdots & \ddots & \vdots \\
u(x_n,t_1) & u(x_n,t_2) & \cdots & u(x_n,t_m) \\
v(x_1,t_1) & v(x_1,t_2) & \cdots & v(x_1,t_m) \\
v(x_2,t_1) & v(x_2,t_2) & \cdots & v(x_2,t_m) \\
\vdots & \vdots & \ddots & \vdots \\
v(x_n,t_1) & v(x_n,t_2) & \cdots & v(x_n,t_m)
\end{bmatrix}, \tag{6.19}
$$

where $n = 2n_x$, since there are two measurements for each spatial location. Note that we are not making any assumptions concerning the spatial or temporal resolution of our data, or in what quantity or quantities are being measured. The only implicit assumption that we are making in forming $\boldsymbol{X}$ is that we measure the same quantities for each snapshot. We next subtract from each column of $\boldsymbol{X}$ a base condition $\boldsymbol{y}_0$. Most typically, this is the mean of all of the columns in $\boldsymbol{X}$, though it need not always be.

If $\boldsymbol{X}$ represents the transient response of a system to a given input, it would make sense to consider the steady or average state as $t \to \infty$ instead of the mean of the early-time data. For notational convenience, we will refer to this base state that is subtracted from the data as the "mean" in any case. Denoting the kth column of $\boldsymbol{X}$ by $\boldsymbol{y}_k(x)$, we now form the mean-subtracted matrix

$$\boldsymbol{Y} = \begin{bmatrix} \boldsymbol{y}_1' & \boldsymbol{y}_2' & \cdots & \boldsymbol{y}_m' \end{bmatrix}, \tag{6.20}$$

where $\boldsymbol{y}_j' = \boldsymbol{y}_j - \boldsymbol{y}_0$ is the jth mean-subtracted snapshot. Most simply, the POD can be found from taking the SVD of Y. In particular, if we have

$$\boldsymbol{Y} = \boldsymbol{\Phi\Sigma\Psi}^*, \tag{6.21}$$

then POD modes are given by the columns $\phi_j(x)$ of $\boldsymbol{\Phi}$, which evolve in time via the corresponding coefficients $\boldsymbol{a}_j(t) = \sigma_j \psi_j(t)^*$ (and for the case of real data with a standard inner product, $\psi_j(t)^* = \psi_j(t)^T$). That is, with reference to (6.1), a POD expansion of the data is given by the dyadic expansion of the SVD as described in Section 6.2.2:

$$\boldsymbol{y}_k(x) = \sum_{j=1}^{q} \phi_j(x)\sigma_j\psi_j^*(t_k). \tag{6.22}$$

Since we have arranged our data such that spatial location depends only on the row number, and temporal location depends only on the column, the left and right singular vectors depend only on space and time, respectively, allowing this direct use of the SVD to obtain this separation of variables in our data. The optimality property of the SVD described in Section 6.2.2 means that a truncation of this sum to r terms gives the closest rank-r approximation to the full data set.

When assembling the data matrix in (6.19), we chose to place the measurements of the v-component of velocity below all of the measurements of u, though we could have also arranged the data in other ways, for example, with alternating rows containing the u and v components of velocity. From the permutation properties of the SVD discussed in Section 6.2.2, changing the arrangement does not affect the decomposition, aside from permuting the entries in the POD modes to the appropriate locations. Similarly, we can rearrange the columns of the data matrix without affecting the POD modes at all. Indeed, this fact can be used to demonstrate that POD does not require time-resolved data, given that shuffling the order of time-resolved snapshots can give a non-time-resolved sequence of snapshots.

From Section 6.2.2, the right singular vectors of $\boldsymbol{Y}$ are eigenvectors of the matrix $\boldsymbol{Y}^*\boldsymbol{Y}$. This $m \times m$ matrix can be interpreted as the correlation matrix between all snapshots:

$$\boldsymbol{Y}^*\boldsymbol{Y} = \begin{bmatrix} \langle \boldsymbol{y}_1', \boldsymbol{y}_1' \rangle_x & \langle \boldsymbol{y}_2', \boldsymbol{y}_1' \rangle_x & \cdots & \langle \boldsymbol{y}_m', \boldsymbol{y}_1' \rangle_x \\ \langle \boldsymbol{y}_1', \boldsymbol{y}_2' \rangle_x & \langle \boldsymbol{y}_2', \boldsymbol{y}_2' \rangle_x & \cdots & \langle \boldsymbol{y}_m', \boldsymbol{y}_2' \rangle_x \\ \vdots & \vdots & \ddots & \vdots \\ \langle \boldsymbol{y}_1', \boldsymbol{y}_m' \rangle_x & \langle \boldsymbol{y}_2', \boldsymbol{y}_m' \rangle_x & \cdots & \langle \boldsymbol{y}_m', \boldsymbol{y}_m' \rangle_x \end{bmatrix}, \tag{6.23}$$

where $\langle \cdot, \cdot \rangle_x$ denotes a spatial inner product. Similarly, if we let $z_j'(t)$ be a row vector that is the jth row of $\boldsymbol{Y}$, we have

$$\boldsymbol{Y}\boldsymbol{Y}^* = \begin{bmatrix} \langle z_1', z_1' \rangle_t & \langle z_1', z_2' \rangle_t & \cdots & \langle z_1', z_n' \rangle_t \\ \langle z_2', z_1' \rangle_t & \langle z_2', z_2' \rangle_t & \cdots & \langle z_2', z_n' \rangle_t \\ \vdots & \vdots & \ddots & \vdots \\ \langle z_n', z_1' \rangle_t & \langle z_n', z_2' \rangle_t & \cdots & \langle z_n', z_n' \rangle_t \end{bmatrix}, \tag{6.24}$$

where $\langle \cdot, \cdot \rangle_t$ is an inner product in the time dimension. Again from Section 6.2.2, the POD modes ϕ_j are eigenvectors of the matrix $\boldsymbol{Y}\boldsymbol{Y}^*$. This $n \times n$ matrix can be interpreted as the time-correlation matrix between pairwise rows in $\boldsymbol{Y}$.

Note that the two matrices $\boldsymbol{Y}\boldsymbol{Y}^*$ and $\boldsymbol{Y}^*\boldsymbol{Y}$ are generally of different sizes, with $\boldsymbol{Y}\boldsymbol{Y}^*$ being $n \times n$ and $\boldsymbol{Y}^*\boldsymbol{Y}$ being $m \times m$. Since the SVD of $\boldsymbol{Y}$ is related to the eigendecompositions of these square matrices, it might be easiest to compute and work with the smaller of these two matrices. For example, if one had very large snapshots but comparatively few of them, then $m \ll n$, and so it can be more tractable to compute the (full or partial) eigendecomposition of $\boldsymbol{Y}^*\boldsymbol{Y}$ to obtain the POD coefficients $\boldsymbol{a}_j(t)$. The POD modes can then be computed using $\phi_j = \sigma_j^{-1}\boldsymbol{Y}\boldsymbol{a}_j$, following the properties of the SVD in Section 6.2.2. This is precisely the "method of snapshots" for computing POD introduced in Sirovich (1987). Conversely, if $n \ll m$, then one can instead start by computing an eigendecomposition of $\boldsymbol{Y}\boldsymbol{Y}^*$.

6.4 The Proper Orthogonal Decomposition of Continuous Systems

Up until this point, we have viewed the POD as being something that is computed from a given data set. In reality, however, if our data comes from measurements of a given system, then the POD should ideally be independent of the exact data that we have collected and used for analysis. That is to say, we can alternatively think of POD as being a property of a system, which we hope to accurately compute from data. Looking back at the inner products of discrete data that feature in (6.23) and (6.24), these can be viewed as discrete approximations to the integrals

$$\left\langle \boldsymbol{y}_j', \boldsymbol{y}_k' \right\rangle_x = \int_{\Omega_x} \boldsymbol{y}_1'(x, t_j)\boldsymbol{y}_2'(x, t_k)dx, \tag{6.25}$$

$$\left\langle z_j', z_k' \right\rangle_t = \int_{\Omega_t} z_1'(x_j, t) z_2'(x_k, t)dt, \tag{6.26}$$

where the first equation is the spatial integral over the appropriate number of spatial dimensions, and Ω_x and Ω_t define the spatial and temporal domains of interest. Note that a desire to correctly approximate these integrals can lead to the need to include integration weights for discrete data as described in Section 6.2.2.[3] Note that,

[3] To be more general, we also could have included an additional weight function in these integrals, but we omit this for simplicity.

depending on the context, sometimes it is most convenient to divide the integrals in (6.25) and (6.26) by the size of the domains of integration, so that the inner products become averages/expectations. More generally, we can think about (6.25) and (6.26) as continuous functions of time and space, respectively (rather than being computed for the discrete indices j and k). We can also be slightly more general, and think of these integrals as computing an expected value of some stochastic process. With this in mind, we can define the spatial correlation function

$$C(x,x') = \mathbb{E}[\bar{\mathbf{y}}'(x,t)\mathbf{y}'(x',t)], \tag{6.27}$$

where $\mathbb{E}$ is the expectation, taken over time and/or ensembles. The POD in a continuous setting amounts to finding modes $\phi_j(x)$ that are eigenfunctions of the integral equation

$$\int_{\Omega_x} C(x,x')\phi_j(x')dx' = \sigma_j^2\phi_j(x). \tag{6.28}$$

As is perhaps indicated by the notation used, the eigenvalues σ_j^2 are real and positive, and this equation is the continuous equivalent to (6.12). Note that it can be shown that many of the same properties of the POD discussed in the discrete, finite-dimensional context, such as orthogonality and optimality, also extend to equivalent concepts in the continuous setting. In cases where the system is spatially homogenous (i.e., is invariant under translations), $C(x,x') = C(x-x')$ is only a function of $x-x'$, and it can readily be shown from (6.28) that the POD modes become Fourier modes. To see this, if we change the variable of integration to $r = x - x'$, (6.28) can be expressed as

$$\int_{\Omega_r} C(r)\frac{\phi_j(x+r)}{\phi_j(x)}dr = \sigma_j^2. \tag{6.29}$$

The right-hand side of (6.29) is independent of x, so eigenfunctions ϕ_j must possess the property that $\frac{\phi_j(x+r)}{\phi_j(x)}$ is independent of x. We can verify that Fourier modes (i.e., $\exp(2\pi ikx)$) satisfy this property. In the case where the domain is finite in the direction on homogeneity, periodic boundary conditions determine the permissible values of the spatial wavenumbers, k. Similarly, if the system is homogeneous in only certain directions (such as the streamwise and spanwise directions in channel flow), then POD modes become Fourier modes in these directions of spatial homogeneity.

Rather than only looking at the spatial correlation tensor, we can also consider the space-time correlation tensor

$$C(x,x',t,t') = \mathbb{E}[\bar{\mathbf{y}}'(x,t)\mathbf{y}'(x',t')], \tag{6.30}$$

where now the expectation can be thought of as an ensemble average across different realizations of a stochastic field for all pairs of space-time coordinates (x,t) and (x',t'). An equivalent to (6.28) in this space-time setting is then given by

$$\int_{\Omega_x,\Omega_t} C(x,x',t,t')\tilde{\phi}_j(x',t')dx'dt' = \sigma_j^2\tilde{\phi}_j(x,t). \tag{6.31}$$

Note in particular that the eigenfunctions of this equation are now functions of both space and time, which we distinguish from the space-only modes that we have dealt with so far using a $\tilde{\cdot}$. Most typically, systems modeled using this approach are stationary in time, meaning that $C(x, x', t, t') = C(x, x;, t - t')$ is only a function of the difference in time $\tau = t - t'$. Similar to the case of spatial homogeneity, this means that the temporal dependence on these modes become Fourier modes. This allows for a temporal Fourier transform of (6.31), giving

$$\int_{\Omega_x} S(x, x', f)\hat{\phi}_j(x', f)dx' = \sigma_j(f)^2\hat{\phi}_j(x, f), \tag{6.32}$$

where the cross-spectral density tensor $S(x, x', f)$ is obtained from the temporal Fourier transforms:

$$S(x, x', f) = \int_{\infty}^{\infty} C(x, x', \tau)\exp(-i2\pi f\tau)d\tau. \tag{6.33}$$

This space-time or spectral POD was, in fact, the original formulation of POD as described in Lumley (1967) and Lumley (1970). However, the "space-only" POD has been much more commonly used since then, likely due to two main reasons: it is easy to formulate and compute from data, and it is readily amenable for the construction of projection-based reduced-order models (e.g., Holmes et al. 2012), as will be discussed in Section 6.5. Indeed, the difference between the original and most commonly used formulations has perhaps been under appreciated in the community (including by the author). Recently, the work of Towne et al. (2018) has brought the "spectral" version of POD back to prominence, though an exposition of the method and its close connections to other modeling approaches, and in the development of practical algorithms for its implementation (Schmidt & Towne 2019).

6.5 The Proper Orthogonal Decomposition for Projection-Based Modeling

Perhaps the most common usage of the POD involves using a truncated basis of POD modes to obtain a subspace that differential equations describing the physics of a system may be projected onto. For example, if we consider the incompressible Navier–Stokes momentum equation

$$\frac{\partial \boldsymbol{u}}{\partial t} = -(\boldsymbol{u} \cdot \nabla)\boldsymbol{u} + Re^{-1}\Delta\boldsymbol{u} - \nabla p, \tag{6.34}$$

then taking the spatial inner product of both sides of this equation with a POD mode $\phi_j(x)$ gives

$$\left\langle \frac{\partial \boldsymbol{u}}{\partial t}, \phi_j \right\rangle_x = -\left\langle (\boldsymbol{u} \cdot \nabla)\boldsymbol{u}, \phi_j \right\rangle_x + Re^{-1}\left\langle \Delta\boldsymbol{u}, \phi_j \right\rangle_x - \left\langle \nabla p, \phi_j \right\rangle_x. \tag{6.35}$$

Approximating the full velocity field with a set of POD modes (as in (6.1)), we obtain

$$\dot{\boldsymbol{a}} = L\boldsymbol{a} + Q(\boldsymbol{a}, \boldsymbol{a}) + \boldsymbol{f}, \tag{6.36}$$

where L is a linear operator, Q is a bilinear operator, and f is a vector, each defined based on the identified spatial POD modes by

$$L_{ij} = \nu \left\langle \Delta\phi_j, \phi_i \right\rangle_x, \quad Q_{ijk} = -\left\langle (\phi_j \cdot \nabla)\phi_k, \phi_i \right\rangle_x, \quad f = -\left\langle \nabla p, \phi_i \right\rangle_x. \tag{6.37}$$

This gives a means of approximating the Navier–Stokes equations by a set of nonlinear ordinary differential equations, which is typically of much lower dimension than the full discretized equations.

6.6 Extensions of the Proper Orthogonal Decomposition

There are many modifications and extensions of the basic POD definition and algorithm that can improve on its suitability, accuracy, and efficiency in various applications. If being used for reduced-order modeling as described in Section 6.5, a subspace based on POD modes is optimal from an energetic sense, but is not necessarily the best choice for retaining all features of dynamic importance. For the case of linear systems, one can seek to define the subspace and direction of projection such that they best preserve the system dynamics (quantified by the observability and controllability Gramians). This is known as balanced truncation (Moore 1981), and it was shown by Rowley (2005) that these subspaces can be identified from POD using inner products that are weighted by the observability and controllability Gramians (see also the similar method developed in Willcox and Peraire (2002)). Further discussion of notions of these concepts are given in Chapter 10.

Several other variants of POD account for other desired qualities in spatio-temporal decompositions, such as finding a balance between energetic optimality and frequency localization (Sieber et al. 2016), and seeking scale separation (Mendez et al. 2019). Additional variants aim to reparameterize the spatio-temporal domain with translations, rotations, and/or rescalings to make certain data/systems (often those dominated by convection) more amenable to low-dimensional approximation (Rowley & Marsden 2000, Rowley et al. 2003, Mojgani & Balajewicz 2017, Mowlavi & Sapsis 2018, Reiss et al. 2018, Black et al. 2019, Mendible et al. 2020).

Aside from these variants, computation of POD can require overcoming several practical challenges involving the acquisition and processing of data of sufficient quantity and quality to ensure converged results. For further discussion of such practical considerations for experimental data in particular, see Chapter 9.

6.7 Examples

This section will present two simple examples to show how the POD can be applied, and its outputs interpreted. Both of these examples come with accompanying Python code, available on the book's website.[4] The example considered here is similar to an

[4] www.datadrivenfluidmechanics.com/download/book/chapter6.zip

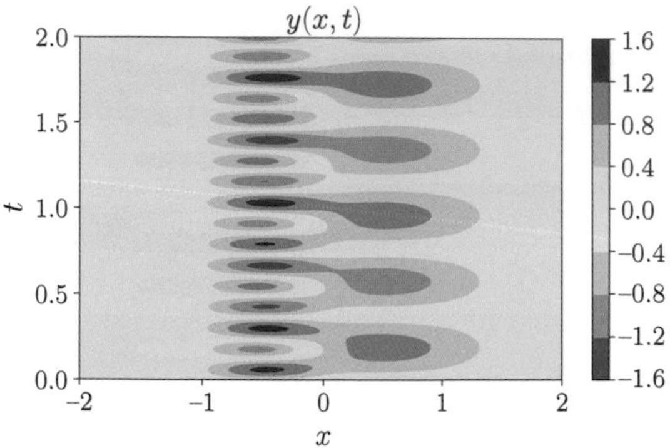

Figure 6.1 Surface plot of data from (6.38).

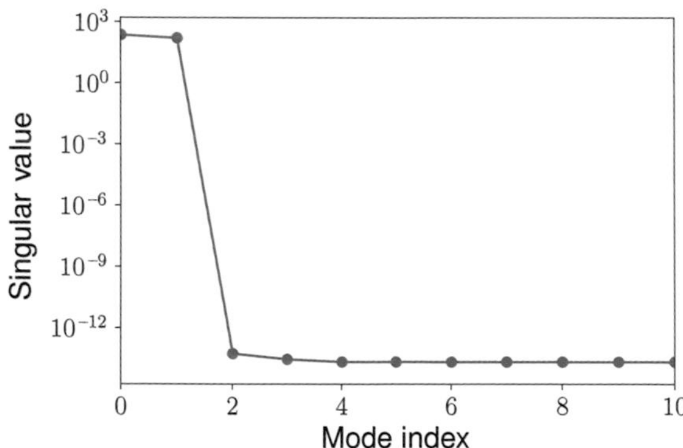

Figure 6.2 Singular values corresponding to each POD mode for data from (6.38).

example considered in (Tu et al. 2014). We consider data that is sampled from the function

$$\mathbf{y}(x,t) = c_1 \exp\left[\frac{(x-x_1)^2}{2s_1^2}\right] \sin(2\pi f_1 t) + c_2 \exp\left[\frac{(x-x_2)^2}{2s_2^2}\right] \sin(2\pi f_2 t), \qquad (6.38)$$

where we choose the parameters $c_1 = 1$, $x_1 = 0.5$, $s_1 = 0.6$, $f_1 = 1.3$, $c_2 = 1.2$, $x_2 = -0.5$, $s_2 = 0.3$, and $f_2 = 4.1$. We collect data in the range $x \in [-2, 2]$ and $t \in [0, 50]$. A visualization of this function over four time units is shown in Figure 6.1.

We perform POD by compiling the data in a matrix and performing a SVD, using the procedure discussed in Section 6.3. The resulting singular values, spatial POD modes, and corresponding temporal coefficients are shown in Figures 6.2 through 6.4, respectively. The leading two singular values are much larger than the rest (which are essentially zero), due to the fact that the data comes from a function that is the sum of two terms (with each the product of a spatial function and a temporal function).

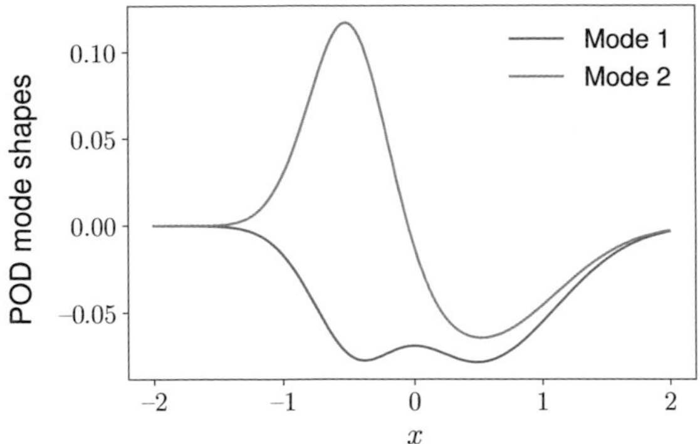

Figure 6.3 POD mode shapes computed from data from (6.38).

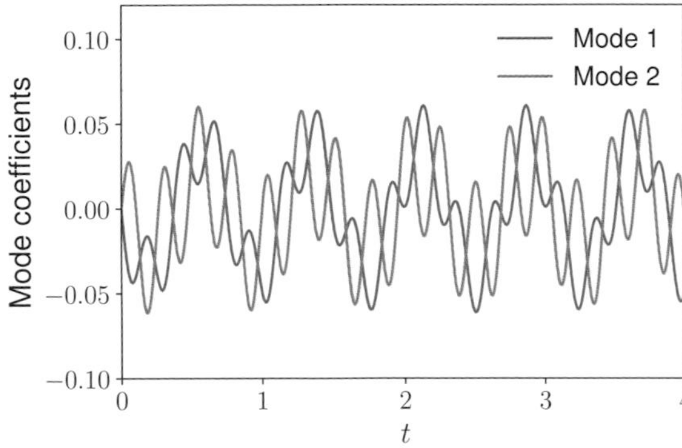

Figure 6.4 POD mode coefficients for data from (6.38).

However, the leading two POD modes do not cleanly partition the two component functions, and instead each mode is a mix of the two. This is due to the fact that the SVD is "greedy" in the sense that the first set of singular vectors captures as much energy as possible, which in this case means taking part of each of the two terms in (6.38).

In Figure 6.5, we show the results of applying spectral POD to the same data set, which we see is able to distinguish between the two spatial functions corresponding to different temporal frequencies, both of which show up as peaks in the corresponding spectrum, shown in Figure 6.6. Note that here a simple Fourier transform in time would also suffice, since the data is entirely deterministic and free of noise. Here we could also separate out the frequency content using dynamic mode decomposition on appropriately arranged data.

The next example that we consider involves data from a fluid simulation, considering two-dimensional flow over a circular cylinder at a Reynolds number of 60.

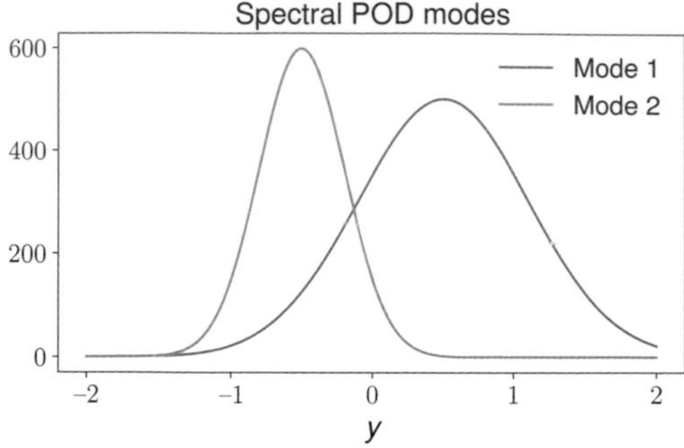

Figure 6.5 Spectral POD mode shapes computed from data generated from (6.38).

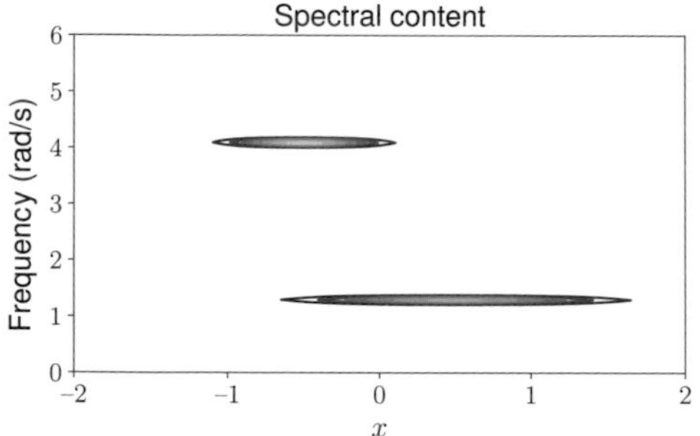

Figure 6.6 Spectral POD energy content of data generated from (6.38).

A similar example has already been introduced and discussed in Chapter 1. We consider 41 snapshots of data collected over approximately eight convective times (being the time taken for fluid in the freestream to travel one cylinder diameter). The data is collected when the system is in a transient state, close to the unstable equilibrium solution, but with oscillations (corresponding to vortex shedding) that are slowly growing in time.

The accompanying Python code shows how the data for both the streamwise and transverse velocity components can be arranged as in (6.19); how POD can be performed via an SVD; and how the resulting modes and mode coefficients can be visualized. The first two modes are shown in Figures 6.7 and 6.8, with their accompanying time-varying coefficients shown in Figure 6.9. We see that these leading two modes are similar in structure, featuring oscillatory structures in the streamwise (horizontal) direction, which have a phase difference between modes 1 and 2 (with a similar phase difference observed for the mode coefficients in time).

Mode 1, u

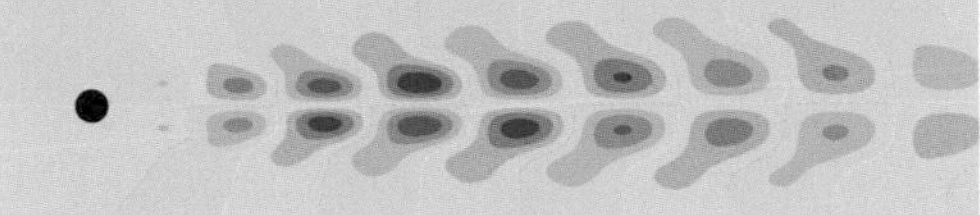

Mode 1, v

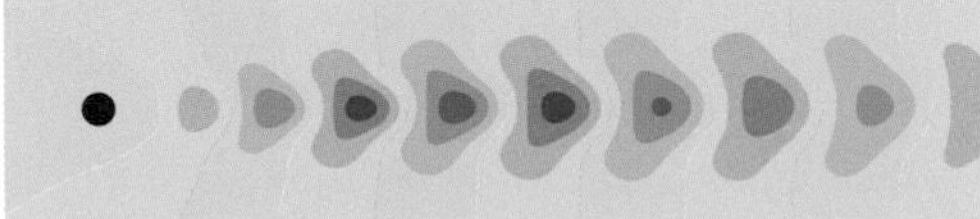

Figure 6.7 Streamwise (u) and transverse (v) velocity components of the leading POD mode for flow over a cylinder in a transient regime.

Mode 2, u

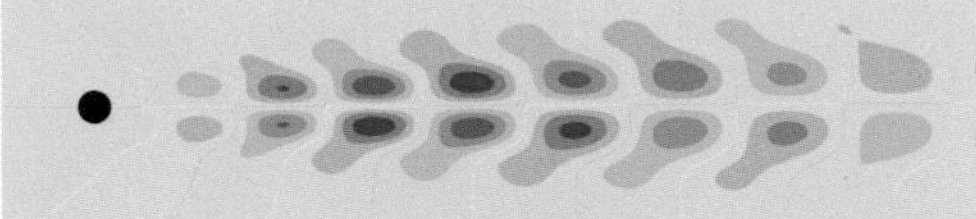

Mode 2, v

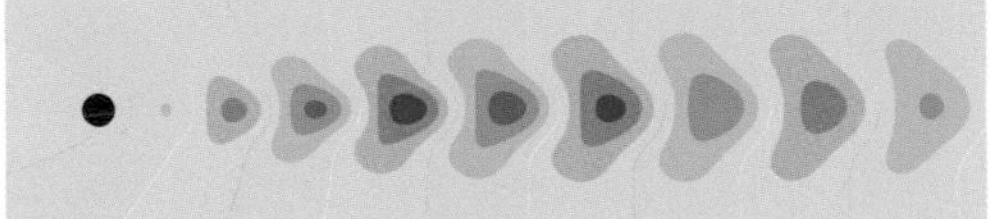

Figure 6.8 Streamwise (u) and transverse (v) velocity components of the second POD mode for flow over a cylinder in a transient regime.

This phase difference between both the streamwise-oscillating structures in space and coefficients in time is approximately $\pi/2$. This can be reasoned from the fact that two purely oscillatory signals at the same frequency will only be orthogonal if they are exactly a quarter-cycle out of phase. These mode shapes, which can be thought of as standing waves, allow for the representation of structures that convect downstream over time.

Lower-energy modes (not shown) include modes that oscillate at harmonic frequencies of these leading modes, as well as modes that capture the distortion of the base flow as the system moves between the unstable equilibrium and limit cycle of vortex shedding (as discussed in Chapter 1). Note that had we considered data for a different section of the transient (or on the limit cycle), these leading two POD modes would be slightly different, though would share the same qualitative features. In other words, on this transient trajectory, the system is not statistically stationary, so the POD that we obtain is a function of the data set that we use.

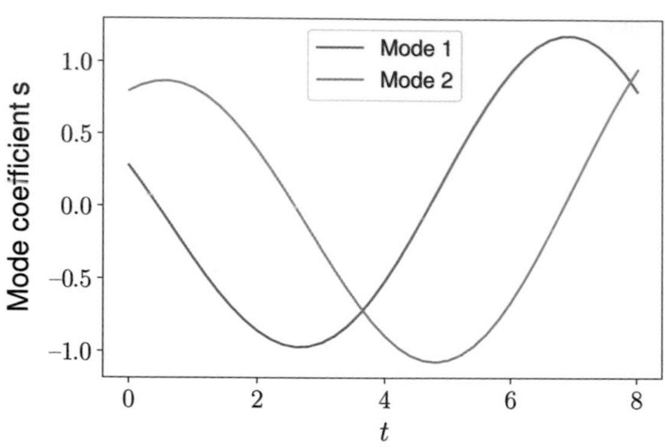

Figure 6.9 Time-varying coefficients corresponding to the POD modes shown in Figures 6.7 and 6.8.

7 The Dynamic Mode Decomposition: From Koopman Theory to Applications

P. J. Schmid

The dynamic mode decomposition (DMD) is a data-driven method for flow-field analysis, closely linked to a powerful mathematical framework for dynamical systems known as Koopman theory. It provides approximations to the key elements of Koopman analysis and has spawned many developments from its original formulation. This chapter gives a brief derivation and outline of this decomposition method, presents three applications to experimental and numerical data sequences, and touches on some of the many exciting extensions and generalizations.

7.1 Introduction

Nonlinear dynamical systems are a common mathematical concept to describe many evolution processes in the physical, biological, and social sciences. The motion of fluids across all spatial and temporal scales is represented by a large-scale dynamical system that arises from a discretization of a partial differential equation. This differential equation, in turn, is derived from first principles, such as the conservation of mass, momentum, energy, and other appropriate flow markers. Our understanding of physical processes is thus closely linked to the analysis of nonlinear dynamical systems and to the tools available for this task. Over many decades, the mathematical methods at our disposal have been dominated by linear techniques. Relying on linear superposition, a complex process can be deconstructed into subprocesses that can be studied in isolation. Furthermore, linear techniques can be applied locally near special points of the solution space and furnish information about the disturbance behavior in their vicinity.

At the same time, with the rise of new experimental techniques and massively parallel computers, the amount of data within our reach is staggering, and the detail of this data is impressive. The data-rich environment has spawned many methods to extract essential information from large data files, to describe physical processes and to quantify output variables of common interest. Besides statistical methodology, data decomposition techniques have dominated the field of data-driven analysis. These techniques aim at a factorization of the full spatio-temporal observations into subunits of decreasing coherence. For each of the available decompositions, the precise meaning of "coherence" has to be defined anew, and often has to be tailored to the specific circumstances of the data and/or flow configuration.

The data-centric approach carries both advantages and disadvantages. It is simpler to justify, particularly in an experimental setting, since the gathered data – by definition – is observable; model-based computational approaches, on the other hand, can suffer from robustness issues. Still, even in data-driven analysis, we can encounter hidden (and sometimes high-dimensional) variables that are essential from a physical point of view, but are difficult to measure or absent in our data. Data are often noisy, which can pose challenges to the subsequent analysis (see also Chapter 13). The uncertainty in the data has to be taken into account and balanced against both further approximations in the decomposition. In particular, the uncertainty in the output variables has to be determined and accounted for in the final interpretation of the results.

Many common techniques in describing processes, based on the data they produce, are based on spectral analysis (in the most general sense of the word). This involves a transformation into another basis, during which some of the independent variables (time, space) will be replaced by their dual equivalents (frequency, wavenumber). Ideally, this transformation introduces a sparse, localized, and hierarchical description of the physical process (contained in the data sequence) in the new basis – and allows a more concentrated representation of a potentially complex process.

Among various decompositions, modal decompositions (in a general sense) are most common (Lumley 1970, Berkooz et al. 1993, Holmes et al. 2012, Taira et al. 2017). They are the data-driven equivalent of a separation-of-variables approach: we factorize the data into purely spatial components (the modes), purely temporal components (the dynamics), and the associated amplitudes (the spectrum); see also Chapter 8 for additional material on linear decompositions. This procedure is predicated on a linear assumption: we can recover the original signals by a superposition of (some of) the identified mode–dynamics–amplitude triplets. For this hierarchical breakup and reassembly, a great many decompositions are available; for nonlinear systems, or a nonlinear analysis, the options are far more limited.

A promising way forward seeks to postulate and design a coordinate transformation, under which the nonlinear dynamics reduces to a linear one. This approach has been studied from a mathematical point of view for many decades, but as an exact transformation has found only limited applicability to a few nonlinear partial differential equations within a model-based setting.

Koopman analysis falls within this latter category, as it proposes a nonlinear coordinate transformation that embeds a nonlinear finite-dimensional dynamical system into an equivalent linear system, albeit of infinite dimensions. The remainder of this chapter will motivate data analysis from this point of view (Lasota & Mackey 1994, Mezić 2005, Budišić et al. 2012, Mezić 2013) and present a computational framework for the decomposition of data sequences generated by a nonlinear system. The link to Koopman theory is also developed in Chapter 10, and the reader is urged to consult this material.

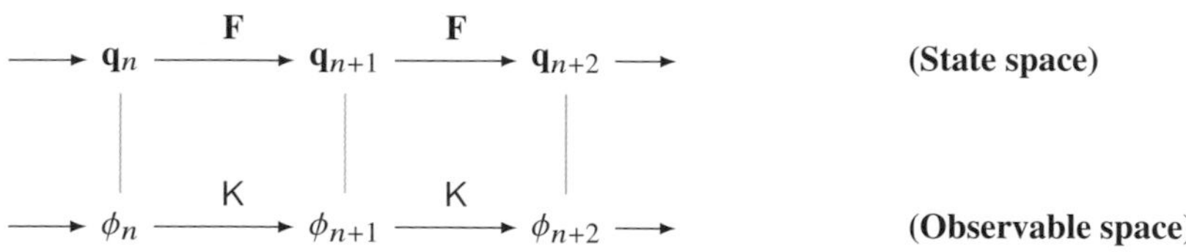

Figure 7.1 Schematic of the Koopman idea. We have used the abbreviaton $\phi_n = \phi(\mathbf{q}_n)$.

7.2 The Koopman Idea

Koopman analysis is a mathematical formalism for dynamical systems theory. Its core principle is a coordinate transformation that embeds a nonlinear dynamical system in an equivalent linear dynamical system of observables. As such it presents an alternative to the classical viewpoint of dynamical systems (Guckenheimer & Holmes 1983), given by a state vector $\mathbf{q}$ and a governing equation for its evolution $\mathbf{f}$ or a discrete map $\mathbf{F}$ which together give the dynamical system

$$\frac{d\mathbf{q}}{dt} = \mathbf{f}(\mathbf{q}) \qquad \text{(continuous)}, \tag{7.1a}$$

$$\mathbf{q}_{n+1} = \mathbf{F}(\mathbf{q}_n) \qquad \text{(discrete)}. \tag{7.1b}$$

Dynamical systems of this form have been analyzed by a variety of mathematical techniques and are part of the standard mathematical curriculum (Guckenheimer & Holmes 1983). Equilibrium points and their stability can be computed and analyzed, limit cycles, Poincaré maps, attractors of various dimensions, and bifurcations between them can be defined and determined. This type of analysis is based on the geometry of phase space, but encounters limitations when the dimensionality of the state vector becomes exceedingly large.

Koopman analysis takes a different and data-driven approach, arguing that the observed data are simply function evaluations of an underlying state. Concentrating on the observed data, Koopman analysis provides tools and a mathematical foundation for the analysis of the state-vector evolution. To this end, we introduce an operator K that maps the observables over one time-step (for the discrete case). We have

$$\mathsf{K}\phi(\mathbf{q}_n) = \phi(\mathbf{q}_{n+1}) = \phi(\mathbf{F}(\mathbf{q}_n)), \tag{7.2}$$

defining the Koopman operator K. It is straightforward to show that K is a **linear** operator, and satisfies a semi-group property, that is, closedness and associativity under its binary operation. Further advancement in time is given by a multiple application of the Koopman operator K or powers thereof. We can think of the Koopman formulation as a lifting procedure of the dynamics from a state-space description to an observable-space description. Figure 7.1 gives a sketch of the two viewpoints. While the dynamics in state-space is finite-dimensional but governed by a nonlinear operator $\mathbf{F}$, the dynamics in the observable space can be described by a linear operator K, but is infinite-dimensional in nature.

The choice of observables under which our nonlinear dynamical system is transformed into a linear system is the key issue in Koopman analysis. In effect, we are looking for a coordinate transformation or an embedding $\phi(\mathbf{q})$ that renders the evolution in ϕ linear. Two examples shall illustrate the main ideas of finding these Koopman embeddings (Tu et al. 2014, Brunton & Kutz 2019).

The first example (Tu et al. 2014) is given by the two-dimensional nonlinear map

$$x_{n+1} = a x_n, \tag{7.3a}$$

$$y_{n+1} = b(y_n - x_n^2). \tag{7.3b}$$

We choose the observable vector $\phi_n = (x_n, y_n, x_n^2)$. Simple algebra confirms that a linear mapping K links the observable vector ϕ_n to its successive realization ϕ_{n+1} according to

$$\underbrace{\begin{pmatrix} x_{n+1} \\ y_{n+1} \\ x_{n+1}^2 \end{pmatrix}}_{\phi_{n+1}} = \underbrace{\begin{pmatrix} a & 0 & 0 \\ 0 & b & -b \\ 0 & 0 & a^2 \end{pmatrix}}_{\mathsf{K}} \underbrace{\begin{pmatrix} x_n \\ y_n \\ x_n^2 \end{pmatrix}}_{\phi_n}. \tag{7.4}$$

We see that a three-dimensional (rather than infinite-dimensional) vector of observables, inspired by the nonlinearity in the original mapping, transforms the nonlinear two-dimensional system into a linear one.

A second, even simpler, example (Brunton & Kutz 2019) is the logistic map $x_{n+1} = \beta x_n(1 - x_n)$. Based on the success of the previous example, we choose the observable $\phi_n = (x_n, x_n^2)$. When deriving a mapping for this choice of observable, we see that terms of cubic and quartic order in x_n appear. These terms are not part of the observable vector but, as we subsequently include them, even higher-order terms will appear – leading to a truly infinite system. In essence, by choosing powers of the state vector x_n, we outrun the span of the observable basis. Consequently, closure (as accomplished in the first example) cannot be reached. A more systematic manner of computing Koopman embeddings is needed. Simply considering higher powers of the observables, a technique related to Carleman linearization (Carleman 1932), is generally not ensuring a mapping onto a linear dynamical system. See also the chapter on nonlinear dynamics for additional details.

As the Koopman operator is defined as the linear mapping between two successive observations, it seems prudent to consider its eigenfunctions, that is,

$$\mathsf{K}\boldsymbol{\Phi} = \boldsymbol{\Phi}\boldsymbol{\Lambda}, \tag{7.5}$$

and express our observables in this Koopman eigenvector basis according to $\phi = \boldsymbol{\Phi}\xi$. Applying discrete time-stepping then produces a sequence of observables as

$$\phi(\mathbf{q}_k) = \mathsf{K}^k \phi(\mathbf{q}_0) = \mathsf{K}^k \boldsymbol{\Phi}\xi = \boldsymbol{\Phi}\xi\boldsymbol{\Lambda}^k. \tag{7.6}$$

The choice of the Koopman eigenvector basis facilitates closure in the observable basis, as it represents the invariant subspace of K. A truncation of the eigenvector basis has to be considered with care, as the full closure relies on the infinite nature of the observable basis. We also note that $\boldsymbol{\Phi}$ is dependent on $\mathbf{q}_0$.

The above representation holds the key to establishing a computational technique to compute approximations to the Koopman eigenfunctions directly from a snapshot sequence of observables.

Using expression (7.6) and the abbreviation $\phi_i = \phi(\mathbf{q}_i)$, we can establish the snapshot matrix of observables as

$$\begin{bmatrix} | & | & & | \\ \phi_0 & \phi_1 & \cdots & \phi_{n-1} \\ | & | & & | \end{bmatrix} = \mathbf{\Phi} \left[\mathbf{I}\, \mathbf{\Lambda}\, \cdots\, \mathbf{\Lambda}^{n-1} \right] (\mathbf{I} \otimes \xi) \; = \mathbf{\Phi}\, \mathrm{diag}(\xi)\, \mathsf{C},$$

where $\otimes$ denotes the Kronecker product, $\mathbf{I}$ stands for the $n \times n$ identity matrix, and the matrix C represents the (discrete) temporal dynamics.

In what follows, we consider the original data sequence as observables. Juxtaposing the previous final expression with a general decomposition of a data matrix D, consisting of a temporal sequence of measurements/observable of our dynamic process arranged in columns, we seek

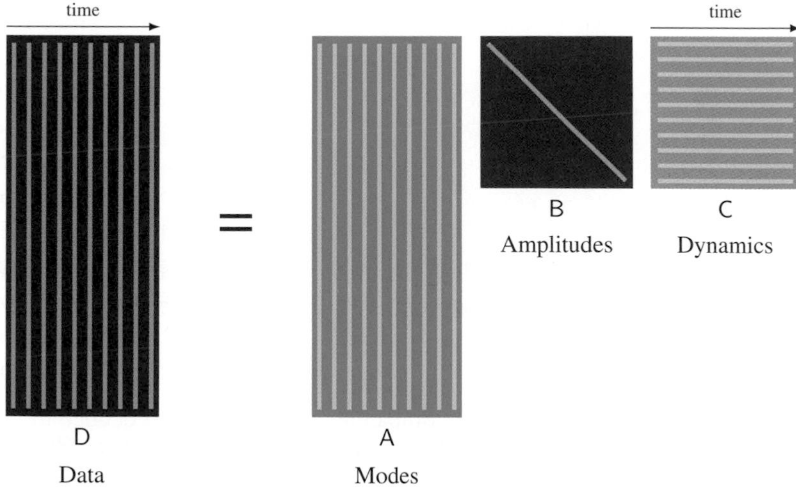

from which we can identify the matrix A as containing the Koopman eigenvectors $\mathbf{\Phi}$, the matrix B as containing the amplitudes, and the matrix C as containing the temporal dynamics. The diagonality of B decouples the decomposition into rank-one components. In addition, expression (7.6) establishes the matrix C as a Vandermonde matrix, containing the eigenvalues of K, with $\mathrm{diag}(\{\lambda\}) = \Lambda$. We have

$$\mathsf{C} = \begin{pmatrix} 1 & \lambda_1 & \lambda_1^2 & \cdots & \lambda_1^{n-1} \\ 1 & \lambda_2 & \lambda_2^2 & \cdots & \lambda_2^{n-1} \\ \vdots & \vdots & \vdots & & \vdots \\ 1 & \lambda_n & \lambda_n^2 & \cdots & \lambda_n^{n-1} \end{pmatrix}, \tag{7.7}$$

where n denotes the number of snapshots (columns) in the data matrix D. In other words, given a data matrix $\mathsf{D} \in \mathbb{C}^{m\times n}$ we have to find a decomposition into three factors, a general matrix $\mathsf{A} \in \mathbb{C}^{m\times n}$, a diagonal matrix $\mathsf{B} \in \mathbb{C}^{n\times n}$, and a Vandermonde matrix $\mathsf{C} \in \mathbb{C}^{n\times n}$.

Computing Koopman Eigenvectors from Data

The decomposition of a rectangular matrix into three factors is a common procedure in numerical analysis. Constraints on the three factors have to be imposed to arrive at a unique decomposition. For example, the constraints of an orthogonal A- and C-matrix and a diagonal B-matrix yield the familiar singular value decomposition (SVD) which is related to the proper orthogonal decomposition from Chapter 6. Other constraints will result in less familiar factorizations.

To advance on our decomposition, we use a relationship between Vandermonde and companion matrices: Vandermonde matrices diagonalize companion matrices. We can write

$$\mathsf{S} = \mathsf{C}^{-1}\mathsf{B}\mathsf{C}, \tag{7.8}$$

with S as a companion matrix, that is, a matrix consisting of a subdiagonal of ones and a nonzero final column:

$$\mathsf{S} = \begin{pmatrix} & & & & a_0 \\ 1 & & & & a_1 \\ & 1 & & & a_2 \\ & & \ddots & & \vdots \\ & & & 1 & a_{n-1} \end{pmatrix}. \tag{7.9}$$

We can replace the product BC by the product CS and recast our decomposition as $\mathsf{D} = \mathsf{ACS}$. We rename the product AC as D' and write

$$\mathsf{D} = \mathsf{D}'\mathsf{S}. \tag{7.10}$$

We take advantage of the companion structure of S to deduce the matrix D' and the last column of matrix D. The subdiagonal of ones links the $j+1$th column of D' to the jth column of D for $j = 1,\ldots,n-1$. This part of the companion matrix S is thus responsible for an index shift for the columns of the two matrices D and D'. In other words, the columns in D consist mostly of columns of D'. The last column of S (containing the coefficients a_j) states that the nth column of D is a linear combination of all the columns of D'. In the end, the columns of D are the backshifted columns of D with the final column a linear combination of all columns of D'.

7.3 The Dynamic Mode Decomposition

The transformation of our Koopman decomposition from a Vandermonde-based into a companion-based factorization now provides a first algorithm to compute approximations of the Koopman eigenvectors and eigenvalues. This decomposition is known as the DMD (Schmid 2010, Schmid et al. 2011, Kutz et al. 2016).

Starting with a sequence of $n + 1$ snapshots, sampled (for simplicity) at equispaced intervals Δt in time, we break the full sequence into two subsequences, each with n columns, according to

$$\mathsf{D}_1^{n+1} \quad \rightarrow \quad \begin{cases} \mathsf{D}_1^n, \\ \\ \mathsf{D}_2^{n+1}, \end{cases} \tag{7.11}$$

where we have introduced the notation D_i^j to represent a sequence of measurement from snapshot i to snapshot j, that is, $\mathsf{D}_i^j = \{\mathbf{d}_i, \mathbf{d}_{i+1}, \ldots, \mathbf{d}_j\}$, with $\mathbf{d}$ denoting the measurement (observable) vector. Equation (7.10) then suggests to express the last column of D_2^{n+1}, that is, the observable vector $\mathbf{d}_{n+1}$ as a linear combination of all columns of D_1^n. We accomplish this by formulating a least-squares problem, and using a QR-decomposition of D_1^n, we arrive at

$$\mathsf{S} = \mathsf{R}^{-1}\mathsf{Q}^H\mathsf{D}_2^{n+1} \quad \text{with} \quad \mathsf{QR} = \mathsf{D}_1^n, \tag{7.12}$$

that is, the companion matrix S. The superscript ^{H} denotes the Hermitian (conjugate transpose) operation. We then have the relation

$$\mathsf{KD}_1^n = \mathsf{D}_2^{n+1} \approx \mathsf{D}_1^n\mathsf{S}. \tag{7.13}$$

This latter expression states that the action of the Koopman operator K on the data set of observables D_1^n, pushing each snapshot forward over one time step Δt, can be expressed on the basis of all gathered snapshots using the companion matrix S (Rowley et al. 2009). While K scales with the (high) dimensionality of the observer vectors, the matrix S scales with number of gathered snapshots. Equation (7.13) is reminiscent of an Arnoldi iteration, where the action of a large matrix on an orthonormal set of vectors (from a Krylov subspace sequence) is expressed on the same basis. As a consequence, any spectral information we wish to gather about the operator K can be gathered, in an approximate manner, from the operator S. For example, a subset of eigenvalues of K can be approximated by the eigenvalues of S.

Determining S via a least-squares approximation, as proposed earlier, can quickly become ill-conditioned. For long data sequences, or data sequences that are characterized by a noticeable amount of noise or uncertainty, the matrix D_1^n can become rank-deficient. This is particularly true when data redundancy or near data redundancy arises. For example, sampling a wake flow past a bluff body will eventually fail to produce new data; instead, the repeated shedding of coherent structures will cause the matrix to quickly decline in rank, once a few full shedding cycles have been sampled. In this case, the matrix R from the QR-decomposition of D_1^n is no longer invertible. For this reason, a more robust algorithm must be implemented.

While keeping with the general premise of the above least-squares problem, we switch to solving it for the rank-deficient case, using a singular-value decomposition (Schmid 2010). We then have

1 given a snapshot sequence $\{\mathbf{d}_1, \mathbf{d}_2, \ldots, \mathbf{d}_{n+1}\}$ sampled equispaced in time with Δt

2 $\mathsf{D}_1^n \leftarrow \{\mathbf{d}_1, \mathbf{d}_2, \ldots, \mathbf{d}_n\}$

3 $\mathsf{D}_2^{n+1} \leftarrow \{\mathbf{d}_2, \mathbf{d}_3, \ldots, \mathbf{d}_{n+1}\}$

4 $\mathsf{U\Sigma V}^H \leftarrow \mathsf{D}_1^n$

5 $\tilde{\mathsf{S}} \leftarrow \mathsf{U}^H \mathsf{D}_2^{n+1} \mathsf{V\Sigma}_\epsilon^+$

6 $[\mathsf{X}, \Lambda] \leftarrow \mathrm{eig}(\tilde{\mathsf{S}})$

7 $\mu_j \leftarrow \log(\Lambda_{jj})/\Delta t$

8 $\Phi_j \leftarrow \mathsf{UX}_{(:,j)}$

Figure 7.2 SVD-based dynamic mode decomposition algorithm. The matrix Φ represents the matrix A in our general decomposition $\mathsf{D} = \mathsf{ABC}$. The eigenvalues Λ can be recast into the Vandermonde matrix C of our general decomposition.

$$\tilde{\mathsf{S}} = \mathsf{U}^H \mathsf{D}_2^{n+1} \mathsf{V\Sigma}_\epsilon^+ \quad \text{with} \quad \mathsf{U\Sigma V}^H = \mathsf{D}_1^n, \qquad \Sigma_\epsilon^+ = \mathrm{diag}\left(\left\{ \begin{array}{lll} 1/\sigma_j & \text{for} & \sigma_j \geq \epsilon \\ 0 & \text{for} & \sigma_j < \epsilon \end{array} \right. \right), \tag{7.14}$$

which applies a Moore–Penrose pseudo-inverse to the least-squares problem. We have introduced a threshold ϵ in the pseudo-inverse to signify a cutoff value for considering the processed data matrix as rank-deficient. In expression (7.14) for $\tilde{\mathsf{S}}$ we have also projected the dynamics on the singular vectors contained in U, which are equivalent to the POD-modes of D_1^n.

We can then state the simple algorithm for computing the DMD from a sequence of data snapshots, and with it the approximate spectral properties of the Koopman operator K.

This algorithm accommodates data sequences with a potential degree of redundancy or near-redundancy, leading to a rank deficiency when processing the data matrix in step 4 of the algorithm (see Figure 7.2). When forming the pseudo-inverse in step 5, a threshold value ϵ has to be specified and principal vectors of the SVD associated with singular values below this threshold can be discarded. The logarithmic mapping in step 7 is commonly applied in hydrodynamic applications to transform from a discrete to a continuous time variable. The above algorithm is a rather robust technique for extracting dynamic modes (and approximate Koopman eigenvectors) directly from a sequence of observable data. It should be mentioned that an additional SVD of D_2^{n+1} will produce exact eigenvectors of the high-dimensionial operator for the case of D_1^n and D_2^{n+1} not spanning the same vector space (Tu et al. 2014). Extensions and generalizations of the above core algorithm (regarding robustness, accuracy, efficiency and applicability) will be discussed in Section 7.4.

7.4 Sparsity Promotion for Finding Amplitudes

Once the dynamic structure and their time dependence (contained in the Koopman eigenvalues λ_j) have been identified, we are still left with the computation of the amplitudes, that is, the diagonal matrix B of our original decomposition. These amplitudes will provide information about the importance and dominance of the identified structures, as they are present in the processed data sequence. Ideally, structures and their dynamics that are key components in the overall data sequence should be identified by a larger amplitude, while smaller amplitudes should be attached to less dominant processes or flow features. For data from high-performance simulations, this desirable relation between amplitude and relevance can often be realized; for experimental data and data with a significant amount of noise, it cannot be guaranteed. In particular, when the processed data sequence is marred by outliers or other inaccuracies that are fairly localized in time, we may encounter intermittent structures that are characterized by a large decay rate. This large decay rate accounts for the fact that the associated flow feature disappears quickly over the time horizon of the sampled data. If the intermittent structure is of sufficient size, a simple amplitude algorithm (matching the identified dynamic modes to the original data sequence) would assign a sizeable amplitude to it, thus giving it importance among the extracted modes and processes. For a robust and physically meaningful assignment of amplitudes to identified dynamic processes, a more subtle balance between the reproduction of the original data sequence and the number of participating modes must be struck.

Mathematically and computationally, the recovery of the amplitudes is a nontrivial step. It is best formulated as a mixed-norm optimization problem. The first component of the optimization is responsible for enforcing data fidelity of the recovered process. In other words, we choose the amplitudes of the modes and their dynamics to best reproduce the original data sequence. This is accomplished by minimizing the Frobenius norm of the modal expansion and the data sequence (Jovanović et al. 2014). We denote the vector of amplitudes by $\mathbf{b}$, that is, the diagonal elements of the matrix B above. We have

$$\|\mathsf{D}_1^n - \mathbf{\Phi}\mathrm{diag}(\mathbf{b})\mathsf{C}\|_F \quad \rightarrow \quad \min, \tag{7.15}$$

with $\mathbf{\Phi}$ as the matrix containing the computed dynamic modes Φ_j and C as the Vandermonde matrix containing powers of the identified eigenvalues $\lambda_j = \Lambda_{jj}$. Using a SVD of D_1^n, the definition of the matrix $\mathbf{\Phi}$, and some algebra, the optimization problem above can be reformulated into the quadratic form

$$\mathcal{J}(\mathbf{b}) = \mathbf{b}^H \mathsf{P}\mathbf{b} - \mathbf{q}^H\mathbf{b} - \mathbf{b}^H\mathbf{q} + s \quad \rightarrow \quad \min, \tag{7.16}$$

with

$$\mathsf{P} = (\mathsf{X}^H\mathsf{X}) \odot (\mathsf{C}\mathsf{C}^H)^*, \tag{7.17a}$$

$$\mathbf{q}^* = \mathrm{diag}(\mathsf{C}\mathsf{V}\Sigma\mathsf{X}), \tag{7.17b}$$

$$s = \mathrm{trace}(\Sigma^2), \tag{7.17c}$$

where * denotes the complex conjugate operation and $\odot$ stands for the elementwise (Hadamard) product. The amplitude vector $\mathbf{b}_*$ that optimizes the match between the dynamic mode expansion and the original data sequence is simply given as the solution to this quadratic optimization problem, namely,

$$\mathbf{b}_* = \mathsf{P}^{-1}\mathbf{q}. \tag{7.18}$$

This amplitude distribution only minimizes the reconstruction error of the DMD, taking all available modes – whether physically relevant or linked to outliers in the data – into account. For a more robust and physically meaningful reconstruction, we have to restrict the number of modes being considered. We will do this by juxtaposing the sparsity of the amplitude vector and the achieved data reconstruction error (Jovanović et al. 2014). These two objectives are in conflict with each other: a very sparse amplitude vector consisting of only a few nonzero amplitudes will not produce a small data reconstruction error, while a minimal reconstruction error will not be achieved by only a few amplitudes. A compromise has to be found that produces an acceptable reconstruction error with as sparse an amplitude vector as can be managed.

The sparsity of a vector is commonly measured by its ℓ_0-norm or cardinality. The ℓ_0-norm of a vector simply counts the number of its nonzero components, and is strictly speaking not a norm (but a quasi-norm). Optimizations based on ℓ_0-norms are combinatorial in nature and difficult to implement. To this end, we substitute the ℓ_1-norm of the amplitude vector (i.e., the sum of the absolute values of its components) as a proxy for its sparsity, and augment the cost functional $\mathcal{J}(\mathbf{b})$ to read

$$\mathcal{J}(\mathbf{b}) + \gamma\|\mathbf{b}\|_1 \tag{7.19}$$

with a user-specified parameter γ that quantifies the trade-off between the reconstruction error $\mathcal{J}$ and the sparsity of the amplitude vector $\|\mathbf{b}\|_1$. This regularization step is reminiscent of Lasso regression, where a least-squares fit is combined with an L_1-penalty (Tibshirani 1996). Minimizing the above cost functional yields a lower-dimensional representation of the full data matrix of the form.

In the augmented cost functional, γ is a positive parameter that quantifies our focus on the sparsity of the amplitude vector $\mathbf{b}$. Larger values of γ produce sparser solutions, at the expense of a larger data reconstruction error (see Figure 7.3).

As the value of γ increases, the amplitude vector $\mathbf{b}$ becomes progressively sparser. There is no straightforward link between the number of nonzero amplitudes (cardinality of $\mathbf{b}$) and the required value of γ to reach this outcome. In practice, one has to experiment with different values of γ to determine the value for a desired number of nonzero amplitudes. We notice, however, that the optimization procedure involved only matrices that scale with the number of snapshots; multiple optimizations with different values of γ should therefore be computationally feasible.

After we have reached the desired number of amplitudes, while keeping the reconstruction error minimal, we lock the sparsity structure of the amplitude vector and solve, once again, an optimization problem, where we optimize the reconstruction

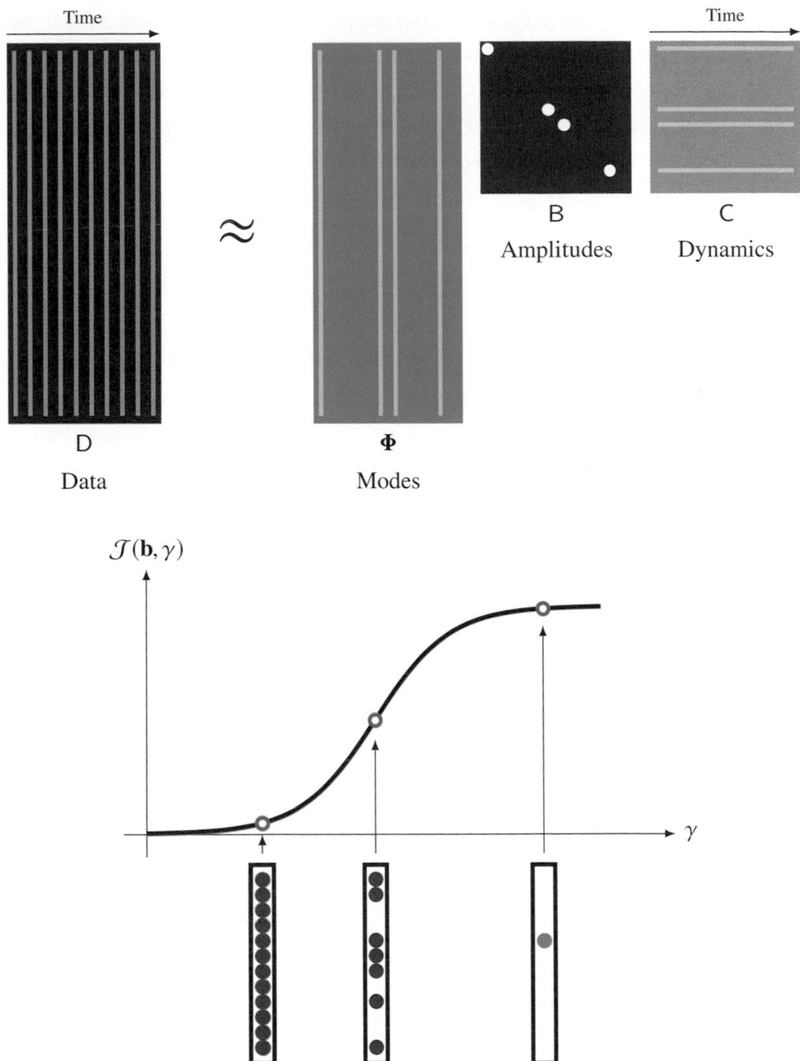

Figure 7.3 Cost functional versus the user-specified parameter γ. Low values of γ result in lower reconstruction error, but fuller amplitude vectors; larger values of γ lead to sparser amplitude vectors at the expense of larger reconstruction errors.

error subject to enforcing the identified sparsity structure explicitly. This results in the convex optimization problem (Jovanović et al. 2014)

$$\mathcal{J}(\mathbf{b}) \quad \rightarrow \quad \min, \tag{7.20a}$$

$$\text{subject to } \mathsf{E}^H \mathbf{b} = 0, \tag{7.20b}$$

where the columns of E consist of unit vectors that identify the zero entries of the final amplitude vector **b**, found from the previous mixed-norm optimization.

The mixed-norm optimization problem can be solved by common numerical techniques, such as ADMM (alternating direction method of multipliers) (Boyd et al. 2011), split Bregman iteration (Goldstein & Osher 2009), or IRLS (iteratively reweighted least-squares) (Daubechies et al. 2010).

In the supplemental material, a three-step MATLAB® code will analyze a snapshot sequence of PIV data (flow through a cylinder bundle) and extract relevant flow processes via a sparsity-promoting DMD.

7.5 Applications

A few selected applications shall demonstrate the DMD on experimental and numerical data sets. The examples are meant to illustrate the algorithm and showcase different manners of utilizing the flexibility of the methodology in analyzing fluid behavior.

Example 1: Tomographic TR-PIV of a water jet

As a first example, we consider experimental data from tomographic time-resolved PIV measurements of a water jet (Schmid et al. 2012). The jet emerges from a round nozzle of diameter 10 mm into an octagonal water tank that allows optimal access for tomographic imaging. The exit velocity of the jet is $U = 0.5$ m/s, resulting in a Reynolds number of $Re = 5000$. Neutrally buoyant particles are used to produce light-scatter images, which are recorded and processed by the tomographic imaging system. Snapshots are acquired at a rate of 1 kHz in a three-dimensional measurement domain of 50 mm × 50 mm × 32 mm. A representative snapshot from the experiment is shown in Figure 7.4, visualized by velocity vectors in the axial center plane.

The data matrix D is formed by a sequence of snapshots (sampled at a frequency of 1 kHz), each consisting of $107 \times 62 \times 62$ three-dimensional velocity vectors. We process $n = 40$ snapshots. The DMD, with sparsity promotion for the amplitude recovery, is then applied to the data matrix.

The DMD-spectrum Λ_{jj} is shown in Figure 7.4, superimposed on the unit disk. Four eigenvalues are associated with modal structures of significant amplitude; they represent two distinct structure with Strouhal number $St = 0.374$ and $St = 0.691$. The amplitude distribution versus the frequencies confirms this finding: two marked peaks at the aforementioned Strouhal numbers are clearly visible above a background amplitude distribution. It appears that the jet dynamics captured by the tomo-PIV data sequence is well described by two principal structures. These two structures are displayed in Figure 7.5, in a two-dimensional slide through the center plane, and as iso-contours of the Q-criterion. We identify the higher Strouhal number ($St = 0.691$) as the primary instability of the jet, as its spatial support is closer to the nozzle exit. At the lower Strouhal number ($St = 0.374$), an instability is identified that results in the merging of nearly axisymmetric vortex rings, halving the characteristic axial scale and halving the temporal frequency (Strouhal number) in the process. In the three-dimensional visualizations, the onset of an azimuthal breakdown of the vortex rings (via a secondary instability) is clearly visible.

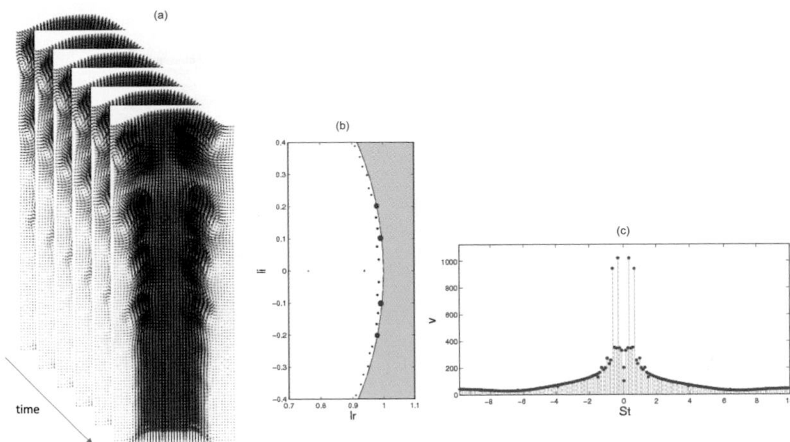

Figure 7.4 Tomographic TR-PIV of a water jet. (a) Representative snapshot (visualized in a two-dimensional slice through the center plane). (b) The DMD spectrum (superimposed on the unit disk). (c) Amplitude distribution versus identified frequencies. (a) and (b) from Schmid, Violato & Scarano (2012), reprinted by permission from Springer Nature.

Example 2: Composite analysis of an acoustically active jet

The second example is based on data from numerical simulations of the axisymmetric compressible Navier–Stokes equations. We consider a jet exiting from a round nozzle in quiescent air. With a Mach number of $Ma = 0.9$, the shear layers forming between the exiting jet and the quiescent freestream cause acoustic radiation. We are interested in analyzing the jet flow and identify structures that are responsible for the sound radiated into the freestream.

To this end, we generate and process observables that account for each of the dominant physical processes: hydrodynamic instabilities and acoustic sound radiation. The hydrodynamics is represented by the vorticity $\nabla \times \mathbf{u}$ of the flow field, acting as a substitute for instability structures. The acoustic component is captured by the dilatation, that is, the divergence $\nabla \cdot \mathbf{u}$ of the velocity field. Other combinations or markers for the respective physical processes are conceivable. Our two choices are then combined into an observable vector according to

$$\mathbf{d}_i = \begin{bmatrix} \nabla \times \mathbf{u} \\ \hline \nabla \cdot \mathbf{u} \end{bmatrix}_i, \tag{7.21}$$

and the data matrix is formed by equispaced snapshots of synchronized vorticity-dilatation fields. The underlying premise of this composite data processing is that the

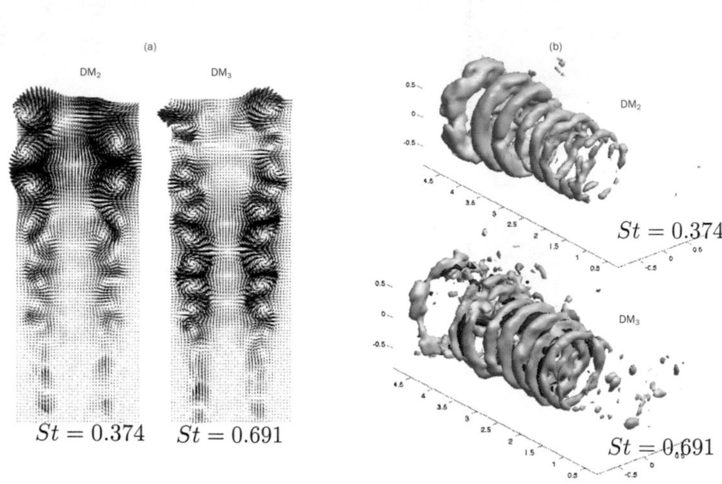

Figure 7.5 Tomographic TR-PIV of a water jet. Visualization of the two dominant dynamic modes, (a) visualized by velocity vectors in a two-dimensional slide through the center plane, and (b) visualized by iso-contours of the Q-criterion.

Koopman operator propagating this observed flow state will contain information about the interplay between the two processes encapsulated in the chosen data. Eigenvectors of the Koopman operator that have significant activity in both components may point toward structures that are "synced" within our data stream and hence may have a physical link, as they emerge with an identical temporal dynamics.

This processing of composite data is a rather compelling tool to analyze data from multi-physics processes. While it cannot establish a cause–effect relationship between the processed fields, it can identify structures (and their frequencies) that are "in resonance" or arise conjointly. For example, fluid-structure problems, shock-boundary layer interaction, or the link between coherent structures and their wall-shear stress footprint can be treated using this composite multivariable DMD technique.

In our case, a DMD analysis of the gathered data sequence from axisymmetric simulations of the compressible jet reveals a spectrum of eigenvalues (not shown) with two characteristic, but broad, peaks in frequency. The first, lower-frequency peaks are dominated by modes with mostly hydrodynamic (vorticity) components, which concentrate in the center of the jet or on the developing instabilities of the downstream outer shear layer (see Figure 7.6); the same structures also underscore the importance of the collapse of the potential core, which appears on the centerline at an axial distance of $x \approx 12$ in the simulations. No significant or very little acoustic activity in the freestream is visible.

Sampling from the higher-frequency peak, a different picture emerges (see Figure 7.7). While the hydrodynamic component still concentrates on the center of

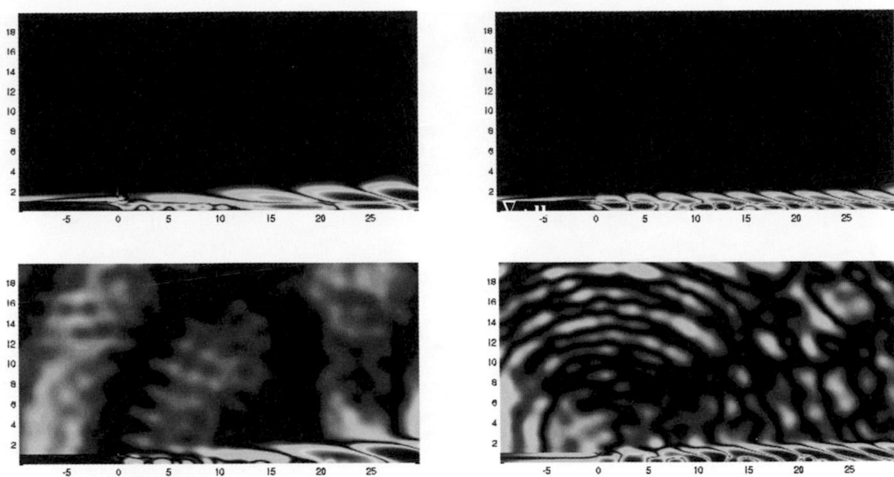

Figure 7.6 Composite analysis of a compressible axisymmetric jet at $Ma = 0.9$. Two dynamic modes (left and right) from the lower-frequency peak, with the top corresponding to the hydrodynamic vorticity component and the bottom visualizing the acoustic dilatation component.

the jet, the sound component of the dynamic modes is clearly visible and captures the acoustic activity in the freestream. While the first mode (on the left) is dominated by a nearly omnidirectional radiation of sound, the second mode (on the right) displays a preferred directionality toward the upstream region. Especially these latter structures are often responsible for feedback loops that magnify and sustain acoustic activity and cause concentrated peaks in pressure sound spectrograms.

This example demonstrates the versatility of a Koopman perspective on dynamical systems, as well as the flexibility of the DMD to explore physical processes from a data-driven viewpoint. A traditional Poincaré approach based on a state-space (rather than observable-space) formulation would be difficult to carry out with the same ease.

Example 3: Streamline-based analysis of a jet in cross-flow

The final example further extends the capabilities of Koopman/DMD analysis to treat configurations where the appropriate evolution direction is not the time coordinate. In our derivation of the DMD, we have assumed the time coordinate as our evolution direction. Consequently, we have stacked our data matrix such that each column represents a different snapshot in time. The Vandermonde matrix C and the approximate eigenvalues of the Koopman operator thus represent a temporal dynamics; the associated modes, captured in $\mathbf{\Phi}$, describe spatial structures. While this division into

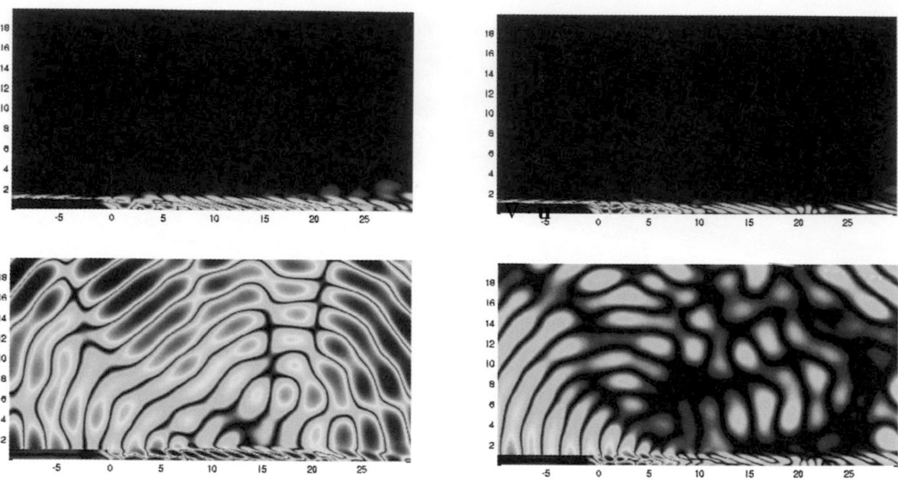

Figure 7.7 Composite analysis of a compressible axisymmetric jet at $Ma = 0.9$. Two dynamic modes (left and right) from the higher-frequency peak, with the top corresponding to the hydrodynamic vorticity component and the bottom visualizing the acoustic dilatation component. Significant acoustic activity is observed in both cases.

temporal dynamics and spatial structure may be appropriate for many applications, there are just as many configurations where an alternative dynamics-structure division seems more fitting. An example is given by a jet in cross-flow.

A jet in cross-flow (or transverse jet) is generated when a flow exiting an orifice on a wall is subjected to a second flow parallel to the wall. Situations like this arise, for example, in smoke stacks or every time a fluid is injected into a cross-stream, such as in fuel injectors. The corresponding flow is rather complex and develops a great many instabilities that interact in a nontrivial manner. Among the main instabilities is the roll-up of the vortex sheet, which initially axisymmetric at the orifice quickly deforms into two counterrotating vortex pairs that follow a bent trajectory. Along this bent, rolled up vortex sheet instabilities develop quickly: initiated a short distance from the orifice, vortex ring-like structures amplify before they break down into smaller and less coherent fluid elements.

The description of the principal instability for a jet in cross-flow clearly shows that the evolution of the instability is more aptly expressed as a *spatial* evolution along the *bent* vortex sheet of the base flow. This realization immediately sways the analysis of this physical process toward a data-driven (Koopman) approach, rather

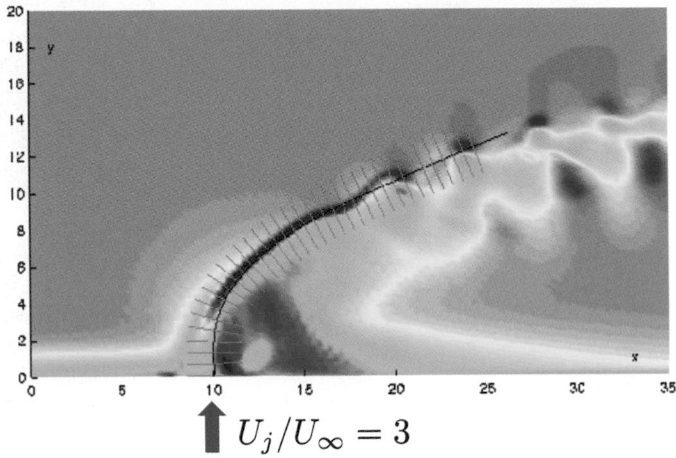

Figure 7.8 Streamline-based analysis of a jet in cross-flow. A base streamline, emanating from the jet nozzle, serves as the spatial evolution direction. Flow fields are sampled in n plane orthogonal to the streamline and equispaced in arclength along the streamline. Figure reproduced with permission from Henningson (2010).

than a traditional state-space description that would require the formulation of the governing equations for the spatial evolution of disturbances along curved paths.

In our data-driven approach, we simply have to sample the flow fields in planes orthogonal to a chosen evolution direction and stack the data matrix accordingly. To this end, we determine a streamline emanating from the center of the orifice and following the base flow. This streamline will serve as our spatial evolution direction. We then partition the streamline into $n = 110$ equispaced (in arclength) intervals between the location of instability onset and the location of ultimate instability breakdown. For each point along the streamline, we define a plane normal to its tangent, and interpolate the three-dimensional velocity field into this plane. These planes (when reshaped into vectors) will form the columns of our data matrix D. Figure 7.8 illustrates the aforementioned procedure.

Once we have established the evolution direction of our observables and formed the data matrix, a standard DMD/Koopman analysis can be performed. It is worth noting that the "spatial" direction, that is, the row direction of the data matrix, contains the coordinates in the cross-planes (normal to the base-flow streamline tangent) as well as the time coordinate. In order to reduce the size of the problem, we Fourier transform in time and focus on the Strouhal number (nondimensional frequency) associated with the Kelvin–Helmholtz instability of the vortex sheet.

Evaluating the dynamic modes from the analysis of our data matrix, we capture the spatial breakdown of the vortex sheet. The identified instabilities are localized on the base-flow counterrotating vortex sheet and describe vortical structures of increasing complexity and scale; see Figure 7.9.

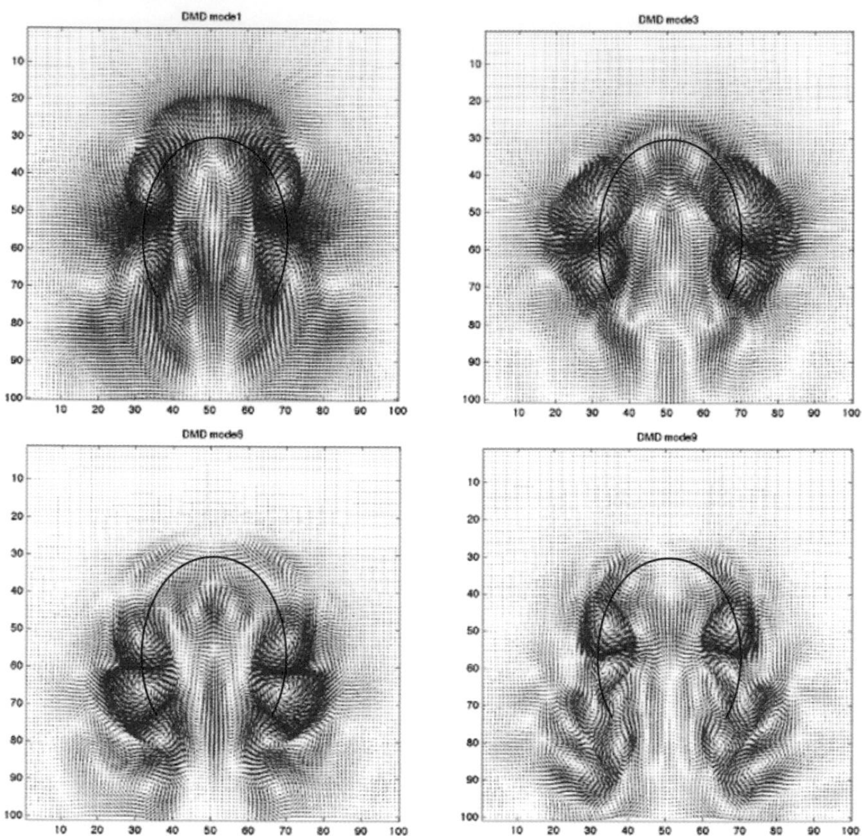

Figure 7.9 Streamline-based analysis of a jet in cross-flow. Dominant dynamic modes displaying the structures causing the spatial breakdown of the base vortex sheet along the base streamline. The approximate location of the base vortex sheet is indicated in red. Reproduced with permission from Henningson (2010).

The dynamic modes consist of vortical structures that align with the counterrotating vortex sheet of the base flow. Higher-order modes display more vortical elements that progressively concentrate on the flanks of the base-flow vortex sheet. This is consistent with the findings from direct numerical simulations as well as physical experiments. The amplitude distribution of the modes (not shown) shows a broad peak, which suggests that a superposition of a few dominant modes may best capture the pertinent dynamics of the spatial instability and may provide a reduced-order description of the Kelvin–Helmholtz-type breakdown of the base-flow's counterrotating vortex sheet.

This final example demonstrates the capability of Koopman/DMD analysis to identify coherent fluid motion directly from data snapshots in situations where the relevant evolution direction is not aligned with a coordinate direction, but is instead given by a curved base-flow streamline. With a projection and rearrangement of the data fields, the dynamic characteristics of complex instabilities can be readily computed.

7.6 Extensions and Generalizations

Since its introduction, the DMD and the underlying Koopman analysis have gained in popularity as a tool for quantitative flow analysis. Both fields have spawned a great deal of improvements, extensions, and generalizations – both from an algorithmic and conceptual point of view. Only a few of them shall be mentioned here.

The choice of observables is a key issue in Koopman analysis. In the above exposition, we use the raw snapshots, assuming that their evolution contains the necessary information to embed the dynamics in a linear fashion. However, extended DMD (Williams et al. 2015) aims at explicitly including observables that are higher powers of the original snapshots. This follows the original idea of supplementing the observables by selected nonlinearities. Alas, extended DMD quickly becomes computationally expensive; a reduction in dimensionality using a kernel trick is possible (Williams et al. 2015), but only provides temporary relief.

It has quickly been noted that, while equispaced snapshots are advantageous in the motivation of the DMD algorithm, they are not required for a more general formulation. In fact, only the snapshots in equal column position in D_1^n and D_2^{n+1} have to be separated by a constant Δt. Within either snapshot sequence, non-equispaced data can be accommodated.

A SVD is a core element of the DMD algorithm, which allows for a great deal of improvements from an algorithmic standpoint. An incremental formulation of the SVD allows the design of an efficient DMD algorithm for streaming data (Hemati et al. 2017), where new data snapshots get incorporated into an already existing decomposition by a perturbative adjustment of the modes and spectrum. This type of algorithm permits the analysis of experimental data in real time. In addition, recent developments in randomized algorithms for the SVD can also be incorporated into the overall DMD algorithm and lead to higher efficiencies, while sacrificing little in terms of accuracy.

Using delay-embedding techniques (related to the Ruelle-Takens methodology), the predictive horizon of a reduced description of a dynamical system via DMD can be substantially enhanced. In essence, the data sequence is transformed into a phase-space description and decomposed there. The data matrix is block-Hankelized and processed by the standard DMD algorithm (Arbabi & Mezić 2017, Brunton et al. 2017).

Finally, the recent popularity of machine learning and artificial neural network techniques has also influenced data decompositions and model reduction methods. By proposing a neural net in the form of an auto-encoder, a nonlinear Koopman embedding can be learned directly from the data (Takeishi et al. 2017, Lusch et al. 2018, Mardt et al. 2018, Otto & Rowley 2019, Yeung et al. 2017), by imposing linearity of the mapping over one time step between the final step of the encoder portion and the first layer of the decoder part. More advances in this direction are expected over the coming years.

7.7 Summary and Conclusions

DMD is a simple, yet effective data-driven tool to extract quantitative information about fluid flows from sequences of measurements. In its simplest form, it is straightforward to implement and use. Nonetheless, it is based on a sound, sophisticated and rich mathematical formalism – Koopman analysis – that provides guidance and support for further developments, applications, and interpretations. Many extensions and improvements of the original algorithm are available, and a wide range of applications (to numerical and experimental data) can be found in a growing body of literature.

8 Generalized and Multiscale Modal Analysis

M. A. Mendez

This chapter describes modal decompositions in the framework of matrix factorizations. We highlight the differences between classic space-time decompositions and 2D discrete transforms and discuss the general architecture underpinning *any* decomposition. This setting is then used to derive simple algorithms that complete *any* linear decomposition from its spatial or temporal structures (bases). Discrete Fourier transform (DFT), proper orthogonal decomposition (POD), dynamic mode decomposition (DMD), and eigenfunction expansions (EF) are formulated in this framework and compared in a simple exercise. Finally, this generalization is used to analyze the impact of spectral constraints on the classical POD, and to derive the multiscale proper orthogonal decomposition (mPOD). This decomposition combines multi-resolution analysis (MRA) and POD. This chapter contains four exercises and two tutorial test cases. The PYTHON scripts associated to these are provided in the book's website.[1]

8.1 The Main Theme

Modal analysis aims at decomposing a data set as a linear combination of elementary contributions called *modes*. This provides the foundations to many areas of applied mathematics, including pattern recognition and machine learning, data compression, filtering, and model order reduction. A mode is the representation of the data along an element of a *basis*, and the decomposition is the *projection* of the data set onto the basis (see Chapter 4). A basis maps the data onto a different space from the one in which it is sampled, with the hope that this new space is capable of better highlighting the features of interest for the task at hand. This is the central theme of this chapter: modal decompositions are ways of representing the *same* information in different bases.

There is no "perfect" basis. Although some decompositions are more flexible than others, they all have their merit for a specific task. In other words, every basis can highlight or "isolate" a specific pattern in the data. Depending on how these features are used, one enters into various areas of applied mathematics – all relevant in fluid mechanics – as discussed in different chapters of this book.

In data compression or model order reduction, for example, one aims at representing the data with the least number of modes, to distill its "most important" features.

[1] www.datadrivenfluidmechanics.com/download/book/chapter8.zip

We thus look for the smallest basis, that is, the smallest subspace that can still handle all the information we need. Yet, the definition of "most important" strongly depends on the application and the experience of the user. In some settings, it is essential to identify (and retain) information contained within a specific range of frequencies. In others, it might be necessary to focus on the information that is localized within a particular location in space or time; in still others, it might be essential to extract some "coherency" or "energy" or "variance" contribution. More generally, one might be interested in a combination of the previous, and the tools described in this chapter give the reader full control over the full spectrum of options.

8.1.1 The Scope of Modal Decompositions

Data-driven modal decompositions were historically driven in fluid mechanics by the quest for identifying (and objectively define) coherence structures in turbulent flows (see Chapter 2). In parallel to this, another critical driver has always been reduced-order modeling, that is the quest for identifying the "right" basis onto which to project partial differential equations (PDEs). If such a basis consists of a few carefully chosen modes, the projected system preserves the relevant dynamics and can be simulated at a fraction of the computational cost. This opens the path to fast analysis and prediction, optimization, system identification, control, and more modeling: reducing the dimensionality of a system, we facilitate regression problems such as, for example, the derivation of new models for turbulence.

In a book on data science and machine learning, this chapter could be titled *linear dimensionality reduction*; this is a fundamental tool to simplify classification or regression problems. A complementary chapter should include nonlinear decompositions based on kernels or artificial neural networks (ANN), or cluster-based methods. These methods were initially developed in computer vision, data science, and statistics, and are now revolutionizing the toolbox of the fluid dynamicists, as discussed in Chapters 1, 3, and 14.

This chapter is about *linear* decompositions in which a *linear combination* of modes describes the data. Each mode might feature a nonlinear dynamic,[2] but their combination is linear, with coefficients determined using inner products (see Chapter 4). In this chapter, we assume that the data are sampled in space and time; hence, every mode has a spatial structure and a temporal structure. These structures represent, respectively, a basis for the discrete space and the discrete time. Furthermore, the decompositions presented consider data sets composed of a single signal. Extended methods that construct hybrid bases combining different signals (e.g., velocities and temperatures) are described in Chapter 9.

8.1.2 How Many Decompositions?

Modal decompositions can be classified into data-independent and data-driven. In the first category, the basis of the decomposition is constructed regardless of the data set

[2] In the sense that its evolution might not be described by a linear equation.

at hand and is defined by the size of the data and the basis construction criteria. This is the case of the DFT (see Chapter 4) or the discrete wavelet transforms (DWT, see Chapter 5). In the first, structures are harmonics with a frequency that is an integer multiple of a fundamental one; in the second, structures are obtained by scaling and shifting a template basis (mother and father wavelets).

In data-driven decompositions, the basis is tailored to the data. The most classic examples are the POD (see Chapter 6) and the DMD (see Chapter 7). Both POD and DMD have many variants from which we can identify two categories of data-driven decomposition: those arising from the POD are "energy-based;" those arising from the DMD are "frequency-based."

The POD basis is obtained from the eigenvalue decomposition of the temporal or the spatial correlation matrices. This is dictated by a constrained optimization problem that maximizes the energy (i.e., variance) along its basis, constrained to be *orthogonal*, such that the error produced by an approximation of rank $\tilde{r} < R$ is the least possible. Variants of the POD can be constructed from different choices of the inner product or in the use of different averaging procedures in the computation of the correlations. Examples of the first variants are proposed by Maurel et al. (2001), Rowley et al. (2004), Lumley and Poje (1997), where multiple quantities are involved in the inner product. Examples of the second variants are proposed by Citriniti and George (2000) and Towne et al. (2018), where the correlation matrix is computed in the frequency domain using time averaging over short windows, following the popular Welch's periodogram method (Welch 1967).

Within the frequency-based decompositions, the DMD basis is constructed assuming that a linear dynamical system can well approximate the data. The DMD is thus essentially a system identification procedure that aims at extracting the eigenvalues of the linear system that best fit the data. Variants of the DMD propose different methods to compute these eigenvalues. Examples are the sparsity promoting DMD (Jovanović et al. 2014), the optimized DMD (Chen et al. 2012), or the randomized DMD (Erichson et al. 2019), while higher-order formulations have been proposed by Le Clainche and Vega (2017). Although the DMD represents the most popular formalism in fluid mechanics for such linear system identification process, analogous formulations (with slightly different algorithms) were introduced in the late 1980s in climatology under the names of principal oscillation patterns (POP, see Hasselmann 1988, von Storch & Xu 1990) or linear inverse modeling (LIM, see Penland & Magorian 1993, Penland 1996).

Both "energy-based" and "frequency-based" methods have limits, illustrated in some of the proposed exercises of this chapter. These limits motivate the need for hybrid decompositions that mix the constraints of energy optimality and spectral purity. Examples of such methods for stationary data sets are the spectral POD proposed by Sieber et al. (2016), the multiresolution DMD (Kutz et al. 2016), the recursive DMD (Noack et al. 2016), or the Cronos–Koopman analysis (Cammilleri et al. 2013). A hybrid method that does not hinge on the stationary assumption is the mPOD (see Mendez et al. 2018, Mendez et al. 2019).

All the decompositions mentioned thus far have a common underlying architecture. This chapter presents this architecture and the formulation of the mPOD. This decomposition offers the most general formalism, unifying the energy-based and the frequency-based approaches.

8.2 The General Architecture

All the modal decomposition introduced in Section 8.1 can be written as a special kind of matrix factorization. This view allows for defining a general algorithm for modal decomposition that is presented in Section 8.3. First, Section 8.2.1 briefly reviews the notation followed throughout the chapter while Sections 8.2.2 and 8.2.3 put this factorization in a more general context of 2D transforms. Section 8.3 briefly discusses the link between discrete and continuous domain, which is essential to render all decompositions statistically convergent to grid-independent results.

8.2.1 A Note on Notation and Style

The matrix to be factorized in this chapter is denoted as $\boldsymbol{D}(\mathbf{x_i}, t_k) = \boldsymbol{D}[\mathbf{i}, k]$ and is assumed to be a collection of samples of a real quantity (e.g., gray scale levels in a set of images, pressure fields in a CFD simulation or deformation fields in solid mechanics) along a spatial discretization $\mathbf{x_i}$ and a time discretization t_k. The notation introduced in Chapter 4 is maintained for continuous and discrete signals. The boldface notation in the spatial discretization indicates that this can be high-dimensional, for example, $\mathbf{x_i} = (x_i, y_j)$ in a 2D domain and the boldface index $\mathbf{i}$ denotes a linear matrix index.

The matrix linear index is important when we transform a spatial realization which has the form of a matrix (e.g., a scalar pressure field $p[i, j]$ with $i \in [0, n_x - 1]$ and $j \in [0, n_y - 1]$) into a vector (i.e., $p[\mathbf{i}]$ with $\mathbf{i} \in [0, n_s - 1]$, being $n_s = n_x\, n_y$). This index accesses entries of a matrix in a different way depending of whether the flattening is performed column-wise or row-wise. For example, consider the case of a matrix $\boldsymbol{A} \in \mathbb{C}^{3\times 3}$. The column-wise and the row-wise matrix indices are[3]

$$\text{column-wise } \mathbf{i} : \; \boldsymbol{A} = \begin{bmatrix} 0 & 3 & 6 \\ 1 & 4 & 7 \\ 2 & 5 & 8 \end{bmatrix}, \qquad \text{row-wise } \mathbf{i} : \; \boldsymbol{A} = \begin{bmatrix} 0 & 1 & 2 \\ 3 & 4 & 5 \\ 6 & 7 & 8 \end{bmatrix}.$$

In what follows, we consider a column-wise flattening.

All the material presented in this chapter assumes a 2D space domain, the generalization to higher or lower dimensions being trivial. Moreover, we assume that space and time domains are sampled on uniform meshes; the generalization to nonuniform mesh requires extra care in the normalization process, as discussed

[3] Recall that here we use a Python-like indexing. Hence the first entry is 0 and not 1.

in Section 8.2.5. The space domain is sampled over a grid $(x_i, y_i) \in \mathbb{R}^{n_x \times n_y}$, with $x_i = i\Delta x, i \in [0, n_x - 1]$ and $y_j = j\Delta x$, $y \in [0, n_y - 1]$.

In a vector quantity, for example, a velocity field $\boldsymbol{U}[u(\mathbf{x_i}), v(\mathbf{x_i})]$, we consider that the reshaping stacks all the components one below the other producing a state vector of size $n_s = n_C\, n_x\, n_y$, where $n_C = 2$ is the number of velocity components. Therefore, the snapshot of the data at a time t_k is a vector $\mathbf{d}_k[\mathbf{i}] \in \mathbb{R}^{n_s \times 1}$, while the temporal evolution of the data at a location $\mathbf{i}$ is a vector $\mathbf{d_i}[k] \in \mathbb{R}^{n_t \times 1}$. We then have $\boldsymbol{D}[\mathbf{i}, k = c] = \mathbf{d}_c[\mathbf{k}] \in \mathbb{R}^{n_s \times 1}$ and $\boldsymbol{D}[\mathbf{i} = \mathbf{c}, k] = \mathbf{d_c}[l] \in \mathbb{R}^{1 \times n_t}$.

8.2.2 Projections in Space *or* Time

Every linear operation on a discrete 1D signal, represented as a column vector, can be carried out via matrix multiplication. This is true for convolutions, changes of bases (i.e., linear transforms), and filtering. Understanding the matrix formalism of these operations brings at least three benefits. First, the notation is simplified by replacing summations on sequences with matrix-vector multiplications. Second, the efficiency and compactness of computer codes for signal processing are largely augmented,[4] by avoiding the nested "for" loops implied in the summation notation. Third, the linear algebra representation of these operations offers a valuable geometrical interpretation.

Consider the problem of decomposing a column vector $\boldsymbol{u} \in \mathbb{R}^{n_u \times 1}$ into a linear combination of basis vectors $\{\mathbf{b}_1, \mathbf{b}_2, \ldots, \mathbf{b}_{n_b}\}$ of equal dimensions. This means

$$u[k] = \boldsymbol{u} = \sum_{r=1}^{n_u} c_r\, \boldsymbol{b}_r \Longleftrightarrow \mathbf{u} = \boldsymbol{B}\, \mathbf{u}_B, \tag{8.1}$$

where $\boldsymbol{B} = [\mathbf{b}_1, \mathbf{b}_2, \ldots \mathbf{b}_{n_B}] \in \mathbb{C}^{n_u \times n_b}$ is the basis matrix having all the elements of the basis along its columns, and $\mathbf{u}_B = [c_1, c_1, \ldots, c_{n_b}]^T$ is the set of coefficients in the linear combination, that is, the representation of the vector in the new basis. Computing the transform of a vector with respect to a basis $\boldsymbol{B}$ means solving a linear system of algebraic equations. Such a system can have no solution, one solution, or infinite solutions depending on n_b and n_u.

If $n_b < n_u$, as it is the case in model-order reduction, the system is *overdetermined* and *there is no solution*.[5] In this case, we look for the approximated solution $\tilde{\mathbf{u}}_B$ that is obtained by projecting $\mathbf{u}$ onto the column space of $\boldsymbol{B}$. This is provided by the well-known least-squares approximation, which gives[6]

$$\mathbf{u} = \boldsymbol{B}\, \mathbf{u}_B \Longrightarrow \tilde{\mathbf{u}}_B = (\boldsymbol{B}^\dagger \boldsymbol{B})^{-1} \boldsymbol{B}^\dagger \mathbf{u} \Longrightarrow \tilde{\mathbf{u}} = \boldsymbol{B}(\boldsymbol{B}^\dagger \boldsymbol{B})^{-1} \boldsymbol{B}^\dagger \mathbf{u} = \mathcal{P}_B \mathbf{u}\,. \tag{8.2}$$

The least-squares solution minimizes $||\mathbf{u} - \boldsymbol{B}\, \mathbf{u}_B||_2 = ||\mathbf{e}||_2$; the minimization imposes that the error vector $\mathbf{e}$ is orthogonal to the column space of $\boldsymbol{B}$. In the machine learning terminology, the matrix $\mathcal{P}_B = \boldsymbol{B}(\boldsymbol{B}^\dagger \boldsymbol{B})^{-1} \boldsymbol{B}^\dagger$ is an *autoencoder* that maps a

4 Especially in interpreted languages such as MATLAB or Python.

5 Unless the $\boldsymbol{u}$ lays in the range of the column space of $\boldsymbol{B}$ in which case a unique solution exists. This is rarely the case in practice.

6 Multiply the system by $\boldsymbol{B}^\dagger$ and then invert the resulting $\boldsymbol{B}^\dagger \boldsymbol{B}$.

signal in $\mathbb{R}^{n_u \times 1}$ to $\mathbb{R}^{n_b \times 1}$ (this is an *encoding*) and then back to $\mathbb{R}^{n_u \times 1}$ (this is the *decoding*).

Because the underlying linear system has no solution, the linear encoding does not generally admit an inverse if $n_b < n_u$: it is not possible to retrieve $\mathbf{u}$ from $\tilde{\mathbf{u}}_B$– that is, the autoencoding loses information. A special case occurs if $n_b = n_u$. Under the assumption that the columns are linearly independent– that is, $\boldsymbol{B}^{-1}$ exists – there is only one solution. It is easy to show that[7] this yields $\mathcal{P}_B = \mathbf{I}$: in this case $\tilde{\mathbf{u}} = \mathbf{u}$, and the tilde is removed because the autoencoding is lossless. This is fundamental in filtering applications for which the basis matrix is usually square: a signal is first projected onto a certain basis (e.g., Fourier or wavelets), manipulated, and then projected back.

The last possibility, that is $n_b > n_u$, results in an *underdetermined* system. In this case, there are *infinite solutions*. Among these, it is common practice to consider the one that leads to the least energy in the projected domain, that is, such that $min(||\mathbf{u}_B||)$. This approach, which also yields a reversible projection, is known as the least norm solution and reads

$$\mathbf{u} = \boldsymbol{B}\,\mathbf{u}_B \iff \mathbf{u}_B = \boldsymbol{B}^{\dagger}(\boldsymbol{B}\boldsymbol{B}^{\dagger})^{-1}\mathbf{u}\,. \tag{8.3}$$

It is now interesting to apply these notions to decompositions (projections) in space and time domains. Considering first the projection in the space domain, let the signal in (8.2) be a column of the data set matrix, that is, $\mathbf{u} := \mathbf{d}_k[\mathbf{i}] \in \mathbb{R}^{n_s \times 1}$. Let $\boldsymbol{\Phi} = [\boldsymbol{\phi}_1, \boldsymbol{\phi}_2, \dots, \boldsymbol{\phi}_{n_\phi}] \in \mathbb{C}^{n_s \times n_\phi}$ denote the spatial basis. Because the matrix multiplication acts independently on the columns of $\boldsymbol{D}$, (8.1) is

$$\mathbf{d}_k = \boldsymbol{\Phi}\,\mathbf{d}_\phi \Rightarrow \tilde{\mathbf{d}}_{k\phi} = (\boldsymbol{\Phi}^{\dagger}\boldsymbol{\Phi})^{-1}\boldsymbol{\Phi}^{\dagger}\mathbf{d}_k \;\; ; \; \boldsymbol{D} = \boldsymbol{\Phi}\boldsymbol{D}_\phi \;\Rightarrow \tilde{\boldsymbol{D}}_\phi = (\boldsymbol{\Phi}^{\dagger}\boldsymbol{\Phi})^{-1}\boldsymbol{\Phi}^{\dagger}\boldsymbol{D}. \tag{8.4}$$

Here the transformed vector is $\tilde{\mathbf{d}}_{k\phi}$ while the matrix $\tilde{\boldsymbol{D}}_\phi$ collects all the transformed vectors, that is, the coefficients of the linear combinations of basis elements $\{\boldsymbol{\phi}_1, \boldsymbol{\phi}_2, \dots, \boldsymbol{\phi}_R\}$ that represents a given snapshot $\boldsymbol{d}_k$.

The same reasoning holds for transforms in the time domain. In this case, let the signal be a row of the data set matrix, that is, $\mathbf{u} := \mathbf{d}_\mathbf{i}^T[k] \in \mathbb{R}^{n_t \times 1}$. Defining the temporal basis matrix as $\boldsymbol{\Psi} = [\boldsymbol{\psi}_1, \boldsymbol{\psi}_2, \dots, \boldsymbol{\psi}_{n_\psi}] \in \mathbb{C}^{n_t \times n_\psi}$ and handling the transpositions with care,[8] the analogous of (8.4) in the time domain reads

$$\mathbf{d}_\mathbf{i}^T = \boldsymbol{\Psi}\,\mathbf{d}_\psi^T \Rightarrow \tilde{\mathbf{d}}_{\mathbf{i}\psi}^T = (\boldsymbol{\Psi}^{\dagger}\boldsymbol{\Psi})^{-1}\boldsymbol{\Psi}^{\dagger}\mathbf{d}_\mathbf{i}^T \;\; ; \; \boldsymbol{D} = \boldsymbol{D}_\psi\boldsymbol{\Psi}^T \;\Rightarrow \tilde{\boldsymbol{D}}_\psi = \boldsymbol{D}\overline{\boldsymbol{\Psi}}(\boldsymbol{\Psi}^{\dagger}\boldsymbol{\Psi})^{-1}\,. \tag{8.5}$$

Here the transformed vectors are $\tilde{\mathbf{d}}_{\mathbf{i}\psi}$ and the matrix $\tilde{\boldsymbol{D}}_\psi$ collects in its columns all the transformed vectors, that is, the coefficients of the linear combinations of basis elements $\{\boldsymbol{\psi}_1, \boldsymbol{\psi}_2, \dots, \boldsymbol{\psi}_R\}$ that represent the data evolution at a location $\mathbf{d}_\mathbf{i}[k]$.

Note that approximations can be obtained along space and time domains as

$$\tilde{\boldsymbol{D}}_\phi = \boldsymbol{\Phi}(\boldsymbol{\Phi}^{\dagger}\boldsymbol{\Phi})^{-1}\boldsymbol{\Phi}^{\dagger}\boldsymbol{D} \quad \text{and} \quad \tilde{\boldsymbol{D}}_\psi = \boldsymbol{D}\overline{\boldsymbol{\Psi}}(\boldsymbol{\Psi}^{\dagger}\boldsymbol{\Psi})^{-1}\boldsymbol{\Psi}^T. \tag{8.6}$$

[7] Use the distributive property of the inversion to show that $\mathcal{P}_B = \boldsymbol{B}(\boldsymbol{B}^{\dagger}\boldsymbol{B})^{-1}\boldsymbol{B}^{\dagger} = \boldsymbol{B}\boldsymbol{B}^{-1}\left(\boldsymbol{B}^{\dagger}\right)^{-1}\boldsymbol{B} = \boldsymbol{I}$.

[8] More generally, in a matrix multiplication $\boldsymbol{A}\boldsymbol{B}$, the matrix $\boldsymbol{A}$ is acting on the columns of $\boldsymbol{B}$. The same action along the rows of $\boldsymbol{B}$ is obtained by $\boldsymbol{A}\boldsymbol{B}^T$.

These approximations are exact (the projections are reversible) if the bases are complete, that is, the matrices $\mathbf{\Phi}$ and $\mathbf{\Psi}$ are square[9] ($n_\phi = n_s$ and $n_\psi = n_t$). In these cases, autoencoding is loseless: $\tilde{\boldsymbol{D}}_\phi = \boldsymbol{D}$ and $\tilde{\boldsymbol{D}}_\psi = \boldsymbol{D}$.

It is left as an exercise to see how (8.4)–(8.6) simplify if the bases in the space and time are also orthonormal, that is, the inner products yield $\mathbf{\Phi}^\dagger\mathbf{\Phi} = \mathbf{I}$ and $\mathbf{\Psi}^\dagger\mathbf{\Psi} = \mathbf{I}$ regardless of the number of basis elements[10] n_ϕ and n_ψ. On the other hand, from the fact that (8.6) is *exact* only for *complete* bases, one can see that $\mathbf{\Phi}\mathbf{\Phi}^\dagger = \mathbf{I}$ *only if* $n_\phi = n_s$ and $\mathbf{\Psi}\mathbf{\Psi}^\dagger = \mathbf{I}$ *only if* $n_\psi = n_t$.

8.2.3 Projections in Space *and* Time

Let us now consider 2D transforms of the data set matrix. These are the discrete version of 2D transforms, which in this case act on the columns and the rows of the matrix independently. Let $\boldsymbol{D}_{\phi\psi}$ be the 2D transform of the matrix $\boldsymbol{D}$ taking $\mathbf{\Phi} = [\boldsymbol{\phi}_1, \boldsymbol{\phi}_2, \ldots, \boldsymbol{\phi}_{n_\phi}] \in \mathbb{C}^{n_s \times n_\phi}$ and $\mathbf{\Psi} = [\boldsymbol{\psi}_1, \boldsymbol{\psi}_2, \ldots, \boldsymbol{\psi}_{n_\psi}] \in \mathbb{C}^{n_t \times n_\psi}$ as basis matrices for the columns and the rows, respectively. The general 2D transform pair is

$$\boldsymbol{D} = \mathbf{\Phi}\boldsymbol{D}_{\phi\psi}\mathbf{\Psi}^T \Leftrightarrow \tilde{\boldsymbol{D}}_{\phi\psi} = (\mathbf{\Phi}^\dagger\mathbf{\Phi})^{-1}\mathbf{\Phi}^\dagger\,\boldsymbol{D}\,\overline{\mathbf{\Psi}}(\mathbf{\Psi}^\dagger\mathbf{\Psi})^{-1}\,. \tag{8.7}$$

A classic example of this kind of transforms is the 2D Fourier transform in which case the bases matrices are

$$\mathbf{\Phi}_{\mathcal{F}}[m,n] = \frac{1}{\sqrt{n_s}}e^{2\pi \mathrm{j}\, m\, n/n_s} \text{ and } \mathbf{\Psi}_{\mathcal{F}}[m,n] = \frac{1}{\sqrt{n_t}}e^{2\pi \mathrm{j}\, m\, n/n_t}\,. \tag{8.8}$$

These are the *Fourier matrices*, which are orthonormal. If these are also square (i.e., the number of frequency bins equals n_s and n_t, respectively), these matrices are also symmetric. Then, (8.7) is reversible and the 2D discrete Fourier pair is[11]

$$\boldsymbol{D} = \mathbf{\Phi}_{\mathcal{F}}\widehat{\boldsymbol{D}}\mathbf{\Psi}_{\mathcal{F}} \Longleftrightarrow \widehat{\boldsymbol{D}} = \overline{\mathbf{\Phi}}_{\mathcal{F}}\boldsymbol{D}\,\overline{\mathbf{\Psi}}_{\mathcal{F}}\,. \tag{8.9}$$

The (discrete) Fourier transform plays an essential role in Section 8.4. Let us use a wide hat $\widehat{\ }$ to indicate the Fourier transform of a matrix, whether this is carried out along its rows, columns, or both.

Different decompositions can be derived by taking different bases for the rows and for the columns. This is often done in the 2D discrete wavelet transform (see Chapter 5), as one might desire to highlight different features along different directions. For example, in the literature of image processing, we define *horizontal details* as those features that have high frequencies over the column space and low frequencies over the row space. We can thus construct a basis (i.e., design a filter) that enhances or reduces these details.

Decompositions of this form are common and useful in image processing. For reduced-order modeling, however, these have a substantial disadvantage: the number

[9] And, of course, have full rank.

[10] If $\mathbf{\Psi}^\dagger\mathbf{\Psi} = \mathbf{I}$, then taking the conjugation on both sides gives $\mathbf{\Psi}^T\overline{\mathbf{\Psi}} = \mathbf{I}$ while taking the transposition on both sides gives $\overline{\mathbf{\Psi}}\mathbf{\Psi}^T = \mathbf{I}$.

[11] Observe that transpositions are removed because of the symmetry of the bases.

of basis elements (i.e., its "modes") is $n_s \times n_t$. While most of the data have a sparse representation in a 2D Fourier or wavelet bases,[12] this representation is still inefficient. The inefficiency stems from the lack of a "variable separation": the entry $\boldsymbol{D}_{\phi\psi}[m,n]$ measures the correlation with a basis matrix constructed from the *mth* basis element of the column space and the *nth* basis element of the row space. Even in the ideal case of orthonormal bases on both columns and rows (i.e., space and time), the decomposition in (8.7) requires a double summation:

$$\boldsymbol{D}[\mathbf{i},k] = \sum_{m=0}^{n_s-1} \sum_{n=0}^{n_t-1} \boldsymbol{D}_{\phi\psi}[m,n]\boldsymbol{\phi}_m[\mathbf{i}]\,\boldsymbol{\psi}_n[k]. \tag{8.10}$$

In modal analysis, we seek a transformation that renders $\boldsymbol{D}_{\phi\psi}$ *diagonal* and that has no more than $R = min(n_s, n_t)$ modes. We seek separation of variables and hence enter data-driven modal analysis from Section 8.2.4.

8.2.4 The Fundamental Factorization

We impose that $\boldsymbol{D}_{\phi\psi}$ is diagonal. In this case, the transformed matrix is denoted as $\boldsymbol{\Sigma} = \text{diag}(\sigma_1, \ldots, \sigma_R) \in \mathbb{R}^{R\times R}$ and the entries σ_r are referred to as *mode amplitudes*. The summation in (8.10) and the factorization in (8.7) become

$$\boldsymbol{D} = \boldsymbol{\Phi}\,\boldsymbol{\Sigma}\,\boldsymbol{\Psi}^T \Longleftrightarrow \boldsymbol{D}[\mathbf{i},k] = \sum_{r=0}^{R-1} \sigma_r \boldsymbol{\phi}_m[\mathbf{i}]\,\boldsymbol{\psi}_n[k] \Longleftrightarrow \boldsymbol{D} = \sum_{r=0}^{R-1} \sigma_r \boldsymbol{\phi}_r\,\boldsymbol{\psi}_r^T\,. \tag{8.11}$$

Each of the terms $\sigma_r \boldsymbol{\phi}_r\,\boldsymbol{\psi}_r^T$ produces a matrix of unitary rank, which we call *mode*. Note that the rank of a (full rank) rectangular matrix is $rank(\boldsymbol{D}) = min(n_s, n_t)$, but in general the number of relevant modal contributions is not associated to the rank.[13] We obtain approximations of the data set by zeroing some of the entries along the diagonal of $\boldsymbol{\Sigma}$. As in Section 8.2.3 we use tildes to denote approximations. Moreover, since infinite decompositions could be obtained by dividing the diagonal entries $\sigma_r = \boldsymbol{\Sigma}[r,r]$ by the length of the corresponding basis matrices, we here assume that $||\boldsymbol{\phi}_r|| = ||\boldsymbol{\psi}_r|| = 1$ for all $r \in [0, \ldots, R-1]$.

The reader has undoubtedly recognized that the factorization in (8.11) has the same structure of the singular value decomposition (SVD) introduced in Chapter 6. The SVD is a very special case of (8.11), but there are *infinite* other possibilities, depending on the choice of the basis. Nevertheless, such a choice has an important constraint that must be discussed here.

Assume that the spatial structures $\boldsymbol{\Phi}$ are given. Projecting on the left gives

$$(\boldsymbol{\Phi}^\dagger\boldsymbol{\Phi})^{-1}\boldsymbol{\Phi}^\dagger\boldsymbol{D} = \boldsymbol{\Sigma}\,\boldsymbol{\Psi}^T = \boldsymbol{D}_\phi = \begin{bmatrix} \sigma_1\psi_1[1] & \sigma_1\psi_1[2] & \ldots & \sigma_1\psi_1[n_t] \\ \sigma_2\psi_2[1] & \sigma_2\psi_2[2] & \ldots & \sigma_2\psi_2[n_t] \\ \vdots & \vdots & \vdots & \vdots \\ \sigma_R\psi_R[1] & \sigma_R\psi_R[2] & \ldots & \sigma_R\psi_R[n_t] \end{bmatrix}. \tag{8.12}$$

[12] That is, most of the entries in $\boldsymbol{D}_\phi$, $\boldsymbol{D}_\psi$ or $\boldsymbol{D}_{\phi\psi}$ are zero or almost zero.

[13] For example, we might add five modes and obtain a matrix of rank four. The number of modes equals the rank of the matrix only if these modes are orthogonal in both space and time, that is, only if these are POD modes.

This is equivalent to (8.4): this is a transform along the columns of $\boldsymbol{D}$. The transformed snapshot, that is, the set of coefficients in the linear combination of $\boldsymbol{\phi}$'s, changes from snapshot to snapshot – they evolve in time. The key difference with respect to a 2D transform is that the evolution of the structure in the basis element $\boldsymbol{\phi}_r$ only depends on the corresponding temporal structure $\boldsymbol{\psi}_r$. Moreover, notice that $\sigma_r = ||\sigma_r \psi_r||$, since $||\psi_r|| = 1$. This explains why the amplitudes are real quantities by construction: they represent the norm of the rth column $\boldsymbol{D}_\phi$, which we denote as $\boldsymbol{D}_\phi[:,r]$ following a PYTHON notation.

The same observation holds if the temporal structures $\boldsymbol{\Psi}$ are given. In this case, projecting on the right gives

$$\boldsymbol{D}\overline{\boldsymbol{\Psi}}(\boldsymbol{\Psi}^\dagger\boldsymbol{\Psi})^{-1} = \boldsymbol{\Phi\Sigma} = \boldsymbol{D}_\psi = \begin{bmatrix} \sigma_1\phi_1[1] & \sigma_2\phi_1[1] & \dots & \sigma_R\phi_R[1] \\ \sigma_1\phi_2[2] & \sigma_2\phi_2[2] & \dots & \sigma_R\phi_R[2] \\ \vdots & \vdots & \vdots & \vdots \\ \sigma_1\phi_R[n_s] & \sigma_2\phi_R[n_s] & \dots & \sigma_R\phi_R[n_s] \end{bmatrix}. \tag{8.13}$$

This is equivalent to (8.5): this is a transform along the rows of $\boldsymbol{D}$. The set of coefficients in the linear combination of $\boldsymbol{\psi}$'s are spatially distributed according to the corresponding spatial structures $\boldsymbol{\phi}$'s. As before, we see that the amplitudes of the modes can also be computed as $\sigma_r = ||\sigma_R\boldsymbol{\phi}_r|| = ||\boldsymbol{D}_\psi[:,r]||$.

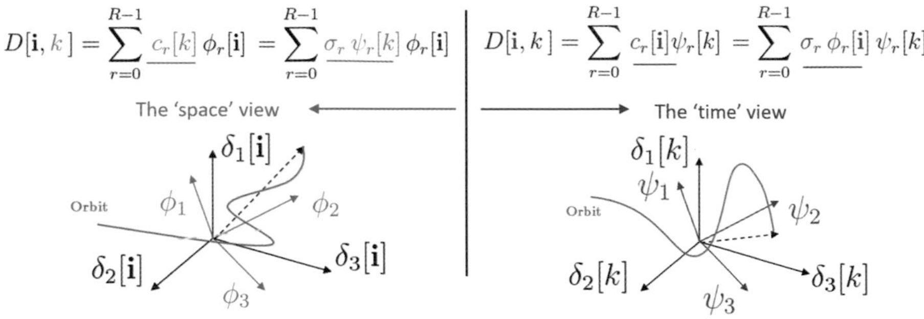

Figure 8.1 The "space view" and the "time-view" of modal analysis. In the "space view," every snapshot $\mathbf{d}_k[\mathbf{i}]$ is a linear combination of basis elements $\boldsymbol{\phi}$'s. The coefficients of this combination evolve in time according to the associated $\boldsymbol{\psi}$'s. In the "time-view," every temporal evolution $\mathbf{d}_\mathbf{i}[k]$ is a linear combination of basis elements $\boldsymbol{\psi}$'s. The coefficients of this combination are spatially distributed according to the associated $\boldsymbol{\phi}$'s.

The space-time symmetry is further elucidated in Figure 8.1, considering also the cases in which $\boldsymbol{\Phi} := \mathbf{I}/\sqrt{n_s}$ or $\boldsymbol{\Psi} := \mathbf{I}/\sqrt{n_t}$. In the "space view" on the left, we follow the data in time from a specific spatial basis. If this basis is the set of impulses $\delta_k[\mathbf{i}]$, the temporal evolutions along each element of the basis is given by the time evolution $\mathbf{d}_\mathbf{i}[k]$. In the "time view" on the right, we analyze the spatial distribution from a specific temporal basis. If this basis is the set of impulses $\delta_\mathbf{i}[k]$, the spatial distribution of each member of the basis (i.e., every instant) is given by the snapshot $\mathbf{d}_k[\mathbf{i}]$ itself.

With these views in mind, the reader should understand the most important observations of this section: *in the decomposition in* (8.11) *it is not possible to impose both basis matrices* $\boldsymbol{\Phi}$ *and* $\boldsymbol{\Psi}$*– given one of the two, the other is univocally determined.* This observation also leads to the formulations of two general algorithms to compute this factorization given one of the two bases. These are listed below:

Algorithm 1: Projection from $\boldsymbol{\Phi}$

Input: $\boldsymbol{D} \in \mathbb{R}^{n_s \times n_t}$ and $\boldsymbol{\Phi} \in \mathbb{R}^{n_s \times R}$

Output: $\boldsymbol{\Sigma} \in \mathbb{R}^{R \times R}$ and $\boldsymbol{\Psi} \in \mathbb{R}^{n_t \times R}$

Project $\tilde{\boldsymbol{D}}_\phi = (\boldsymbol{\Phi}^\dagger \boldsymbol{\Phi})^{-1} \boldsymbol{\Phi}^\dagger \boldsymbol{D}$ (eq. 12)

From (4.12) it is $\boldsymbol{D}_\phi = \boldsymbol{\Sigma} \boldsymbol{\Psi}^T$, hence :

for $r <= R$, r=r++, $\sigma_r = ||\boldsymbol{D}_\phi[r,:]||_2$; $\psi_r = \boldsymbol{D}_\phi[r,:]/\sigma_r$

Assembly $\boldsymbol{\Sigma} = \text{diag}(\sigma_0, \sigma_1, \dots \sigma_R) \in \mathbb{R}^{R \times R}$ and $\boldsymbol{\Psi} = [\psi_0, \psi_1, \dots \psi_R] \in \mathbb{C}^{n_t \times R}$

Algorithm 2: Projection from $\boldsymbol{\Psi}$

Input: $\boldsymbol{D} \in \mathbb{R}^{n_s \times n_t}$ and $\boldsymbol{\Psi} \in \mathbb{R}^{n_t \times R}$

Output: $\boldsymbol{\Sigma} \in \mathbb{R}^{R \times R}$ and $\boldsymbol{\Phi} \in \mathbb{R}^{n_s \times R}$

Project $\tilde{\boldsymbol{D}}_\psi = \boldsymbol{D} \overline{\boldsymbol{\Psi}} (\boldsymbol{\Psi}^\dagger \boldsymbol{\Psi})^{-1} = \boldsymbol{\Phi} \boldsymbol{\Sigma}$ (eq. 13)

From (4.12) it is $\boldsymbol{D}_\psi = \boldsymbol{\Phi} \boldsymbol{\Sigma}$, hence :

for $r <= R$, r=r++, $\sigma_r = ||\boldsymbol{D}_\psi[r,:]||_2$; $\phi_r = \boldsymbol{D}_\phi[r,:]/\sigma_r$

Assembly $\boldsymbol{\Sigma} = \text{diag}(\sigma_0, \sigma_1, \dots \sigma_R) \in \mathbb{R}^{R \times R}$ and $\boldsymbol{\Phi} = [\phi_0, \phi_1, \dots \phi_R] \in \mathbb{C}^{n_s \times R}$

Finally, note that these algorithms are the most general ones to complete a decomposition from its spatial or its temporal structures. Such a level of generality highlights the common structure but also leads to the least efficient approach: every decomposition offers valuable shortcuts that we discuss in Section 8.3.

8.2.5 Amplitudes and Energies

The normalization of a data-driven decomposition cannot be performed beforehand, because the basis is initially unknown. We here analyze the consequences of imposing a unitary norm for the bases vectors $\boldsymbol{\phi}$ and $\boldsymbol{\psi}$ and the link of this operation with the continuous domain. Using the continuous inner product (see Chapter 4), the energies of continuous basis element in space and time are

$$||\phi_r(\mathbf{x})||^2 = \frac{1}{\Omega} \int_\Omega \phi_r(\mathbf{x})\, \overline{\phi}_r(\mathbf{x})\, d\Omega \quad \text{and} \quad ||\psi_r(t)||^2 = \frac{1}{T} \int_T \psi_r(t) \overline{\psi}_r(t)\, dT \,, \tag{8.14}$$

having taken the spacial basis in a domain[14] Ω, and the temporal basis in a domain T. If a Cartesian and uniform mesh is used in space and time, and if a right end point integration scheme is considered, these are approximated by their discrete equivalent

[14] This could be an area in a 2D domain or a volume in a 3D domain. In this sense, the bases in the continuous domain are always considered in terms of "densities."

$$||\phi_r(\mathbf{x})||^2 = \frac{1}{n_s}||\boldsymbol{\phi}_r||^2 = \frac{1}{n_s}\boldsymbol{\phi}_r^\dagger\boldsymbol{\phi} \quad \text{and} \quad ||\psi_r(t)||^2 = \frac{1}{n_t}||\boldsymbol{\psi}_r||^2 = \frac{1}{n_t}\boldsymbol{\psi}_r^\dagger\boldsymbol{\psi}\,, \tag{8.15}$$

where the norms are simply Euclidean norms $||\boldsymbol{a}|| = \boldsymbol{a}^\dagger\boldsymbol{a}$. For nonuniform meshes or more sophisticated integration schemes, weighted inner products must be introduced (see Chapter 6).

Consider now an approximation of the data using only one mode. This is a matrix $\tilde{\boldsymbol{D}}[\mathbf{i},k] = \sigma_r\phi_r[\mathbf{i}]\psi_r[k]$ of unitary rank. The *total* energy associated to this mode, assuming that this is the discrete version of a continuous space-time evolution, is

$$\mathcal{E}\{\tilde{D}[\mathbf{x},t]\} = \frac{1}{\Omega T}\int_T\int_\Omega \tilde{D}^2(\mathbf{x},t)d\Omega dt \approx \frac{1}{n_t n_s}\sum_{\mathbf{i}=0}^{n_s}\sum_{k=0}^{n_t}\tilde{\boldsymbol{D}}^2[\mathbf{i},k]\,. \tag{8.16}$$

The summation on the right is the Frobenious norm,[15] hence we have

$$\mathcal{E}\{\tilde{D}[\mathbf{x},t]\} \approx \frac{1}{n_t n_s}\mathrm{tr}\{\tilde{\boldsymbol{D}}^\dagger\tilde{\boldsymbol{D}}\} = \frac{\sigma_r^2}{n_t n_s}\mathrm{tr}\{\boldsymbol{\psi}_r\boldsymbol{\phi}_r^\dagger\boldsymbol{\phi}_r\boldsymbol{\psi}_r^\dagger\} = \frac{\sigma_r^2}{n_t n_s}. \tag{8.17}$$

Observe that $\boldsymbol{\phi}_r^\dagger\boldsymbol{\phi}_r = 1$ because of the unitary length of the spatial structures and $\mathrm{tr}\{\boldsymbol{\psi}_r\boldsymbol{\psi}_r^T\} = \boldsymbol{\psi}_r^\dagger\boldsymbol{\psi}_r = 1$ for the same constraint on the temporal structures. *We thus conclude that the square of the normalized amplitude $\hat{\sigma}_r = \sigma_r/\sqrt{n_t n_s}$ is mesh-independent approximation of the energy contribution of each mode.*

This statement is extremely important but requires great care: *in general, summing the energy contribution of each mode does not equal the energy in the entire data.* This occurs only if *both the temporal and the spatial structures are orthonormal*, that is, only for the POD (see Chapter 6). For any other decomposition, the energy of an approximation with $\tilde{r} < R$ modes is

$$\mathcal{E}\{\tilde{D}[\mathbf{x},t]\} = \frac{1}{n_s n_t}\mathrm{tr}\left\{\tilde{\boldsymbol{\Psi}}\,\tilde{\boldsymbol{\Sigma}}\,\tilde{\boldsymbol{\Phi}}^\dagger\;\tilde{\boldsymbol{\Phi}}\,\tilde{\boldsymbol{\Sigma}}\,\tilde{\boldsymbol{\Psi}}^\dagger\right\} = \frac{1}{n_s n_t}\mathrm{tr}\left\{\tilde{\boldsymbol{\Phi}}\tilde{\boldsymbol{\Sigma}}\tilde{\boldsymbol{\Psi}}^\dagger\;\tilde{\boldsymbol{\Psi}}\,\tilde{\boldsymbol{\Sigma}}\,\tilde{\boldsymbol{\Phi}}^\dagger\right\}. \tag{8.18}$$

The convergence of a decomposition indicates how efficiently this can represent the data set, that is, how quickly the norm of the approximation converges to the norm of the data as $\tilde{r} \to R$. Convergence can be either measured in terms of l^2 norm or in terms of Frobenius norm. Using the l^2 norm, the convergence can be measured as

$$E(\tilde{r}) = \frac{||\boldsymbol{D} - \tilde{\boldsymbol{D}}(\tilde{r})||}{||\boldsymbol{D}||} = \frac{||\boldsymbol{D} - \tilde{\boldsymbol{\Phi}}\tilde{\boldsymbol{\Sigma}}\tilde{\boldsymbol{\Psi}}||_2}{||\boldsymbol{D}||}. \tag{8.19}$$

8.3 Common Decompositions

We here describe how common decompositions fit in the factorization framework. We adopt a "time-view," that is, we identify every decomposition from its temporal structure $\boldsymbol{\Psi}$ and complete the factorization following Algorithm 2 from Section 8.2.

[15] Recall: the Frobenious norm of a matrix $\boldsymbol{A} \in \mathbb{C}^{n_s \times n_t}$ can be written as $||\boldsymbol{A}||_F = \mathrm{tr}\{\boldsymbol{A}^\dagger\boldsymbol{A}\} = \mathrm{tr}\{\boldsymbol{A}\boldsymbol{A}^\dagger\}$.

8.3.1 The Delta Decomposition

This decomposition is useless, but conceptually important. Take the basis of impulses $\mathbf{\Psi}_\delta = \mathbf{I}/\sqrt{n_t}$ as the temporal basis. This basis is orthonormal and we distinguish this decomposition from the others using the subscript δ.

This decomposition generalizes the convolution representation of a signal introduced in Chapter 4. The general temporal structure is $\psi_r[k] = \boldsymbol{\delta}_r[k] = \delta[k-r]$. The decomposition of the snapshot $\boldsymbol{d}_k[\mathbf{i}]$ becomes

$$\boldsymbol{d}_k[\mathbf{i}] = \sum_{r=0}^{R-1} \sigma_{\delta\, r}\boldsymbol{\phi}_{\delta r}[\mathbf{i}]\psi_{\delta r}[k] = \sum_{r=0}^{R-1} \sigma_{\delta\, r}\boldsymbol{\phi}_{\delta r}[\mathbf{i}]\frac{\delta[k-r]}{\sqrt{n_t}} = \frac{\sigma_{\delta\, k}\boldsymbol{\phi}_{\delta k}[\mathbf{i}]}{\sqrt{n_t}}. \tag{8.20}$$

Therefore, the spatial basis corresponds to the normalized snapshots:

$$\boldsymbol{\phi}_{\delta k}[\mathbf{i}] = \frac{\boldsymbol{d}_k[\mathbf{i}]}{||\boldsymbol{d}_k[\mathbf{i}]||} \quad \text{and} \quad \sigma_{\delta\, k} = ||\boldsymbol{d}_k[\mathbf{i}]||\sqrt{n_t} = ||\boldsymbol{d}_k(\mathbf{x_i})||\sqrt{n_t\, n_s}. \tag{8.21}$$

This decomposition yields the highest possible temporal localization: large amplitudes $\sigma_{\delta r}$ correspond to a snapshot with large energy content. A strong amplitude decay indicates that some snapshots have much lower energy content than others. In a data set in which a strong impulsive event occurs, the POD temporal structure may approach a delta-like form: when this happens, the associated spatial structure tends to the corresponding normalized snapshot. A similar limit is obtained in a DWT (see Chapter 5) if the finest scale is pushed to its limit. In a Haar wavelet transform, for example, the finest scale corresponds to elements of the form $\boldsymbol{w}_r = (\boldsymbol{\delta}_r + \boldsymbol{\delta}_{r+1})/\sqrt{2}$. The spatial structures associated to these are normalized averages of the snapshots $\mathbf{d}_k$ and $\mathbf{d}_{k+1}$. Similarly, the spatial structure associated with the basis of the form $\boldsymbol{\psi}_r = 1/\sqrt{n_t}$ is the normalized time average of the data.

8.3.2 The Discrete Fourier Transform

We here refer to the DFT of the data set in time, that is, this is equivalent to performing the DFT of $\boldsymbol{D}$ row-wise as in (8.5), followed by a normalization to compute the amplitudes. In the DFT we set $\mathbf{\Psi} = \mathbf{\Psi}_{\mathcal{F}}$ and we use the subscript $\mathcal{F}$ to distinguish it from the others. This matrix is defined in (8.8). This basis is orthonormal and symmetric. Using (8.13) and the general Algorithm 2 gives

$$\boldsymbol{D}\overline{\mathbf{\Psi}}_{\mathcal{F}} = \mathbf{\Phi}_{\mathcal{F}}\mathbf{\Sigma}_{\mathcal{F}} = \widehat{\boldsymbol{D}} \Longleftrightarrow \sigma_{\mathcal{F}r} = ||\boldsymbol{\phi}_r\sigma_r|| = ||\boldsymbol{D}\overline{\boldsymbol{\psi}}_r||, \quad \text{that is, } \mathbf{\Phi}_{\mathcal{F}} = \widehat{\boldsymbol{D}}\mathbf{\Sigma}_{\mathcal{F}}^{-1}. \tag{8.22}$$

The first step on the left is the row-wise Fourier transform of $\boldsymbol{D}$, denoted as $\widehat{\boldsymbol{D}}$. The second step is the normalization, which can be computed along the rows of $\widehat{\boldsymbol{D}}$, and in the last step we see that the spatial structures correspond to the normalized distribution of a given entry of the Fourier spectra. We thus see that each entry of the spatial basis tells how much the data at a specific location correlates with the harmonic associated with the discrete frequency r. The computation of the digital frequencies from the sampling frequency of the data is described in Chapter 4. In the same section, it was recalled that the DFT implicitly assumes that the data is periodic of period $T = n_t\Delta t$.

From a computational point of view, the multiplication by $\mathbf{\Psi}_{\mathcal{F}}$ can be efficiently carried out using the FFT algorithm. Note that FFT routines from MATLAB or PYTHON define the Fourier basis matrix as the conjugate of (8.8). This decomposition is, by definition, the one with highest frequency resolution. As described in Chapter 6, this comes at the price of no time localization, since this temporal basis is of infinite duration. An important property of the Fourier basis is that its elements are eigenvectors of circulant matrices (see Gray 2005).

This property links the DFT and the POD in stationary data, and it is worth a brief discussion. An intuitive derivation follows from the notions introduced in Chapter 4, specifically the convolution theorem in the discrete domain and the link between impulse response of an LTI system and frequency transfer functions.

Hence, let $\mathbf{u} := u[k] \in \mathbb{R}^{n_t \times 1}$ with $k \in [0, n_t - 1]$ be a discrete signal, represented by a column vector of appropriate size, which is introduced as an input to a SISO system. Let $\mathbf{y} := y[k] \in \mathbb{R}^{n_t \times 1}$ be the corresponding output. Let $h[k]$ be the *impulse response* of the LTI system, assuming that this is of finite duration and that it has at most n_t nonzero entries. The input–output link in the time domain is provided by the convolution sum (see Chapter 4). If both input and output are periodic, the convolution can be written using a circulant matric $\boldsymbol{C}_h$ and a matrix multiplication:

$$y[k] = \sum_{l=-\infty}^{l=\infty} u[k]h[k-l] \rightarrow \mathbf{y} = \boldsymbol{C}_h \mathbf{u},$$

$$\begin{bmatrix} y[0] \\ y[1] \\ y[2] \\ \vdots \\ y[n-1] \end{bmatrix} = \begin{bmatrix} h[0] & h[n-1] & h[n_t-2] & \dots & h[1] \\ h[1] & h[0] & h[n_t-1] & \dots & h[2] \\ h[2] & h[1] & h[0] & \ddots & \vdots \\ \vdots & h[2] & h[1] & \ddots & h[n-1] \\ h[n-1] & \dots & h[2] & h[1] & h[0] \end{bmatrix} \begin{bmatrix} u[0] \\ u[1] \\ u[2] \\ \vdots \\ u[n-1] \end{bmatrix}. \quad (8.23)$$

On the other hand, we know from the convolution theorem that the same can be achieved in the frequency domain using an entry-by-entry multiplication of the Fourier transform of the signals and the one of the impulse response. In terms of matrix multiplications, these are, respectively,

$$\widehat{\mathbf{u}} = \overline{\mathbf{\Psi}}_{\mathcal{F}}\mathbf{u}, \quad \widehat{\mathbf{y}} = \overline{\mathbf{\Psi}}_{\mathcal{F}}\mathbf{y}, \text{and } \widehat{\mathbf{H}} = \overline{\mathbf{\Psi}}_{\mathcal{F}}\mathbf{h}, \quad (8.24)$$

where $\mathbf{h} := h[k] \in \mathbb{R}^{n_t \times 1}$ is the impulse response arranged as a column vector. To perform the entry-by-entry matrix multiplication, we define a diagonal matrix $\boldsymbol{H} = \text{diag}(\widehat{\mathbf{H}})$. Hence the operation in the frequency domain becomes $\widehat{\mathbf{y}} = \boldsymbol{H}\,\widehat{\mathbf{u}}$. Introducing the Fourier transforms in (8.24) and moving back to the time domain, we get

$$\widehat{\mathbf{y}} = \boldsymbol{H}\,\widehat{\mathbf{u}} \rightarrow \overline{\mathbf{\Psi}}_{\mathcal{F}}\mathbf{y} = \boldsymbol{H}\,\overline{\mathbf{\Psi}}_{\mathcal{F}}\mathbf{u} \rightarrow \mathbf{y} = \mathbf{\Psi}_{\mathcal{F}}\boldsymbol{H}\,\overline{\mathbf{\Psi}}_{\mathcal{F}}\mathbf{u} \quad \text{hence } \boldsymbol{C}_h = \mathbf{\Psi}_{\mathcal{F}}\boldsymbol{H}\,\overline{\mathbf{\Psi}}_{\mathcal{F}}. \quad (8.25)$$

The last equation on the right is obtained by direct comparison with (8.24) and is extremely important: this is an eigenvalue decomposition. Each $\widehat{\mathbf{H}}[n]$ entry of the frequency response vector is an eigenvalue of $\boldsymbol{C}_h$ and the Fourier basis element $\psi_{\mathcal{F}n}$ (corresponding to the frequency f_n) is the associated eigenvector.

Of great interest is the case in which the impulse response is even, that is, $h[k] = h[-k]$. In this case the circulant matrix $\boldsymbol{C}_h$ becomes also symmetric and its eigenvalues are real: this is the case of zero-phase (noncausal) filters. In this special case, also the eigenvectors can be taken as real harmonics: they could be either sinusoidals or cosinusoidals, that is, the basis of the discrete sine transform (DST) or the discrete cosine transform (DCT) (see Strang 2007). In a completely different context, this eigenvalue decomposition also appears in the POD of a special class of signals. Because of its importance, the reader is encouraged to pause and practice with the following exercise.

Exercise 1: The DFT and the Diagonalization of Circulant Matrices

Consider the discrete signal **u** from Exercise 5 in Chapter 4. First, compute the DFT via matrix multiplication and compare that result with the one from numpy's FFT routine. Then prepare the impulse response $h[k]$ of a low-pass filter of order $N = 211$ using the windowing method and a Hamming window. Then implement this filter using the matrix multiplication form of the convolution in (8.23) and the matrix multiplication in the frequency domain in (8.25). Compare the results with any of the previous implementations from Chapter 4. Finally, construct the convolution matrix from (8.25) and test the validity of the eigenvalue decomposition previously introduced.

8.3.3 The Proper Orthogonal Decomposition

The POD is introduced in Chapter 6; we here only focus on its special properties in relation to Algorithm 2.

The temporal structures of the POD are eigenvectors of the temporal correlation matrix, which in the notation of this chapter reads

$$\boldsymbol{K} = \boldsymbol{D}^{\dagger}\boldsymbol{D} = \boldsymbol{\Psi}_{\mathcal{P}}\boldsymbol{\Lambda}\boldsymbol{\Psi}_{\mathcal{P}}^{T}, \quad \text{that is,} \quad \boldsymbol{K}[i,j] = \sum_{r=0}^{n_t-1} \sigma_{\mathcal{P}}^2\, \psi_{\mathcal{P}}[i]\psi_{\mathcal{P}}[j]. \tag{8.26}$$

The subscript $\mathcal{P}$ is used to distinguish the POD. The notation on the right is based on an outer product representation of the eigenvalue decomposition of symmetric matrices; this will be useful in Section 8.4. The first key feature of the decomposition is that the eigenvalues of $\boldsymbol{K}$ are linked to the POD amplitudes as $\boldsymbol{\Lambda} = \boldsymbol{\Sigma}_{\mathcal{P}}^2$. Hence the diagonalization in (8.26) *also provides the POD amplitudes* and the normalization in line 4 of Algorithm 2 is not needed. Introducing $\boldsymbol{\Psi}_{\mathcal{P}}$ in this algorithm gives the Sirovinch formulation[16] described in Chapter 6. Observe that since the POD amplitudes are the singular values of the data set matrix, it is possible to compute the convergence in (8.19) without computing norms:

[16] Actually a much less efficient version: the projection step in line 1 of the algorithm could simply be $\tilde{\boldsymbol{D}} = \boldsymbol{D}\boldsymbol{\Psi}_{\mathcal{P}}$.

$$E(\tilde{r}) = \frac{||\boldsymbol{D} - \tilde{\boldsymbol{D}}(\tilde{r})||_2}{||\boldsymbol{D}||_2} = \sqrt{\frac{\sum_{r=\tilde{r}}^{R-1} \sigma_{\mathcal{P}r}^2}{\sum_{r=0}^{R-1} \sigma_{\mathcal{P}r}^2}}. \tag{8.27}$$

For the following discussion, two observations are of interest. The first is that introducing the POD factorization (i.e., the SVD) in (8.26), we have

$$\boldsymbol{K} = \boldsymbol{\Psi}_{\mathcal{P}}\boldsymbol{\Sigma}_{\mathcal{P}}^{-1}\boldsymbol{\Phi}_{\mathcal{P}}^T\boldsymbol{\Phi}_{\mathcal{P}}\boldsymbol{\Sigma}_{\mathcal{P}}^{-1}\boldsymbol{\Psi}_{\mathcal{P}}^T \quad \text{i.e.,} \quad \boldsymbol{\Lambda} = \boldsymbol{\Sigma}_{\mathcal{P}}^{-1}\boldsymbol{\Phi}_{\mathcal{P}}^T\boldsymbol{\Phi}_{\mathcal{P}}\boldsymbol{\Sigma}_{\mathcal{P}}^{-1}. \tag{8.28}$$

The last step arises from a direct comparison with (8.26). Because $\boldsymbol{\Lambda}$ is diagonal, we see that we must have $\boldsymbol{\Phi}_{\mathcal{P}}^T\boldsymbol{\Phi}_{\mathcal{P}} = \mathbf{I}$, that is, the spatial structures *are also orthonormal.* We know this from Chapter 6, where it was shown that these are eigenvectors of the spatial correlation matrix. The key observation is that the reverse must also be true: *every decomposition that has orthonormal temporal and spatial structure is a POD.* This brings us to the second observation on the uniqueness of the POD. In a data set that leads to modes of equal energetic importance, the amplitudes of the associated POD modes tend to be equal. This means repeated eigenvalues of $\boldsymbol{K}$ and thus nonunique POD. In the extreme case of a purely random data set, it is easy to see that the POD modes are all equal (see Mendez et al. 2017), and there are infinite possible PODs. The impact of noise in a POD decomposition is further discussed in Chapter 9.

Finally, observe that the POD is based on error minimization (or, equivalently, amplitude maximization) and has no constraints on the frequency content in its temporal structures $\boldsymbol{\Psi}_{\mathcal{P}}$. However, a special case occurs in an ideally stationary process. In such a process, the temporal correlations are invariant with respect to time delays and solely depend on the time lag considered in the correlation. Hence the correlation $\boldsymbol{K}[1,4] = \mathbf{d}_1^T\,\mathbf{d}_4$ is equal to $\boldsymbol{K}[4,7] = \mathbf{d}_4^T\,\mathbf{d}_7$ or $\boldsymbol{K}[11,14] = \mathbf{d}_{11}^T\,\mathbf{d}_{14}$, for example.

In other words, the temporal correlation matrix $\boldsymbol{K}$ becomes circulant, like the matrix $\mathbf{C}_h$ in (8.23) for the convolution. Therefore, its eigenvectors are harmonics: we conclude that *the POD of an ideally stationary data set is either a DCT or a DST.*[17] This property is the essence of the spectral POD proposed by Sieber et al. (2016), which introduces an ingenious FIR filter along the diagonal of $\boldsymbol{K}$ to reach a compromise between the energy optimality of the POD modes and the spectral purity of Fourier modes. Depending on the strength of this filter, the SPOD offers an important bridge between the two decompositions.

8.3.4 The Dynamic Mode Decomposition

The DMD is introduced in Chapter 7 and we here focus on its link with the Algorithm 2. The DMD aims at fitting a linear dynamical system to the data, writing every snapshot as

[17] Depending on whether the temporal average has been removed. The mean is accounted for in a DCT but not in a DST.

$$\mathbf{d}_{\mathbf{i}}[k+1] = \sum_{r=0}^{n_s-1} a_r \psi_{\mathcal{D}r}[\mathbf{i}] e^{-p_r t_k} = \sum_{r=0}^{n_s-1} a_r \psi_{\mathcal{D}r}[\mathbf{i}] \lambda_r^{k-1} \, . \tag{8.29}$$

The subscript $\mathcal{D}$ is used to distinguish the DMD. In Chapter 4, we have seen that the complex exponentials are eigenfunctions of linear dynamical systems; Chapter 10 describes their state-space representation. Observe that the DMD is a sort of Z-transform in which the basis elements only include the *poles* of the systems (see Chapter 4), that is, the eigenvalues of the propagating matrix $\boldsymbol{A}$ in the state-space representation in Chapter 10. Such decomposition is natural for a homogeneous linear system, while the extension of the DMD for the forced system is proposed by Proctor et al. (2016). This extension makes the DMD an extremely powerful system identification tool that can be combined with the linear control methods introduced in Chapter 10.

Many algorithms have been developed to compute the eigenvalues (λ_r's) from the data set (see Chapter 7). Once these are computed, the matrix containing the temporal structures of the DMD can be constructed,

$$\boldsymbol{\Psi}_{\mathcal{D}} = \begin{bmatrix} 1 & 1 & 1 & \dots & 1 \\ \lambda_1 & \lambda_2 & \lambda_3 & \dots & \lambda_{n_t} \\ \lambda_1^2 & \lambda_2^2 & \lambda_3^2 & \dots & \lambda_{n_t}^2 \\ \vdots & \vdots & \ddots & \vdots & \\ \lambda_1^{n_t} & \lambda_2^{n_t} & \lambda_3^{n_t} & \dots & \lambda_{n_t}^{n_t} \end{bmatrix} \in \mathbb{C}^{n_t \times n_t} \, , \tag{8.30}$$

and the decomposition completed following Algorithm 2.

From a dynamical system perspective, we know from Chapter 4 that a system is stable if all its poles are within the unit circle, hence if all the λ's have modulus $|\lambda| \leq 1$. From an algebraic point of view, we note that $\boldsymbol{\Psi}_{\mathcal{D}}$ differs from the temporal structures of the other decompositions in two important ways. First, it is generally not orthonormal: the full projection must be considered in line 1 of Algorithm 2. Second, its inverse might not exist: convergence is not guaranteed.

Different communities have different ways of dealing with this lack of convergence. In the fluid dynamics community, more advanced DMD algorithms, such as the sparsity promoting DMD (Jovanović et al. 2014) or the optimized DMD (Chen et al. 2012), have been developed to enforce that all the λ's have unitary modulus. This is done by introducing an optimization problem in which the cost function is defined on the error minimization of the DMD approximation. Consequently, vanishing or diverging decompositions are penalized. Observe that having all the modes *on the unit circle does not necessarily imply that these are DFT modes*: the DMD does not impose any orthogonality condition, which would force the modes to have frequencies that are multiples of a fundamental one. If orthogonality is enforced as a constraint to the optimization, then the only degree of freedom distinguishing DMD and DFT is in the choice of the fundamental tone: while this is $T = n_t \Delta t$ for the DFT, the DMD can choose different values and bypass problems like spectral leakage or windowing (Harris 1978).

In the climatology community in which the DMD is known as POP (see Hasselmann 1988, von Storch & Xu 1990) or LIM (see Penland & Magorian 1993, Penland 1996), the lack of convergence is seen as a natural limit of the linearization process. Accordingly, the results of these decompositions are usually presented in terms of *e-folding time*, which is the time interval within which the temporal structures decrease or increase by a factor e. In time-series analysis, modes with a short e-folding time are modes with shorter predictive capabilities. Variants of the POP/LIM have been presented in the fluid dynamics community under the name of oscillating pattern decomposition (OPD, see Uruba 2012). The link between DMD, POP, and LIM is also discussed by Schmid (2010) and Tu et al. (2014).

Exercise 2: Eigenfunction Expansion and Modal Analysis

Consider a 2D pulsating Poiseuille flow within two infinite plates at a distance $2b$ over a length L, as shown in the left panel in the following figure. Assume that the flow is fully developed ($\partial_x u = 0$) and incompressible ($\partial_x u + \partial_y v = 0$). It is easy to show that the pressure gradient must thus be uniform ($\partial_x p =$ const). If a sinusoidal pressure pulsation is introduced at the inlet, the Navier–Stokes equation governing the velocity profile in the streamwise direction is

$$\rho \partial_t u = -\frac{\Delta p(t)}{L} + \mu \partial_{yy} u \quad \text{with} \quad \Delta p(t) = p_M + p_A \cos(\omega t), \tag{8.31}$$

where ρ and μ are the density and the dynamic viscosity of the fluid. Introducing the reference quantities

$$[y] = b \quad [x] = L \quad [p] = p_M \quad [t] = 1/\omega, \tag{8.32}$$

the problem can be written in dimensionless form as follows:

$$\begin{cases} \partial_{\hat{t}} \hat{u} - \dfrac{1}{\mathcal{W}^2} \partial_{\hat{y}\hat{y}} \hat{u} = -\dfrac{\hat{p}_A}{\mathcal{W}^2} \cos(\hat{t}), & -1 < \hat{y} < 1,\, t \geq 0, \\ \hat{u}(-1,\hat{t}) = \hat{u}(1,\hat{t}) = 0, & t \geq 0, \\ \hat{u}(\hat{y},0) = u_0(\hat{y}), & -1 < \hat{y} < 1\,. \end{cases} \tag{8.33}$$

Here the hat denotes dimensionless variables, that is, $\hat{a} = a/[a]$ is the dimensionless scaling of a variable a with respect to the reference quantity $[a]$. The dimensionless number controlling the response of the profile is the *Womersley* number $\mathcal{W} = b\sqrt{\omega \rho / \mu}$.

Equation (8.33) is a linear parabolic PDE with homogeneous Dirichlet boundary conditions on both sides of the domain. A variable separated solution can be found by expanding the velocity profile in terms of eigenfunctions of the Laplacian operator. Given the boundary conditions in (8.32), these are cosine functions. The solution via EF expansion (see Mendez & J.M-Buchlin 2016) is

$$\hat{u}(\hat{y},\hat{t}) = \sum_{n=1}^{\infty} \phi_{\mathcal{E}n}(\hat{y})\, \sigma_{\mathcal{E}n}\, \psi_{\mathcal{E}n}(\hat{t}), \tag{8.34}$$

with

$$\begin{aligned} \phi_{\mathcal{E}n} &= \cos\left[(2n-1)\frac{\pi}{2}\hat{y}\right], \\ \sigma_{\mathcal{E}n} &= \frac{16\hat{p}_A}{(2n-1)\pi\sqrt{16\mathcal{W}^4+(2n-1)^4\pi^4}}, \\ \psi_{\mathcal{E}n} &= (-1)^n \cos\left[\hat{t} - \tan^{-1}\left(\frac{\mathcal{W}}{[(2n-1)\pi]}\right)^2\right]. \end{aligned} \tag{8.35}$$

Note that the spatial structures are orthogonal but the temporal ones are not. The amplitudes of the modes decay as $\propto 1/(2n-1)^3$ when $\mathcal{W} \to 0$ and as $\propto 1/(2n-1)$ when $\mathcal{W} \to \infty$. Finally, note that all the amplitudes tend to $\sigma_{\mathcal{E}n} \to 0$ as $\mathcal{W} \to \infty$: as the frequency of the perturbation increases, the oscillations in the velocity profile are attenuated. Consider a case with $\mathcal{W} = 10$ and $\hat{p}_a = 60$. Several dimensionless velocity profiles during the oscillations are shown in the left panel of the figure.

In this exercise, the reader should compare the DFT, the DMD, and the POD with the eigenfunction solution. Assume that the space discretization consists of $n_y = 2000$ while the time discretization consists of $n_t = 200$ with a dimensionless sampling frequency of $\hat{f}_s = 10$.

First, construct the discrete data set from (8.34) to (8.35) by setting it in terms of the canonical factorization in (8.11). Then, prepare a function that implements Algorithm 2 described in Section 8.2.4 and use this algorithm to compute the DFT, POD, and DMD. Finally, plot the amplitude decay of all the decompositions and show the first three dominant structures in space and time for each.

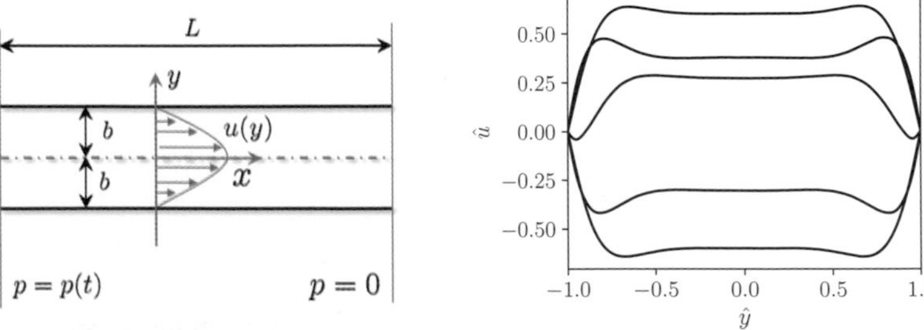

(a) Schematic of the 2D pulsating Poiseuille flow considered in this exercise. (b) Snapshots of the velocity profile during the pulsation.

8.4 The Multiscale Proper Orthogonal Decomposition

This section is composed of three parts. We first analyze in Section 8.4.1 the impact of frequency constraints on the POD. In particular, we are interested in what happens if we filter a data set before computing the POD. We will see that if a frequency is removed from the data, it is also removed from the temporal structures of its POD. We extend this result to a decomposition of the data via MRA (see Chapters 4 and 5). The MRA uses a filter bank to break the data into *scales*, and we here see how to keep the PODs of all scales *mutually orthogonal*. Then, the POD bases of each scale can be assembled into a single orthonormal basis. That is the basis of the mPOD – the mPOD, presented in Section 8.4.2. Finally, the mPOD algorithm is described in Section 8.4.3.

8.4.1 Frequency-Constrained Proper Orthogonal Decomposition

We compute the POD of a data set that has first been filtered. *Are the POD modes also filtered? Can we obtain the same modes by filtering the data or filtering the temporal correlation matrix? How are these linked to the POD modes of the original data?* This section answers these questions.

First, we need some definitions. Let the filter have a finite impulse response[18] $h \in \mathbb{R}^{n_t \times 1}$. Following Section 8.3.2, the frequency response in the discrete domain is a complex vector $\widehat{\mathbf{H}} = \mathbf{\Psi}_{\mathcal{F}} \boldsymbol{h} \in \mathbb{C}^{n_t \times 1}$. This vector can be arranged in two possible matrices, both useful for the following discussion. The first is linked to the diagonalization of circulant matrices, and is introduced in (8.25). The second consists of a row-wise replication of the same vector. These are $\boldsymbol{H} \in \mathbb{C}^{n_t \times n_t}$ and $\mathcal{H} = \mathbf{1} \cdot \widehat{\mathbf{H}}^T \in \mathbb{C}^{n_s \times n_t}$, with $\mathbf{1} \in \mathbb{R}^{n_s \times 1}$ a column vector of ones.[19] To avoid confusion, we here show both matrices[20]:

$$\boldsymbol{H} = \begin{bmatrix} \widehat{\mathbf{H}}[0] & 0 & 0 & 0 \\ 0 & \widehat{\mathbf{H}}[1] & 0 & 0 \\ \vdots & \vdots & \ddots & \vdots \\ 0 & 0 & 0 & \widehat{\mathbf{H}}[n_t] \end{bmatrix} \qquad \mathcal{H} = \begin{bmatrix} \widehat{\mathbf{H}}[0] & \widehat{\mathbf{H}}[1] & \dots & \widehat{\mathbf{H}}[n_t] \\ \widehat{\mathbf{H}}[0] & \widehat{\mathbf{H}}[1] & \dots & \widehat{\mathbf{H}}[n_t] \\ \vdots & \vdots & \vdots & \vdots \\ \widehat{\mathbf{H}}[0] & \widehat{\mathbf{H}}[1] & \dots & \widehat{\mathbf{H}}[n_t] \end{bmatrix}. \tag{8.36}$$

Recall that $\boldsymbol{H}$ is the eigenvalue matrix of the circulant convolution matrix $\boldsymbol{C}_h$ in (8.23). Assuming that the filtering in the time domain is performed in all the spatial location equally, the associated matrix multiplications are

$$\boldsymbol{D}_{\mathcal{H}} = \boldsymbol{D}\,\boldsymbol{C}_h = \overbrace{\underbrace{\boldsymbol{D}\overline{\mathbf{\Psi}}_{\mathcal{F}}}_{\widehat{\boldsymbol{D}}}\,\boldsymbol{H}}^{\widehat{\boldsymbol{D}}_{\mathcal{H}}}\,\mathbf{\Psi}_{\mathcal{F}} \quad \text{or} \quad \boldsymbol{D}_{\mathcal{H}} = \overbrace{\Big[\underbrace{\big(\boldsymbol{D}\,\overline{\mathbf{\Psi}}_{\mathcal{F}}\big)}_{\widehat{\boldsymbol{D}}} \odot \mathcal{H}\Big]}^{\widehat{\boldsymbol{D}}_{\mathcal{H}}}\,\mathbf{\Psi}_{\mathcal{F}}, \tag{8.37}$$

[18] Which thus implies that the filter cannot be "ideal," as discussed in Chapter 4.

[19] To be normalized by $1/\sqrt{n_s}$ to keep it of unitary length.

[20] Note that the right multiplication by $\boldsymbol{H}$ is equivalent to the Hadamard product by $\mathcal{H}$.

where $\odot$ is the Hadamard product, that is, the entry-by-entry multiplication between two matrices. Observe that $\widehat{\boldsymbol{D}}_{\mathcal{H}}\boldsymbol{\Psi}_{\mathcal{F}}$ is the inverse Fourier transform of $\widehat{\boldsymbol{D}}_{\mathcal{H}}$ (i.e., the frequency spectra of the filtered data) while $\widehat{\boldsymbol{D}} = \boldsymbol{D}\boldsymbol{\Psi}_{\mathcal{F}}$ is the Fourier transform of $\boldsymbol{D}$ along its rows (i.e., in the time domain).

The representation on the left is a direct consequence of the link between convolution theorem and eigenvalue decomposition of circulant matrices introduced in (8.25). The representation on the right opens to an intuitive and graphical representation of the filtering process that is worth discussing briefly.

First, we introduce the cross-spectral density matrix $\boldsymbol{K}_{\mathcal{F}}$. This collects the inner product between the frequency spectra of the data evolution; it is the frequency domain analogous of the temporal correlation matrix $\boldsymbol{K}$. These matrices are linked:

$$\boldsymbol{K}_{\mathcal{F}} = \widehat{\boldsymbol{D}}^{\dagger}\,\widehat{\boldsymbol{D}} = \boldsymbol{\Psi}_{\mathcal{F}}\left[\boldsymbol{D}^{\dagger}\,\boldsymbol{D}\right]\overline{\boldsymbol{\Psi}}_{\mathcal{F}} = \boldsymbol{\Psi}_{\mathcal{F}}\boldsymbol{K}\,\overline{\boldsymbol{\Psi}}_{\mathcal{F}} \Longleftrightarrow \boldsymbol{K} = \overline{\boldsymbol{\Psi}}_{\mathcal{F}}\boldsymbol{K}_{\mathcal{F}}\boldsymbol{\Psi}_{\mathcal{F}}. \tag{8.38}$$

Since $\boldsymbol{\Psi}_{\mathcal{F}}\overline{\boldsymbol{\Psi}}_{\mathcal{F}} = \mathbf{I}$, the equation on the right is a similarity transform, hence $\boldsymbol{K}_{\mathcal{F}}$ and $\boldsymbol{K}$ share the same eigenvalues. Introduce the eigenvalue decomposition of $\boldsymbol{K}$ in (8.26):

$$\boldsymbol{K}_{\mathcal{F}} = \boldsymbol{\Psi}_{\mathcal{F}}\left(\boldsymbol{\Psi}_{\mathcal{P}}\boldsymbol{\Sigma}^2_{\mathcal{P}}\boldsymbol{\Psi}^T_{\mathcal{P}}\right)\overline{\boldsymbol{\Psi}}_{\mathcal{F}} = \overline{\widehat{\boldsymbol{\Psi}}}_{\mathcal{P}}\boldsymbol{\Sigma}^2_{\mathcal{P}}\overline{\widehat{\boldsymbol{\Psi}}}^{\dagger}_{\mathcal{P}}, \quad \text{that is,} \quad \boldsymbol{K}_{\mathcal{F}}[i,j] = \sum_{r=0}^{n_t-1}\sigma^2_{\mathcal{P}}\,\overline{\widehat{\boldsymbol{\psi}}}_{\mathcal{P}}[i]\widehat{\boldsymbol{\psi}}_{\mathcal{P}}[j]. \tag{8.39}$$

We thus see that the eigenvectors of $\boldsymbol{K}_{\mathcal{F}}$ are the conjugate of the Fourier transform of the eigenvectors of $\boldsymbol{K}$. The outer product notation on the right shows that the diagonal of this matrix contains the sum of the power spectra of the temporal structures of all the modes. This is the sum of positive real quantities.

Consider now the cross-spectral density matrix of the filtered data in (8.36) and use the distributive property of the Hadamard product to get

$$\boldsymbol{K}_{\mathcal{F}} = \widehat{\boldsymbol{D}}^{\dagger}_{\mathcal{H}}\widehat{\boldsymbol{D}}_{\mathcal{H}} = \left(\widehat{\boldsymbol{D}}\odot\mathcal{H}\right)^{\dagger}\left(\widehat{\boldsymbol{D}}\odot\mathcal{H}\right) = \left(\widehat{\boldsymbol{D}}^{\dagger}\widehat{\boldsymbol{D}}\right)\odot\left(\mathcal{H}^{\dagger}\odot\mathcal{H}\right) = \boldsymbol{K}_{\mathcal{F}}\odot\underline{\mathcal{H}}, \tag{8.40}$$

having introduced the 2D symmetric frequency transfer function $\underline{\mathcal{H}} = \mathcal{H}^{\dagger}\mathcal{H}$. The figure in Exercise 2 gives a pictorial representation of the magnitude of such 2D frequency transfer function for ideal low-pass, band-pass, and high-pass filters. In a low-pass filter, for example, the 2D frequency transfer function is unitary within a square $\mathcal{H}[i,j] \neq 0$, for all $i,j \in [-i_c, i_c]\times[-i_c, i_c]$ and approximately zero everywhere else.

We thus see that the filtering of the data constraints the spectral content of the POD modes: writing the cross-spectral density of the filtered data in terms of its eigenvectors as in (8.38), we find

$$\boldsymbol{K}_{\mathcal{F}\mathcal{H}} = \boldsymbol{K}_{\mathcal{F}}\odot\underline{\mathcal{H}} \Longleftrightarrow \boldsymbol{K}_{\mathcal{F}\mathcal{H}}[i,j] = \sum_{r=0}^{n_t-1}\sigma^2_{\mathcal{P}\mathcal{H}}\,\overline{\widehat{\boldsymbol{\psi}}}_{\mathcal{P}\mathcal{H}}[i]\widehat{\boldsymbol{\psi}}_{\mathcal{P}\mathcal{H}}[j]. \tag{8.41}$$

Since $\boldsymbol{K}_{\mathcal{F}\mathcal{H}} \approx 0$ outside the band-pass range of the 2D filter, and since the entries along its diagonals are a summation of positive quantities, we conclude that *frequencies removed from the data cannot be present in any of the POD modes.* Moreover, we can compute the POD modes of the filtered data from the filtered

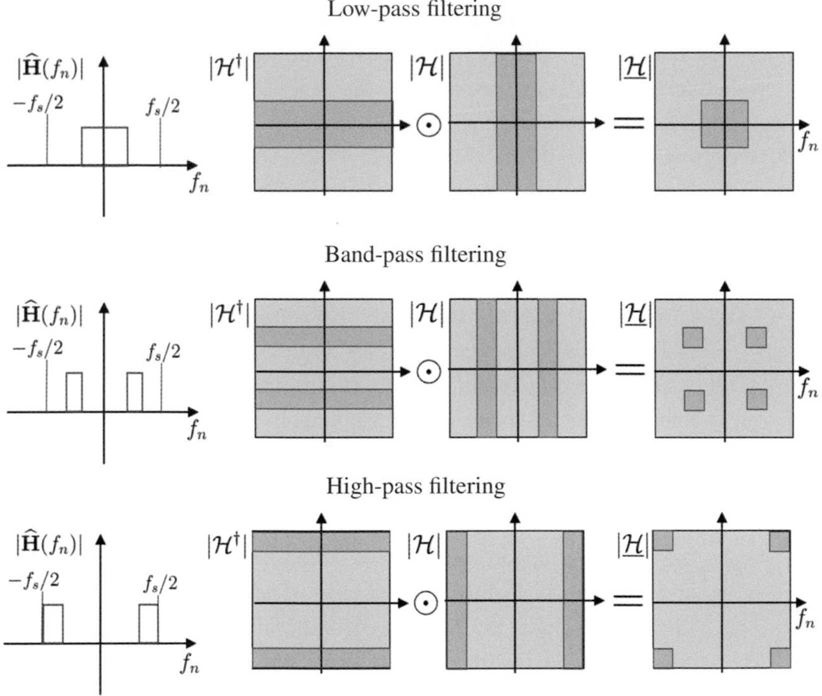

Figure 8.2 Pictorial representation of the magnitudes of the frequency transfer functions of the filters considered in this section. The figure on the left shows the magnitude of the 1D frequency response $\widehat{\mathbf{H}}$. On the right, the figure shows how 2D filters $\underline{\mathcal{H}} = \mathcal{H}^\dagger \mathcal{H}$ are constructed from the extended response $\mathcal{H}$. The red region represents the passband portion, that is, within which $|\mathcal{H}| \approx 1$ while in the gray area it is $|\mathcal{H}| \approx 0$.

cross-spectral matrix $\boldsymbol{K}_{\mathcal{F}\mathcal{H}}$ or, more conveniently, from the temporal correlation matrix $\boldsymbol{K}_{\mathcal{H}}$. The link between these matrices can be written as

$$\boldsymbol{K}_{\mathcal{H}} = \boldsymbol{D}^\dagger_{\mathcal{H}} \boldsymbol{D}_{\mathcal{H}} = \overline{\boldsymbol{\Psi}}_{\mathcal{F}} \boldsymbol{K}_{\mathcal{F}\mathcal{H}} \boldsymbol{\Psi}_{\mathcal{F}} = \overline{\boldsymbol{\Psi}}_{\mathcal{F}} \widehat{\boldsymbol{K}}_{\mathcal{H}} \overline{\boldsymbol{\Psi}}_{\mathcal{F}} \iff \widehat{\boldsymbol{K}}_{\mathcal{H}} = \boldsymbol{P}_\pi \boldsymbol{K}_{\mathcal{F}\mathcal{H}}, \tag{8.42}$$

where the definition of 2D Fourier transform in (8.8) is introduced and $\boldsymbol{P}_\pi = \boldsymbol{\Psi}_{\mathcal{F}} \boldsymbol{\Psi}_{\mathcal{F}}$ is the permutation matrix obtained by multiplying the Fourier matrix twice[21]:

$$\boldsymbol{P}_\pi = \boldsymbol{\Psi}_{\mathcal{F}} \boldsymbol{\Psi}_{\mathcal{F}} = \overline{\boldsymbol{\Psi}}_{\mathcal{F}} \overline{\boldsymbol{\Psi}}_{\mathcal{F}} = \begin{bmatrix} 1 & 0 & \dots & 0 & 0 \\ 0 & 0 & & 0 & 1 \\ 0 & 0 & & 1 & \vdots \\ \vdots & \vdots & \iddots & & \vdots \\ 0 & 1 & 0 & \dots & 0 \end{bmatrix}. \tag{8.43}$$

Finally, to answer the last question from this section, use the multiplication by the diagonal matrix $\boldsymbol{H}$ in (8.39) instead of the Hadamard product:

$$\boldsymbol{K}_{\mathcal{F}\mathcal{H}} = \left(\widehat{\boldsymbol{D}}\boldsymbol{H}\right)^\dagger \left(\widehat{\boldsymbol{D}}\boldsymbol{H}\right) = \boldsymbol{H}^\dagger \widehat{\boldsymbol{D}}^\dagger \boldsymbol{D} \boldsymbol{H} = \boldsymbol{H}^\dagger \boldsymbol{K}_{\mathcal{F}} \boldsymbol{H}\,. \tag{8.44}$$

[21] A funny question arose after reading this: what happens if the Fourier transform is performed four times?

Since $\boldsymbol{H}^{\dagger}\boldsymbol{H} \neq \mathbf{I}$, this is not a similarity transform. Hence the eigenvalues of these two matrices are different: there is no simple link between $\sigma_{\mathcal{P}}$ and the $\sigma_{\mathcal{P}\mathcal{H}}$, nor between the POD of the data and the POD of the filtered data. We do not attempt to build such a connection: it suffices to notice that the POD of the filtered data is the optimal basis for the portion of data whose spectra fall within the band-pass region of the filter. In the terminology of MRA (see Chapter 4), we call this a *scale*.

Exercise 3: The POD of Filtered Data

Consider the test case from Exercise 2 but assume that the forcing is composed of two terms. These are sinusoids associated to a Womersley number of $\mathcal{W} = 1$ and $\mathcal{W} = 4$, modulated by two windows We refer to these terms as F_1 and F_2, respectively. As the problem is linear, we can sum the response of the velocity profile to each term. The response of the velocity profile at the centerline, for each of the terms, is shown in Figure 8.4.

In this exercise, we want to analyze the response of the velocity profile to the perturbation at the highest frequency. First, perform a standard POD and analyze the resulting structures. Then, filter the data using a high-pass filter that isolates the forcing F_2 and compute the POD again. Compute this decomposition by filtering the data and by filtering the correlation matrix. Compare the results. As an optional exercise, perform DMD and DFT using the codes from Exercise 2: is it possible to separate the contribution of each forcing term with these decompositions? The answer is no. Show it!

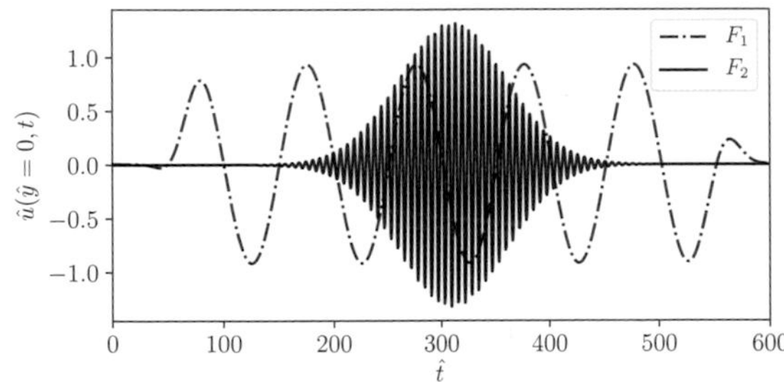

Time evolution of the velocity profile at the center of the channel when each of the two forcing terms F_1 and F_2 is active.

8.4.2 From Frequency Constraints to Multi-resolution Analysis

Section 8.4.1 provided the background to generalize the study of the frequency-constrained POD to the mPOD. The mPOD constructs a set of orthogonal modes that have no frequency overlapping, that is, no common frequency content within specific frequency bandwidths.

To illustrate the general idea, let us consider a data set $\boldsymbol{D} \in \mathbb{R}^{n_s \times n_t}$, with $n_t < n_s$. We now use MRA to break the data set into scales. With no loss of generalities, let us consider three scales; these are identified by three filters with band-pass bandwidths $\Delta f_1 = [0, f_1]$, $\Delta f_2 = [f_1, f_2]$ and $\Delta f_3 = [f_2, f_s/2]$. We lump this information in a frequency splitting vector $F_V = [f_1, f_2, f_3]$. With three scales, the temporal correlation matrix has nine partitions:

$$\boldsymbol{K} = \boldsymbol{D}^\dagger \boldsymbol{D} = \left(\boldsymbol{D}_1 + \boldsymbol{D}_2 + \boldsymbol{D}_3\right)^\dagger \left(\boldsymbol{D}_1 + \boldsymbol{D}_2 + \boldsymbol{D}_3\right) = \sum_{i=1}^{3} \sum_{j=1}^{3} \boldsymbol{D}_i^\dagger \boldsymbol{D}_j \,. \tag{8.45}$$

Following an MRA formulation, we assume that these scales are isolated by filters with complementary frequency response, that is,

$$\boldsymbol{D}_i = \left[(\boldsymbol{D}\overline{\boldsymbol{\Psi}}_{\mathcal{F}}) \odot \boldsymbol{\mathcal{H}}_i\right]\boldsymbol{\Psi}_{\mathcal{F}} \quad \text{with} \quad \sum \boldsymbol{\mathcal{H}}_i \approx \boldsymbol{1} \quad \text{and} \quad \boldsymbol{\mathcal{H}}_i \odot \boldsymbol{\mathcal{H}}_j \approx \boldsymbol{0}. \tag{8.46}$$

This ensures a lossless decomposition of the data. We now use (8.39) and (8.45) in (8.44) to analyze which portion of the cross-spectral density $\boldsymbol{K}_{\mathcal{F}}$ is taken by each of the contributions in (8.44). Figure 8.3 gives a pictorial view of the partitioning, which can be constructed following the graphical representation in Exercise 2. In wavelet terminology, the term $\boldsymbol{\mathcal{H}}_1^\dagger \boldsymbol{\mathcal{H}}_1$ is the *approximation term* at the largest scale. The other "pure" terms $\boldsymbol{\mathcal{H}}_2^\dagger \boldsymbol{\mathcal{H}}_2$ and $\boldsymbol{\mathcal{H}}_3^\dagger \boldsymbol{\mathcal{H}}_3$ are *diagonal details* of the scales 2 and 3. The terms $\boldsymbol{\mathcal{H}}_i^\dagger \boldsymbol{\mathcal{H}}_j$ with $i > j$ are *horizontal details* while those with $j < i$ are *vertical details*.

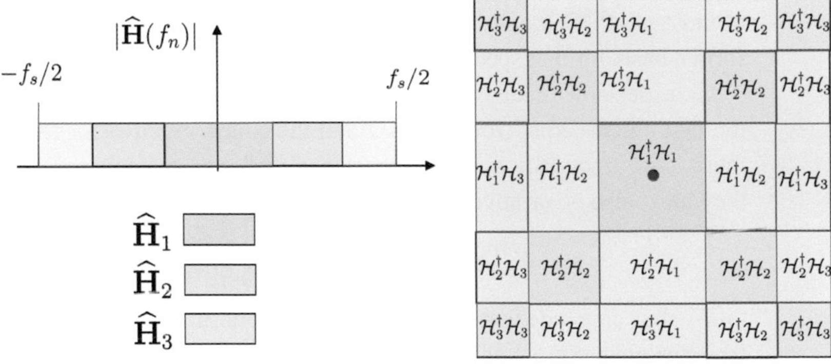

Figure 8.3 Repartition of the cross-spectral density matrix into the nine contributions identified by three scales. The origin is marked with a red circle. The "mixed terms" $\boldsymbol{\mathcal{H}}_i^\dagger \boldsymbol{\mathcal{H}}_j$ with $i \neq j$ are colored in light gray, while the "pure terms" $\boldsymbol{\mathcal{H}}_i^\dagger \boldsymbol{\mathcal{H}}_i$ are colored following the legend on the left, where the 1D transfer functions are also shown.

We have seen in Section 8.4.1 that removing a frequency from the data set removes it from the temporal structures of its POD. There is thus no frequency overlapping[22] between the eigenvectors of the contributions $\boldsymbol{K}_1 = \boldsymbol{D}_1^\dagger \boldsymbol{D}_1$, $\boldsymbol{K}_2 = \boldsymbol{D}_2^\dagger \boldsymbol{D}_2$, $\boldsymbol{K}_3 = \boldsymbol{D}_3^\dagger \boldsymbol{D}_3$, which we denote as $\boldsymbol{K}_m = \boldsymbol{\Psi}_{\mathcal{H}m} \boldsymbol{\Lambda}_{\mathcal{H}m} \boldsymbol{\Psi}_{\mathcal{H}m}^T$. However, the same is not necessarily true for the correlations of the "mixed terms" $\boldsymbol{D}_i^\dagger \boldsymbol{D}_j$ with $i \neq j$: the filters $\boldsymbol{\mathcal{H}}_i^\dagger \boldsymbol{\mathcal{H}}_j$ with $i \neq j$ leave the diagonals of $\boldsymbol{K}_{\mathcal{F}}$ unaltered and thus the spectral constraints are much

[22] At least in case of ideal filters, which are perfectly complementary.

weaker: it is possible that eigenvectors from $\boldsymbol{D}_2^\dagger \boldsymbol{D}_3$, for example, have frequencies that are already present among the eigenvectors of $\boldsymbol{D}_2^\dagger \boldsymbol{D}_2$ or $\boldsymbol{D}_3^\dagger \boldsymbol{D}_3$.

We opt for a drastic approach: remove all the "mixed terms" $\boldsymbol{D}_i^\dagger \boldsymbol{D}_j$ with $i \neq j$ and consider the correlation as the sum of M "pure terms",

$$\boldsymbol{K} \approx \sum_{m=1}^{M} \boldsymbol{D}_m^\dagger \boldsymbol{D}_m = \sum_{m=1}^{M} \boldsymbol{\Psi}_{\mathcal{F}} \big[\hat{\boldsymbol{K}} \odot \mathcal{H}_m^\dagger \mathcal{H}_m \big] \boldsymbol{\Psi}_{\mathcal{F}} = \sum_{m=1}^{M} \boldsymbol{\Psi}_{\mathcal{H}m} \boldsymbol{\Lambda}_{\mathcal{H}m} \boldsymbol{\Psi}_{\mathcal{H}m}^T, \tag{8.47}$$

having introduced (4.41) and the eigenvalue decomposition of each term.

This is a fundamental step to link the mPOD to POD and DFT. Such simplification results in a loss of information in the temporal correlation matrix, but it still ensures a complete reconstruction of the data set. It remains to be seen if a basis constructed by combining all the eigenvectors of different scales can form a complete and orthonormal basis for n_t.

Consider first each of these scales independently. Let us assume that the number of nonzero frequencies in each scale, namely the number of frequency bins in the band-pass region, is n_m. If the frequency transfer functions are complementary, we have $\sum n_m = n_t$. If each scale contains at most n_m frequencies, every temporal evolution of the data within that specific scale can be written as a linear combination of n_m Fourier modes. Therefore, every contribution $\boldsymbol{D}_m$ is at most of $rank(\boldsymbol{D}_m) = n_m$ and hence has at most n_m POD modes. These are clearly orthonormal, that is, $\boldsymbol{\Psi}_{\mathcal{H}m}^T \boldsymbol{\Psi}_{\mathcal{H}m} = \mathbf{I}$ and form a basis for that specific scale.

Consider now modes from different scales, say for example $\boldsymbol{\psi}_{\mathcal{H}2}^{(1)}$ and $\boldsymbol{\psi}_{\mathcal{H}3}^{(4)}$, that is, the first eigenvectors from scales 2 and the fourth eigenvector from scale 3. Because there is no frequency overlapping, their Fourier transforms are orthogonal, and thus they must also be orthogonal:

$$\widehat{\boldsymbol{\psi}}_{\mathcal{H}2}^{(1)\dagger} \widehat{\boldsymbol{\psi}}_{\mathcal{H}4}^{(4)} = 0 = \boldsymbol{\psi}_{\mathcal{H}2}^{(1)\dagger} \overline{\boldsymbol{\Psi}}_{\mathcal{F}} \boldsymbol{\Psi}_{\mathcal{F}} \boldsymbol{\psi}_{\mathcal{H}4}^{(4)} = \boldsymbol{\psi}_{\mathcal{H}2}^{(1)\dagger} \boldsymbol{\psi}_{\mathcal{H}4}^{(4)}. \tag{8.48}$$

We thus conclude that a complete and orthonormal basis for $\mathbb{R}^{n_t}$ can be constructed from the eigenvectors of the various scales. That is the mPOD basis.

We close this section by highlighting how the mPOD connects POD and DFT or DMD. In case of no frequency partitioning, the mPOD is a POD: the temporal structures can span the entire frequency range and are derived, as described in Section 8.2.4 and Chapter 6, under the constraint of optimal approximation for any given number of modes. Introducing the frequency partitioning and the approximation in (8.46), we identify modes that are optimal only within the frequency bandwidth of each scale.

The optimality of the full basis is lost as modes from different scales are not allowed to share the same frequency bandwidth. As we introduce finer and finer partitioning, each mode is limited within a narrower frequency bandwidth. At the limit for $n_m = 1$ and $M = n_t$, every mode is allowed to have only one frequency. The approximation in (8.46) forces the spectra of the temporal correlation matrix to be diagonal, that is, the correlation matrix is approximated as a Toeplitz circulant matrix. Accordingly, the mPOD tends toward the DFT or DMD depending on the boundary conditions used in the filtering of $\boldsymbol{K}$. If periodicity is assumed, the DFT is produced. If periodicity is not assumed, the decomposition selects harmonic modes whose frequency is not necessarily a multiple of the observation time, and a DMD is recovered.

8.4.3 The Multiscale Proper Orthogonal Decomposition Algorithm

The pseudo-code for computing the mPOD listed in Algorithm 3. This is composed of eight steps.

Algorithm 3: Algorithm for computing the multiscale POD.

Input: $\boldsymbol{D} \in \mathbb{R}^{n_s \times n_t}$, $F_V \in \mathbb{R}^{M \times 1}$

Output: $\boldsymbol{\Psi}_\mathcal{M} \in \mathbb{R}^{n_t \times R}$, $\boldsymbol{\Sigma}_\mathcal{M} \in \mathbb{R}^{R \times R}$, and $\boldsymbol{\Phi}_\mathcal{M} \in \mathbb{R}^{n_s \times R}$

1 Compute temporal correlation matrix $\boldsymbol{K} = \boldsymbol{D}^\dagger \boldsymbol{D}$.

2 Prepare the filter bank $\boldsymbol{H}_1, \boldsymbol{H}_2, \ldots, \boldsymbol{H}_M$.

3 Decompose $\boldsymbol{K}$ into $\boldsymbol{K}_1, \boldsymbol{K}_2, \ldots, \boldsymbol{K}_m$ contributions.

4 Compute the temporal structures of each scale $\boldsymbol{K} = \boldsymbol{\Psi}_m \boldsymbol{\Lambda}_m \boldsymbol{\Psi}_m^T$.

5 Assembly basis $\boldsymbol{\Psi}^0 = [\boldsymbol{\Psi}_1, \ldots, \boldsymbol{\Psi}_M]$ and $\lambda^0 = [\text{diag}(\boldsymbol{\Lambda}_1), \ldots, \text{diag}(\boldsymbol{\Lambda}_M)]$.

6 Prepare permutation $\boldsymbol{P}_\Lambda$, sorting λ^0. Then shuffle columns $\boldsymbol{\Psi}^1 = \boldsymbol{\Psi}^0 \boldsymbol{P}_\Lambda$.

7 Enforce orthogonality $\boldsymbol{\Psi}^1 = \boldsymbol{\Psi}_\mathcal{M} \boldsymbol{R} \rightarrow \boldsymbol{\Psi}_\mathcal{M} = \boldsymbol{\Psi}^1 \boldsymbol{R}^{-1}$.

8 Complete the decomposition using Algorithm 2 $\boldsymbol{\Phi}_\mathcal{M} \rightarrow$ from $\boldsymbol{D}, \boldsymbol{\Psi}_\mathcal{M}$.

The first step computes the temporal correlation matrix, as in the POD. The second step prepares the filter bank, according to the introduced frequency vector F_V. The third step computes the MRA of the temporal correlation matrix; the fourth step computes the eigenvectors of each.[23]

At this stage, we can proceed with the preparation of a single basis. First, in step 5, all the temporal structures and associated eigenvalues are collected into a single matrix $\boldsymbol{\Psi}^0$ and a single vector of eigenvalues λ^0. These eigenvalues give an estimation of the relative importance of each term. This is sorted in descending order and the information is used to permute the columns of $\boldsymbol{\Psi}^0$ using an appropriate permutation matrix $\boldsymbol{P}_\Lambda$.

If the filters were ideal, the resulting temporal basis $\boldsymbol{\Psi}^1 = \boldsymbol{\Psi}^0 \boldsymbol{P}_\Lambda$ would be completed. However, to compensate for the nonideal frequency response of the filters, a reduced QR factorization is used in step 7 to enforce the orthonormality of the mPOD basis. The result of the orthogonalization procedure is the matrix of the temporal structure of the mPOD. The final step is the projection to compute the spatial structures that is common to every decomposition and that can be computed using[24] Algorithm 2.

Note that special attention should be given to the filtering process in the third step. If this is done as in Exercise 4, it is implicitly assumed that the matrix $\boldsymbol{K}$ is

[23] *Can we by-pass this step? Can we compute all the $\boldsymbol{\Psi}_m$'s from one single diagonalization (of a properly filtered matrix)?* These were two brilliant questions by Bo B. Watz, development engineer at Dantec Dynamics, who attended the lecture series. The answer is that it is possible, in principle, to compute the basis from one clever diagonalization. However, the filtering is challenging: that was the starting point of new exciting developments toward the *fast mPOD*.

[24] Observe that a simplified version could be used since the temporal structure is orthonormal!

periodic. When such an assumption is incorrect, edge effects could appear after the filtering process. The cure for these effects is described in classical textbooks on image processing (see Gonzalez & Woods 2017) and include *reflections* or *extrapolations*. To limit the scope of this chapter, we do not address these methods here.[25] The reader should nevertheless be able to construct the mPOD algorithm using all the codes developed for the previous exercises.

Exercise 4: Your own mPOD Algorithm

Combine the codes from the previous exercises to build your own function for computing the mPOD, following Algorithm 3. This function should compute $\mathbf{\Psi}_{\mathcal{M}}$, while the decomposition can be completed using Algorithm 2.
Assume that filters are constructed using Hamming windows, and the filter order is an input parameter. Test your function with the previous exercise and show the structures of the first two mPOD modes. Compare their amplitudes with the ones of the POD modes.

8.5 Tutorial Test Cases

In addition to the coding exercises, this chapter includes two tutorial test cases from time-resolved particle image velocimetry measurements. The first data set collects the velocity field of a planar impinging gas jet in stationary conditions; the second is the flow past a cylinder in transient conditions. These are described in Mendez, Balabane and Buchlin (2019) and Mendez et al. (2020). Other examples of applications of the mPOD on experimental data can be found in Mendez et al. (2018), Mendez et al. (2019) and Esposito et al. (2021). More exercises and related codes can be found in `https://github.com/mendezVKI/MODULO`, together with an executable with graphical user interface (GUI) developed by Ninni and Mendez (2021).

8.5.1 Test Case 1

To download the data set and store it in a local directory, run the following script

```
import urllib.request
print('Downloading Data for Tutorial 1...')
url = 'https://osf.io/c28de/download'
urllib.request.urlretrieve(url, 'Ex_4_TR_PIV_Jet.zip')
print('Download Completed! I prepare data Folder')
# Unzip the file
from zipfile import ZipFile
String='Ex_4_TR_PIV_Jet.zip'
zf = ZipFile(String,'r'); zf.extractall('./'); zf.close()
```

[25] Interested readers can find the related codes at `https://github.com/mendezVKI/MODULO`

The data set consists of $n_t = 2000$ velocity fields, sampled at $f_s = 2$ kHz over a grid of 60×114 points. The spatial resolution is approximately $\Delta x = 0.33$ mm. The script provided in the book's website[26] describes how to manipulate this data set, plot a time step, and more. For the mPOD of this test case, the frequency splitting vector used in Mendez, Balabane and Buchlin (2019) is constructed in terms of Strouhal number $St = f H/U_J$, that is, the dimensionless frequency computed from the advection time H/U_J, with H the standoff distance and U_J the mean velocity of the jet at the nozzle outlet. These are H = 4mm and U_J = 6.5 m/s.

The reader is encouraged to compare the mPOD results with those achievable from other decompositions. The spatial structures and the spectra of the temporal structures in the first five modes are produced by the script TUT_1.PY. These are shown in Figure 8.4 for the dominant mPOD mode in the scale $St = 0.1 - 0.2$. This mode isolates the roll-like structures produced by the evolution of the shear layer instability downstream the potential core of the jet. Recall that the velocity components are stacked into a single vector.

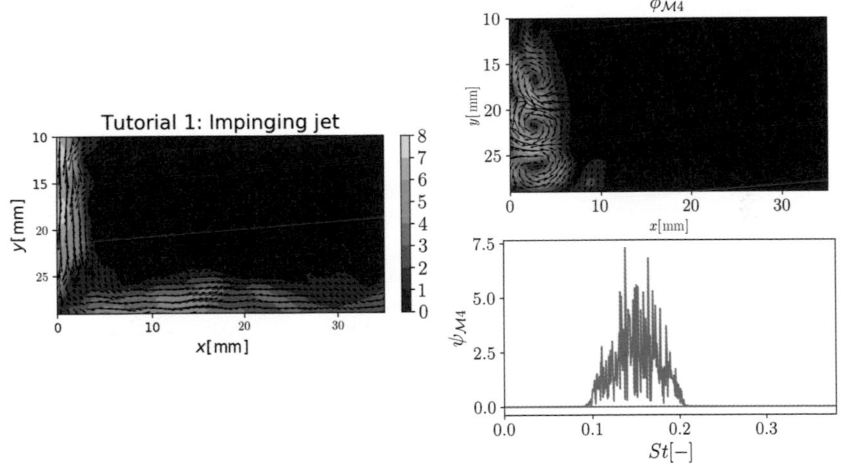

Figure 8.4 On the left: example snapshot from the first tutorial test case on the TR-PIV of an impinging gas jet. Colormap in m/s. On the right: example of spatial structure (top) and frequency content (bottom) of the temporal structures of a mPOD mode.

8.5.2 Test Case 2

To download this data set and store it into a local directory, run the following script

```
import urllib.request
print('Downloading Data for Tutorial 2...')
url = 'https://osf.io/qa8d4/download'
urllib.request.urlretrieve(url, 'Ex_5_TR_PIV_Cylinder.zip')
print('Download Completed! I prepare data Folder')
```

[26] www.datadrivenfluidmechanics.com/chapter8

```
# Unzip the file
from zipfile import ZipFile
String='Ex_5_TR_PIV_Cylinder.zip'
zf = ZipFile(String,'r');
zf.extractall('./DATA_CYLINDER'); zf.close()
```

The experimental setup and configuration are extensively described in Mendez et al. (2020). This data set is computationally much more demanding than the previous, and consists of $n_t = 13{,}200$ velocity fields sampled at $f_s = 3$ kHz over a grid of 71×30 points. The spatial resolution is approximately $\Delta x = 0.85$ mm. A plot of a time step is shown in Figure 8.5

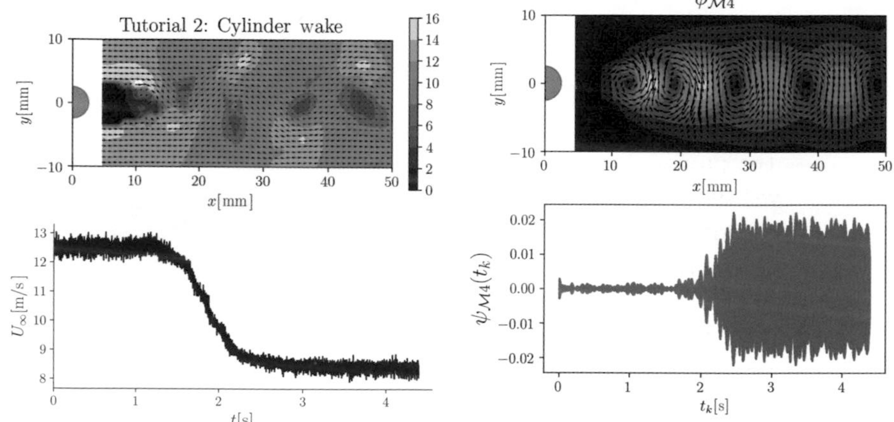

Figure 8.5 Top left: a snapshot of the velocity field past a cylinder, obtained via TR-PIV. Colormap in m/s. Bottom left: evolution of the free-stream velocity as a function of time, sampled from the top left corner of the field. Right: example of spatial structure (top) and frequency content (bottom) of the temporal structures of an mPOD mode.

As for the previous tutorial, a function is used to extract all the information about the grid (in this case stored on a different file) and the velocity field.

This test case is characterized by a large-scale variation of the free-stream velocity. A plot of the velocity magnitude in the main stream is shown in Figure 8.5. In the first 1.5 s, the free-stream velocity is at approximately $U_\infty \approx 12$ m/s. Between $t = 1.5$ s and $t = 2.5$ s, this drops down to $U_\infty \approx 8$ m/s. The variation of the flow velocity is sufficiently low to let the vortex shedding adapt, and hence preserve an approximately constant Strouhal number of $St = fdU_\infty \approx 0.19$, with $d = 5$ mm the diameter of the cylinder. Consequently, the vortex shedding varies from $f \approx 459$ Hz to $f \approx 303$ Hz. Interestingly, the POD assigns the entire evolution to a single pair of modes, and it is hence not possible to analyze the vortex structures in the shedding for the two phases at approximately constant velocity, nor to distinguish them from the flow organization during the transitory phase.

The mPOD can be used to identify modes related to these three phases. Three scales are chosen in the exercise. The first, in the range $\Delta f = [0 - 10]$ Hz, is designed to isolate the large-scale motion of the flow, hence the variation of the free-stream velocity. The scale with $\Delta f = [290–320]$ Hz is designed to capture the vortex shedding when the velocity is at $U_\infty \approx 8$m/s, while the scale with $\Delta f = [430–470]$ Hz identifies the shedding when the velocity is at $U_\infty \approx 12$ m/s.

The spatial structure and the temporal evolution of an mPOD mode are also shown in Figure 8.5. This mode is clearly associated to the second stationary conditions. Other modes capture the first part while others are focused on the transitory phase: besides allowing for spectral localization, the MRA structure of the decomposition also allows for providing time localization capabilities.

8.6 What's Next?

This chapter presented a general matrix factorization framework to formulate *any* linear decomposition. This general framework allowed for deriving the mPOD, a formulation that allows us to bridge energy-based and frequency-based formulations. The proposed exercises and the extensive collection of codes listed in the book's website should allow the reader to efficiently use these decompositions on his/her own data and provide a good starting point to develop computational proficiency on this subject.

It is important to note that this chapter has only focused on the mathematical framework of *linear decompositions*: Every decomposition provides its own way of representing the data with respect to a certain basis. From here, the reader might be interested in two possible directions. On the representation side, while linear dimensionality reduction has been at the heart of an extensive body of literature in fluid mechanics, nonlinear methods are now offering a new set of tools as showcased in Chapter 1. On the reduced-order modeling side, whether a low dimensional representation is capable of reducing the complexity of a dynamical system depends on whether the projection (or the nonlinear mapping) is able to preserve the most dynamically relevant phenomena in the new representation. These challenges are described in Chapter 14.

9 Good Practice and Applications of Data-Driven Modal Analysis

A. Ianiro

This chapter[1] develops a theoretical background for the definition of good practice (number of samples and sampling time versus the effect of noise) for data-driven modal analysis and shows an example application to a channel-flow data set. Snapshot proper orthogonal decomposition (POD) is employed as a benchmark allowing to move the discussion toward the convergence of the flow statistics and the effect of measurement noise on the temporal correlation matrix.

Beyond the estimation of the effect of noise on the estimated modes, this chapter also presents several application examples showing how hidden information can be extracted from a well-converged POD. In particular, it is shown how temporal modes of non-time-resolved data can provide detailed phase information and how extended POD modes can provide a linear stochastic estimation of correlated events. This is useful for applications that involve flow sensing and for the study of convection problems.

9.1 Introduction

When analyzing flow data sets, both from experiments and simulations, the researcher is challenged by important questions about the needed size and completeness of the data set.

The basic question that the researcher might ask him/herself is *how many samples do I need to store to obtain a reliable modal analysis?* Considering, for instance, the shedding phenomenon of the wake of a cylinder as discussed in Chapter 1, it is rather intuitive that it is needed to correctly sample several phases of the shedding period and to acquire enough samples per phase to achieve a satisfactory convergence of the flow statistics and thus of the correlation matrix. Unfortunately, data storage, simulation or experiment duration, and processing time might limit the amount of samples stored. The question previously identified can be further subdivided into two questions: *while acquiring a data set, what should be the acquisition frequency? Moreover, given a*

[1] Andrea Ianiro acknowledges the support of his colleagues and friends Prof. S. Discetti and Prof. M. Raiola. The majority of the original concepts summarized in the present chapter is the result of a fruitful collaboration at UC3M in the last eight years. The application of the turbulent heat transfer in a pipe is the result of a collaboration with Dr. A. Antoranz, Prof. O. Flores and Prof. M. García-Villalba. A. Antoranz, O. Flores and M. García-Villalba are kindly acknowledged for sharing the turbulent-pipe data set.

certain acquisition frequency, how many samples should be acquired? Answers to these questions can be obtained the from basic statistical data characterization theory described in books on experimental methods (Discetti & Ianiro 2017) and sits on the theoretical arguments presented in Chapters 4 and 5.

While the two previous questions are of rather straightforward answer for numerical simulations (which have only negligible truncation errors), experimental data are often affected by uncertainties. Measurement uncertainty includes bias and random parts. While the bias contribution mostly affects the estimated values of mode amplitude and energy content, random uncertainties have a more complicated effect on the correlation matrix. This fact raises a more intriguing question, that is, *how does measurement noise affect the convergence of the modes?* A seminal work on the topic was presented by Venturi (2006); following Venturi's work, a simple practical criterion for POD truncation was proposed by Raiola et al. (2015). Although the results are derived for POD, all the modal decomposition approaches based on the singular value decomposition of a snapshot matrix are affected by the noise perturbation of the time-correlation matrix.

With a special focus on particle image velocimetry (PIV) data, despite the continuous improvement of the available hardware, it is often not possible to perform measurements at frame rates sufficiently high to obtain a characterization of the flow dynamics. The researcher should wonder *whether it could be possible to extract information about the modes dynamics*. Although temporal sampling will not satisfy the Nyquist theorem, the phase relation between modes might be preserved, and Lissajous curves of the temporal modes can allow retrieving this information. Knowing dynamical information about some specific modes, both from theory or from additional experiments, it can then be possible to identify the flow dynamics. The suggested approach is explained by means of an application to the wake behind a couple of tandem cylinders in ground effect (Raiola et al. 2016).

The temporal modes, as shown in for example Chapters 6 and 8, are a basis for the temporal evolution of the data and can be used as a basis to project and decompose other synchronized data sets. This decomposition, referred to as extended proper orthogonal decomposition (herein EPOD) and first proposed by Borée (2003), allows us to ascertain correlations between different flow quantities and/or flow measurements (Picard & Delville 2000, Duwig & Iudiciani 2010).

This results in an excellent opportunity to retrieve temporal information from synchronized time-resolved point measurements (for instance, from hot wires or other high-repetition-rate pressure probes) and non-time-resolved flow-field measurements (Tinney et al. 2008, Baars & Tinney 2014, Hosseini et al. 2015, Discetti et al. 2018, Discetti et al. 2019). Even if recent works have proven the superior performances of neural networks for the purpose of sensor-based flow reconstructions (Güemes et al. 2019, Guastoni et al. 2020), the author believes that EPOD represents a founding brick of machine learning-based flow sensing, and a description is provided herein.

For what concerns the correlation between different synchronized flow quantities, EPOD is especially well suited for the analysis of combustion or heat transport problems (Antoranz et al. 2018), allowing us to obtain composite modes, for example, modes of temperature and velocity that provide a clear description of the scalar transport mechanisms. A MATLAB® exercise focusing on the analysis of a data set from the work by Antoranz et al. (2018) is proposed at the end of the chapter.

9.1.1 A Brief Recap of the Snapshot Proper Orthogonal Decomposition Procedure

Even if already introduced in Chapter 6, it is here worth briefly recalling the procedure employed for the derivation of the snapshot POD. Consider a data set of n_s snapshot of discrete measurements of a certain quantity a evaluated at n_x equally spaced measurement points x_i. Snapshot POD, introduced in Chapter 6, requires us to assemble the data set into a matrix $\boldsymbol{A}$ in which each column represents a snapshot of all the data measured at a given instance in time t_j, and each row consists of all the instances of a given spatial point x_i:

$$\boldsymbol{A} = \begin{bmatrix} a_{x_1,t_1} & a_{x_1,t_2} & \cdots & a_{x_1,t_{n_S}} \\ a_{x_2,t_1} & a_{x_2,t_2} & \cdots & a_{x_2,t_{n_S}} \\ \vdots & \vdots & \ddots & \vdots \\ a_{x_{n_x},t_1} & a_{x_{n_x},t_2} & \cdots & a_{x_{n_x},t_{n_S}} \end{bmatrix}. \tag{9.1}$$

Following the notation of Chapter 6, it is then useful to subtract a base condition $\boldsymbol{y_0}$ to each snapshot, commonly chosen to be the mean of all the columns of $\boldsymbol{A}$, obtaining a matrix $\boldsymbol{Y}$ composed of vectors $\boldsymbol{y_j}'$. The POD can be found from the singular value decomposition of $\boldsymbol{Y} = \boldsymbol{\Phi\Sigma\Psi}^*$. If $n_x > n_s$, a computationally more efficient approach requires the solution of the eigenvalue problem for the temporal correlation matrix $\boldsymbol{Y}^*\boldsymbol{Y}$,

$$\boldsymbol{Y}^*\boldsymbol{Y} = \begin{bmatrix} \langle \boldsymbol{y}_1', \boldsymbol{y}_1' \rangle & \langle \boldsymbol{y}_1', \boldsymbol{y}_2' \rangle & \cdots & \langle \boldsymbol{y}_1', \boldsymbol{y}_{n_s}' \rangle \\ \langle \boldsymbol{y}_2', \boldsymbol{y}_1' \rangle & \langle \boldsymbol{y}_2', \boldsymbol{y}_2' \rangle & \cdots & \langle \boldsymbol{y}_2', \boldsymbol{y}_{n_s}' \rangle \\ \vdots & \vdots & \ddots & \vdots \\ \langle \boldsymbol{y}_{n_s}', \boldsymbol{y}_1' \rangle & \langle \boldsymbol{n}_s, \boldsymbol{y}_2' \rangle & \cdots & \langle \boldsymbol{y}_{n_s}', \boldsymbol{y}_{n_s}' \rangle \end{bmatrix} = \boldsymbol{\Psi\Lambda\Psi}^*, \tag{9.2}$$

where $\boldsymbol{Y}^*$ is the conjugate transpose of $\boldsymbol{Y}$ and $\left\langle \boldsymbol{y}_i', \boldsymbol{y}_j' \right\rangle$ is the scalar product of the vectors $\boldsymbol{y_i}$ and $\boldsymbol{y_j}$ estimated at time instants t_i and t_j. The solution of the eigenvalue problem of the cross-correlation matrix $\boldsymbol{Y}^*\boldsymbol{Y}$ returns as right singular vectors the rows of $\boldsymbol{\Psi}^*$, that is, the temporal modes of the POD, while the diagonal eigenvalue matrix $\boldsymbol{\Lambda} = \boldsymbol{\Sigma}^2$ contains the variance of each mode. The matrix $\boldsymbol{\Psi}$ can be used as a basis for the projection of the matrix $\boldsymbol{Y}$ in order to obtain the spatial modes, which are the columns of the matrix $\boldsymbol{\Phi}$. Spatial modes are thus a weighted average of the snapshots through the matrix product:

$$\boldsymbol{\Phi} = \boldsymbol{Y\Psi\Sigma}^{-1}. \tag{9.3}$$

9.2 Data Set Size and Richness

9.2.1 Effect of the Number of Samples on the Convergence of the Statistics

The answer to the question "*how many samples are needed to obtain a well-converged modal decomposition?*" depends on the data set. It is rather intuitive that the statistical properties of laminar flows can be estimated with much fewer samples than those of a turbulent flow. The solution of the eigenvalue problem of the correlation matrix, employed for the snapshot POD, provides modes that are ordered by their contribution to the flow-field variance. It is thus intuitive that a reliable modal decomposition requires us to have access to a data set that is sufficiently large and rich to achieve a good statistical convergence for what concerns both the estimation of the mean quantities and of their variance (i.e., turbulent kinetic energy if we are analyzing a velocity field).

Considering n independent samples of a flow field $u(x, y)$, without noise content, it is possible to estimate the mean and the variance of a flow field as

$$\begin{aligned} \overline{u} &\approx \frac{\sum_{i=1}^{n} u_i}{n}, \\ \overline{u'^2} &\approx \frac{\sum_{i=1}^{n} (u_i - \overline{u})^2}{n-1}. \end{aligned} \tag{9.4}$$

The two relations in (9.4) tend to equalities for large numbers of samples. According to Everitt and Skrondal (2002), the standard error of a statistic is the standard deviation σ of its sampling distribution. Following Kenney and Keeping (1951), it can be derived that the standard errors of the mean and the variance are equal to

$$\begin{aligned} \sigma_{\overline{u}} &= \frac{\sqrt{\overline{u'^2}}}{\sqrt{n}}, \\ \sigma_{\overline{u'^2}} &= \overline{u'^2}\sqrt{\frac{(n-1)^2}{n^3}\gamma_u - \frac{(n-1)(n-3)}{n^3}}, \end{aligned} \tag{9.5}$$

where γ_u is the kurtosis, that is, the fourth standardized statistical moment of u. For a normal distribution, γ_u is equal to three that results in having $\sigma_{\overline{u'^2}} = \sqrt[2]{\frac{2(n-1)}{n^2}}\overline{u'^2}$. Both mean and variance reach statistical convergence with a dependence to the square root of n; thus doubling the number of elements employed for the calculation of the statistics allows us to reduce the convergence error by approximately 30%. For what concerns the number n:

- For the estimation of the standard error of the mean-flow field, when averaging over time, n is equal to the number of snapshots n_s.
- For the estimation of the standard error of the temporal correlation matrix, n is the number of element of a column of $\boldsymbol{Y}$, n_x.

In fact, the elements of the diagonal of the correlation matrix contain the variance of a certain column of **Y**, which is evaluated along all this elements. Considering average values of the turbulent intensity below 10%, this results in an uncertainty of approximately 1% of the mean velocity and in a 1% of relative uncertainty for the elements of the covariance matrix if a total of 100 independent snapshots of a size of 100×100 elements is employed. The trace of the matrix Λ contains the cumulative variance of the flow field along the entire data set; thus, its uncertainty should be further reduced of a factor $\sqrt{n_s}$; however, it has to be remarked that the variance content associated to any mode is a fraction of the trace of lambda and that modes are ordered by decreasing variance content. The uncertainty of the variance associated to each mode thus affects in a different way the modes depending on their larger or lower energy content. It is intuitive that flow field with a sparse eigenspectrum, that is, with few modes containing a large amount of the variance, enables a modal analysis with a very low relative uncertainty even with a small number of snapshots while the uncertainty increases significantly for more spectrally rich data sets.

Regarding the question, *while acquiring a data set, what should be the acquisition frequency and how many periods should be acquired?* it is a bit more complicated to provide a rigorous answer. In case of a purely periodic phenomenon, such as the wake of a cylinder at a relatively low Reynolds number, exactly sampling an integer number of periods with any sampling frequency should provide a statistically converged modal analysis; statistically converged modes should be also obtained sampling one shedding period at frequency which is an integer multiple of the shedding frequency. The application of such a criterion, however, is not possible for the majority of the flows, especially when the information about the spectral content of the flow is not known a priori.

A more general derivation about the required number of samples (and thus the required sampling time in case of time-resolved data sets) can be obtained considering the effect of the data autocorrelation on the standard errors. The analysis that leads to (9.5) is based on the assumption that the snapshots used for the estimation of the mean and of the correlation matrix are statistically independent. However, when dealing with fluid-flow data sets, sometimes the data sequences are time resolved or spatial information is somehow correlated (consider the typical case of PIV interrogation windows with overlap, which will have a high degree of correlation, typically up to 75%). Under these conditions of correlated data, for a moderate to large data set, the standard errors in (9.5) are larger, of a factor f (Bence 1995):

$$f = \sqrt{\frac{1+\rho}{1-\rho}}, \qquad \text{where } \rho = \frac{\sum_{i=2}^{n} u_{i-1} \cdot u_i}{\sum_{i=2}^{n} u_i^2}. \tag{9.6}$$

The autocorrelation coefficient of the data ρ is an indicator of how much the number of samples should be increased in order to attain statistical convergence, for example, for time-resolved fields in which ρ is likely to be of the order of 0.8, a nine-time-larger number of samples is required to obtain the same statistical convergence as if the samples were statistically independent. This last statement is equivalent to saying that, if we observe a convective flow with a certain convective velocity, to obtain a good convergence we should make sure that the total sampling time is at least one order of magnitude larger than the time to convect from one measurement point to another (in order to obtain at least 100 independent samples for each measurement point). However, due to spatial correlation of the observed flow features this sampling time should be, in principle, even larger and should be one or two orders of magnitude larger than the characteristic convective time of the largest flow features analyzed.

9.2.2 The Effect of Noise

When dealing with experimental data, measurements are affected by errors, typically classified into bias and random error. Considering the case of PIV, while the bias error is mainly responsible for a signal attenuation (in particular, PIV interrogation windows behave as a low-pass filter) random noise affects almost in the same way low-frequency and high-frequency features in a certain measured flow (velocity, pressure, or temperature) field. With a special focus on PIV, several works in the literature have attempted to quantify the effect of noise on the modes, and notable examples are certainly the works by Venturi (2006), Raiola et al. (2015), and Epps and Krivitzky (2019).

Raiola et al. (2015) provide a simple yet useful estimation criterion to quantify which POD modes are mostly affected by noise and to choose a suitable number of modes to use in an optimal low-order reconstruction. The analysis is based on the assumption that random error is Gaussian and the snapshot matrix containing the measured field $\boldsymbol{Y}$ is the sum of the "true" snapshot matrix $\tilde{\boldsymbol{Y}}$ without noise plus the error snapshot matrix $\boldsymbol{E}$, that is, $\boldsymbol{Y} = \tilde{\boldsymbol{Y}} + \boldsymbol{E}$. From this assumption, it is possible to build the snapshot covariance matrix $\boldsymbol{Y}^*\boldsymbol{Y}$ as the sum of the four terms:

$$\boldsymbol{Y}^*\boldsymbol{Y} = \tilde{\boldsymbol{Y}}^*\tilde{\boldsymbol{Y}} + \boldsymbol{E}^*\tilde{\boldsymbol{Y}} + \tilde{\boldsymbol{Y}}^*\boldsymbol{E} + \boldsymbol{E}^*\boldsymbol{E}. \tag{9.7}$$

Assuming that the noise has a Gaussian distribution with standard deviation σ_e and is identically distributed among the snapshots, it is possible to approximate the noise covariance matrix as diagonal matrix with the diagonal elements equal to the square of the noise standard deviation times the number of the elements of the snapshots (Huang et al. 2005), that is, $\boldsymbol{E}^*\boldsymbol{E} = n_x\sigma_e^2\boldsymbol{I}$.[2] It is also reasonable (although not always true) to assume that the noise and the true field are not correlated, so that $\boldsymbol{E}^*\tilde{\boldsymbol{Y}} + \tilde{\boldsymbol{Y}}^*\boldsymbol{E}$ can be neglected in (9.7). This yields

$$\boldsymbol{Y}^*\boldsymbol{Y} \approx \tilde{\boldsymbol{Y}}^*\tilde{\boldsymbol{Y}} + n_x\sigma_e^2\boldsymbol{I}. \tag{9.8}$$

According to Marčenko and Pastur (1967), this approximation is still valid when the matrix $\boldsymbol{Y}$ is rectangular, even for a non-Gaussian error that is independent on the flow field and is identically distributed among the snapshots so that all the eigenvalues of $\boldsymbol{E}^*\boldsymbol{E}$ are in a neighborhood of $n_x\sigma_e^2$. This last assumption is equivalent to the assumption that the eigenvalues of the covariance matrix of the random error have a constant spectra distribution.

[2] Note that according to (9.4) the unbiased estimation of $\sigma_e \approx \sqrt{\frac{\sum_{i=1}^{n_x} e_i^2}{n_x - 1}}$. However, when dealing with the total energy corresponding to a certain eigenvector, this scales with n_x.

Comparing the solution of the eigenvalue problem of the measured and true covariance matrices, it is possible to write that

$$\begin{aligned} \boldsymbol{\Psi}^* \boldsymbol{Y}^* \boldsymbol{Y} &= \boldsymbol{\Psi}^* \left(\tilde{\boldsymbol{Y}}^* \tilde{\boldsymbol{Y}} + n_x \sigma_e^2 \boldsymbol{I} \right) = \tilde{\boldsymbol{\Lambda}} \boldsymbol{\Psi}^*, \\ \tilde{\boldsymbol{\Psi}}^* \tilde{\boldsymbol{Y}}^* \tilde{\boldsymbol{Y}} &= \tilde{\boldsymbol{\Lambda}} \tilde{\boldsymbol{\Psi}}^*. \end{aligned} \tag{9.9}$$

If the random-error eigenvalues are small with respect to the difference between two successive eigenvalues $\tilde{\lambda}_i$ and $\tilde{\lambda}_{i+1}$ of $\tilde{\boldsymbol{Y}}^* \tilde{\boldsymbol{Y}}$, it is possible to write that $\lambda_i = \tilde{\lambda}_i + n_x \sigma_e^2$ and that the ith temporal mode of $\boldsymbol{Y}$ is approximately equal to the ith temporal mode of $\tilde{\boldsymbol{Y}}$ (Venturi 2006). The perturbation of the eigenvectors increases with an increasing mode number as the variance content of the ith mode $\tilde{\lambda}_i$ approaches the value of $n_x \sigma_e^2$. These relationships can be accurately derived from matrix perturbation theory, along with their bounds, and are a common assumption in perturbed principal component analysis (PCA) applications (Huang et al. 2005). However, such a simplified description can be useful and is certainly sufficient to draw some conclusions and identify some practical insights:

- If a data set can be represented with a small number of modes, it is possible to obtain very high signal-to-noise ratios employing a low-order reconstruction with a reduced number of modes, that is, $\boldsymbol{Y_k} = \boldsymbol{\Phi} \boldsymbol{\Sigma_k} \boldsymbol{\Psi}^*$. $\boldsymbol{\Sigma_k}$ is the singular value matrix keeping only the first k elements of the diagonal. Consider, for instance, a data set of 100 snapshots in which the standard deviation of the noise is 1% of the reference velocity while the turbulent fluctuations are 10% and assuming that 90% of the turbulent fluctuations are reconstructed with 10 modes. A low-order reconstruction with only 10 modes allows us to reconstruct fluctuating fields with a signal-to-noise ratio of 90, clearly larger than the initial SNR = 10.
- Since the eigenvalues of the covariance matrix are representative of the total variance in the data set, they will increase linearly with an increasing number of samples, assuming that the flow is statistically stationary. The measurement noise, instead, spreads almost uniformly among all the modes keeping a constant contribution equal to $n_x \sigma_e^2$. It is thus possible to increase the ratio $\frac{\tilde{\lambda}_i}{n_x \sigma_e^2}$ and improve the quality of the modal decomposition simply increasing the number of snapshots employed. This result is very appealing since this relation is linear!
- While it is possible to obtain a theoretical demonstration of the existence of an optimum number of modes (Raiola et al. 2015), often the noise variance is unknown and cannot be determined a priori without previous knowledge of the spectral content of the noise-free data set. The optimality property of POD ensures that the most energetically relevant features are contained in the first modes and that the energy content rapidly decays along the eigenspectrum. As discussed here, the contribution of the noise should instead be practically constant. After a certain mode number, thus, the variance content will remain almost constant, being mainly ascribed to noise. This idea returns an operative criterion, inspired by the scree plot test (Cattell 1966), which consists in looking for an elbow in the plot of the cumulative variance content versus the mode number.

9.2.3 An Example: Noisy Measurements in a Turbulent Channel

The effect of noise on POD modes, along with the choice of a suitable number of modes to obtain an optimal low-order reconstruction, was assessed against noisy fields obtained from a turbulent channel flow (Raiola et al. 2015). A total of 8 000 turbulent fields were extracted from the Johns Hopkins turbulence database (Li et al. 2008) in streamwise/wall-normal planes with a size of $h \times h$ with h being the channel half height. These fields were employed to generate synthetic PIV particle image couples, which were then processed to obtain "measured" flow fields as in a PIV experiment. For this data set, PIV images allowed us to obtain a known level of random noise while the amount of snapshots and measurement points were sufficiently large to ensure statistical convergence. It was shown that employing the scree plot criterion would allow us to find an optimum number of modes maximizing the true signal and removing most of the noise. The reconstruction error was defined as the Frobenius norm of the difference between the low-order reconstruction of the data set $\boldsymbol{Y}_i$ with respect to the measured data set $\boldsymbol{Y}$. The scree plot method was employed looking for $F \approx 1$ with F being the ratio of the backward and forward derivative of the reconstruction error versus the number of modes; that is,

$$\begin{aligned} \delta_{RM}(i) &= \frac{1}{n_x} ||\boldsymbol{Y} - \boldsymbol{Y}_i||_F, \\ F(i) &= \frac{\delta_{RM}(i+1) - \delta_{RM}(i)}{\delta_{RM}(i) - \delta_{RM}(i-1)}. \end{aligned} \tag{9.10}$$

Figure 9.1 reports the evolution of the reconstruction error and of the parameter F versus the number of modes and shows that for $F > 0.999$ the reconstruction error with respect to the true fields $\delta_{RT}(i) = \frac{1}{n_x}||\tilde{\boldsymbol{Y}} - \boldsymbol{Y}_i||_F$ is minimum, thus this criterion can be employed to identify an optimal number of modes k_F to mitigate noise contamination of PIV data. This result is further confirmed from a visual inspection of Figure 9.2, which shows that two vorticity patterns of similar magnitude are measured by PIV, one present in the real flow field and the other one being an artifact. The low-order reconstruction with the optimal number of modes conserves the true vortex while the artifact is removed.

9.3 Extracting Phase Information

When dealing with flow-field measurements, experimental data sets are often sampled at frame rates that are not sufficient to correctly describe the dynamics of the problem under study. However, even when it is not possible to actually measure the frequency of the flow unsteady phenomena, the POD temporal modes allow us to determine relevant phase information. In case of phenomena characterized by dominant periodic features (such as the shedding in the wake of a bluff body), these periodic features can be usually represented with a compact subset of modes. In fact, a plot of the correlation-matrix eigenvalues λ_i versus mode number usually shows a quite clear spectral separation between first modes, accounting for the dominant periodic features,

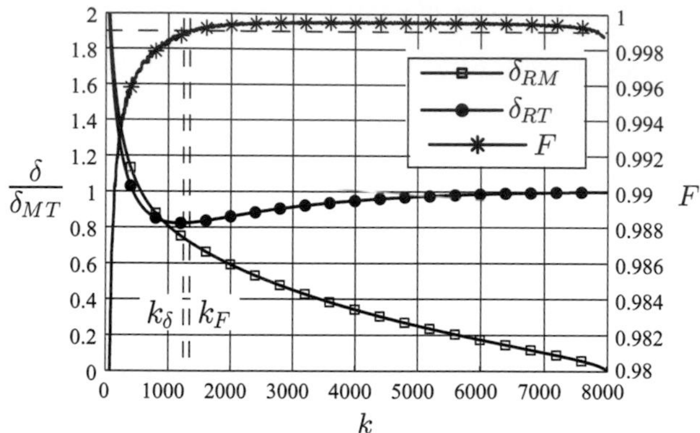

Figure 9.1 Reconstruction errors with respect to true and measured fields (δ_{RT} and δ_{RM}) (left axis) and F (right axis) versus the number of modes used in the reconstruction. (From Raiola, Discetti & Ianiro (2015), reproduced by permission from Springer Nature.)

and the following modes that describe the smaller-scale turbulent features. This separation in the eigenspectra can be exploited to effectively extract periodic flow features (Perrin et al. 2007) as detailed in the following.

Considering a temporally homogeneous flow or a temporally periodic phenomenon (with period τ), the temporal correlation matrix is a function of solely the temporal separation between two snapshots, that is, the temporal correlation between the time t and the time t' is only a function of $t - t'$, that is, $R_t(t,t') = R_t(t - t')$. The eigenfunctions of the temporal correlation matrix must thus be Fourier modes as discussed in Chapter 6. In case of a shedding-dominated phenomenon in which the most energetic flow features are characterized by a statistical periodicity with a dominant frequency, it is safe to assume that the first POD modes, apart from being orthogonal, also align to a Fourier decomposition of the field and show a strong harmonic relation. Therefore, the POD temporal modes with larger eigenvalues, obtained solving the eigenvalue problem of the temporal correlation matrix, can unveil their relative phase information.

If we consider cases such as that of the vortices shed in the wake of a bluff body or the vortices developed due to shear-layer instabilities in a jet, a traveling wave is described by two high-energy modes (which are often found to be the first two modes). Both modes have to share the same periodicity, that is, the shedding period τ, thus, according to the orthogonality of the POD temporal modes, it is possible to assume that the low-order reconstruction employing only the first two POD modes would coincide with the decomposition (Raiola et al. 2016)

$$\boldsymbol{a}_2(t) = \boldsymbol{y}_0 + \psi_1(\vartheta)\sigma_1\boldsymbol{\phi}_1 + \psi_2(\vartheta)\sigma_2\boldsymbol{\phi}_2, \tag{9.11}$$

where $\sigma_i = \sqrt{\lambda_i}$ and $\vartheta = \frac{2\pi t}{\tau}$ is the period phase. The functions $\psi_1(\vartheta)$ and $\psi_2(\vartheta)$ will likely be two sinusoidal functions and must be in phase quadrature, that is, $\psi_1 \propto \sin(\vartheta)$ and $\psi_2 \propto \sin(\vartheta + \pi/2)$. The scatter plot of the temporal modes $\boldsymbol{\psi}_1$ and $\boldsymbol{\psi}_2$ distributes

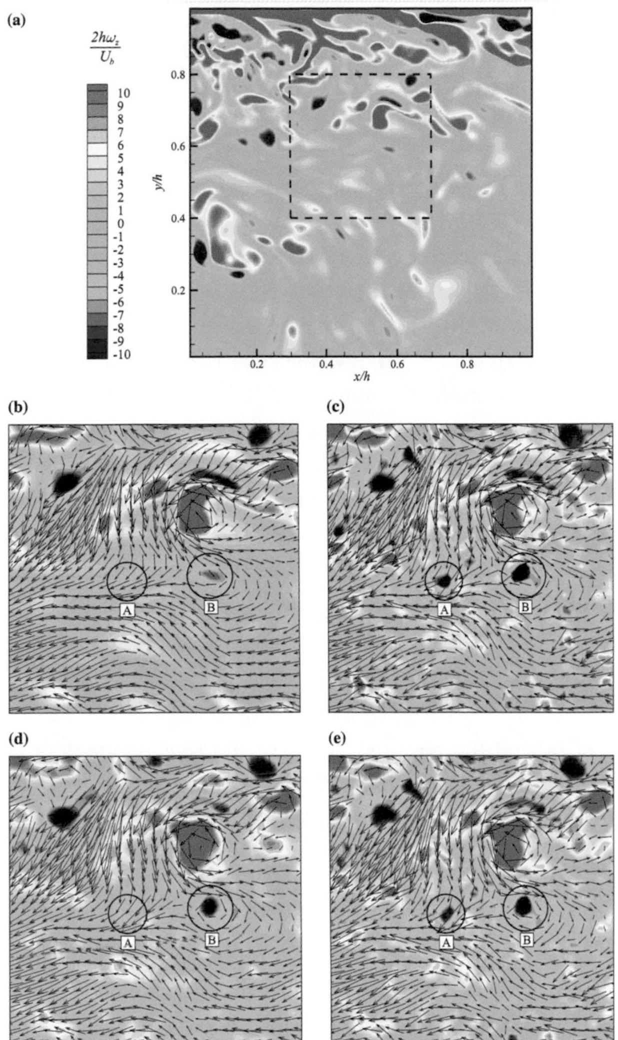

Figure 9.2 Instantaneous fluctuating vorticity (ω_z) field. (a) The DNS field used for this benchmark. Magnified view of the: (b) DNS field, (c) measured field, (d) field reconstructed with 1300 modes, (e) field reconstructed with 3000 modes. (From Raiola, Discetti & Ianiro (2015), reproduced by permission from Springer Nature.)

in the neighborhood of a circle if $\psi_1(\vartheta)$ and $\psi_2(\vartheta)$ are sinusoidal functions. Since temporal-mode vectors have a unitary norm and zero mean, the scatter plot will be in the neighborhood of a goniometric circle (with radius 1) if the temporal modes are multiplied by the factor $\sqrt{n_s/2}$.

In general, the scatter plot of the time coefficients might unlock information on the phase and frequency relation between the first and other modes, often corresponding to higher-order harmonics, thus shedding light on the interconnection between the different flow features highlighted by the modal analysis. Assuming that the ith POD mode is harmonically related and phase shifted with respect to the first mode, it is possible to write that

$$\sqrt{n_s/2}\psi_i(\vartheta) = \sin(\alpha_i\vartheta + \delta_i), \tag{9.12}$$

where α_i is a positive integer and δ_i is the phase shift of the ith mode with respect to the first mode.

In order to ascertain these frequency relations, the scatter plot of the POD temporal modes ψ_1 and ψ_i should be observed in search of Lissajous curves. A Lissajous curve is the graph of a system of parametric equations of the type $x = A\sin(\alpha t + \delta)$ and $y = B\sin(\beta t)$, which can describe even very complex harmonic motions. Visually, the ratio α/β determines the number of "lobes" of the figure. The procedure to identify the harmonic relation and the phase shift for higher order modes relies on the simplifying assumption that the first two modes are at the same frequency with a $\pi/2$ phase shift. It is possible to extract the period phase from the time coefficients of the first two modes, that is, for any snapshot $tan(\vartheta) = \psi_1/\psi_2$. Subsequently, the positive integer α_i and the phase shift δ_i, which characterize the harmonic relation, can be extracted from the solution of the optimization problem

$$\underset{\substack{\alpha_i\in\mathbb{N}\\ \delta_i\in\mathbb{R}}}{\operatorname{argmin}}\left(\sqrt{n_s/2}\,\psi_i - \sin(\alpha_i\vartheta + \delta_i)\right), \tag{9.13}$$

where α_i and the phase shift δ_i are the free parameters to identify. An example of this procedure is given in the following, where it is applied to the analysis of the flow features in the wake of two tandem cylinders located near a wall.

9.3.1 An Example: The Wake of Two Cylinders in Ground Effect

The work by Raiola et al. (2016) reports a PIV wind tunnel study in the wake of cylinders near a wall. The experimental arrangement is sketched in Figure 9.3. The cylinders had a diameter of 32 mm. The acquisition was performed with a 2Mpixels camera (1600 × 1200 pixels array) with a spatial resolution of about 7.2 pixels/mm while the sampling frequency of the PIV measurements was 10 Hz and the Reynolds number was equal to 4.9×10^3, thus the sampling frequency was not sufficient to sample the characteristic frequencies involved in the problem. An ensemble of 2 000 image couples was acquired for each experiment, obtaining a total of 2 000 flow fields. The interrogation strategy was an iterative multistep image deformation algorithm, with final interrogation windows of 16 × 16 pixels with 50% overlap (i.e., final vector spacing of 8 pixels, corresponding to $0.035D$). The ensemble size both in term of snapshots and number of points enables a sufficient convergence of the statistics (errors well below 1% of the variance both in time and space).

The results of the modal analysis are reported herein for the case of two cylinders with a relative distance $L = 1.5D$ and a gap from the wall $G = 1D$. Under these conditions the vortex shedding for the upstream cylinder is completely suppressed, and the two cylinders act as a single bluff body (Zdravkovich 1987). At the same time, the ground affects the pressure distributions, as well as the flow development.

As reported in Figure 9.4, the first two modes account for almost 60% of the variance, and the scatter plots of the temporal modes (multiplied times $\sqrt{n_s/2}$) return a perfect goniometric circle confirming that these two modes are representative of a traveling pattern. The asymmetry of the first POD mode highlights the strong interaction of the wall boundary layer with the shedding in the wake, evident from the presence of near-wall vorticity visualized in the first (and second mode, not reported here for brevity). Higher-order modes are mostly characterized by intermittent release of vorticity at a spatial and temporal frequency that is a multiple of the principal shedding frequency as shown from the contour plot of the fourth spatial mode and outlined from the Lissajous curves in the scatter plot of the fourth modes against the first one. This is an asymmetric mode that models the cross-wise oscillation of the shedding wake.

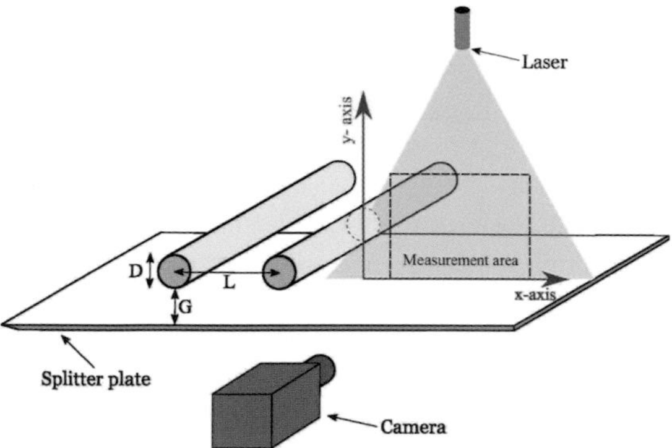

Figure 9.3 Experimental setup for PIV measurements in the wake of two tandem cylinders in ground effect. (Reprinted from Raiola, Ianiro & Discetti (2016), copyright 2016, with permission from Elsevier.)

9.4 Extended Proper Orthogonal Decomposition

Until this point, this chapter has been centered on the analysis of a given data set $\boldsymbol{A}$. However, the availability of multiple data sets, for example, $\boldsymbol{A}$ and $\boldsymbol{B}$, are rather frequent. Subtracting the base conditions $\boldsymbol{y_{0,A}}$ to $\boldsymbol{A}$ and $\boldsymbol{y_{0,B}}$ to $\boldsymbol{B}$, it is possible to obtain $\boldsymbol{Y_A}$ and $\boldsymbol{Y_B}$. Given these two data sets, the most straightforward task would be to perform two separate modal analyses, as for instance performed by Mallor et al. (2019). However, if the measured quantities are synchronized, it is tempting to perform a combined analysis to extract information not only about the modes of a certain data set, for example, $\boldsymbol{A}$, but also about the features of the second data set ($\boldsymbol{B}$) that correlates with these modes.

To this purpose, the reader must remember that the POD spatial modes can be obtained from a projection of the snapshot matrix onto the temporal-mode matrix as in (9.3). This projection approach can be used to extend the POD to other quantities, in a generalized approach named Extended POD (Borée 2003). The columns of $\boldsymbol{\Psi_A}$, that is, the POD temporal modes, form a basis in the $\mathbb{R}^{n_s}$ vector space; thus, it is possible to use $\boldsymbol{\Psi_A}$ as a basis for the projection also of the matrix $\boldsymbol{Y_B}$, provided that

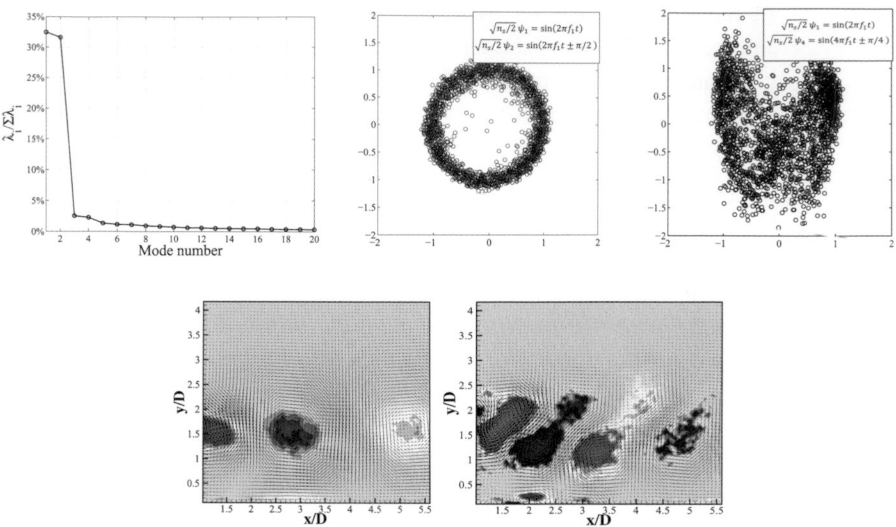

Figure 9.4 Modal analysis results for two tandem cylinders with $L/D = 1.5$ and $G/D = 1$. On the top, the correlation matrix eigenvalues are reported along with the scatter plots of the first temporal mode against the second and the fourth modes (temporal modes are multiplied by $\sqrt{n_S/2}$). At the bottom of the figure, the spatial modes 1 and 4 are reported, highlighting the fact that the first and the fourth modes have an even spatial frequency ratio with a phase shift of $\pi/4$. (Adapted from Raiola, Ianiro & Discetti (2016), copyright 2016, with permission from Elsevier.)

the available number of snapshots of $\boldsymbol{A}$ and $\boldsymbol{B}$ is the same. Consequently, while in (9.3) the projection was used to estimate the spatial modes $\boldsymbol{\Phi_A}$ of $\boldsymbol{A}$, it is possible to define the EPOD modes of $\boldsymbol{B}$ as

$$\boldsymbol{Y_B} = \boldsymbol{\Phi_{B_A}} \boldsymbol{\Sigma_{B_A}} \boldsymbol{\Psi_A^*} \rightarrow \boldsymbol{\Phi_{B_A}} \boldsymbol{\Sigma_{B_A}} = \boldsymbol{Y_B} \boldsymbol{\Psi_A}, \tag{9.14}$$

where the subscripts A and B refer to the quantities A and B, respectively. The columns of $\boldsymbol{\Phi_{B_A}}$ are thus the spatial modes of $\boldsymbol{Y_B}$ obtained from the projection on the temporal modes of $\boldsymbol{Y_A}$. The data sets $\boldsymbol{A}$ and $\boldsymbol{B}$ have to be captured/generated in the same time reference frame; however, they might represent different physical quantities, and the snapshots of $\boldsymbol{A}$ and $\boldsymbol{B}$ can have different numbers of elements.

While the POD temporal modes $\boldsymbol{\psi_{A_i}}$ are estimated from the solution of the eigenvalue problem of the correlation matrix $\boldsymbol{Y_A^* Y_A}$ and are optimal for $\boldsymbol{Y_A}$, the extended spatial modes $\boldsymbol{\phi_{B_{A,i}}}$ are not optimal nor are necessarily ordered by their variance content $\sigma_{B_i}^2$. The value of $\sigma_{B_i}^2$ that can be computed as the square root of the variance of the columns of $\boldsymbol{\Phi_{B_A}} \boldsymbol{\Sigma_{B_A}}$ in order to obtain $\boldsymbol{\phi_{B_{A,i}}}$ vectors of unitary norm.

It has to be remarked that the projection of the snapshot matrix $\boldsymbol{B}$ on the temporal mode matrix $\boldsymbol{\Psi_A}$ is energy preserving since $\boldsymbol{\Psi_A}$ is a basis in the $\mathbb{R}^{n_s}$ vector space. Due to the orthogonality of the temporal modes $\boldsymbol{\psi_{A_i}}$ this has the consequence that EPOD decomposition, although not being optimal is orthogonal. The orthogonality principle has the advantage to guarantee that each EPOD mode $\sigma_{B_{A,i}} \boldsymbol{\phi_{B_{A,i}}}$ contains only the part of $\boldsymbol{B}$ that is correlated with the contribution of the ith POD mode to $\boldsymbol{A}$. This concept is closely connected with the idea of the linear stochastic estimation (LSE), first developed by Adrian (1975), a technique that is typically employed to estimate an unknown quantity given a known quantity. As the name suggests, this technique attempts to statistically draw a linear relation between a known quantity and another quantity that has to be estimated. The EPOD can be considered as the LSE of the spatial modes $\sigma_{B_{A,i}}\boldsymbol{\phi}_{\boldsymbol{B}_{A,i}}$, given the spatial modes $\boldsymbol{\phi}_{A_i}$. All the properties of the LSE can therefore be extended to the EPOD. In particular, it must be noted that, while the relation between velocity fields and other quantities may be formally nonlinear, the LSE may still prove to be adequate due to the small magnitude of second-order terms, as shown by Adrian et al. (1989) for the case of homogeneous turbulence. Several attempts to determinate the velocity modes correlated to the POD modes of other quantities have been reported in the literature. Two examples are reported in the following: the first making use of the extended POD to reconstruct time-resolved fields from time resolved hot-wire measurements and non-time-resolved PIV measurements and the second one that analyzes the thermal transport in a pipe with nonhomogeneous heating.

9.4.1 An Application: Time-Resolved Flow Sensing with the Extended Proper Orthogonal Decomposition

In the works by Discetti et al. (2018) and Discetti et al. (2019), we have employed a combination of non-time-resolved field measurements and time-resolved point measurements for the estimation of time-resolved turbulent fields, as a reduced-order reconstruction with a limited subset of modes. The theoretical concepts behind this approach have been developed through a vast literature (Ewing & Citriniti 1999, Tinney et al. 2008, Baars & Tinney 2014, Hosseini et al. 2015, Discetti et al. 2018, 2019) and essentially grounded in the fact that EPOD can be employed for linear stochastic estimation, that is, we can estimate the mode time coefficients of a certain quantity with the known time coefficients of the modes of another synchronized quantity.

Consider the two synchronized data sets: one at low repetition rate with PIV and the other one at a high repetition rate with hot-wire anemometry. Due to the different repetition rates, the number of time samples for the hot wire is much larger than that for PIV, although fewer points are available in space. Employing only the time instants at which both PIV and hot-wire measurements are available, it is possible to build two matrices with synchronized snapshots: one named $\boldsymbol{Y}_P$ (where P stands for PIV) and another named $\boldsymbol{Y}_H$ (where H stands for hot wire). The eigenvalue decomposition of the correlation matrices of both quantities allows us to identify the matrices of the temporal modes $\boldsymbol{\Psi}_P$ and $\boldsymbol{\Psi}_H$. The product $\boldsymbol{\Psi}_P^*\boldsymbol{\Psi}_H$ returns a new matrix $\boldsymbol{\Xi} = \boldsymbol{\Psi}_P^*\boldsymbol{\Psi}_H$ that contains in its element i, j the correlation between the ith mode of the first data set (PIV) and the jth mode of the second data set (HW).

The correlation information contained in the matrix $\boldsymbol{\Xi}$ can be used to estimate the PIV flow fields at those time instants when hot-wire measurements are available, even if PIV was not sampling. At any time instant t, the availability of a hot-wire snapshot

$\boldsymbol{y}_H(t)$ allows us to estimate the vector containing the time coefficients of the hot-wire spatial modes, $\boldsymbol{\psi}_H(t)$, through a simple projection:

$$\boldsymbol{\psi}_H(t) = \boldsymbol{\Sigma}_H^{-1}\boldsymbol{\Phi}_H^* \boldsymbol{y}_H(t). \tag{9.15}$$

Given $\boldsymbol{\psi}_H(t)$, it is possible to estimate the time coefficients of the PIV modes. The vector $\boldsymbol{\psi}_P(t)^+$, containing the estimated time coefficients of all the PIV spatial modes, is obtained following (9.16), and the PIV snapshot can be reconstructed as in (9.17):

$$\boldsymbol{\psi}_P^+(t) = \boldsymbol{\Xi}\boldsymbol{\psi}_H(t), \tag{9.16}$$

$$\boldsymbol{y}_P^+(t) = \boldsymbol{\Phi}_P\boldsymbol{\Sigma}_P\boldsymbol{\psi}_P^+(t). \tag{9.17}$$

The reader might notice that the success of such a procedure would require a significant number of hot-wire probes in order to avoid a significant difference between the dimensionality of PIV and hot-wire snapshot matrices. Even if some works in the literature have employed large amounts of probes (Kerhervé et al. 2017), this is not easily available. The most common approach passes through the definition of pseudo-snapshots $\boldsymbol{a}_{\boldsymbol{H}_i}$ in which several measurements are taken at the same "instant" as for PIV adding "virtual" probes. Virtual probes are obtained using time-shifted probe data with the support of the Taylor hypothesis of uniform convection, converting the hot-wire temporal information into spatial information.

It is worth noticing that the columns of both $\boldsymbol{\Psi}_{\boldsymbol{P}}$ and $\boldsymbol{\Psi}_{\boldsymbol{H}}$ are orthonormal vectors and form two bases in the $\mathbb{R}^{n_s}$ vector space; consequently $\boldsymbol{\Xi}$ is also composed by columns forming a basis. This implies that the process is energy preserving and that the higher-order modes that might be mutually uncorrelated are taken into account for the estimation in (9.16). Since all rows/columns of $\boldsymbol{\Xi}$ have unitary norm, if a certain ith probe mode (jth field mode) is uncorrelated with all the field modes (the snapshot modes), the ith row (jth column) of $\boldsymbol{\Xi}$ has to be composed of randomly distributed elements, with unitary norm and zero mean, thus standard deviation equal to $1/\sqrt{n_t}$. This reasoning allows us to filter the matrix $\boldsymbol{\Xi}$ removing all the elements $|\Xi_{i,j}| < 1/\sqrt{n_t}$ (Discetti et al. 2018). This approach has allowed us to obtain a reconstruction of the behavior of large-scale and very-large-scale motions high-Re turbulent flows. An example of experiment is reported in Figure 9.5 in which PIV measurments are synchronized with five hot-wire probes in the large-scale pipe-flow facility CICLoPE (Univ. of Bologna, Italy). Figure 9.5, right-hand side, reports a comparison between the PIV estimated fields and the hot-wire measurements that shows a significant coherence and low noise level, despite the significant spectral richness of the flow (Discetti et al. 2019).

As a further extension of this work, aiming at the identification of a suitable sensing tool for the detection of large-scale and very-large-scale motions in a turbulent wall-bounded flow, we have recently explored the performances of convolutional neural networks (Güemes et al. 2019, Guastoni et al. 2020), obtaining better results than EPOD. This highlights that machine learning tools have the capability to enhance and eventually replace traditional data analysis tools in the field of fluid mechanics.

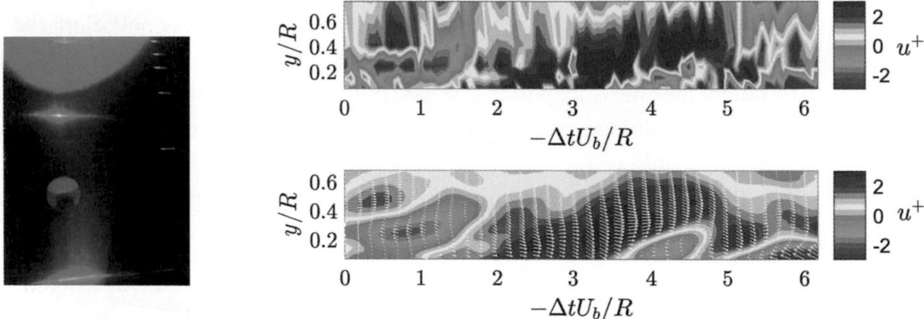

Figure 9.5 Left: Image of the experimental arrangement with PIV image plane and hot-wire rake located downstream. Right: Contour of the evolution of the streamwise velocity component obtained using Taylor's hypothesis, $Re_\tau = 9500$. Top: hot-wire data. Bottom: Estimated fields. The velocity contours are shown in inner units, that is, $u^+ = u/u_\tau$. (Reprinted from Discetti et al. (2019), copyright 2019, with permission from Elsevier.)

9.4.2 An Exercise: The Turbulent Heat Transfer in a Pipe Flow

In the work by Antoranz et al. (2018), we analyzed the thermal transport in a turbulent pipe flow with nonhomogeneous heating. This configuration is representative of the receivers of thermal solar power systems in which mirrors concentrate solar radiation toward a receiver made of pipes through which a molten nitrate salt is flowing. In this application, it is of foremost importance that the received thermal input is convected along the whole pipe section in order to avoid that uneven temperature distributions cause thermal stresses in the pipe and/or chemical reactions in the molten nitrate salts employed as working fluid.

The data set analyzed in the paper was generated with direct numerical simulations of a fully developed turbulent pipe flow, with circumferentially varying heat flux boundary conditions. Details about the simulations are available in the works by Antoranz et al. (2015) and Antoranz et al. (2018). The boundary conditions at the pipe wall were nonslip conditions for the velocity while the heat flux had a sinusoidal distribution for azimuthal cohordinate θ between 0 and π and was set to zero for $\pi < \theta < 2\pi$. Exploiting the statistical homogeneity of the fully developed pipe flow, a total of 4 500 snaphsots were extracted taking radial planes at 50 streamwise locations of the pipe at 90 time instants. Available quantities were three velocity components (u, v, w), and the temperature T. In the following, the case at $Re_\tau = 180$ and $Pr = 0.7$ is analyzed.

Snapshot POD is performed on the temperature snapshot matrix, and extended velocity modes (projected on the temperature temporal basis $\boldsymbol{\Psi}_T$) are estimated. Figure 9.6, on the top, reports the normalized variance content of the temperature modes and of the velocity extended modes. Although it should be clear to the reader that the POD temperature modes are optimal for the reconstruction of the temperature variance, it is still surprising that the majority of the thermal variance (and thus of the turbulent thermal transport) is correlated only with a little amount of the velocity variance. Looking at the cumulative sum of the variance, 80% of the temperature variance is correlated with only 40% of the velocity variance. The fact that the turbulent thermal transport is ascribed only to a limited amount of the velocity variance suggests possible approaches for heat transfer enhancement, such as the use of vortex generators able to produce a flow pattern similar to the first extended velocity mode (see Figure 9.6, bottom) to improve the temperature uniformity in the pipe section.

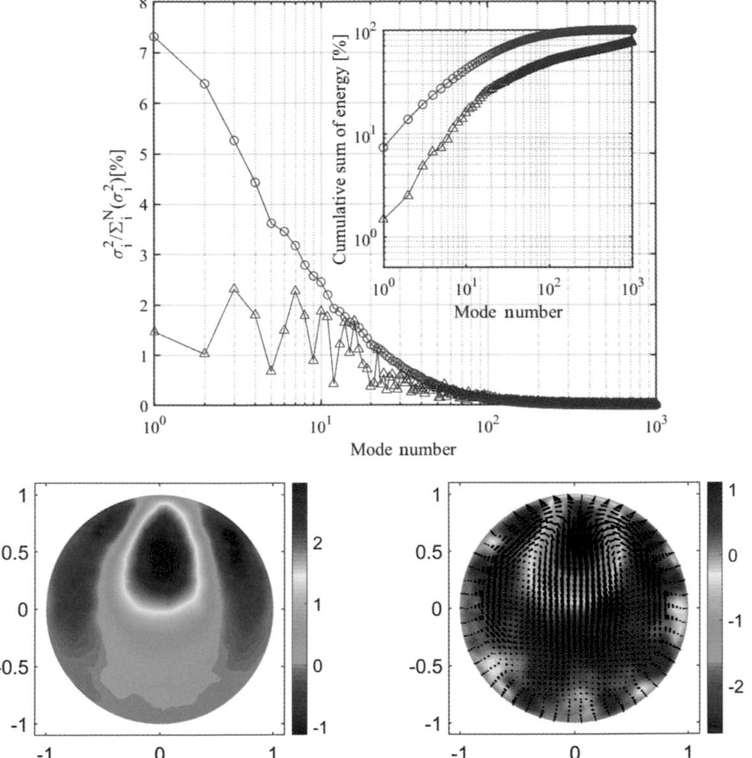

Figure 9.6 Results of the EPOD analysis in a turbulent pipe with nonhomogeneous heat flux. Top: Variance content (cumulative variance content in the inset) of temperature modes (blue) and velocity extended modes (red). Bottom left: First temperature mode. Bottom Right: First extended velocity mode with contour of the streamwise velocity component and in-plane velocity vectors.

As an exercise, a few lines of the MATLAB®code used to generate the results presented in Figure 9.6 are provided. For the present data set, the data from the numerical simulation are not available on a uniform Cartesian mesh, hence requiring us to take into account the different flow areas corresponding to each point of the snapshot matrix. Since the spatial mesh is not spatially uniform in the $x-y$ plane, the temporal correlation matrix of a generic quantity g should be computed taking into account the weights corresponding to the different areas of all grid points (see also Chapter 6 on the weighted SVD).

A data set named *case*1.*m* is provided, containing temperature and velocity snapshot matrices, along with geometrical information about the grid points of the snapshots. A total amount of $n_s = 2275$ snapshots is available with a mesh made of $n_r = 69$ radial locations r and $n_\theta = 100$ equally spaced tangential locations. It has to be noted, here, that a limited number of snapshots is provided, in order to enable quick calculations on the laptops of the students attending the lecture series; however, the turbulent pipe flow is characterized by a rich spectrum that causes a relatively poor convergence of the modal analysis for higher mode numbers.

```
function C = corr_matrix_polar(g,r)
nr = size(g,1);
ntheta = size(g,2);
ns = size(g,3);
%weights calculation
%The first point is in the center of the domain and
    ↪ has
%area dr*dr
dr = diff(r);
weight = 0*r;
weight (1) = dr(1)/2;
weight(2:nr-1) = (dr(1:nr-2)+dr(2:nr-1))/2;
weight(nr) = dr(end)/2;
temp = g.*repmat((2*pi/ntheta*r.*weight).^0.5,[1,nt,
    ↪ ns]);
% All the points have area r*dr*2*pi/ntheta.
temp = reshape(temp,nr*ntheta,ns);
%Area weighted correlation matrix
C = temp'*temp/pi;
% Correlation matrix has to be divided by
% the total area pi*R^2 (here R=1).
```

Once the correlation matrix has been computed, it is possible to estimate the temperature modes and the extended velocity modes for u, v, and w based on the temperature temporal modes $\boldsymbol{\Psi}_{\boldsymbol{T}}$:

```
%building snapshot matrices
nzplanes = size(T,3);
T = reshape(T,nr*ntheta,ns);
u = reshape(u,nr*ntheta,ns);
v = reshape(v,nr*ntheta,ns);
w = reshape(w,nr*ntheta,ns);
%SVD of the temperature covariance matrix
[Psi_T lmd_T Psistar_T]=svd(C_T);
%Calculating spatial modes
SigmaPhi_T=(T)*Psi_T;
SigmaPhi_x_T=(u)*Psi_T;
SigmaPhi_y_T=(v)*Psi_T;
SigmaPhi_z_T=(w)*Psi_T;
```

To calculate the variance content σ_i^2 of each mode $\boldsymbol{\phi_i}$, it is again necessary to take into account the different flow areas corresponding to each point:

```
function sigma = get_sigma(sigmaphi,r,nr,nt)
sigmaphi = reshape(sigmaphi,nr,ntheta);
dr = diff(r);
weight = 0*r;
weight (1) = dr(1)/2;
weight(2:nr-1) = (dr(1:nr-2)+dr(2:nr-1))/2;
```

```
weight(nr) = dr(end)/2;
temp = sigmaphi.*repmat((2*pi/ntheta*r.*weight)
    ↪ .^0.5,[1,nt]);
temp = reshape(temp,nr*ntheta,1);
sigma = (temp'*temp/pi)^0.5;
```

Part IV

Dynamical Systems

10 Linear Dynamical Systems and Control

S. Dawson

This chapter introduces a suite of ideas and tools from linear control theory, which can be used to modify the behavior of linear dynamical systems using feedback control. We use a simple example problem to motivate these methods and explicitly demonstrate how they can be applied. The system that we focus on is a linearization of unstable fluid flow over a bluff body. This system has already been discussed in Chapter 1, and here we start with the same data set used in Chapter 6 as an example for computing the proper orthogonal decomposition (POD). We start with a brief review of linear systems in the time and frequency domain before introducing the example system and embarking on a journey through different approaches to control it. In particular, we introduce proportional, integral, and derivative (PID) control, pole placement, and full-state linear quadratic regulator (LQR) optimal control. Due to the breadth of topics covered, which might typically take one or several semester-long courses to cover, we will not be able to provide a full theoretical background behind each method but will focus on the main ideas and the practical implementation of these methods. We also briefly discuss further extensions and a number of applications of linear control theory.

10.1 Linear Systems

In many applications in science and engineering, we are interested in analyzing the properties of systems that evolve in time. These temporal dynamics are often due both to the internal dynamics of the system itself, and to the effect of external inputs on the system, which might be known or unknown, controllable or uncontrollable, and either beneficial or harmful to the desired behavior of the system.

The simplest dynamical system is one that is linear, with time evolution that can be represented using a system of linear ordinary differential equations,

$$\dot{\boldsymbol{x}}(t) = \boldsymbol{A}\boldsymbol{x}(t), \tag{10.1}$$

where $\boldsymbol{x}(t) \in \mathbb{R}^n$ is a column vector consisting of each state in the system, $\dot{\boldsymbol{x}}(t)$ is the time derivative of this quantity, and $\boldsymbol{A}$ is an $n \times n$ matrix, which we assume to have real entries. Recall that this system is asymptotically stable if the eigenvalues of $\boldsymbol{A}$ (also referred to as the poles of the system) are located in the left half of the complex plane. Accounting for inputs $\boldsymbol{u}(t)$ and outputs $\boldsymbol{y}(t)$, the full system can be specified by matrices $(\boldsymbol{A}, \boldsymbol{B}, \boldsymbol{C}, \boldsymbol{D})$, with

$$\begin{aligned} \dot{\boldsymbol{x}}(t) &= \boldsymbol{A}\boldsymbol{x}(t) + \boldsymbol{B}\boldsymbol{u}(t), \\ \boldsymbol{y}(t) &= \boldsymbol{C}\boldsymbol{x}(t) + \boldsymbol{D}\boldsymbol{u}(t). \end{aligned} \tag{10.2}$$

The state-space system in (10.2) can be expressed as a transfer function in the Laplace (frequency) domain by

$$P(s) = C(sI - A)^{-1}B + D. \tag{10.3}$$

$P(s)$ can also be interpreted as the ratio $\frac{Y(s)}{U(s)}$, where $U(s)$ and $Y(s)$ are the Laplace transforms of the input, $\boldsymbol{u}(t)$ and output, $\boldsymbol{y}(t)$, respectively.[1] The eigenvalues of $\boldsymbol{A}$ in (10.2) correspond to the values of s where the term $(sI - A)^{-1}$ in (10.3) (which is known as the resolvent) is not defined, due to $(sI - A)$ being non-invertible. Further discussion of the stability of linear systems, and their analysis through the resolvent, can be found in Chapter 13. A linear system of the form given by (10.2) or (10.3) is the "plant" that we will ultimately be seeking to control. We can represent this system and its inputs and outputs schematically as

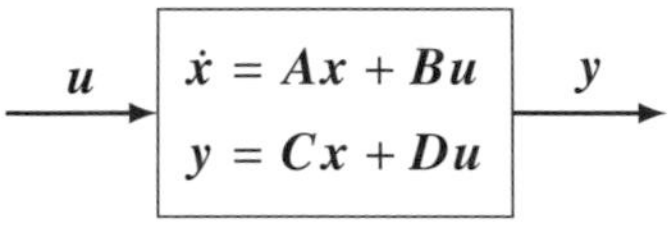

$$\xrightarrow{\;\boldsymbol{u}\;} \boxed{\begin{aligned} \dot{\boldsymbol{x}} &= \boldsymbol{A}\boldsymbol{x} + \boldsymbol{B}\boldsymbol{u} \\ \boldsymbol{y} &= \boldsymbol{C}\boldsymbol{x} + \boldsymbol{D}\boldsymbol{u} \end{aligned}} \xrightarrow{\;\boldsymbol{y}\;}$$

The properties of linear systems in both the time and frequency domain are discussed in further detail in Chapter 4.

To illustrate the various concepts and ideas that will be covered in this lecture, we will focus on a specific example, coming from a relatively simple system in fluid mechanics. We will look at two-dimensional flow over a circular cylinder at a Reynolds number of 60, particularly focusing on the dynamics of the system as it evolves in time away from its unstable equilibrium solution, and toward its vortex shedding limit cycle. The dynamics of this system are discussed in further detail in Chapter 1. The specific data set that we use is the same one utilized in Chapter 6 for an example for computing the POD.

To obtain a linear model of the form indicated in (10.1), we take data from a short region in time, identify the two leading POD modes in this region (see Chapter 6 for a detailed discussion of the POD), and then identify a linear model that captures the dynamics of the evolution of the corresponding POD coefficients during this short region of time. This model identification is done via the dynamic mode decomposition algorithm, which is discussed in detail in Chapter 7. For additional details concerning the implementation of this system identification method, as well as the various control techniques that will be applied to this identified system, see the accompanying Python code (available on the book's website[2]). The leading two POD modes, along with the evolution of their corresponding coefficients in time, are shown in Figures 10.1

[1] In the case where there are multiple inputs or outputs, the components of $G(s)$ are given by the ratios between the Laplace transforms of each pair of inputs and outputs.

[2] `www.datadrivenfluidmechanics.com/download/book/chapter10.zip`

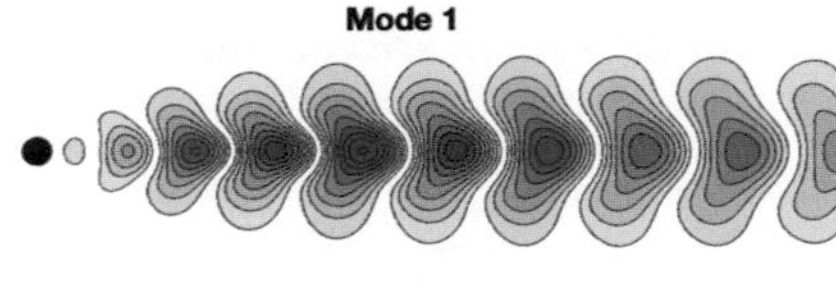

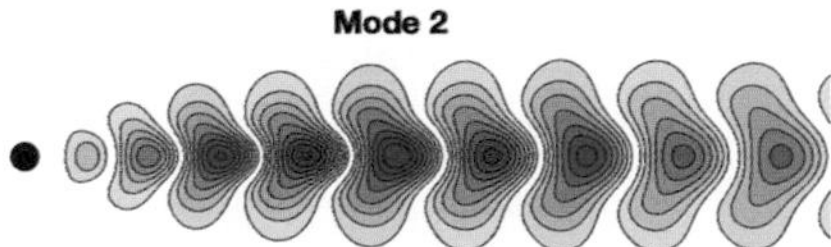

Figure 10.1 The first two POD modes (transverse velocity component) for flow over a circular cylinder at a Reynolds number of 60, on a trajectory going between the unstable equilibrium and limit cycle of the system.

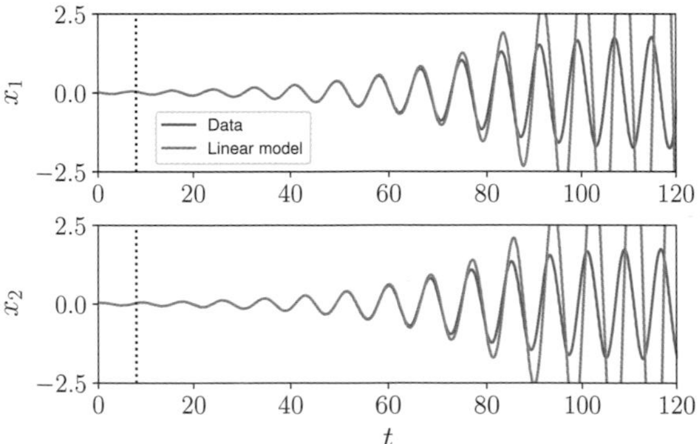

Figure 10.2 Evolution of the first two POD mode coefficients, corresponding to modes shown in Figure 10.1. Shown are both the actual coeffiicents, and those predicted by the identified linear dynamics (10.4). Only the data to the left of the vertical line is used for the identification of the linear model.

and 10.2, respectively. As well as showing the true evolution of the POD coefficients (which will be our system states x_1 and x_2), we also show the evolution predicted by the identified linear model. The linear model accurately captures the true behavior over a region in time substantially larger than that used for system identification, though is unable to capture the eventual saturation of the coefficient amplitudes as the true system converges toward the limit cycle. Note that the portion of the data used for system identification is the same as that used to compute POD for the example in Chapter 6. Using feedback control, we might hope to be able to keep the oscillations seen in Figure 10.2 to small enough amplitudes such that the linear model remains accurate, and that controllers designed using this linear model remain effective.

For this system, we obtain the linearized dynamics

$$\boldsymbol{A} = \begin{bmatrix} 0.0438 & -0.7439 \\ 0.7373 & 0.0527 \end{bmatrix}. \tag{10.4}$$

This system has two unstable eigenvalues at $0.0483 \pm 0.7406i$, which is expected, since the envelope of the POD coefficient amplitudes plotted in Figure 10.2 is growing in time. The fact that these eigenvalues come as a complex conjugate pair is also expected, since the dynamics of the system are oscillatory.

Note that this linear system is a simplification of the true dynamics not just because we are using a linear model, but also because we are neglecting other states of the system, corresponding to lower-energy, truncated POD modes. For example, identifying a linear system with more states could model (at least in a linear sense) the slow modification of the "mean" as the amplitude of the oscillations increases, as well as dynamics with frequency content at harmonics of the primary oscillation frequency. These dynamics are discussed in further detail in Chapter 14.

We will explore various choices for the $\boldsymbol{B}$ and $\boldsymbol{C}$ matrices in Sections 10.2–10.5, but will typically choose to directly observe and control one of the system states. This also represents a simplification of what measurement and control might look like for the true system in practice, where localized sensors and actuators might be more feasible than directly measuring or controlling an entire POD mode.

10.2 Proportional, Integral, and Derivative Feedback Control

Suppose now that we want to manipulate our system to achieve a desired output, $\boldsymbol{y}(t)$. For the example described, perhaps we want to stabilize the unstable system, and thus eventually drive all components of $\boldsymbol{y}(t)$ to zero. For an unstable system such as this, this objective is impossible without the availability of real-time information about the state of the system, and how it is responding to attempts at control. It would be equivalent, for example, trying to balance an inverted pendulum without any visual or tactile measurements.

What we want to do then, is to develop a means for utilizing knowledge of the system output $\boldsymbol{y}(t)$ in order to determine appropriate system inputs to achieve the desired control objective. This general concept is known as *feedback control*. For example, we could link the output to the input using the following arrangement:

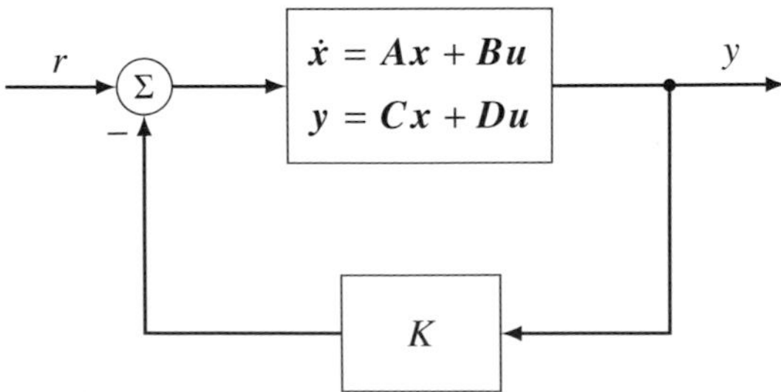

Here we refer to K as the *controller*, which could simply be a constant, or might have its own internal dynamics.

In this section, we will assume that we have a single input to our system (so $\boldsymbol{B}$ has one column), and a single output (so $\boldsymbol{C}$ has one row), and will always assume that there is no direct feedthrough from the input to the output (so $\boldsymbol{D} = 0$). It will be helpful to look at our system in the frequency domain, which we can obtain from a state-space

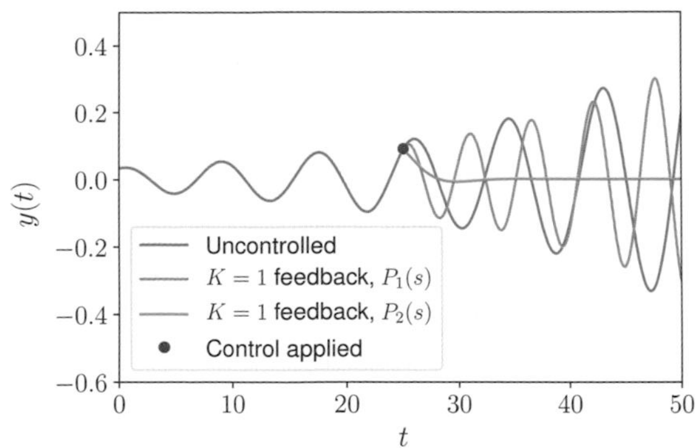

Figure 10.3 Effect of applying proportional feedback control with $K = 1$ to the two systems, P_1 and P_2.

system using (10.3). We will explore two different input and output combinations. First, we have the case where the input uses a different state variable as the output (system P_1), using the input and output matrices

$$\boldsymbol{B}_1 = \begin{bmatrix} 1 \\ 0 \end{bmatrix}, \quad \boldsymbol{C}_1 = \begin{bmatrix} 0 & 1 \end{bmatrix}, \tag{10.5}$$

giving the transfer function (see (10.3))

$$P_1(s) = \frac{0.7373}{s^2 - 0.09654s + 0.5508}. \tag{10.6}$$

Second, we use the input and output matrices

$$\boldsymbol{B}_2 = \begin{bmatrix} 0 \\ 1 \end{bmatrix}, \quad \boldsymbol{C}_2 = \begin{bmatrix} 0 & 1 \end{bmatrix}, \tag{10.7}$$

giving the transfer function

$$P_2(s) = \frac{s - 0.04381}{s^2 - 0.09654s + 0.5508}. \tag{10.8}$$

Note that the denominators of the transfer functions $P_1(s)$ and $P_2(s)$ are the same quadratic polynomial. This is as expected, since the roots of this polynomial are the poles of the system, which are in turn given by the eigenvalues of $\boldsymbol{A}$, which is the same for both systems.

To begin with, and before doing any further analysis of this system, suppose that we try to control our systems with the simplest feedback control, where K from our schematic diagram being a multiplicative constant. This is known as *proportional control*. The results of applying this control to our system with $K = 1$ is shown in Figure 10.3, where we have the reference set at $r = 0$. Here, we allow the system to evolve without any inputs, before turning on control at a time $t = 15$. Note that here we are applying control to the unstable linearized system, rather than the full nonlinear system. We notice that the system P_2 seems to be stabilized, while P_1 remains unstable.

How might we have predicted this observed behavior? With $r = 0$, the system input is given by law $\boldsymbol{u} = -K\boldsymbol{y} = -K\boldsymbol{C}\boldsymbol{x}$. This means that the system with feedback evolves according to

$$\dot{\boldsymbol{x}} = (\boldsymbol{A} - \boldsymbol{B}K\boldsymbol{C})\boldsymbol{x}. \tag{10.9}$$

The stability of this system is determined by the eigenvalues of the matrix $\boldsymbol{A} - \boldsymbol{B}K\boldsymbol{C}$. In the frequency domain, it is easy to show that the transfer function of the controlled system is given by

$$G_{CL} = \frac{P}{1 + PK}, \tag{10.10}$$

where K is the transfer function for the controller in the frequency domain (which is a constant for proportional control). Note that more generally, a closed-loop transfer function is given by

$$G_{CL} = \frac{G_{OL}(s)}{1 + G_{FB}(s)}, \tag{10.11}$$

where G_{OL} denotes the open-loop transfer function between r and y, and G_{FB} is the transfer function going around the feedback loop. Going back to (10.10), here the poles of the closed-loop system will be the solutions to

$$1 + PK = 0, \tag{10.12}$$

which can be expressed as the roots of a *characteristic polynomial.* To study the systems P_1 and P_2 in more detail, consider a more general system of the form

$$P(s) = \frac{a_1 s + a_0}{s^2 + b_1 s + b_0}, \tag{10.13}$$

for which $P_1(s)$ and $P_2(s)$ are both special cases. From (10.12), the closed-loop poles of this system for control with a constant K can be found to be the solutions to the polynomial

$$s^2 + (b_1 + a_1 K)s + (b_0 + a_0 K) = 0. \tag{10.14}$$

These solutions are given by

$$s = -\frac{b_1 + a_1 K}{2} \pm \frac{\sqrt{(b_1 + K)^2 - 4(b_0 + a_0 K)}}{2}. \tag{10.15}$$

From this, we can start to understand why proportional control with $K = 1$ worked for the system P_2, but not for P_1. The fact that P_2 has a nonzero a_1 term means that the control is able to shift the real component of the poles by $-a_1 K/2$, thus making them more stable (assuming that $a_1 > 0$). Note also that when the argument of the square root is negative, this term will not affect the real component of the poles. Conversely, if $a_1 = 0$ as is the case with P_1, the control only affects the square root term in (10.15). If the argument of the square root is negative, then this does not change the real component, and thus the stability, of the system. On the other hand, if the argument of the square root is positive, then one pole will become more stable, but the other becomes more unstable as the feedback gain either increases or decreases. From this, we can conclude that proportional feedback is unable to stabilize P_1.

As an aside, note that the dynamics of this system can be related to those of a spring-mass/damper system, which also has a quadratic denominator in its transfer function. In this analolgy, the negative coefficient of the linear term in the denominator of P_1 and P_2 corresponds to a negative damping, which is what leads to the system being unstable. Furthermore, (10.14) shows that K can only affect the damping term when a is nonzero, and otherwise only affects the stiffness-to-mass ratio, giving another interpretation for why this proportional control cannot stabilize P_1.

What sort of feedback would work for P_1? From (10.12), we see that we could achieve the same behavior of the poles of the closed-loop system if we added a term to our controller to mimic the first-order term in the numerator of P_2. For example, we could let

$$K = a_1 s + a_0, \tag{10.16}$$

to achieve the same closed-loop pole behavior as we found with P_2 using $K = 1$. These systems are not exactly equivalent due to the zero being in the feedback part of the loop, but the control results are very similar to those shown for P_2 in Figure 10.3. While this control strategy results in a stable system, the fact that the zero in P_2 is in the right half-plane is typically not desirable, as it can result in phenomena such as the system response initially moving in the wrong direction from the desired state. A stable system with zeros in the right half-plane is known as a non-minimum-phase system.

More generally, control of the form given in (10.16) is called proportional-derivative (or PD) control. Our original objective was to implement control to drive the system output to zero (or to any other desired output signal). While it looks like we are successful in doing this visually in Figure 10.3, the only way that this will be exact in reality is for the denominator of the closed-loop transfer function to not have a constant term. To achieve this, we could consider adding an "integrator" to our control design. Generally, combining proportional (P), integral (I), and derivative (D) control is called *PID*, as is a very commonly used classical control technique, which relies on being able to choose appropriate gains in the control law

$$K(s) = k_p + \frac{k_i}{s} + k_p s. \tag{10.17}$$

Note that adding an integrator can also destabilize the system, so care must be taken when choosing the gains k_p, k_i, and k_d used in PID control.

The analysis that we have performed in this section is tractable for this relatively simple system, but quickly becomes unmanageable for more complex systems with higher order numerators and denominators in their transfer functions. This motivates the formulation of general rules to predict the effect of feedback controllers without requiring us to explicitly solve for the roots of high-order polynomial equations. Such rules will be discussed in Section 10.3.

10.3 The Root Locus Plot

This section introduces a tool that will allow us to predict in advance what the effect of a given control strategy will be. In particular, it turns out that there are several simple rules that can allow us to sketch where the poles and zeros go as the controller gain increases. The *root locus* (developed by Evans (1948)) is a tool to see graphically how the poles of a closed-loop system move as a parameter k (which most typically is a controller gain) is varied. In particular, the root locus plot shows the locations of all poles of a system as a function of k, where we can write the characteristic equation for the poles of the closed-loop system as

$$1 + kG(s) = 0. \tag{10.18}$$

In the case of proportional control, $G(s)$ is simply the plant $P(s)$ and k is the proportional controller gain, though note that root locus plots can be constructed for other forms of control, so long as the closed-loop poles can be expressed in the form given by (10.18). Note in particular the similarity between (10.18) and (10.12). While it is easy to make a root locus plot in MATLAB or Python, it is also possible to draw these plots relatively accurately by hand, just by following a few simple rules (and importantly, without needing to explicitly solve (10.18) by hand for all values of k. We start by assuming that we know where the n poles (denoted as p_j) and m zeros (z_j) of G are located. It is also important to know $d = n - m$, which is the relative degree of G. From this, we have the following rules:

1. The branches of a root locus start at the poles of G (i.e., when no control is applied, or equivalently where $k = 0$), and end either at zeros of the open-loop system, or at infinity (with a total of d ending at ∞).
2. A point on the real axis is included in the root locus if there is an odd total number of poles and zeros to the right of it.
3. As k goes to ∞, the d branches of the root locus that go to ∞ asymptote to straight lines with equal angles between them, with those angles (from the positive real axis) given for $k = 0, 1, \ldots, d-1$ by

$$\theta_k = \frac{\pi + 2k\pi}{d}. \tag{10.19}$$

Moreover, the asymptotes originate at a common center, given by

$$c_0 = d^{-1}\left(\sum_{j=1}^{n} p_j - \sum_{j=1}^{m} z_j\right). \tag{10.20}$$

If $d = 0$, then all branches of the root locus end up at the zeros of the open-loop system as k goes to ∞.

These rules typically give enough information to roughly sketch how the poles of the closed-loop system will move as the controller gain varies. This can be particularly useful for determining the best choice of control law for a given system. Note that there are more complex rules that can enhance the accuracy of a hand-drawn root

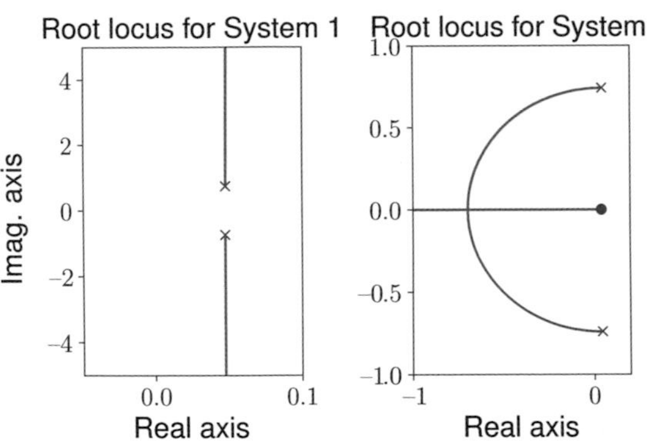

Figure 10.4 Root locus plots for $P_1(s)$ and $P_2(s)$ with proportional control.

locus plot, but we will not discuss them here. Root locus plots of the systems $P_1(s)$ and $P_2(s)$ are shown in Figure 10.4. These plots are consistent with our findings in Section 10.2, where it was found that proportional control could stabilize $P_2(s)$ but not $P_1(s)$.

Note also that there are many other tools available to assist in designing controllers for linear systems, such as Bode plots, Nyquist plots, and other auxiliary systems that can help predict the behavior of the controlled system when it is subject to sensor noise and disturbances. Details for these methods can be found in textbooks such as Aström and Murray (2010) and Franklin et al. (1994).

10.4 Controllability and Observability

So far, we have looked at a few different control strategies, and were fortunately able to find an approach that succeeded in stabilizing both of our systems. It is natural to wonder if there are cases where this is an entirely futile exercise, in the sense that no possible control strategy could work, given the ways in which we can manipulate the state of our system, $\boldsymbol{x}$, through the input $\boldsymbol{u}$. It turns out, there is a relatively simple way to determine whether a system is controllable in this sense. This can be done through the use of the controllability matrix

$$C = \begin{bmatrix} \boldsymbol{B} & \boldsymbol{AB} & \boldsymbol{A}^2\boldsymbol{B} & \cdots & \boldsymbol{A}^{n-1}\boldsymbol{B} \end{bmatrix}. \tag{10.21}$$

In particular, we say that a system is *controllable* if the controllability matrix has a rank of n (i.e., is full rank). Note that the C has n rows and $n \times n_i$ columns, where n_i is the number of inputs. Along with how controllable a system is, we can also analyze whether the outputs from the system are sufficient for us to be able to estimate the full state of the system, through the observability matrix

$$O = \begin{bmatrix} C \\ CA \\ CA^2 \\ \vdots \\ CA^{n-1} \end{bmatrix}. \tag{10.22}$$

We say that a system is *observable* if the observability matrix has rank n. O has n columns and $n \times n_o$ rows, where n_o is the number of outputs. It is easy for us to verify that the state-space versions of systems P_1 and P_2 are both controllable and observable, and indeed this is the case for any choice of input and output channels (i.e., for $\boldsymbol{B} = [0\ 1]^T$ or $[1\ 0]^T$, and $\boldsymbol{C} = [0\ 1]$ or $[1\ 0]$).

It turns out that we can also talk about controllability and observability not just as binary notions, but can also study how controllable a system is through related objects called the controllability and observability Gramians (Zhou & Doyle 1998), which are discussed in Chapter 12.

10.5 Full-State Feedback

Thus far, we have only considered control using the system output, which we have taken to be a scalar quantity. For the following sections, it will be useful to consider the case where we have access to (or somewhat equivalently, have a method of estimating) the full state of the system $\boldsymbol{x}$. This can be represented with the following control diagram, which will be useful in the following sections:

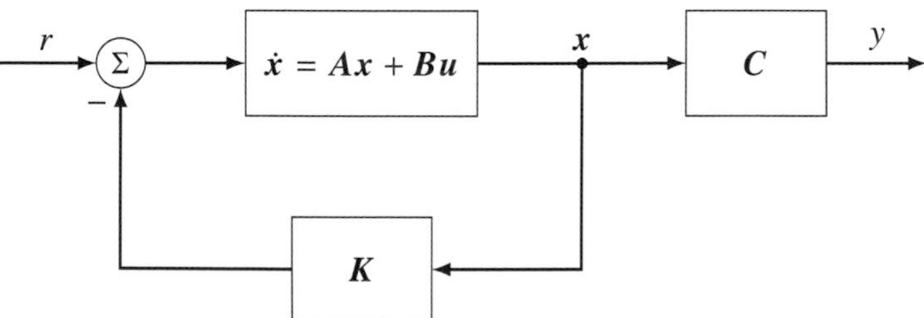

We now consider two different methods for controller design using full-state feedback: pole placement in Section 10.5.1, and optimal control using LQR in Section 10.5.2.

10.5.1 Pole Placement

When using full-state feedback with a control law given by

$$\boldsymbol{u} = -\boldsymbol{K}\boldsymbol{x}, \tag{10.23}$$

the dynamics of the closed-loop system can be written as

$$\dot{\boldsymbol{x}} = (\boldsymbol{A} - \boldsymbol{B}\boldsymbol{K})\boldsymbol{x}, \tag{10.24}$$

which is the same as (10.9) with $\boldsymbol{C} = \boldsymbol{I}$. If we have desired closed-loop poles, then this can be achieved by finding the components of $\boldsymbol{K}$ such that $\boldsymbol{A} - \boldsymbol{B}\boldsymbol{K}$ has eigenvalues at these locations. If the system is controllable, then this can always be achieved. For example, if we wish to place poles at $p = -0.5 \pm 0.5i$ for our systems P_1 or P_2 with full-state feedback, then we find that $\boldsymbol{K} = [1.0965 \;\; 0.0095]^T$. Pole placement can be computed using the command `K = place(A,B,p)` in either MATLAB or Python.

10.5.2 Optimal Control

So far, the design of controllers has seemed somewhat ad hoc, and has relied upon tuning parameters based on insight and desired properties. This motivates the question: is there a way to define a controller that is, in some sense, the "best"? At least for some definition of "best," then this is something that we can indeed do. For simplicity, assume that we have full-state feedback. The idea is that, when applying control, we want to balance a desire to drive the system to a specified controlled state (e.g., to zero), while at the same time ensuring that we are limiting the exertion of our controller (i.e., penalizing the use of too much input energy). Mathematically, one way to account for both of these objectives is by constructing a cost function of the form

$$J = \int_{t=0}^{\infty} \left(\boldsymbol{x}^T \boldsymbol{Q} \boldsymbol{x} + \boldsymbol{u}^T \boldsymbol{R} \boldsymbol{u} \right) dt, \tag{10.25}$$

where $\boldsymbol{Q}$ and $\boldsymbol{R}$ are positive definite matrices that define the penalty associated with nonzero states and control effort, respectively. A controller that is designed to minimize (10.25) is called an LQR controller. Mathematically, this control law that minimizes (10.25) is given by

$$\boldsymbol{K} = \boldsymbol{R}^{-1}\boldsymbol{B}^T\boldsymbol{M}, \tag{10.26}$$

where $\boldsymbol{M}$ is the unique positive definite matrix that satisfies the algebraic Riccati equation

$$\boldsymbol{A}^T\boldsymbol{M} + \boldsymbol{M}\boldsymbol{A} - \boldsymbol{M}\boldsymbol{B}\boldsymbol{R}^{-1}\boldsymbol{B}^T\boldsymbol{M} + \boldsymbol{Q} = 0. \tag{10.27}$$

This might look like a difficult problem to solve, but once again there are commands in MATLAB and Python libraries that will output the gain matrix K that optimizes (10.25). As well as being "optimal" in this sense, it turns out that the resulting controller also satisfies certain robustness properties (e.g., Stengel 1994).

To test this method, we design LQR controllers for the system P_1, using $\boldsymbol{Q} = \boldsymbol{I}$ and three different choices of $\boldsymbol{R}$ (10, 1, and 0.1), which prescribe how aggressive the resulting controller is in achieving the desired state. The characteristics of these feedback systems are explored in Figures 10.5 and 10.6. We see in Figure 10.5 that all three controllers stabilize the system, though the locations of the poles are noticeably

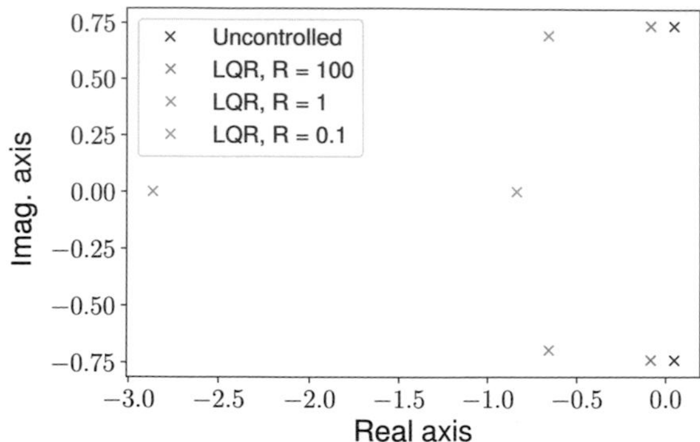

Figure 10.5 Poles and system response for implementing full-state feedback control with LQR using three choices of control effort penalties.

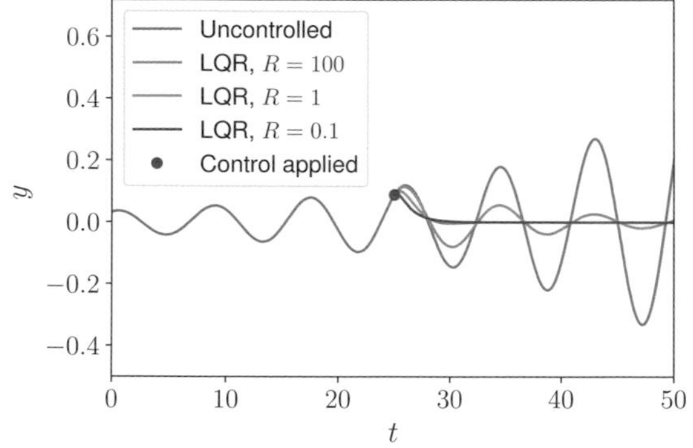

Figure 10.6 Poles and system response for implementing full-state feedback control with LQR using three choices of control effort penalties.

different for each case. Roughly speaking, larger $\boldsymbol{R}$ values penalize large amounts of controller work, which tends to result in closed-loop systems that are "less different" from the initial uncontrolled system, which is consistent with what is observed. Similarly, we see in Figure 10.6 that the larger $\boldsymbol{R}$ value results in a controlled system that takes longer to converge to the desired steady state, while the more aggressive controllers (with smaller $\boldsymbol{R}$ values) converge to this state more rapidly, though with larger control effort (which is not shown explicitly).

10.6 Additional Control Techniques and Considerations

As mentioned at the outset, there are numerous methods in linear control theory that we have not touched on in this lecture. For a more comprehensive treatment of linear control theory, standard references such as Aström and Murray (2010), Franklin et al. (1994), and Skogestad and Postlethwaite (2007) are a good place to start, with the latter in particular covering more advanced topics. In this final section of this chapter, we briefly discuss additional ideas and techniques related to feedback control.

10.6.1 State Estimation

We observed in Section 10.5.2 that control could be achieved "optimally" if using measurements of the full state. But is any of this useful if we do not have access to the full state? As it turns out, if our system is observable (as described in Section 10.4), then we can design an estimator that can reconstruct an estimate of the full state from the system outputs, which can then be used for state feedback, such as with an LQR controller. The combined estimator and controller system, known as a linear-quadratic-Gaussian (LQG) control system, will have poles at the locations of the LQR subsystem, as well as those from the estimator subsystem. This property leads to the separation principle, which says that if we can separately design a full-state feedback system that is stable, along with an estimator that is stable, then the overall system is also guaranteed to be stable.

10.6.2 Robust Control

There are many additional methods for controller design that extend beyond the scope of what has been discussed here. In practice, factors such as measurement noise, external disturbances, time delays, and modeling uncertainties can significantly influence the performance of feedback control strategies. There are several common metrics for determining how *robust* a controller will be to such effects, the most common of which are concepts such as the *phase margin* and *gain margin* of a feedback system. Roughly speaking, these are measures of how close a system is to becoming unstable. There exist more advanced controller design techniques that explicitly seek to produce feedback systems that are robust to disturbances, such as H_∞ methods (e.g., Skogestad & Postlethwaite 2007).

10.6.3 Nonlinear Control

This chapter has only considered the control of systems that are (or are assumed to be) linear. While almost all real-life systems are nonlinear, linear methods can still be very effective, subject to caveats that might relate to how close we stay to a location about which a nonlinear system is linearized. However, there are certainly cases where linear methods are ineffective or undesirable, thus requiring more sophisticated nonlinear techniques. Extensions include gain scheduling and linear parameter varying control (White et al. 2013), feedback linearization, and backstepping (Khalil & Grizzle 2002).

10.6.4 Application of Feedback Control in Fluids

The ability to manipulate fluid flows to achieve desired behavior has the potential to substantially reduce energy requirements for transportation of fluids, transportation of vehicles through fluids, and to improve the reliability and safety of such engineering tasks. The past several decades have seen substantial growth in control-theoretic methods (those discussed in this chapter, and beyond) to enable the design of control strategies that achieve such purposes. For example, many of the methods discussed in this section (and related methods) have been directly applied to fluids systems for the control of boundary layers (Bagheri et al. 2009, Semeraro et al. 2011, 2013, Belson et al. 2013, Leclercq et al. 2019), flow over backward-facing steps (Hervé et al. 2012) and cavities (Rowley & Williams 2006, Barbagallo et al. 2009, Illingworth et al. 2012), bluff-body wakes (Bergmann et al. 2005, Illingworth et al. 2014, Brackston et al. 2016, Illingworth 2016, Flinois & Morgans 2016), airfoils (Magill et al. 2003, Ahuja & Rowley 2010, Kerstens et al. 2011, Brunton et al. 2014), and wall-bounded turbulence (Choi et al. 1994, Kim 2011, Luhar et al. 2014, Toedtli et al. 2019). Beyond methods based on linear control theory, recent developments in applications of cluster (Kaiser et al. 2017a, Nair et al. 2019), network (Nair et al. 2018), reinforcement learning (Rabault et al. 2019, see also Chapter 18), genetic programming (Debien et al. 2016, Minelli et al. 2020, see also Chapter 17), and other machine learning control methodologies (Gautier et al. 2015, Duriez et al. 2017) have expanded the set of tools available for controlling fluid flows. For more comprehensive reviews on methods and applications of flow control, see, for example, Kim and Bewley (2007), Gad-el Hak (2007), Noack et al. (2011), Brunton and Noack (2015), Rowley and Dawson (2017), and Brunton et al. (2020).

11 Nonlinear Dynamical Systems

S. Brunton

Dynamical systems provide a mathematical framework to describe the evolution of complex physical systems, such as unsteady fluid dynamics. The field of dynamical systems lives at the interface of many fields of mathematics, including linear algebra, differential equations, topology, geometry, and numerical analysis. Modern dynamical systems began with the seminal work of Poincaré on the chaotic motion of planets, and it may be viewed as the culmination of hundreds of years of mathematical modeling with differential equations. This chapter presents a modern perspective on dynamical systems in the context of current goals and open challenges in fluid dynamics. Data-driven dynamical systems is a rapidly evolving field, and therefore, we focus on a mix of established and emerging methods that are driving current developments. In particular, we will focus on the three key challenges of (1) high-dimensionality, (2) discovering dynamics from data, and (3) finding data-driven representations that make nonlinear systems amenable to linear analysis.

11.1 Introduction

In recent decades, the field of dynamical systems has grown rapidly, with analytical derivations and first principle models giving way to data-driven approaches. Thus, machine learning and big data are driving a paradigm shift in the analysis and understanding of dynamical systems in science and engineering. This trend is particularly evident in the field of fluid dynamics, which is one of the original big data fields.

In addition, the classical geometric and statistical perspectives on dynamical systems are being complemented by a third *operator-theoretic* perspective, based on the evolution of measurements of the system. This so-called *Koopman* operator theory is poised to capitalize on the increasing availability of measurement data from complex systems. Moreover, Koopman theory provides a path to identify intrinsic coordinate systems to represent nonlinear dynamics in a linear framework. Obtaining linear representations of strongly nonlinear systems has the potential to revolutionize our ability to predict and control these systems (Figure 11.1). See Chapters 10 and 7 for overviews of linear systems and Koopman theory, respectively.

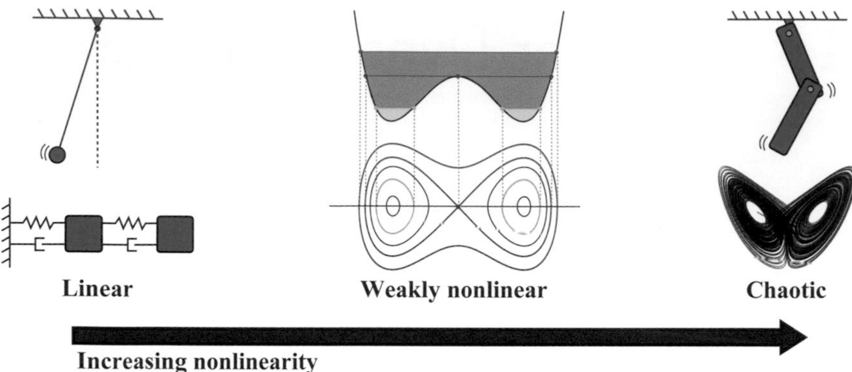

Figure 11.1 Schematic depiction of increasing nonlinearity in dynamical systems. Behavior goes from periodic to quasiperiodic, and eventually becomes chaotic. Reproduced from Brunton et al. (2017).

11.2 Dynamical Systems

11.2.1 Continuous-Time Systems

Throughout this chapter, we will consider dynamical systems of the form

$$\frac{d}{dt}\mathbf{x}(t) = \mathbf{f}(\mathbf{x}(t), t; \boldsymbol{\beta}), \tag{11.1}$$

where $\boldsymbol{x}$ is the state of the system and $\mathbf{f}$ is a vector field that possibly depends on the state $\boldsymbol{x}$, time t, and a set of parameters $\boldsymbol{\beta}$.

Finite-dimensional representations of the Navier–Stokes equations take this form in which case the state $\mathbf{x}$ may contain the flow-field variables on a discretized grid, or a finite number of Fourier coefficients. These dynamics are typically high-dimensional, often requiring millions or billions of degrees of freedom. It is also possible to obtain low-dimensional dynamical systems that describe a fluid flow in terms of the dynamics of the amplitudes of a small set of modes, for example, by Galerkin projection of the governing equations onto a truncated set of Proper Orthogonal Decomposition (POD) modes. In these projected models, the dynamics in (11.1) have quadratic nonlinearities due to the convective nonlinearity $(\mathbf{u} \cdot \nabla)\mathbf{u}$ in the Navier–Stokes equations.

The Lorenz (1963) system

$$\dot{x} = \sigma(y - x), \tag{11.2a}$$

$$\dot{y} = x(\rho - z) - y, \tag{11.2b}$$

$$\dot{z} = xy - \beta z \tag{11.2c}$$

is one of the earliest nonlinear reduced-order models of an unsteady fluid system, as it was developed to model chaotic Rayleigh–Benard convection. The x, y, and z variables denote the amplitudes of three spatial modes, posited by Saltzman (1962), onto which the Navier–Stokes equations are Galerkin projected. In this case, the state vector is $\boldsymbol{x} = \begin{bmatrix} x & y & z \end{bmatrix}^{\mathrm{T}}$ and the parameter vector is $\boldsymbol{\beta} = \begin{bmatrix} \sigma & \rho & \beta \end{bmatrix}^{\mathrm{T}}$. A trajectory

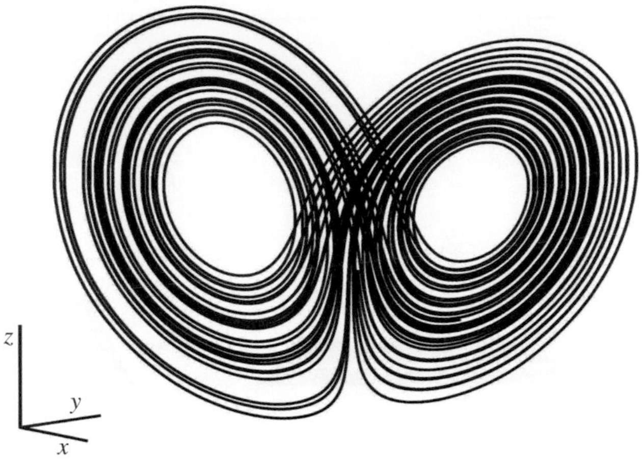

Figure 11.2 Chaotic trajectory of the Lorenz system from (11.2). From Brunton and Kutz (2019), reproduced with permission of the Licensor through PLSclear.

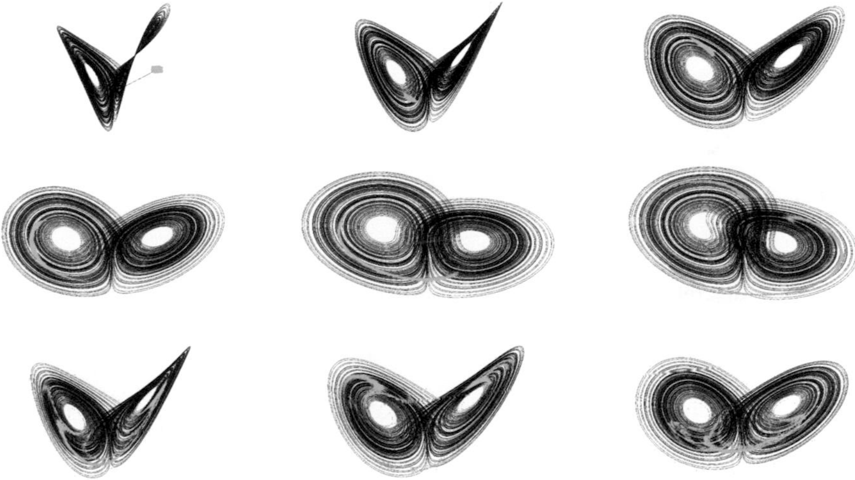

Figure 11.3 Depiction of uncertainty in the Lorenz system. A cube of initial conditions is evolved along the flow. Although the trajectories stay close together for some time, after a while, they begin to spread along the attractor.

of the Lorenz system is shown in Figure 11.2, with parameters $\sigma = 10$, $\rho = 28$, and $\beta = 8/3$.

The Lorenz system is among the simplest and most well-studied dynamical systems that exhibit chaos, which is characterized as a sensitive dependence on initial conditions. Two trajectories with nearby initial conditions will rapidly diverge in behavior, and after long periods, only statistical statements can be made. This sensitivity is depicted in Figure 11.3, where a cube of initial conditions is evolved along the flow of the dynamical system, showing that after some time the initial conditions become spread out along the attractor. This type of chaotic *mixing* is characteristic of turbulence.

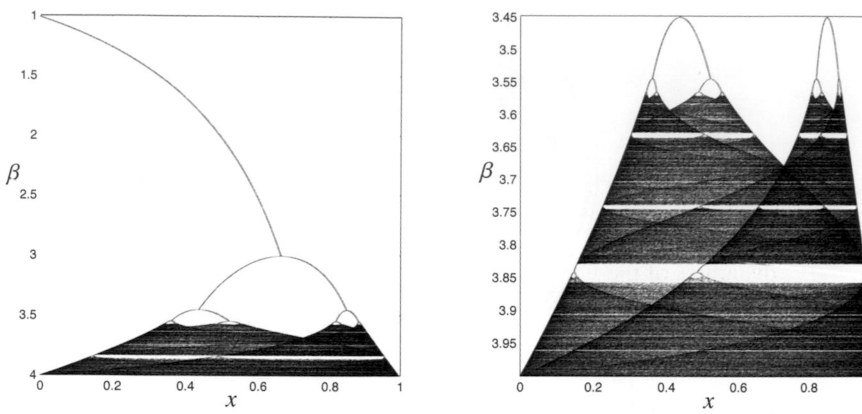

Figure 11.4 Attracting sets of the logistic map for varying parameter β. From Brunton and Kutz (2019), reproduced with permission of the Licensor through PLSclear.

We will often consider the simpler case of an autonomous system without time dependence or parameters:

$$\frac{d}{dt}\mathbf{x}(t) = \mathbf{f}(\mathbf{x}(t)). \tag{11.3}$$

11.2.2 Discrete-Time Systems

In addition to the aforementioned continuous-time dynamics, we will also consider discrete-time dynamical systems

$$\mathbf{x}_{k+1} = \mathbf{F}(\mathbf{x}_k). \tag{11.4}$$

Discrete-time dynamics are often more natural when considering experimental data and simulations, as data is typically generated discretely in time.

Also known as a *map*, the discrete-time dynamics are more general than the continuous-time formulation in (11.3), encompassing discontinuous and hybrid systems as well.

As an example, consider the logistic map, which is a simple model to explore the complexity of population dynamics:

$$x_{k+1} = \beta x_k(1 - x_k). \tag{11.5}$$

As the parameter β is increased, the attracting set becomes increasingly complex, shown in Figure 11.4. A series of period-doubling bifurcations occur until the attracting set becomes fractal. The complexity of turbulence is also often understood as the result of a sequence of period-doubling Hopf bifurcations.

Discrete-time dynamics may be induced from continuous-time dynamics, where $\mathbf{x}_k$ is obtained by sampling the trajectory in (11.3) discretely in time, so that $\mathbf{x}_k = \mathbf{x}(k\Delta t)$. The discrete-time propagator $\mathbf{F}_{\Delta t}$ is now parameterized by the time step Δt. For an arbitrary time t, the *flow map* $\mathbf{F}_t$ is defined as

$$\mathbf{F}_t(\mathbf{x}(t_0)) = \mathbf{x}(t_0) + \int_{t_0}^{t_0+t} \mathbf{f}(\mathbf{x}(\tau))\, d\tau. \tag{11.6}$$

11.2.3 Linear Dynamics and Spectral Decomposition

As explored in Chapter 10, whenever possible it is desirable to work with linear dynamics of the form

$$\frac{d}{dt}\boldsymbol{x} = \mathbf{A}\boldsymbol{x}. \tag{11.7}$$

Even for nonlinear systems, it is often possible to understand the behavior near fixed points in terms of locally linearized dynamics.

Linear dynamical systems admit closed-form solutions, and there are a wealth of techniques for the analysis, prediction, numerical simulation, estimation, and control of such systems. The solution of (11.7) is given by

$$\boldsymbol{x}(t_0 + t) = e^{\mathbf{A}t}\boldsymbol{x}(t_0). \tag{11.8}$$

The dynamics are entirely characterized by the eigenvalues and eigenvectors of the matrix $\mathbf{A}$, given by the *spectral decomposition* (eigendecomposition) of $\mathbf{A}$:

$$\mathbf{A}\mathbf{T} = \mathbf{T}\mathbf{\Lambda}. \tag{11.9}$$

When $\mathbf{A}$ has n distinct eigenvalues, then $\mathbf{\Lambda}$ is a diagonal matrix containing the eigenvalues λ_j and $\mathbf{T}$ is a matrix whose columns are the linearly independent eigenvectors $\boldsymbol{\xi}_j$ associated with eigenvalues λ_j. In this case, it is possible to write $\mathbf{A} = \mathbf{T}\mathbf{\Lambda}\mathbf{T}^{-1}$, and the solution in (11.8) becomes

$$\boldsymbol{x}(t_0 + t) = \mathbf{T}e^{\mathbf{\Lambda}t}\mathbf{T}^{-1}\boldsymbol{x}(t_0). \tag{11.10}$$

More generally, in the case of repeated eigenvalues, the matrix $\mathbf{\Lambda}$ will consist of Jordan blocks (Perko 2013). Note that the continuous-time system gives rise to a discrete-time dynamical system, with $\mathbf{F}_t$ given by the solution map $\exp(\mathbf{A}t)$ in (11.8). In this case, the discrete-time eigenvalues are given by $e^{\lambda t}$.

For non-normal operators $\mathbf{A}$, the eigenvectors may be nearly parallel. In this case, even with stable eigenvalues, the system may exhibit *transient* growth, where an initial condition first grows before eventually converging to the origin. Transient growth is observed in many shear flows, and the excursion from the origin often results in perturbation amplitudes that are large enough to excite nonlinear dynamics. For example, pipe flow is linearly stable at all Reynolds numbers, but non-normality and transient energy growth results in sustained turbulence at finite Reynolds numbers.

The matrix $\mathbf{T}^{-1}$ defines a transformation, $z = \mathbf{T}^{-1}\boldsymbol{x}$, into intrinsic eigenvector coordinates, z, where the dynamics become decoupled:

$$\frac{d}{dt}z = \mathbf{\Lambda}z. \tag{11.11}$$

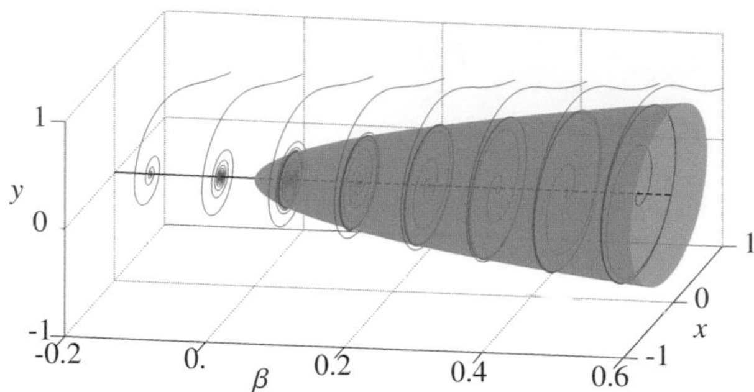

Figure 11.5 Schematic of a Hopf bifurcation. Adapted from Brunton, Proctor and Kutz (2016a).

In other words, each coordinate, z_j, only depends on itself, with dynamics given by

$$\frac{d}{dt} z_j = \lambda_j z_j. \tag{11.12}$$

Thus, it is highly desirable to work with linear systems, since it is possible to easily transform the system into eigenvector coordinates where the dynamics become decoupled. No such closed-form solution or simple linear change of coordinates exists in general for nonlinear systems, motivating many of the directions described in this chapter.

11.2.4 Bifurcations

We have already seen bifurcations that occur in the logistic map. Bifurcations are particularly important in fluid dynamic systems. Small changes in physical parameters, such as the Reynolds number, may drive qualitative changes in the behavior of the resulting dynamics. For example, in the flow past a stationary circular cylinder, when the Reynolds number increases past 47, the flow goes from a steady laminar solution to unsteady, periodic vortex shedding (see Chapter 1). This *Hopf bifurcation* is shown in Figure 11.5, and is given by the following dynamical system:

$$\frac{dx}{dt} = \beta x - \omega y - Ax\left(x^2 + y^2\right), \tag{11.13a}$$

$$\frac{dy}{dt} = \omega x + \beta y - Ay\left(x^2 + y^2\right). \tag{11.13b}$$

As β goes from negative to positive values, a single stable fixed point at the origin becomes unstable, and a stable limit cycle emerges.

As a simpler example, we often consider the one-dimensional *pitchfork bifurcation*, shown in Figure 11.6 and given by the following dynamical system:

$$\frac{dx}{dt} = \beta x - x^3. \tag{11.14}$$

For negative values of the parameter β, there is a single stable fixed point at the origin. As this parameter increases from negative to positive, this fixed point becomes unstable and two other stable fixed points emerge at $\pm\sqrt{\beta}$.

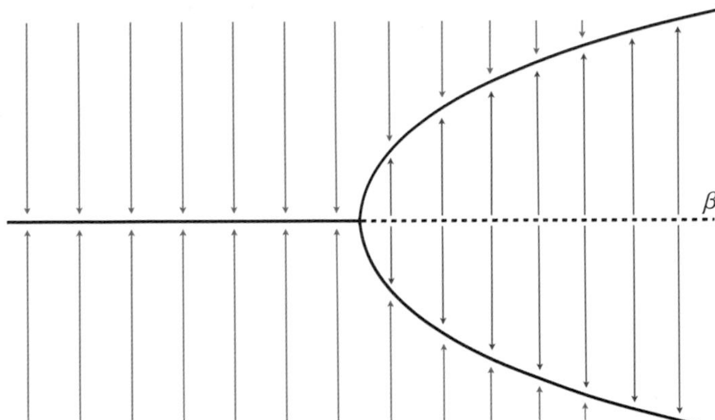

Figure 11.6 Schematic of a pitchfork bifurcation.

It is interesting to note that if we write the Hopf normal form in (11.13) in polar coordinates, with $r^2 = x^2 + y^2$, then the equation for the amplitude r reduces to the pitchfork normal form in (11.14).

11.3 Goals and Challenges in Modern Dynamical Systems

There are a number of goals and uses of dynamical systems models of fluid flows:

1. **Future state prediction.** In many cases, we seek predictions of the future state of a system. Long-time predictions may still be challenging, especially for chaotic systems, such as turbulence, where long-term statistics are a more reasonable goal.
2. **Design and optimization.** The dynamical system may be used as a surrogate model to tune the parameters of a system for improved performance or stability. For example, aircraft and automobile designers modify geometry and control surfaces to improve aerodynamic performance.
3. **Estimation and control.** For many fluid systems, an ultimate goal is to actively control the system through feedback, using measurements of the system to inform actuation to modify the behavior. For high-dimensional systems, such as fluids, it is often necessary to estimate the full state of the system from limited measurements.
4. **Interpretability and physical understanding.** Perhaps a more fundamental goal of dynamical systems modeling is to provide physical insight and interpretability into a system's behavior through analyzing trajectories and solutions to the governing equations of motion. Dynamical systems models of fluids will ideally yield some insight into fundamental mechanisms that drive the flow.

These goals are often challenged by the multiscale and nonlinear nature of fluids. There is also uncertainty in the boundary conditions, parameters, and measurements of the system. These challenges may be summarized as follows:

1. **High-dimensionality.** Fluid flows, and the resulting measurement and simulation data, are often exceedingly high-dimensional. Such systems are expensive to simulate and analyze, especially for iterative design optimization, uncertainty quantification, and real-time control, motivating accurate and efficient surrogate models of lower dimension. Pattern extraction and modal decompositions (Taira et al. 2017, Taira et al. 2019) provide powerful techniques to reduce the dimension of fluid systems. These techniques include both classical linear dimensionality reduction, such as POD and dynamic mode decomposition (DMD) (see Chapters 6 and 7), as well as emerging nonlinear dimensionality reduction techniques from machine learning (see Chapter 3).
2. **Nonlinearity.** Nonlinearity remains a primary challenge in analyzing and controlling fluid dynamics. Even simple quadratic nonlinearity, such as the convection term in the Navier–Stokes equations, significantly complicates analysis and control. Traditional linearization around fixed points make it possible to leverage linear modeling and control techniques, even for nonlinear systems. However, global analysis has remained largely qualitative and computational, limiting the theory of nonlinear prediction, estimation, and control away from fixed points and periodic orbits. To address the issue of nonlinearity, operator-theoretic approaches to dynamical systems are becoming increasingly used. As we will show, it is possible to represent nonlinear dynamical systems in terms of infinite-dimensional but linear operators, such as the Koopman operator from Section 11.4 which advances measurement functions, and the Perron–Frobenius operator which advances probability densities and ensembles through the dynamics.
3. **Unknown dynamics.** Despite knowledge of the governing Navier–Stokes equations, the reduced-order physical mechanisms are often unknown for a particular flow. Traditionally, systems in fluid dynamics were analyzed by making ideal approximations and then deriving simple differential equation models via asymptotic expansions. Dramatic simplifications could often be made by exploiting symmetries and clever coordinate systems. With increasingly complexity, the paradigm is shifting from this classical approach to data-driven methods to discover governing equations. As data become increasingly abundant, and we continue to investigate systems that are not amenable to first-principles analysis, regression and machine learning are becoming vital tools to discover dynamical systems from data. This is the basis of many of modern techniques, including the DMD and data-driven Koopman methods (see Chapter 7), the sparse identification of nonlinear dynamics (SINDy) (see Section 12.3), as well as the use of genetic programming to identify dynamics from data (Bongard & Lipson 2007, Schmidt & Lipson 2009).

11.4 Koopman Theory

Koopman operator theory has recently emerged as an alternative perspective for dynamical systems in terms of the evolution of measurements $g(\boldsymbol{x})$. In 1931, Bernard O. Koopman demonstrated that it is possible to represent a nonlinear dynamical system in terms of an infinite-dimensional linear operator acting on a Hilbert space of measurement functions of the state of the system. Koopman analysis has recently gained renewed interest with the pioneering work of Mezic and collaborators (Mezić & Banaszuk 2004, Mezić 2005, Budišić & Mezić 2009, Budišić et al. 2012, Budišić & Mezić 2012, Lan & Mezić 2013, Mezić 2013). The so-called *Koopman operator* is linear, and its spectral decomposition completely characterizes the behavior of a nonlinear system, analogous to (11.7). However, it is also infinite-dimensional, as there are infinitely many degrees of freedom required to describe the space of all possible measurement functions g of the state. This poses new challenges. Obtaining finite-dimensional, matrix approximations of the Koopman operator is the focus of intense research efforts and holds the promise of enabling globally linear representations of nonlinear dynamical systems. Expressing nonlinear dynamics in a linear framework is appealing because of the wealth of optimal estimation and control techniques available for linear systems and the ability to analytically predict the future state of the system. Obtaining a finite-dimensional approximation of the Koopman operator has been challenging in practice, as it involves identifying a subspace spanned by a subset of eigenfunctions of the Koopman operator.

11.4.1 Mathematical Formulation

The Koopman operator advances measurement functions of the state with the flow of the dynamics. We consider real-valued measurement functions $g : \mathbf{M} \to \mathbb{R}$, which are elements of an infinite-dimensional Hilbert space. The functions g are also commonly known as *observables*, although this may be confused with the unrelated *observability* from control theory. Often, the Hilbert space is given by the Lebesgue square-integrable functions on $\mathbf{M}$; other choices of a measure space are also valid.

The Koopman operator $\mathcal{K}_t$ is an infinite-dimensional linear operator that acts on measurement functions g as

$$\mathcal{K}_t g = g \circ \mathbf{F}_t, \tag{11.15}$$

where $\circ$ is the composition operator. For a discrete-time system with time step Δt, this becomes

$$\mathcal{K}_{\Delta t} g(\mathbf{x}_k) = g(\mathbf{F}_{\Delta t}(\mathbf{x}_k)) = g(\mathbf{x}_{k+1}). \tag{11.16}$$

In other words, the Koopman operator defines an infinite-dimensional linear dynamical system that advances the observation of the state $g_k = g(\mathbf{x}_k)$ to the next time step:

$$g(\mathbf{x}_{k+1}) = \mathcal{K}_{\Delta t} g(\mathbf{x}_k). \tag{11.17}$$

Note that this is true for *any* observable function g and for any state $\mathbf{x}_k$.

The Koopman operator is linear, a property that is inherited from the linearity of the addition operation in function spaces:

$$\mathcal{K}_t\left(\alpha_1 g_1(\boldsymbol{x}) + \alpha_2 g_2(\boldsymbol{x})\right) = \alpha_1 g_1\left(\mathbf{F}_t(\boldsymbol{x})\right) + \alpha_2 g_2\left(\mathbf{F}_t(\boldsymbol{x})\right) \tag{11.18a}$$

$$= \alpha_1 \mathcal{K}_t g_1(\boldsymbol{x}) + \alpha_2 \mathcal{K}_t g_2(\boldsymbol{x}). \tag{11.18b}$$

For sufficiently smooth dynamical systems, it is also possible to define the continuous-time analogue of the Koopman dynamical system in (11.17):

$$\frac{d}{dt} g = \mathcal{K} g. \tag{11.19}$$

The operator $\mathcal{K}$ is the infinitesimal generator of the one-parameter family of transformations $\mathcal{K}_t$ (Abraham et al. 1988). It is defined by its action on an observable function g:

$$\mathcal{K} g = \lim_{t\to 0} \frac{\mathcal{K}_t g - g}{t} = \lim_{t\to 0} \frac{g \circ \mathbf{F}_t - g}{t}. \tag{11.20}$$

The linear dynamical systems in (11.19) and (11.17) are analogous to the dynamical systems in (11.3) and (11.4), respectively. It is important to note that the original state $\mathbf{x}$ may be the observable, and the infinite-dimensional operator $\mathcal{K}_t$ will advance this function. However, the simple representation of the observable $g = \mathbf{x}$ in a chosen basis for Hilbert space may become arbitrarily complex once iterated through the dynamics. In other words, finding a representation for $\mathcal{K}\boldsymbol{x}$ may not be simple or straightforward.

Koopman Eigenfunctions and Intrinsic Coordinates

The Koopman operator is linear, which is appealing, but infinite-dimensional, posing issues for representation and computation. Instead of capturing the evolution of all measurement functions in a Hilbert space, applied Koopman analysis attempts to identify key measurement functions that evolve linearly with the flow of the dynamics. Eigenfunctions of the Koopman operator provide just such a set of special measurements that behave linearly in time. In fact, a primary motivation to adopt the Koopman framework is the ability to simplify the dynamics through the eigendecomposition of the operator.

A discrete-time Koopman eigenfunction $\varphi(\boldsymbol{x})$ corresponding to eigenvalue λ satisfies

$$\varphi(\boldsymbol{x}_{k+1}) = \mathcal{K}_{\Delta t}\varphi(\boldsymbol{x}_k) = \lambda\varphi(\boldsymbol{x}_k). \tag{11.21}$$

In continuous-time, a Koopman eigenfunction $\varphi(\boldsymbol{x})$ satisfies

$$\frac{\mathrm{d}}{\mathrm{d}t}\varphi(\boldsymbol{x}) = \mathcal{K}\varphi(\boldsymbol{x}) = \lambda\varphi(\boldsymbol{x}). \tag{11.22}$$

Obtaining Koopman eigenfunctions from data or from analytic expressions is a central applied challenge in modern dynamical systems. Discovering these eigenfunctions enables globally linear representations of strongly nonlinear systems.

Applying the chain rule to the time derivative of the Koopman eigenfunction $\varphi(\boldsymbol{x})$ yields

$$\frac{\mathrm{d}}{\mathrm{d}t}\varphi(\boldsymbol{x}) = \nabla\varphi(\boldsymbol{x}) \cdot \dot{\boldsymbol{x}} = \nabla\varphi(\boldsymbol{x}) \cdot \mathbf{f}(\boldsymbol{x}). \tag{11.23}$$

Combined with (11.22), this results in a partial differential equation (PDE) for the eigenfunction $\varphi(\boldsymbol{x})$:

$$\nabla\varphi(\boldsymbol{x}) \cdot \mathbf{f}(\boldsymbol{x}) = \lambda\varphi(\boldsymbol{x}). \tag{11.24}$$

With this nonlinear PDE, it is possible to approximate the eigenfunctions, either by solving for the Laurent series or with data via regression, both of which are explored below. This formulation assumes that the dynamics are both continuous and differentiable. The discrete-time dynamics in (11.4) are more general, although in many examples the continuous-time dynamics have a simpler representation than the discrete-time map for long times. For example, the simple Lorenz system has a simple continuous-time representation, yet is generally unrepresentable for even moderately long discrete-time updates.

The key takeaway from (11.21) and (11.22) is that the nonlinear dynamics become completely linear in eigenfunction coordinates, given by $\varphi(\boldsymbol{x})$. As a simple example, any conserved quantity of a dynamical system is a Koopman eigenfunction corresponding to eigenvalue $\lambda = 0$. This establishes a Koopman extension of the famous Noether's theorem (Noether 1918), implying that any symmetry in the governing equations gives rise to a new Koopman eigenfunction with eigenvalue $\lambda = 0$. For example, the Hamiltonian energy function is a Koopman eigenfunction for a conservative system. In addition, the constant function $\varphi = 1$ is always a trivial eigenfunction corresponding to $\lambda = 0$ for every dynamical system.

Eigenvalue Lattices

Interestingly, a set of Koopman eigenfunctions may be used to generate more eigenfunctions. In discrete time, we find that the product of two eigenfunctions $\varphi_1(\boldsymbol{x})$ and $\varphi_2(\boldsymbol{x})$ is also an eigenfunction

$$\mathcal{K}_t\left(\varphi_1(\boldsymbol{x})\varphi_2(\boldsymbol{x})\right) = \varphi_1(\mathbf{F}_t(\boldsymbol{x}))\varphi_2(\mathbf{F}_t(\boldsymbol{x})) \tag{11.25a}$$
$$= \lambda_1\lambda_2\varphi_1(\boldsymbol{x})\varphi_2(\boldsymbol{x}), \tag{11.25b}$$

corresponding to a new eigenvalue $\lambda_1\lambda_2$ given by the product of the two eigenvalues of $\varphi_1(\boldsymbol{x})$ and $\varphi_2(\boldsymbol{x})$.

In continuous time, the relationship becomes

$$\mathcal{K}\left(\varphi_1\varphi_2\right) = \frac{d}{dt}\left(\varphi_1\varphi_2\right) \tag{11.26a}$$
$$= \dot{\varphi}_1\varphi_2 + \varphi_1\dot{\varphi}_2 \tag{11.26b}$$
$$= \lambda_1\varphi_1\varphi_2 + \lambda_2\varphi_1\varphi_2 \tag{11.26c}$$
$$= (\lambda_1 + \lambda_2)\varphi_1\varphi_2. \tag{11.26d}$$

Mathematically, the set of Koopman eigenfunctions forms a commutative monoid under point-wise multiplication; a monoid has the structure of a group, except that the elements do not necessarily have inverses. Thus, depending on the dynamical system, there may be a finite set of *generator* eigenfunction elements that may be used to construct all other eigenfunctions. The corresponding eigenvalues similarly form a lattice, based on the product $\lambda_1\lambda_2$ or sum $\lambda_1 + \lambda_2$, depending on whether the dynamics are in discrete time or continuous time. For example, given a linear system $\dot{x} = \lambda x$, then $\varphi(x) = x$ is an eigenfunction with eigenvalue λ. Moreover, $\varphi^\alpha = x^\alpha$ is also an eigenfunction with eigenvalue $\alpha\lambda$ for any α.

The continuous-time and discrete-time lattices are related in a simple way. If the continuous-time eigenvalues are given by λ, then the corresponding discrete-time eigenvalues are given by $e^{\lambda t}$. Thus, the eigenvalue expressions in (11.25b) and (11.26d) are related as

$$e^{\lambda_1 t} e^{\lambda_2 t} \varphi_1(\boldsymbol{x})\varphi_2(\boldsymbol{x}) = e^{(\lambda_1+\lambda_2)t} \varphi_1(\boldsymbol{x})\varphi_2(\boldsymbol{x}). \tag{11.27}$$

As another simple demonstration of the relationship between continuous-time and discrete-time eigenvalues, consider the continuous-time definition in (11.20) applied to an eigenfunction:

$$\lim_{t\to 0} \frac{\mathcal{K}_t \varphi(\boldsymbol{x}) - \varphi(\boldsymbol{x})}{t} = \lim_{t\to 0} \frac{e^{\lambda t}\varphi(\boldsymbol{x}) - \varphi(\boldsymbol{x})}{t} = \lambda\varphi(\boldsymbol{x}). \tag{11.28}$$

11.4.2 Koopman Mode Decomposition

Until now, we have considered scalar measurements of a system, and we uncovered special *eigen*-measurements that evolve linearly in time. However, we often take multiple measurements of a system. In extreme cases, we may measure the entire state of a high-dimensional spatial system, such as an evolving fluid flow. These measurements may then be arranged in a vector **g**:

$$\mathbf{g}(\boldsymbol{x}) = \begin{bmatrix} g_1(\boldsymbol{x}) \\ g_2(\boldsymbol{x}) \\ \vdots \\ g_p(\boldsymbol{x}) \end{bmatrix}. \tag{11.29}$$

Each of the individual measurements may be expanded in terms of the eigenfunctions $\varphi_j(\boldsymbol{x})$, which provide a basis for Hilbert space:

$$g_i(\boldsymbol{x}) = \sum_{j=1}^{\infty} v_{ij}\varphi_j(\boldsymbol{x}). \tag{11.30}$$

Thus, the vector of observables, **g**, may be similarly expanded:

$$\mathbf{g}(\boldsymbol{x}) = \begin{bmatrix} g_1(\boldsymbol{x}) \\ g_2(\boldsymbol{x}) \\ \vdots \\ g_p(\boldsymbol{x}) \end{bmatrix} = \sum_{j=1}^{\infty} \varphi_j(\boldsymbol{x})\boldsymbol{v}_j, \tag{11.31}$$

where $\boldsymbol{v}_j$ is the jth *Koopman mode* associated with the eigenfunction φ_j.

For conservative dynamical systems, such as those governed by Hamiltonian dynamics, the Koopman operator is unitary. Thus, the Koopman eigenfunctions are orthonormal for conservative systems, and it is possible to compute the Koopman modes $\boldsymbol{v}_j$ directly by projection:

$$\boldsymbol{v}_j = \begin{bmatrix} \langle \varphi_j, g_1 \rangle \\ \langle \varphi_j, g_2 \rangle \\ \vdots \\ \langle \varphi_j, g_p \rangle \end{bmatrix}, \tag{11.32}$$

where $\langle \cdot, \cdot \rangle$ is the standard inner product of functions in Hilbert space. These modes have a physical interpretation in the case of direct spatial measurements of a system, $\mathbf{g}(\boldsymbol{x}) = \boldsymbol{x}$ in which case the modes are coherent *spatial* modes that behave linearly with the same temporal dynamics (i.e., oscillations, possibly with linear growth or decay).

Given the decomposition in (11.31), it is possible to represent the dynamics of the measurements $\mathbf{g}$ as follows:

$$\mathbf{g}(\boldsymbol{x}_k) = \mathcal{K}_{\Delta t}^k \mathbf{g}(\boldsymbol{x}_0) = \mathcal{K}_{\Delta t}^k \sum_{j=0}^{\infty} \varphi_j(\boldsymbol{x}_0)\boldsymbol{v}_j \tag{11.33a}$$

$$= \sum_{j=0}^{\infty} \mathcal{K}_{\Delta t}^k \varphi_j(\boldsymbol{x}_0)\boldsymbol{v}_j \tag{11.33b}$$

$$= \sum_{j=0}^{\infty} \lambda_j^k \varphi_j(\boldsymbol{x}_0)\boldsymbol{v}_j. \tag{11.33c}$$

This sequence of triples $\{(\lambda_j, \varphi_j, \boldsymbol{v}_j)\}_{j=0}^{\infty}$ is known as the *Koopman mode decomposition*, and was introduced in Mezić (2005). The Koopman mode decomposition was later connected to data-driven regression via the dynamic mode decomposition (Rowley et al. 2009).

Invariant Eigenspaces and Finite-Dimensional Models

Instead of capturing the evolution of all measurement functions in a Hilbert space, applied Koopman analysis approximates the evolution on an invariant subspace spanned by a finite set of measurement functions.

A *Koopman-invariant subspace* is defined as the span of a set of functions $\{g_1, g_2, \cdots, g_p\}$ if all functions g in this subspace

$$g = \alpha_1 g_1 + \alpha_2 g_2 + \cdots + \alpha_p g_p \tag{11.34}$$

remain in this subspace after being acted on by the Koopman operator $\mathcal{K}$:

$$\mathcal{K} g = \beta_1 g_1 + \beta_2 g_2 + \cdots + \beta_p g_p. \tag{11.35}$$

It is possible to obtain a finite-dimensional matrix representation of the Koopman operator by restricting it to an invariant subspace spanned by a finite number of functions $\{g_j\}_{j=0}^p$. The matrix representation $\mathbf{K}$ acts on a vector space $\mathbb{R}^p$, with the coordinates given by the values of $g_j(\mathbf{x})$. This induces a finite-dimensional linear system, as in (11.17) and (11.19).

Any finite set of eigenfunctions of the Koopman operator will span an invariant subspace. Discovering these eigenfunction coordinates is, therefore, a central challenge, as they provide intrinsic coordinates along which the dynamics behave linearly. In practice, it is more likely that we will identify an *approximately* invariant subspace, given by a set of functions $\{g_j\}_{j=0}^{p}$, where each of the functions g_j is well approximated by a finite sum of eigenfunctions: $g_j \approx \sum_{k=0}^{p} \alpha_k \varphi_k$.

11.4.3 Examples of Koopman Embeddings

Nonlinear System with Single Fixed Point and a Slow Manifold

Here, we consider an example system with a single fixed point, given by

$$\dot{x}_1 = \mu x_1, \tag{11.36a}$$

$$\dot{x}_2 = \lambda(x_2 - x_1^2). \tag{11.36b}$$

For $\lambda < \mu < 0$, the system exhibits a slow attracting manifold given by $x_2 = x_1^2$. It is possible to augment the state $\boldsymbol{x}$ with the nonlinear measurement $g = x_1^2$, to define a three-dimensional Koopman invariant subspace. In these coordinates, the dynamics become linear:

$$\frac{d}{dt}\begin{bmatrix} y_1 \\ y_2 \\ y_3 \end{bmatrix} = \begin{bmatrix} \mu & 0 & 0 \\ 0 & \lambda & -\lambda \\ 0 & 0 & 2\mu \end{bmatrix} \begin{bmatrix} y_1 \\ y_2 \\ y_3 \end{bmatrix} \quad \text{for} \quad \begin{bmatrix} y_1 \\ y_2 \\ y_3 \end{bmatrix} = \begin{bmatrix} x_1 \\ x_2 \\ x_1^2 \end{bmatrix}. \tag{11.37a}$$

The full three-dimensional Koopman observable vector space is visualized in Figure 11.7. Trajectories that start on the invariant manifold $y_3 = y_1^2$, visualized by the blue surface, are constrained to stay on this manifold. There is a *slow* subspace, spanned by the eigenvectors corresponding to the slow eigenvalues μ and 2μ; this subspace is visualized by the green surface. Finally, there is the original asymptotically attracting manifold of the original system, $y_2 = y_1^2$, which is visualized as the red surface. The blue and red parabolic surfaces always intersect in a parabola that is inclined at a 45° angle in the y_2–y_3 direction. The green surface approaches this 45° inclination as the ratio of fast to slow dynamics becomes increasingly large. In the full three-dimensional Koopman observable space, the dynamics produce a single stable node, with trajectories rapidly attracting onto the green subspace and then slowly approaching the fixed point.

Intrinsic Coordinates Defined by Eigenfunctions of the Koopman Operator

The left eigenvectors of the Koopman operator yield Koopman eigenfunctions (i.e., eigenobservables). The Koopman eigenfunctions of (11.37a) corresponding to eigenvalues μ and λ are

$$\varphi_\mu = x_1, \quad \text{and} \quad \varphi_\lambda = x_2 - bx_1^2 \quad \text{with} \quad b = \frac{\lambda}{\lambda - 2\mu}. \tag{11.38}$$

The constant b in φ_λ captures the fact that for a finite ratio λ/μ, the dynamics only shadow the asymptotically attracting slow manifold $x_2 = x_1^2$, but in fact follow neighboring parabolic trajectories. This is illustrated more clearly by the various surfaces in Figure 11.7 for different ratios λ/μ.

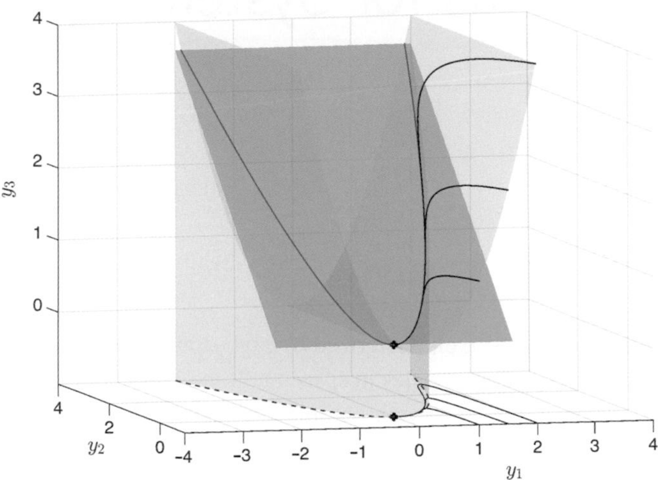

Figure 11.7 Visualization of three-dimensional linear Koopman system from (11.37a) along with projection of dynamics onto the x_1–x_2 plane. The attracting slow manifold is shown in red, the constraint $y_3 = y_1^2$ is shown in blue, and the slow unstable subspace of (11.37a) is shown in green. Black trajectories of the linear Koopman system in **y** project onto trajectories of the full nonlinear system in **x** in the y_1–y_2 plane. Here, $\mu = -0.05$ and $\lambda = 1$. Reproduced from Brunton et al. (2016b).

In this way, a set of intrinsic coordinates may be determined from the observable functions defined by the left eigenvectors of the Koopman operator on an invariant subspace. Explicitly,

$$\varphi_\alpha(\boldsymbol{x}) = \boldsymbol{\xi}_\alpha \boldsymbol{y}(\boldsymbol{x}), \quad \text{where} \quad \boldsymbol{\xi}_\alpha \mathbf{K} = \alpha \boldsymbol{\xi}_\alpha. \tag{11.39}$$

These eigen-observables define observable subspaces that remain invariant under the Koopman operator, even after coordinate transformations. As such, they may be regarded as intrinsic coordinates (Williams et al. 2015) on the Koopman-invariant subspace.

12 Methods for System Identification

S. Brunton

Fluid dynamics are characterized by high-dimensional, nonlinear dynamical systems that exhibit multiple scales in space and time. Simulating these equations is currently too expensive for real-time optimization and control (Bagheri et al. 2009, Fabbiane et al. 2014, Brunton & Noack 2015), and there are efforts to obtain reduced-order models (ROMs) that capture the salient dynamics with a lower-dimensional system. There are two broad approaches to obtain ROMs: first, it is possible to start with a high-dimensional system, such as the discretized Navier–Stokes equations, and project the dynamics onto a low-dimensional subspace identified, for example, using proper orthogonal decomposition (POD; Berkooz et al. 1993, Holmes et al. 2012) and Galerkin projection (Quarteroni & Rozza 2013, Benner et al. 2015). The second approach is to collect data from a simulation or an experiment and identify a low-rank model using data-driven techniques. This approach is typically called system identification, and is often preferred for control design because of the relative ease of implementation. The dynamic mode decomposition (DMD) from Chapter 7 is an example of system identification that is widely used in fluid mechanics.

Most of the system identification techniques discussed here, including DMD from Chapter 7, result in linear models (see Chapter 10). Even for nonlinear systems, such as the Navier–Stokes equations, it is often possible to use linear approaches for closed-loop control. There are numerous examples of successful linear model-based flow control (Bagheri et al. 2009, Fabbiane et al. 2014, Brunton & Noack 2015). Control generally relies on fast, low-latency models for robust real-time performance, especially for fluids where the dynamics often evolve on fast timescales. Therefore, system identification also relies on dimensionality reduction to obtain low-dimensional approximations of the system for use in real-time feedback control.

This chapter explores several system identification techniques. First, we introduce balanced truncation and balanced POD (BPOD), which have been used extensively to obtain linear ROMs in fluid systems. Next, we develop the eigensystem realization algorithm (ERA), which has deep connections to both balanced POD and DMD. Finally, we introduce the sparse identification of the nonlinear dynamics (SINDy) algorithm.

12.1 Balanced Model Reduction

Balanced model reduction refers to an alternative *control-oriented* procedure that results in low-dimensional models for input–output systems. POD is the most common low-dimensional basis used for model reduction, and it orders modes based on the kinetic energy content of the approximated state. Instead of energy, balanced modes are ordered by how much input–output energy they capture, making them useful for feedback control. This section will explore balanced model reduction in the broader context of reduced-order modeling and system identification. Balanced modeling of fluids is based on both data and equations, and is typically used to reduce high-dimensional fluid systems. As with POD-based modeling, the general idea is to determine a hierarchical modal decomposition of the system state that may be truncated at some model order, only keeping the coherent structures that are most important for control.

12.1.1 The Goal of Model Reduction

Consider a high-dimensional system, depicted schematically in Figure 12.1,

$$\frac{d}{dt}\boldsymbol{x} = \mathbf{A}\boldsymbol{x} + \mathbf{B}\boldsymbol{u}, \tag{12.1a}$$

$$\boldsymbol{y} = \mathbf{C}\boldsymbol{x} + \mathbf{D}\boldsymbol{u}, \tag{12.1b}$$

for example from a spatially discretized simulation of a PDE. The primary goal of model reduction is to find a coordinate transformation $\boldsymbol{x} = \boldsymbol{\Psi}\tilde{\mathbf{x}}$ giving rise to a related system $(\tilde{\mathbf{A}}, \tilde{\mathbf{B}}, \tilde{\mathbf{C}}, \tilde{\mathbf{D}})$ with similar input–output characteristics,

$$\frac{d}{dt}\tilde{\mathbf{x}} = \tilde{\mathbf{A}}\tilde{\mathbf{x}} + \tilde{\mathbf{B}}\boldsymbol{u}, \tag{12.2a}$$

$$\boldsymbol{y} = \tilde{\mathbf{C}}\tilde{\mathbf{x}} + \tilde{\mathbf{D}}\boldsymbol{u}, \tag{12.2b}$$

in terms of a state $\tilde{\mathbf{x}} \in \mathbb{R}^r$ with reduced dimension, $r \ll n$. Note that $\boldsymbol{u}$ and $\boldsymbol{y}$ are the same in (12.1) and (12.2) even though the system states are different. Obtaining the projection operator $\boldsymbol{\Psi}$ will be the focus of this section.

As a motivating example, consider the following simplified model:

$$\frac{d}{dt}\begin{bmatrix} x_1 \\ x_2 \end{bmatrix} = \begin{bmatrix} -2 & 0 \\ 0 & -1 \end{bmatrix}\begin{bmatrix} x_1 \\ x_2 \end{bmatrix} + \begin{bmatrix} 1 \\ 10^{-10} \end{bmatrix} u, \tag{12.3a}$$

$$y = \begin{bmatrix} 1 & 10^{-10} \end{bmatrix}\begin{bmatrix} x_1 \\ x_2 \end{bmatrix}. \tag{12.3b}$$

In this case, the state x_2 is barely controllable and barely observable. Simply choosing $\tilde{x} = x_1$ will result in a ROM that faithfully captures the input–output dynamics. Although the choice $\tilde{x} = x_1$ seems intuitive in this extreme case, many model reduction techniques would erroneously favor the state $\tilde{x} = x_2$, since it is more lightly damped. Throughout this section, we will investigate how to accurately and efficiently find the transformation matrix $\boldsymbol{\Psi}$ that best captures the input–output dynamics.

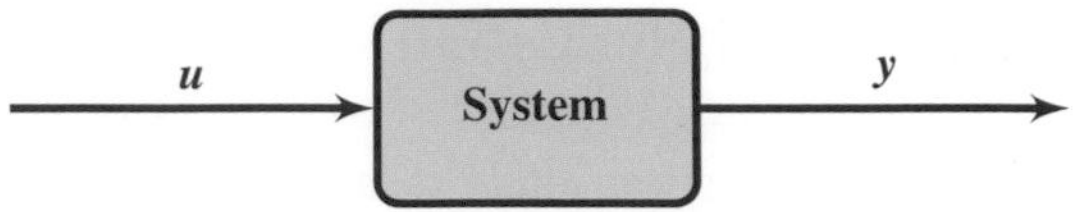

Figure 12.1 Input–output system. A control-oriented reduced-order model will capture the transfer function from $\boldsymbol{u}$ to $\boldsymbol{y}$.

The POD (Berkooz et al. 1993, Holmes et al. 2012) provides a transform matrix $\boldsymbol{\Psi}$, the columns of which are modes that are ordered based on energy content.[1] POD has been widely used to generate ROMs of complex systems, many for control, and it is guaranteed to provide an optimal low-rank basis to capture the maximal energy or variance in a data set. However, it may be the case that the most energetic modes are nearly uncontrollable or unobservable, and therefore may not be relevant for control. Similarly, in many cases the most controllable and observable state directions may have very low energy; for example, acoustic modes typically have very low energy, yet they mediate the dominant input–output dynamics in many fluid systems. The rudder on a ship provides a good analogy: although it accounts for a small amount of the total energy, it is dynamically important for control.

Instead of ordering modes based on energy, it is possible to determine a hierarchy of modes that are most controllable and observable, therefore capturing the most input–output information. These modes give rise to *balanced* models, giving equal weighting to the controllability and observability of a state via a coordinate transformation that makes the controllability and observability Gramians equal and diagonal. These models have been extremely successful, although computing a balanced model using traditional methods is prohibitively expensive for high-dimensional systems. In this section, we describe the balancing procedure, as well as modern methods for efficient computation of balanced models. A computationally efficient suite of algorithms for model reduction and system identification may be found in Belson et al. (2014).

A balanced ROM should map inputs to outputs as faithfully as possible for a given model order r. It is therefore important to introduce an *operator norm* to quantify how similarly (12.1) and (12.2) act on a given set of inputs. Typically, we take the infinity norm of the difference between the transfer functions $\mathbf{G}(s)$ and $\mathbf{G}_r(s)$ obtained from the full system (12.1) and reduced system (12.2), respectively. This norm is given by

$$\|\mathbf{G}\|_\infty \triangleq \max_\omega \sigma_1 \left(\mathbf{G}(i\omega)\right). \tag{12.4}$$

To summarize, we seek a ROM (12.2) of low order, $r \ll n$, so the operator norm $\|\mathbf{G} - \mathbf{G}_r\|_\infty$ is small.

[1] When the training data consists of velocity fields, for example from a high-dimensional discretized fluid system, then the singular values literally indicate the kinetic energy content of the associated mode. It is common to refer to POD modes as being ordered by *energy* content, even in other applications, although *variance* is more technically correct.

12.1.2 Change of Variables in Control Systems

The balanced model reduction problem may be formulated in terms of first finding a coordinate transformation

$$x = \mathbf{T}z, \tag{12.5}$$

that hierarchically orders the states in z in terms of their ability to capture the input–output characteristics of the system. We will begin by considering an invertible transformation $\mathbf{T} \in \mathbb{R}^{n \times n}$, and then provide a method to compute just the first r columns, which will comprise the transformation $\boldsymbol{\Psi}$ in (12.2). Thus, it will be possible to retain only the first r most controllable/observable states, while truncating the rest. This is similar to the change of variables into eigenvector coordinates, except that we emphasize controllability and observability rather than characteristics of the dynamics.

Substituting $\mathbf{T}z$ into (12.1) gives

$$\frac{d}{dt}\mathbf{T}z = \mathbf{A}\mathbf{T}z + \mathbf{B}u, \tag{12.6a}$$

$$y = \mathbf{C}\mathbf{T}z + \mathbf{D}u. \tag{12.6b}$$

Finally, multiplying (12.6a) by $\mathbf{T}^{-1}$ yields

$$\frac{d}{dt}z = \mathbf{T}^{-1}\mathbf{A}\mathbf{T}z + \mathbf{T}^{-1}\mathbf{B}u, \tag{12.7a}$$

$$y = \mathbf{C}\mathbf{T}z + \mathbf{D}u. \tag{12.7b}$$

This results in the transformed equations

$$\frac{d}{dt}z = \hat{\mathbf{A}}z + \hat{\mathbf{B}}u, \tag{12.8a}$$

$$y = \hat{\mathbf{C}}z + \mathbf{D}u, \tag{12.8b}$$

where $\hat{\mathbf{A}} = \mathbf{T}^{-1}\mathbf{A}\mathbf{T}$, $\hat{\mathbf{B}} = \mathbf{T}^{-1}\mathbf{B}$, and $\hat{\mathbf{C}} = \mathbf{C}\mathbf{T}$. Note that when the columns of $\mathbf{T}$ are orthonormal, the change of coordinates becomes

$$\frac{d}{dt}z = \mathbf{T}^{*}\mathbf{A}\mathbf{T}z + \mathbf{T}^{*}\mathbf{B}u, \tag{12.9a}$$

$$y = \mathbf{C}\mathbf{T}z + \mathbf{D}u. \tag{12.9b}$$

Gramians and Coordinate Transformations

The controllability and observability Gramians each establish an inner product on state-space in terms of how controllable or observable a given state is, respectively. As such, Gramians depend on the particular choice of coordinate system and will transform under a change of coordinates. In the coordinate system z given by (12.5), the controllability Gramian becomes

$$\hat{\mathbf{W}}_c = \int_0^\infty e^{\hat{\mathbf{A}}\tau}\hat{\mathbf{B}}\hat{\mathbf{B}}^* e^{\hat{\mathbf{A}}^*\tau}\, d\tau, \tag{12.10a}$$

$$= \int_0^\infty e^{\mathbf{T}^{-1}\mathbf{A}\mathbf{T}\tau}\mathbf{T}^{-1}\mathbf{B}\mathbf{B}^*\mathbf{T}^{-*} e^{\mathbf{T}^*\mathbf{A}^*\mathbf{T}^{-*}\tau}\, d\tau, \tag{12.10b}$$

$$= \int_0^\infty \mathbf{T}^{-1} e^{\mathbf{A}\tau}\mathbf{T}\mathbf{T}^{-1}\mathbf{B}\mathbf{B}^*\mathbf{T}^{-*}\mathbf{T}^* e^{\mathbf{A}^*\tau}\mathbf{T}^{-*}\, d\tau, \tag{12.10c}$$

$$= \mathbf{T}^{-1}\left(\int_0^\infty e^{\mathbf{A}\tau}\mathbf{B}\mathbf{B}^* e^{\mathbf{A}^\tau}\, d\tau\right)\mathbf{T}^{-*}, \tag{12.10d}$$

$$= \mathbf{T}^{-1}\mathbf{W}_c\mathbf{T}^{-*}. \tag{12.10e}$$

Note that here we introduce $\mathbf{T}^{-*} := \left(\mathbf{T}^{-1}\right)^* = (\mathbf{T}^*)^{-1}$. The observability Gramian transforms similarly

$$\hat{\mathbf{W}}_o = \mathbf{T}^*\mathbf{W}_o\mathbf{T}, \tag{12.11}$$

which is an exercise for the reader. Both Gramians transform as tensors (i.e., in terms of the transform matrix $\mathbf{T}$ and its transpose, rather than $\mathbf{T}$ and its inverse), which is consistent with them inducing an inner product on state-space.

Simple Rescaling

This example, modified from Moore (1981), demonstrates the ability to balance a system through a change of coordinates. Consider the system

$$\frac{d}{dt}\begin{bmatrix} x_1 \\ x_2 \end{bmatrix} = \begin{bmatrix} -1 & 0 \\ 0 & -10 \end{bmatrix}\begin{bmatrix} x_1 \\ x_2 \end{bmatrix} + \begin{bmatrix} 10^{-3} \\ 10^3 \end{bmatrix} u, \tag{12.12a}$$

$$y = \begin{bmatrix} 10^3 & 10^{-3} \end{bmatrix}\begin{bmatrix} x_1 \\ x_2 \end{bmatrix}. \tag{12.12b}$$

In this example, the first state x_1 is barely controllable, while the second state is barely observable. However, under the change of coordinates $z_1 = 10^3 x_1$ and $z_2 = 10^{-3} x_2$, the system becomes balanced:

$$\frac{d}{dt}\begin{bmatrix} z_1 \\ z_2 \end{bmatrix} = \begin{bmatrix} -1 & 0 \\ 0 & -10 \end{bmatrix}\begin{bmatrix} z_1 \\ z_2 \end{bmatrix} + \begin{bmatrix} 1 \\ 1 \end{bmatrix} u, \tag{12.13a}$$

$$y = \begin{bmatrix} 1 & 1 \end{bmatrix}\begin{bmatrix} z_1 \\ z_2 \end{bmatrix}. \tag{12.13b}$$

In this example, the coordinate change simply rescales the state $\boldsymbol{x}$. For instance, it may be that the first state had units of millimeters while the second state had units of kilometers. Writing both states in meters balances the dynamics; that is, the controllability and observability Gramians are equal and diagonal.

12.1.3 Balancing Transformations

Now we are ready to derive the balancing coordinate transformation $\mathbf{T}$ that makes the controllability and observability Gramians equal and diagonal:

$$\hat{\mathbf{W}}_c = \hat{\mathbf{W}}_o = \boldsymbol{\Sigma}. \tag{12.14}$$

First, consider the product of the Gramians from (12.10) and (12.11):

$$\hat{\mathbf{W}}_c \hat{\mathbf{W}}_o = \mathbf{T}^{-1}\mathbf{W}_c\mathbf{W}_o\mathbf{T}. \tag{12.15}$$

Plugging in the desired $\hat{\mathbf{W}}_c = \hat{\mathbf{W}}_o = \boldsymbol{\Sigma}$ yields

$$\mathbf{T}^{-1}\mathbf{W}_c\mathbf{W}_o\mathbf{T} = \boldsymbol{\Sigma}^2 \quad \Longrightarrow \quad \mathbf{W}_c\mathbf{W}_o\mathbf{T} = \mathbf{T}\boldsymbol{\Sigma}^2. \tag{12.16}$$

The latter expression in (12.16) is the equation for the eigendecomposition of $\mathbf{W}_c\mathbf{W}_o$, the product of the Gramians in the original coordinates. Thus, the balancing transformation $\mathbf{T}$ is related to the eigendecomposition of $\mathbf{W}_c\mathbf{W}_o$. Equation 12.16 is valid for any scaling of the eigenvectors, and the correct rescaling must be chosen to exactly balance the Gramians. In other words, there are many such transformations $\mathbf{T}$ that make the product $\hat{\mathbf{W}}_c\hat{\mathbf{W}}_o = \boldsymbol{\Sigma}^2$, but where the individual Gramians are not equal (for example diagonal Gramians $\hat{\mathbf{W}}_c = \boldsymbol{\Sigma}_c$ and $\hat{\mathbf{W}}_o = \boldsymbol{\Sigma}_o$ will satisfy (12.16) if $\boldsymbol{\Sigma}_c\boldsymbol{\Sigma}_o = \boldsymbol{\Sigma}^2$).

Below, we will introduce the matrix $\mathbf{S} = \mathbf{T}^{-1}$ to simplify notation.

Scaling Eigenvectors for the Balancing Transformation

To find the correct scaling of eigenvectors to make $\hat{\mathbf{W}}_c = \hat{\mathbf{W}}_o = \boldsymbol{\Sigma}$, first consider the simplified case of balancing the first diagonal element of $\boldsymbol{\Sigma}$. Let $\boldsymbol{\xi}_u$ denote the unscaled first column of $\mathbf{T}$, and let $\boldsymbol{\eta}_u$ denote the unscaled first row of $\mathbf{S} = \mathbf{T}^{-1}$. Then

$$\boldsymbol{\eta}_u\mathbf{W}_c\boldsymbol{\eta}_u^* = \sigma_c, \tag{12.17a}$$

$$\boldsymbol{\xi}_u^*\mathbf{W}_o\boldsymbol{\xi}_u = \sigma_o. \tag{12.17b}$$

The first element of the diagonalized controllability Gramian is thus σ_c, while the first element of the diagonalized observability Gramian is σ_o. If we scale the eigenvector $\boldsymbol{\xi}_u$ by σ_s, then the inverse eigenvector $\boldsymbol{\eta}_u$ is scaled by σ_s^{-1}. Transforming via the new scaled eigenvectors $\boldsymbol{\xi}_s = \sigma_s\boldsymbol{\xi}_u$ and $\boldsymbol{\eta}_s = \sigma_s^{-1}\boldsymbol{\eta}_u$, yields

$$\boldsymbol{\eta}_s\mathbf{W}_c\boldsymbol{\eta}_s^* = \sigma_s^{-2}\sigma_c, \tag{12.18a}$$

$$\boldsymbol{\xi}_s^*\mathbf{W}_o\boldsymbol{\xi}_s = \sigma_s^{2}\sigma_o. \tag{12.18b}$$

Thus, for the two Gramians to be equal,

$$\sigma_s^{-2}\sigma_c = \sigma_s^2\sigma_o \quad \Longrightarrow \quad \sigma_s = \left(\frac{\sigma_c}{\sigma_o}\right)^{1/4}. \tag{12.19}$$

To balance every diagonal entry of the controllability and observability Gramians, we first consider the unscaled eigenvector transformation $\mathbf{T}_u$ from (12.16); the subscript u simply denotes *unscaled*. As an example, we use the standard scaling in most computational software so that the columns of $\mathbf{T}_u$ have unit norm. Then both Gramians are diagonalized, but are not necessarily equal:

$$\mathbf{T}_u^{-1}\mathbf{W}_c\mathbf{T}_u^{-*} = \boldsymbol{\Sigma}_c, \tag{12.20a}$$

$$\mathbf{T}_u^{*}\mathbf{W}_o\mathbf{T}_u = \boldsymbol{\Sigma}_o. \tag{12.20b}$$

The scaling that exactly balances these Gramians is then given by $\boldsymbol{\Sigma}_s = \boldsymbol{\Sigma}_c^{1/4}\boldsymbol{\Sigma}_o^{-1/4}$. Thus, the exact balancing transformation is given by

$$\mathbf{T} = \mathbf{T}_u\boldsymbol{\Sigma}_s. \tag{12.21}$$

It is possible to directly confirm that this transformation balances the Gramians:

$$(\mathbf{T}_u\boldsymbol{\Sigma}_s)^{-1}\mathbf{W}_c(\mathbf{T}_u\boldsymbol{\Sigma}_s)^{-*} = \boldsymbol{\Sigma}_s^{-1}\mathbf{T}_u^{-1}\mathbf{W}_c\mathbf{T}_u^{-*}\boldsymbol{\Sigma}_s^{-1} = \boldsymbol{\Sigma}_s^{-1}\boldsymbol{\Sigma}_c\boldsymbol{\Sigma}_s^{-1} = \boldsymbol{\Sigma}_c^{1/2}\boldsymbol{\Sigma}_o^{1/2}, \tag{12.22a}$$

$$(\mathbf{T}_u\boldsymbol{\Sigma}_s)^{*}\mathbf{W}_o(\mathbf{T}_u\boldsymbol{\Sigma}_s) = \boldsymbol{\Sigma}_s\mathbf{T}_u^{*}\mathbf{W}_o\mathbf{T}_u\boldsymbol{\Sigma}_s = \boldsymbol{\Sigma}_s\boldsymbol{\Sigma}_o\boldsymbol{\Sigma}_s = \boldsymbol{\Sigma}_c^{1/2}\boldsymbol{\Sigma}_o^{1/2}. \tag{12.22b}$$

The manipulations rely on the fact that diagonal matrices commute, so that $\boldsymbol{\Sigma}_c\boldsymbol{\Sigma}_o = \boldsymbol{\Sigma}_o\boldsymbol{\Sigma}_c$, etc.

Example of the Balancing Transform and Gramians

Before confronting the practical challenges associated with accurately and efficiently computing the balancing transformation, it is helpful to consider an illustrative example.

The following example illustrates the balanced realization for a two-dimensional system. First, we generate a system and compute its balanced realization, along with the Gramians for each system. Next, we visualize the Gramians of the unbalanced and balanced systems in Figure 12.2.

To visualize the Gramians in Figure 12.2, we first recall that the distance the system can go in a direction $\boldsymbol{x}$ with a unit actuation input is given by $\boldsymbol{x}^*\mathbf{W}_c\boldsymbol{x}$. Thus, the controllability Gramian may be visualized by plotting $\mathbf{W}_c^{1/2}\boldsymbol{x}$ for $\boldsymbol{x}$ on a sphere with $\|\boldsymbol{x}\| = 1$. The observability Gramian may be similarly visualized.

In this example, we see that the most controllable and observable directions may not be well aligned. However, by a change of coordinates, it is possible to find a new direction that is the most jointly controllable and observable. It is then possible to represent the system in this one-dimensional subspace, while still capturing a significant portion of the input–output energy. If the red and blue Gramians were exactly perpendicular, so that the most controllable direction was the least observable direction, and vice versa, then the balanced Gramian would be a circle. In this case, there is no preferred state direction, and both directions are equally important for the input–output behavior.

12.1.4 Balanced Truncation

We have now shown that it is possible to define a change of coordinates so that the controllability and observability Gramians are equal and diagonal. Moreover, these new coordinates may be ranked hierarchically in terms of their joint controllability

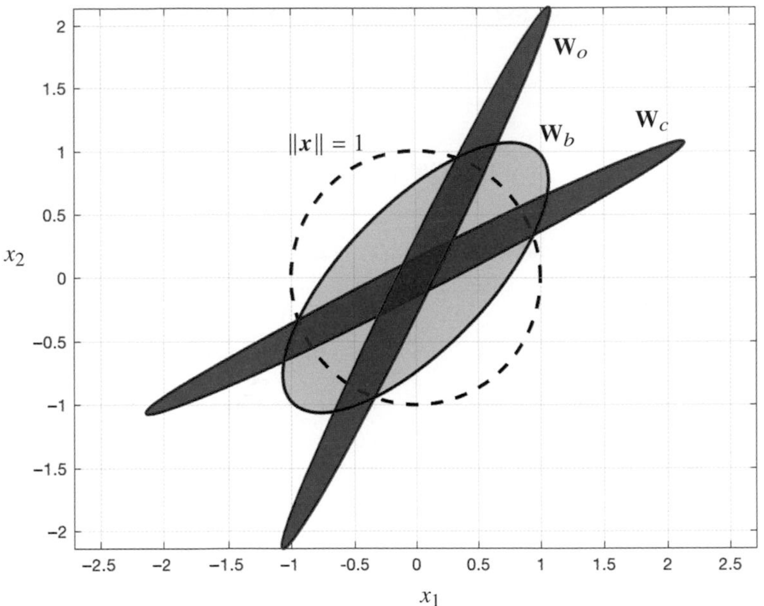

Figure 12.2 Illustration of balancing transformation on Gramians. The reachable set with unit control input is shown in red, given by $\mathbf{W}_c^{1/2}\boldsymbol{x}$ for $\|\boldsymbol{x}\| = 1$. The corresponding observable set is shown in blue. Under the balancing transformation $\mathbf{T}$, the Gramians are equal, shown in purple. From Brunton and Kutz (2019), reproduced with permission of the Licensor through PLSclear.

and observability. It may be possible to truncate these coordinates and keep only the most controllable/observable directions, resulting in a ROM that faithfully captures input–output dynamics.

Given the new coordinates $z = \mathbf{T}^{-1}\boldsymbol{x} \in \mathbb{R}^n$, it is possible to define a reduced-order state $\tilde{\mathbf{x}} \in \mathbb{R}^r$ as

$$z = \begin{bmatrix} \left.\begin{matrix} z_1 \\ \vdots \\ z_r \end{matrix}\right\} \tilde{\mathbf{x}} \\ z_{r+1} \\ \vdots \\ z_n \end{bmatrix} \tag{12.23}$$

in terms of the first r most controllable and observable directions. If we partition the balancing transformation $\mathbf{T}$ and inverse transformation $\mathbf{S} = \mathbf{T}^{-1}$ into the first r modes to be retained and the last $n - r$ modes to be truncated,

$$\mathbf{T} = \begin{bmatrix} \boldsymbol{\Psi} & \mathbf{T}_t \end{bmatrix}, \quad \mathbf{S} = \begin{bmatrix} \boldsymbol{\Phi}^* \\ \mathbf{S}_t \end{bmatrix}, \tag{12.24}$$

then it is possible to rewrite the transformed dynamics in (12.7) as

$$\frac{d}{dt}\left[\begin{array}{c}\tilde{\mathbf{x}}\\ \hline z_t\end{array}\right] = \left[\begin{array}{c|c}\mathbf{\Phi}^*\mathbf{A}\mathbf{\Psi} & \mathbf{\Phi}^*\mathbf{A}\mathbf{T}_t\\ \hline \mathbf{S}_t\mathbf{A}\mathbf{\Psi} & \mathbf{S}_t\mathbf{A}\mathbf{T}_t\end{array}\right]\left[\begin{array}{c}\tilde{\mathbf{x}}\\ \hline z_t\end{array}\right] + \left[\begin{array}{c}\mathbf{\Phi}^*\mathbf{B}\\ \hline \mathbf{S}_t\mathbf{B}\end{array}\right]u, \tag{12.25a}$$

$$y = \left[\begin{array}{c|c}\mathbf{C}\mathbf{\Psi} & \mathbf{C}\mathbf{T}_t\end{array}\right]\left[\begin{array}{c}\tilde{\mathbf{x}}\\ \hline z_t\end{array}\right] + \mathbf{D}u. \tag{12.25b}$$

In balanced truncation, the state z_t is simply truncated (i.e., discarded and set equal to zero), and only the $\tilde{\mathbf{x}}$ equations remain:

$$\frac{d}{dt}\tilde{\mathbf{x}} = \mathbf{\Phi}^*\mathbf{A}\mathbf{\Psi}\tilde{\mathbf{x}} + \mathbf{\Phi}^*\mathbf{B}u, \tag{12.26a}$$

$$y = \mathbf{C}\mathbf{\Psi}\tilde{\mathbf{x}} + \mathbf{D}u. \tag{12.26b}$$

Only the first r columns of $\mathbf{T}$ and $\mathbf{S}^* = \mathbf{T}^{-*}$ are required to construct $\mathbf{\Psi}$ and $\mathbf{\Phi}$, and thus computing the entire balancing transformation $\mathbf{T}$ is unnecessary. The computation of $\mathbf{\Psi}$ and $\mathbf{\Phi}$ without $\mathbf{T}$ will be discussed in the following sections. A key benefit of balanced truncation is the existence of upper and lower bounds on the error of a given order truncation:

$$\textbf{Upper bound:} \quad \|\mathbf{G} - \mathbf{G}_r\|_\infty \leq 2\sum_{j=r+1}^{n}\sigma_j, \tag{12.27a}$$

$$\textbf{Lower bound:} \quad \|\mathbf{G} - \mathbf{G}_r\|_\infty > \sigma_{r+1}, \tag{12.27b}$$

where σ_j is the jth diagonal entry of the balanced Gramians. The diagonal entries of $\mathbf{\Sigma}$ are also known as *Hankel singular values*.

12.1.5 Computing Balanced Realizations

In the previous section, we demonstrated the feasibility of obtaining a coordinate transformation that balances the controllability and observability Gramians. However, the computation of this balancing transformation is nontrivial, and significant work has gone into obtaining accurate and efficient methods, starting with Moore (1981), and continuing with Lall et al. (2002), Willcox and Peraire (2002), and Rowley (2005). For an excellent and complete treatment of balanced realizations and model reduction, see Antoulas (2005).

In practice, computing the Gramians $\mathbf{W}_c$ and $\mathbf{W}_o$ and the eigendecomposition of the product $\mathbf{W}_c\mathbf{W}_o$ in (12.16) may be prohibitively expensive for high-dimensional systems. Instead, the balancing transformation may be approximated from impulse-response data, utilizing the singular value decomposition (SVD) for efficient extraction of the most relevant subspaces.

We will first show that Gramians may be approximated via a snapshot matrix from impulse response experiments/simulations. Then, we will show how the balancing transformation may be obtained from this data.

Empirical Gramians

In practice, computing Gramians via the Lyapunov equation is computationally expensive, with computational complexity of $O(n^3)$. Instead, the Gramians may be approximated by full-state measurements of the discrete-time direct and adjoint systems:

$$\textit{direct}: \quad \boldsymbol{x}_{k+1} = \mathbf{A}_d \boldsymbol{x}_k + \mathbf{B}_d \boldsymbol{u}_k, \tag{12.28a}$$

$$\textit{adjoint}: \quad \boldsymbol{x}_{k+1} = \mathbf{A}_d^* \boldsymbol{x}_k + \mathbf{C}_d^* \boldsymbol{y}_k. \tag{12.28b}$$

Equation (12.28a) is the discrete-time dynamic update equation, and (12.28b) is the adjoint equation. The matrices $\mathbf{A}_d, \mathbf{B}_d$, and $\mathbf{C}_d$ are the discrete-time system matrices. Note that the adjoint equation is generally nonphysical, and must be simulated; thus the methods here apply to analytical equations and simulations, but not to experimental data. An alternative formulation that does not rely on adjoint data, and therefore generalizes to experiments, will be provided in Section 12.2.

Computing the impulse response of the direct and adjoint systems yields the following discrete-time snapshot matrices:

$$\mathcal{C}_d = \begin{bmatrix} \mathbf{B}_d & \mathbf{A}_d\mathbf{B}_d & \cdots & \mathbf{A}_d^{m_c-1}\mathbf{B}_d \end{bmatrix} \qquad \mathcal{O}_d = \begin{bmatrix} \mathbf{C}_d \\ \mathbf{C}_d\mathbf{A}_d \\ \vdots \\ \mathbf{C}_d\mathbf{A}_d^{m_o-1} \end{bmatrix}. \tag{12.29}$$

Note that when $m_c = n$, $\mathcal{C}_d$ is the discrete-time controllability matrix and when $m_o = n$, $\mathcal{O}_d$ is the discrete-time observability matrix; however, we generally consider $m_c, m_o \ll n$. These matrices may also be obtained by sampling the continuous-time direct and adjoint systems at a regular interval Δt.

It is now possible to compute *empirical* Gramians that approximate the true Gramians without solving a Lyapunov equation:

$$\mathbf{W}_c \approx \mathbf{W}_c^e = \mathcal{C}_d \mathcal{C}_d^*, \tag{12.30a}$$

$$\mathbf{W}_o \approx \mathbf{W}_o^e = \mathcal{O}_d^* \mathcal{O}_d. \tag{12.30b}$$

The empirical Gramians essentially comprise a Riemann sum approximation of the integral in the continuous-time Gramians, which becomes exact as the time step of the discrete-time system becomes arbitrarily small and the duration of the impulse response becomes arbitrarily large. In practice, the impulse response snapshots should be collected until the lightly damped transients die out. The method of empirical Gramians is quite efficient, and is widely used (Moore 1981, Lall et al. 1999, Lall et al. 2002, Willcox & Peraire 2002, Rowley 2005). Note that p adjoint impulse responses are required, where p is the number of outputs. This becomes intractable when there are a large number of outputs (e.g., full state measurements), motivating the output projection below.

Balanced POD

Instead of computing the eigendecomposition of $\mathbf{W}_c\mathbf{W}_o$, which is an $n \times n$ matrix, it is possible to compute the balancing transformation via the SVD of the product of the snapshot matrices,

$$\mathcal{O}_d\mathcal{C}_d, \tag{12.31}$$

reminiscent of the method of snapshots (Sirovich 1987). This is the approach taken by Rowley (2005).

First, define the generalized Hankel matrix as the product of the adjoint ($\mathcal{O}_d$) and direct ($\mathcal{C}_d$) snapshot matrices from (12.29), for the discrete-time system:

$$\mathbf{H} = \mathcal{O}_d\mathcal{C}_d = \begin{bmatrix} \mathbf{C}_d \\ \mathbf{C}_d\mathbf{A}_d \\ \vdots \\ \mathbf{C}_d\mathbf{A}_d^{m_o-1} \end{bmatrix} \begin{bmatrix} \mathbf{B}_d & \mathbf{A}_d\mathbf{B}_d & \cdots & \mathbf{A}_d^{m_c-1}\mathbf{B}_d \end{bmatrix} \tag{12.32a}$$

$$= \begin{bmatrix} \mathbf{C}_d\mathbf{B}_d & \mathbf{C}_d\mathbf{A}_d\mathbf{B}_d & \cdots & \mathbf{C}_d\mathbf{A}_d^{m_c-1}\mathbf{B}_d \\ \mathbf{C}_d\mathbf{A}_d\mathbf{B}_d & \mathbf{C}_d\mathbf{A}_d^2\mathbf{B}_d & \cdots & \mathbf{C}_d\mathbf{A}_d^{m_c}\mathbf{B}_d \\ \vdots & \vdots & \ddots & \vdots \\ \mathbf{C}_d\mathbf{A}_d^{m_o-1}\mathbf{B}_d & \mathbf{C}_d\mathbf{A}_d^{m_o}\mathbf{B}_d & \cdots & \mathbf{C}_d\mathbf{A}_d^{m_c+m_o-2}\mathbf{B}_d \end{bmatrix}. \tag{12.32b}$$

Next, we factor $\mathbf{H}$ using the SVD:

$$\mathbf{H} = \mathbf{U}\mathbf{\Sigma}\mathbf{V}^* = \begin{bmatrix} \tilde{\mathbf{U}} & \mathbf{U}_t \end{bmatrix} \begin{bmatrix} \tilde{\mathbf{\Sigma}} & \mathbf{0} \\ \mathbf{0} & \mathbf{\Sigma}_t \end{bmatrix} \begin{bmatrix} \tilde{\mathbf{V}}^* \\ \mathbf{V}_t^* \end{bmatrix} \approx \tilde{\mathbf{U}}\tilde{\mathbf{\Sigma}}\tilde{\mathbf{V}}^*. \tag{12.33}$$

For a given desired model order $r \ll n$, only the first r columns of $\mathbf{U}$ and $\mathbf{V}$ are retained, along with the first $r \times r$ block of $\mathbf{\Sigma}$; the remaining contribution from $\mathbf{U}_t\mathbf{\Sigma}_t\mathbf{V}_t^*$ may be truncated. This yields a bi-orthogonal set of modes given by

$$\textit{Direct modes:} \quad \mathbf{\Psi} = \mathcal{C}_d\tilde{\mathbf{V}}\tilde{\mathbf{\Sigma}}^{-1/2}, \tag{12.34a}$$

$$\textit{Adjoint modes:} \quad \mathbf{\Phi} = \mathcal{O}_d^*\tilde{\mathbf{U}}\tilde{\mathbf{\Sigma}}^{-1/2}. \tag{12.34b}$$

The direct modes $\mathbf{\Psi} \in \mathbb{R}^{n\times r}$ and adjoint modes $\mathbf{\Phi} \in \mathbb{R}^{n\times r}$ are bi-orthogonal, $\mathbf{\Phi}^*\mathbf{\Psi} = \mathbf{I}_{r\times r}$, and Rowley (2005) showed that they establish the change of coordinates that balance the truncated empirical Gramians. Thus, $\mathbf{\Psi}$ approximates the first r-columns of the full $n \times n$ balancing transformation, $\mathbf{T}$, and $\mathbf{\Phi}^*$ approximates the first r-rows of the $n \times n$ inverse balancing transformation, $\mathbf{S} = \mathbf{T}^{-1}$.

Now, it is possible to project the original system onto these modes, yielding a balanced ROM of order r:

$$\tilde{\mathbf{A}} = \mathbf{\Phi}^*\mathbf{A}_d\mathbf{\Psi}, \tag{12.35a}$$

$$\tilde{\mathbf{B}} = \mathbf{\Phi}^*\mathbf{B}_d, \tag{12.35b}$$

$$\tilde{\mathbf{C}} = \mathbf{C}_d\mathbf{\Psi}. \tag{12.35c}$$

It is possible to compute the reduced system dynamics in (12.35a) without having direct access to $\mathbf{A}_d$. In some cases, $\mathbf{A}_d$ may be exceedingly large and unwieldy, and instead it is only possible to evaluate the action of this matrix on an input vector. For example, in many modern fluid dynamics codes the matrix $\mathbf{A}_d$ is not actually represented, but because it is sparse, it is possible to implement efficient routines to multiply this matrix by a vector.

It is important to note that the ROM in (12.35) is formulated in discrete time, as it is based on discrete-time empirical snapshot matrices. However, it is simple to obtain the corresponding continuous-time system. In this example, $\mathbf{D}$ is the same in continuous time and discrete time, and in the full-order and ROMs.

Note that a BPOD model may not exactly satisfy the upper bound from balanced truncation (see (12.27)) due to errors in the empirical Gramians.

Output Projection

Often, in high-dimensional simulations, we assume full-state measurements, so that $p = n$ is exceedingly large. To avoid computing $p = n$ adjoint simulations, it is possible instead to solve an output-projected adjoint equation (Rowley 2005),

$$\boldsymbol{x}_{k+1} = \mathbf{A}_d^* \boldsymbol{x}_k + \mathbf{C}_d^* \tilde{\mathbf{U}} \boldsymbol{y}, \tag{12.36}$$

where $\tilde{\mathbf{U}}$ is a matrix containing the first r singular vectors of $\boldsymbol{C_d}$. Thus, we first identify a low-dimensional POD subspace $\tilde{\mathbf{U}}$ from a direct impulse response, and then only perform adjoint impulse response simulations by exciting these few *POD coefficient* measurements. More generally, if $\boldsymbol{y}$ is high-dimensional but does not measure the full-state, it is possible to use a POD subspace trained on the measurements, given by the first r singular vectors $\tilde{\mathbf{U}}$ of $\mathbf{C}_d \boldsymbol{C_d}$. Adjoint impulse responses may then be performed in these output POD directions.

Data Collection and Stacking

The powers m_c and m_o in (12.32) signify that data must be collected until the matrices $\boldsymbol{C}_d$ and $\boldsymbol{O}_d^*$ are full rank after which the controllable/observable subspaces have been sampled. Unless we collect data until transients decay, the true Gramians are only approximately balanced. Instead, it is possible to collect data until the Hankel matrix is full rank, balance the resulting model, and then truncate. This more efficient approach is developed in Tu and Rowley (2012) and Luchtenburg and Rowley (2011).

The snapshot matrices in (12.29) are generated from impulse response simulations of the direct (12.28a) and adjoint (12.36) systems. These time-series snapshots are then interleaved to form the snapshot matrices.

Historical Note

The balanced POD method described earlier originated with the seminal work of Moore (1981), which provided a data-driven generalization of the minimal realization theory of Ho and Kalman (1965). Until then, minimal realizations were defined in terms of idealized controllable and observable subspaces, which neglected the subtlety of degrees of controllability and observability.

Moore's paper introduced a number of critical concepts that bridged the gap from theory to reality. First, he established a connection between principal component analysis (PCA) and Gramians, showing that information about degrees of controllability and observability may be mined from data via the SVD. Next, Moore showed that a balancing transformation exists that makes the Gramians equal, diagonal, and hierarchically ordered by balanced controllability and observability; moreover, he provides an algorithm to compute this transformation. This sets the stage for principled model reduction, whereby states may be truncated based on their joint controllability and observability. Moore further introduced the notion of an empirical Gramian, although he didn't use this terminology. He also realized that computing $\mathbf{W}_c$ and $\mathbf{W}_o$ directly is less accurate than computing the SVD of the empirical snapshot matrices from the direct and adjoint systems, and he avoided directly computing the eigendecomposition of $\mathbf{W}_c\mathbf{W}_o$ by using these SVD transformations. Lall et al. (2002) generalized this theory to nonlinear systems.

One drawback of Moore's approach is that he computed the entire $n \times n$ balancing transformation, which is not suitable for exceedingly high-dimensional systems. Willcox and Peraire (2002) generalized the method to high-dimensional systems, introducing a variant based on the rank-r decompositions of $\mathbf{W}_c$ and $\mathbf{W}_o$ obtained from the direct and adjoint snapshot matrices. It is then possible to compute the eigendecomposition of $\mathbf{W}_c\mathbf{W}_o$ using efficient eigenvalue solvers without ever actually writing down the full $n \times n$ matrices. However, this approach has the drawback of requiring as many adjoint impulse response simulations as the number of output equations, which may be exceedingly large for full-state measurements. Rowley (2005) addressed this issue by introducing the output projection, discussed earlier, which limits the number of adjoint simulations to the number of relevant POD modes in the data. He also showed that it is possible to use the eigendecomposition of the product $\boldsymbol{O}_d\boldsymbol{C}_d$ The product $\boldsymbol{O}_d\boldsymbol{C}_d$ is often smaller, and these computations may be more accurate.

It is interesting to note that a nearly equivalent formulation was developed 20 years earlier in the field of system identification. The so-called ERA, introduced by Juang and Pappa (1985), obtains equivalent balanced models without the need for adjoint data, making it useful for system identification in experiments. This connection between ERA and BPOD was established by Ma et al. (2011).

12.2 System Identification

In contrast to model reduction, where the system model $(\mathbf{A}, \mathbf{B}, \mathbf{C}, \mathbf{D})$ was known, system identification is purely data-driven. System identification may be thought of as a form of machine learning, where an input–output map of a system is learned from training data in a representation that generalizes to data that was not in the training set. There is a vast literature on methods for system identification (Juang 1994, Ljung 1999), and many of the leading methods are based on a form of dynamic regression that fits models based on data, such as the DMD. For this section, we consider the ERA and observer-Kalman filter identification (OKID) methods because of their connection to balanced model reduction (Moore 1981, Rowley 2005, Ma et al. 2011, Tu et al. 2014) and their successful application in complex systems such as vibration control of aerospace structures and closed-loop flow control (Bagheri, Hoepffner, Schmid & Henningson 2009, Bagheri, Brandt & Henningson 2009, Illingworth et al. 2010). The ERA/OKID procedure is also applicable to multiple-input, multiple-output (MIMO) systems. Other methods include the autoregressive-moving average (ARMA) and autoregressive moving average with exogenous inputs (ARMAX) models (Whittle 1951, Box et al. 2015), the nonlinear autoregressive moving average with exogenous inputs (NARMAX) (Billings 2013) model, and the SINDy method.

12.2.1 Eigensystem Realization Algorithm

The ERA produces low-dimensional linear input–output models from sensor measurements of an impulse response experiment, based on the "minimal realization" theory of Ho and Kalman (1965). The modern theory was developed to identify structural models for various spacecraft (Juang & Pappa 1985), and it has been shown by Ma et al. (2011) that ERA models are equivalent to BPOD models.[2] However, ERA is based entirely on impulse response measurements and does not require prior knowledge of a model. ERA may be equally well applied to model high-dimensional data, such as a fluid vector field, or to model low-dimensional data, such as a scalar time-series of the lift coefficient. In the latter case, where the model is based on partial or incomplete measurements of a high-dimensional state, the effect of latent or hidden variables is included by considering time delays of the measurement. The following mathematical treatment encompasses both cases.

We consider a discrete-time system:

$$\boldsymbol{x}_{k+1} = \mathbf{A}_d \boldsymbol{x}_k + \mathbf{B}_d \boldsymbol{u}_k, \tag{12.37a}$$

$$\boldsymbol{y}_k = \mathbf{C}_d \boldsymbol{x}_k + \mathbf{D}_d \boldsymbol{u}_k. \tag{12.37b}$$

A discrete-time delta function input in the actuation $\boldsymbol{u}$,

$$\boldsymbol{u}_k^\delta \triangleq \boldsymbol{u}^\delta(k\Delta t) = \begin{cases} \mathbf{I}, & k = 0, \\ \mathbf{0}, & k = 1,2,3,\cdots, \end{cases} \tag{12.38}$$

gives rise to a discrete-time impulse response in the sensors $\boldsymbol{y}$:

$$\boldsymbol{y}_k^\delta \triangleq \boldsymbol{y}^\delta(k\Delta t) = \begin{cases} \mathbf{D}_d, & k = 0, \\ \mathbf{C}_d \mathbf{A}_d^{k-1} \mathbf{B}_d, & k = 1,2,3,\cdots. \end{cases} \tag{12.39}$$

In an experiment or simulation, typically q impulse responses are performed, one for each of the q separate input channels. The output responses are collected for each impulsive input, and at a given time step k, the output vector in response to the jth

[2] BPOD and ERA models both balance the empirical Gramians and approximate balanced truncation (Moore 1981) for high-dimensional systems, given a sufficient volume of data.

impulsive input will form the jth column of $\boldsymbol{y}_k^\delta$. Thus, each of the $\boldsymbol{y}_k^\delta$ is a $p \times q$ matrix $\mathbf{C}\mathbf{A}^{k-1}\mathbf{B}$. Note that the system matrices $(\mathbf{A}, \mathbf{B}, \mathbf{C}, \mathbf{D})$ don't actually need to exist, as the following method is purely data-driven.

The Hankel matrix $\mathbf{H}$ from (12.32) is formed by stacking shifted time-series of impulse response measurements into a matrix, as in the Hankel alternative view of the Koopman (HAVOK) method (Brunton et al. 2017):

$$\mathbf{H} = \begin{bmatrix} \boldsymbol{y}_1^\delta & \boldsymbol{y}_2^\delta & \cdots & \boldsymbol{y}_{m_c}^\delta \\ \boldsymbol{y}_2^\delta & \boldsymbol{y}_3^\delta & \cdots & \boldsymbol{y}_{m_c+1}^\delta \\ \vdots & \vdots & \ddots & \vdots \\ \boldsymbol{y}_{m_o}^\delta & \boldsymbol{y}_{m_o+1}^\delta & \cdots & \boldsymbol{y}_{m_c+m_o-1}^\delta \end{bmatrix} \tag{12.40a}$$

$$= \begin{bmatrix} \mathbf{C}_d\mathbf{B}_d & \mathbf{C}_d\mathbf{A}_d\mathbf{B}_d & \cdots & \mathbf{C}_d\mathbf{A}_d^{m_c-1}\mathbf{B}_d \\ \mathbf{C}_d\mathbf{A}_d\mathbf{B}_d & \mathbf{C}_d\mathbf{A}_d^2\mathbf{B}_d & \cdots & \mathbf{C}_d\mathbf{A}_d^{m_c}\mathbf{B}_d \\ \vdots & \vdots & \ddots & \vdots \\ \mathbf{C}_d\mathbf{A}_d^{m_o-1}\mathbf{B}_d & \mathbf{C}_d\mathbf{A}_d^{m_o}\mathbf{B}_d & \cdots & \mathbf{C}_d\mathbf{A}_d^{m_c+m_o-2}\mathbf{B}_d \end{bmatrix}. \tag{12.40b}$$

The matrix $\mathbf{H}$ may be constructed purely from measurements $\boldsymbol{y}^\delta$, without separately constructing $\boldsymbol{O}_d$ and $\boldsymbol{C}_d$. Thus, we do not need access to adjoint equations.

Taking the SVD of the Hankel matrix yields the dominant temporal patterns in the time-series data:

$$\mathbf{H} = \mathbf{U}\boldsymbol{\Sigma}\mathbf{V}^* = \begin{bmatrix} \tilde{\mathbf{U}} & \mathbf{U}_t \end{bmatrix} \begin{bmatrix} \tilde{\boldsymbol{\Sigma}} & \mathbf{0} \\ \mathbf{0} & \boldsymbol{\Sigma}_t \end{bmatrix} \begin{bmatrix} \tilde{\mathbf{V}}^* \\ \mathbf{V}_t^* \end{bmatrix} \approx \tilde{\mathbf{U}}\tilde{\boldsymbol{\Sigma}}\tilde{\mathbf{V}}^*. \tag{12.41}$$

The small singular values in $\boldsymbol{\Sigma}_t$ are truncated, and only the first r singular values in $\tilde{\boldsymbol{\Sigma}}$ are retained. The columns of $\tilde{\mathbf{U}}$ and $\tilde{\mathbf{V}}$ are *eigen*-time-delay coordinates.

Until this point, the ERA algorithm closely resembles the BPOD procedure from Section 12.1. However, we do not require direct access to $\boldsymbol{O}_d$ and $\boldsymbol{C}_d$ or the system $(\mathbf{A}, \mathbf{B}, \mathbf{C}, \mathbf{D})$ to construct the direct and adjoint balancing transformations. Instead, with sensor measurements from an impulse response experiment, it is also possible to create a second, shifted Hankel matrix $\mathbf{H}'$:

$$\mathbf{H}' = \begin{bmatrix} \boldsymbol{y}_2 & \boldsymbol{y}_3^\delta & \cdots & \boldsymbol{y}_{m_c+1}^\delta \\ \boldsymbol{y}_3^\delta & \boldsymbol{y}_4^\delta & \cdots & \boldsymbol{y}_{m_c+2}^\delta \\ \vdots & \vdots & \ddots & \vdots \\ \boldsymbol{y}_{m_o+1}^\delta & \boldsymbol{y}_{m_o+2}^\delta & \cdots & \boldsymbol{y}_{m_c+m_o}^\delta \end{bmatrix} \tag{12.42a}$$

$$= \begin{bmatrix} \mathbf{C}_d\mathbf{A}_d\mathbf{B}_d & \mathbf{C}_d\mathbf{A}_d^2\mathbf{B}_d & \cdots & \mathbf{C}_d\mathbf{A}_d^{m_c}\mathbf{B}_d \\ \mathbf{C}_d\mathbf{A}_d^2\mathbf{B}_d & \mathbf{C}_d\mathbf{A}_d^3\mathbf{B}_d & \cdots & \mathbf{C}_d\mathbf{A}_d^{m_c+1}\mathbf{B}_d \\ \vdots & \vdots & \ddots & \vdots \\ \mathbf{C}_d\mathbf{A}_d^{m_o}\mathbf{B}_d & \mathbf{C}_d\mathbf{A}_d^{m_o+1}\mathbf{B}_d & \cdots & \mathbf{C}_d\mathbf{A}_d^{m_c+m_o-1}\mathbf{B}_d \end{bmatrix} = \boldsymbol{O}_d\mathbf{A}\boldsymbol{C}_d. \tag{12.42b}$$

Based on the matrices $\mathbf{H}$ and $\mathbf{H}'$, we are able to construct a ROM as follows:

$$\tilde{\mathbf{A}} = \tilde{\boldsymbol{\Sigma}}^{-1/2}\tilde{\mathbf{U}}^*\mathbf{H}'\tilde{\mathbf{V}}\tilde{\boldsymbol{\Sigma}}^{-1/2}, \tag{12.43a}$$

$$\tilde{\mathbf{B}} = \tilde{\boldsymbol{\Sigma}}^{1/2}\tilde{\mathbf{V}}^* \begin{bmatrix} \mathbf{I}_p & \mathbf{0} \\ \mathbf{0} & \mathbf{0} \end{bmatrix}, \tag{12.43b}$$

$$\tilde{\mathbf{C}} = \begin{bmatrix} \mathbf{I}_q & \mathbf{0} \\ \mathbf{0} & \mathbf{0} \end{bmatrix} \tilde{\mathbf{U}}\tilde{\boldsymbol{\Sigma}}^{1/2}. \tag{12.43c}$$

Here $\mathbf{I}_p$ is the $p \times p$ identity matrix, which extracts the first p columns, and $\mathbf{I}_q$ is the $q \times q$ identity matrix, which extracts the first q rows. Thus, we express the input–output dynamics in terms of a reduced system with a low-dimensional state $\tilde{\mathbf{x}} \in \mathbb{R}^r$:

$$\tilde{\mathbf{x}}_{k+1} = \tilde{\mathbf{A}}\tilde{\mathbf{x}}_k + \tilde{\mathbf{B}}u, \tag{12.44a}$$

$$y = \tilde{\mathbf{C}}\tilde{\mathbf{x}}_k. \tag{12.44b}$$

$\mathbf{H}$ and $\mathbf{H}'$ are constructed from impulse response simulations/experiments, without the need for storing direct or adjoint snapshots, as in other balanced model reduction techniques. However, if full-state snapshots are available, for example, by collecting velocity fields in simulations or PIV experiments, it is then possible to construct direct modes. These full-state snapshots form $\boldsymbol{C}_d$, and modes can be constructed by

$$\boldsymbol{\Psi} = \boldsymbol{C}_d\tilde{\mathbf{V}}\tilde{\boldsymbol{\Sigma}}^{-1/2}. \tag{12.45}$$

These modes may then be used to approximate the full-state of the high-dimensional system from the low-dimensional model in (12.44) by

$$\boldsymbol{x} \approx \boldsymbol{\Psi}\tilde{\mathbf{x}}. \tag{12.46}$$

If enough data is collected when constructing the Hankel matrix $\mathbf{H}$, then ERA balances the empirical controllability and observability Gramians, $\boldsymbol{O}_d\boldsymbol{O}_d^*$ and $\boldsymbol{C}_d^*\boldsymbol{C}_d$. However, if less data is collected, so that lightly damped transients do not have time to decay, then ERA will only approximately balance the system. It is instead possible to collect just enough data so that the Hankel matrix $\mathbf{H}$ reaches numerical full rank (i.e., so that remaining singular values are below a threshold tolerance), and compute an ERA model. The resulting ERA model will typically have a relatively low order, given by the numerical rank of the controllability and observability subspaces. It may then be possible to apply exact balanced truncation to this smaller model, as is advocated in Tu and Rowley (2012) and Luchtenburg and Rowley (2011).

12.2.2 Observer Kalman Filter Identification

OKID was developed to complement the ERA for lightly damped experimental systems with noise (Juang et al. 1991). In practice, performing isolated impulse response experiments is challenging, and the effect of measurement noise can contaminate results. Moreover, if there is a large separation of timescales, then a tremendous

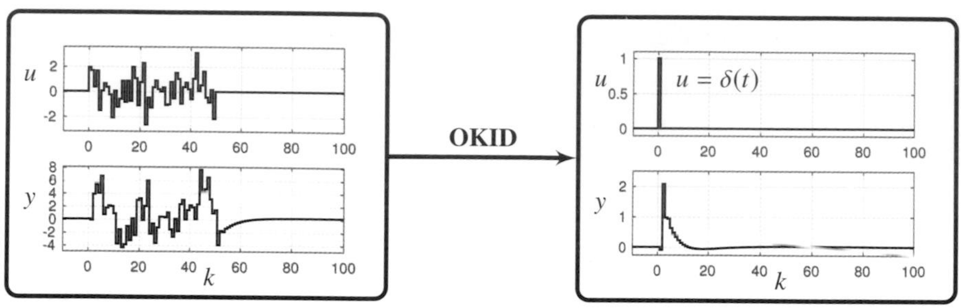

Figure 12.3 Schematic overview of OKID procedure. The output of OKID is an impulse response that can be used for system identification via ERA. From Brunton and Kutz (2019), reproduced with permission of the Licensor through PLSclear.

amount of data must be collected to use ERA. This section poses the general problem of approximating the impulse response from arbitrary input–output data (Figure 12.3). Typically, one would identify ROMs according to the following general procedure:

1. Collect the output in response to a pseudorandom input.
2. This information is passed through the OKID algorithm to obtain the de-noised linear impulse response.
3. The impulse response is passed through the ERA to obtain a reduced-order state-space system.

The output $\boldsymbol{y}_k$ in response to a general input signal $\boldsymbol{u}_k$, for zero initial condition $\boldsymbol{x}_0 = \mathbf{0}$, is given by

$$\boldsymbol{y}_0 = \mathbf{D}_d \boldsymbol{u}_0, \tag{12.47a}$$

$$\boldsymbol{y}_1 = \mathbf{C}_d \mathbf{B}_d \boldsymbol{u}_0 + \mathbf{D}_d \boldsymbol{u}_1, \tag{12.47b}$$

$$\boldsymbol{y}_2 = \mathbf{C}_d \mathbf{A}_d \mathbf{B}_d \boldsymbol{u}_0 + \mathbf{C}_d \mathbf{B}_d \boldsymbol{u}_1 + \mathbf{D}_d \boldsymbol{u}_2, \tag{12.47c}$$

$$\cdots$$

$$\boldsymbol{y}_k = \mathbf{C}_d \mathbf{A}_d^{k-1} \mathbf{B}_d \boldsymbol{u}_0 + \mathbf{C}_d \mathbf{A}_d^{k-2} \mathbf{B}_d \boldsymbol{u}_1 + \cdots + \mathbf{C}_d \mathbf{B}_d \boldsymbol{u}_{k-1} + \mathbf{D}_d \boldsymbol{u}_k. \tag{12.47d}$$

Note that there is no $\mathbf{C}$ term in the expression for $\boldsymbol{y}_0$ since there is zero initial condition $\boldsymbol{x}_0 = \mathbf{0}$. This progression of measurements $\boldsymbol{y}_k$ may be further simplified and expressed in terms of impulse response measurements $\boldsymbol{y}_k^\delta$:

$$\underbrace{\begin{bmatrix} \boldsymbol{y}_0 & \boldsymbol{y}_1 & \cdots & \boldsymbol{y}_m \end{bmatrix}}_{\mathcal{S}} = \underbrace{\begin{bmatrix} \boldsymbol{y}_0^\delta & \boldsymbol{y}_1^\delta & \cdots & \boldsymbol{y}_m^\delta \end{bmatrix}}_{\mathcal{S}^\delta} \underbrace{\begin{bmatrix} \boldsymbol{u}_0 & \boldsymbol{u}_1 & \cdots & \boldsymbol{u}_m \\ \mathbf{0} & \boldsymbol{u}_0 & \cdots & \boldsymbol{u}_{m-1} \\ \vdots & \vdots & \ddots & \vdots \\ \mathbf{0} & \mathbf{0} & \cdots & \boldsymbol{u}_0 \end{bmatrix}}_{\mathcal{B}}. \tag{12.48}$$

It is often possible to invert the matrix of control inputs, $\mathcal{B}$, to solve for the Markov parameters $\mathcal{S}^\delta$. However, $\mathcal{B}$ may either be un-invertible, or inversion may be ill-conditioned. In addition, $\mathcal{B}$ is large for lightly damped systems, making inversion computationally expensive. Finally, noise is not optimally filtered by simply inverting $\mathcal{B}$ to solve for the Markov parameters.

The OKID method addresses each of these issues. Instead of the original discrete-time system, we now introduce an optimal observer system:

$$\hat{\boldsymbol{x}}_{k+1} = \mathbf{A}_d\hat{\boldsymbol{x}}_k + \mathbf{K}_f\left(\boldsymbol{y}_k - \hat{\boldsymbol{y}}_k\right) + \mathbf{B}_d\boldsymbol{u}_k, \tag{12.49a}$$

$$\hat{\boldsymbol{y}}_k = \mathbf{C}_d\hat{\boldsymbol{x}}_k + \mathbf{D}_d\boldsymbol{u}_k, \tag{12.49b}$$

which may be rewritten as:

$$\hat{\boldsymbol{x}}_{k+1} = \underbrace{(\mathbf{A}_d - \mathbf{K}_f\mathbf{C}_d)}_{\bar{\mathbf{A}}_d}\hat{\boldsymbol{x}}_k + \underbrace{\begin{bmatrix}\mathbf{B}_d - \mathbf{K}_f\mathbf{D}_d, & \mathbf{K}_f\end{bmatrix}}_{\bar{\mathbf{B}}_d}\begin{bmatrix}\boldsymbol{u}_k \\ \boldsymbol{y}_k\end{bmatrix}. \tag{12.50}$$

Recall from above that if the system is observable, it is possible to place the poles of $\mathbf{A}_d - \mathbf{K}_f\mathbf{C}_d$ anywhere we like. However, depending on the amount of noise in the measurements, the magnitude of process noise, and uncertainty in our model, there are *optimal* pole locations that are given by the *Kalman filter*. We may now solve for the *observer Markov parameters* $\bar{\mathcal{S}}^\delta$ of the system in (12.50) in terms of measured inputs and outputs according to the following algorithm from Juang et al. (1991):

1. Choose the number of observer Markov parameters to identify, l.

2. Construct the data matrices:

$$\mathcal{S} = \begin{bmatrix}\boldsymbol{y}_0 & \boldsymbol{y}_1 & \cdots & \boldsymbol{y}_l & \cdots & \boldsymbol{y}_m\end{bmatrix}, \tag{12.51}$$

$$\mathcal{V} = \begin{bmatrix}\boldsymbol{u}_0 & \boldsymbol{u}_1 & \cdots & \boldsymbol{u}_l & \cdots & \boldsymbol{u}_m \\ \mathbf{0} & \boldsymbol{v}_0 & \cdots & \boldsymbol{v}_{l-1} & \cdots & \boldsymbol{v}_{m-1} \\ \vdots & \vdots & \ddots & \vdots & \ddots & \vdots \\ \mathbf{0} & \mathbf{0} & \cdots & \boldsymbol{v}_0 & \cdots & \boldsymbol{v}_{m-l}\end{bmatrix}, \tag{12.52}$$

where $\boldsymbol{v}_i = \begin{bmatrix}\boldsymbol{u}_i^T & \boldsymbol{y}_i^T\end{bmatrix}^T$.

The matrix $\mathcal{V}$ resembles $\mathcal{B}$, except that has been augmented with the outputs $\boldsymbol{y}_i$. In this way, we are working with a system that is augmented to include a Kalman filter. We are now identifying the observer Markov parameters of the *augmented* system, $\bar{\mathcal{S}}^\delta$, using the equation $\mathcal{S} = \bar{\mathcal{S}}^\delta\mathcal{V}$. It will be possible to identify these observer Markov parameters from the data and then extract the impulse response (Markov parameters) of the original system.

3. Identify the matrix $\bar{\mathcal{S}}^\delta$ of observer Markov parameters by solving $\mathcal{S} = \bar{\mathcal{S}}^\delta\mathcal{V}$ for $\bar{\mathcal{S}}^\delta$ using the right pseudo-inverse of $\mathcal{V}$ (i.e., SVD).

4. Recover system Markov parameters, $\mathcal{S}^\delta$, from the observer Markov parameters, $\bar{\mathcal{S}}^\delta$:

(a) Order the observer Markov parameters $\bar{\mathcal{S}}^\delta$ as follows:

$$\bar{\mathcal{S}}_0^\delta = \mathbf{D}, \tag{12.53}$$

$$\bar{\mathcal{S}}_k^\delta = \begin{bmatrix}(\bar{\mathcal{S}}^\delta)_k^{(1)} & (\bar{\mathcal{S}}^\delta)_k^{(2)}\end{bmatrix} \quad \text{for } k \geq 1, \tag{12.54}$$

where $(\bar{\mathcal{S}}^\delta)_k^{(1)} \in \mathbb{R}^{q\times p}$, $(\bar{\mathcal{S}}^\delta)_k^{(2)} \in \mathbb{R}^{q\times q}$, and $\boldsymbol{y}_0^\delta = \bar{\mathcal{S}}_0^\delta = \mathbf{D}$.

(b) Reconstruct system Markov parameters:

$$y_k^\delta = (\bar{\mathcal{S}}^\delta)_k^{(1)} + \sum_{i=1}^{k} (\bar{\mathcal{S}}^\delta)_i^{(2)} y_{k-i}^\delta \text{ for } k \geq 1. \tag{12.55}$$

Thus, the OKID method identifies the Markov parameters of a system augmented with an asymptotically stable Kalman filter. The system Markov parameters are extracted from the observer Markov parameters by (12.55). These system Markov parameters approximate the impulse response of the system, and may be used directly as inputs to the ERA algorithm.

ERA/OKID has been widely applied across a range of system identification tasks, including to identify models of aeroelastic structures and fluid dynamic systems. There are numerous extensions of the ERA/OKID methods. For example, there are generalizations for linear parameter varying (LPV) systems and systems linearized about a limit cycle.

12.3 Sparse Identification of Nonlinear Dynamics

Discovering dynamical systems models from data is a central challenge in mathematical physics, with a rich history going back at least as far as the time of Kepler and Newton and the discovery of the laws of planetary motion. Historically, this process relied on a combination of high-quality measurements and expert intuition. With vast quantities of data and increasing computational power, the *automated* discovery of governing equations and dynamical systems is a new and exciting scientific paradigm.

Typically, the form of a candidate model is either constrained via prior knowledge of the governing equations, as in Galerkin projection (Noack et al. 2003, Rowley et al. 2004, Schlegel et al. 2004, Carlberg et al. 2011, Noack et al. 2011, Wang et al. 2012, Balajewicz et al. 2013, Carlberg et al. 2017), or a handful of heuristic models are tested and parameters are optimized to fit data. Alternatively, best-fit linear models may be obtained using DMD or ERA. Simultaneously identifying the nonlinear structure and parameters of a model from data is considerably more challenging, as there are combinatorially many possible model structures.

The SINDy algorithm (Brunton et al. 2016a) bypasses the intractable combinatorial search through all possible model structures, leveraging the fact that many dynamical systems

$$\frac{d}{dt}\mathbf{x} = \mathbf{f}(\mathbf{x}) \tag{12.56}$$

have dynamics $\mathbf{f}$ with only a few active terms in the space of possible right-hand side functions; for example, the Lorenz equations only have a few linear and quadratic interaction terms per equation.

We then seek to approximate $\mathbf{f}$ by a generalized linear model,

$$\mathbf{f}(\boldsymbol{x}) \approx \sum_{k=1}^{p} \theta_k(\boldsymbol{x})\xi_k = \boldsymbol{\Theta}(\boldsymbol{x})\boldsymbol{\xi}, \tag{12.57}$$

with the fewest possible nonzero terms in $\boldsymbol{\xi}$. It is then possible to solve for the relevant terms that are active in the dynamics using sparse regression (Tibshirani 1996, Zou & Hastie 2005, Hastie et al. 2009, James et al. 2013), which penalizes the number of terms in the dynamics and scales well to large problems.

First, time-series data are collected from (12.56) and formed into a data matrix:

$$\mathbf{X} = \begin{bmatrix}\mathbf{x}(t_1) & \mathbf{x}(t_2) & \cdots \mathbf{x}(t_m)\end{bmatrix}^T. \tag{12.58}$$

A similar matrix of derivatives is formed:

$$\dot{\mathbf{X}} = \begin{bmatrix}\dot{\mathbf{x}}(t_1) & \dot{\mathbf{x}}(t_2) & \cdots \dot{\mathbf{x}}(t_m)\end{bmatrix}^T. \tag{12.59}$$

In practice, this may be computed directly from the data in $\mathbf{X}$; for noisy data, the total-variation regularized derivative tends to provide numerically robust derivatives (Chartrand 2011). Alternatively, it is possible to formulate the SINDy algorithm for discrete-time systems $\mathbf{x}_{k+1} = \mathbf{F}(\mathbf{x}_k)$, as in the DMD algorithm, and avoid derivatives entirely.

A library of candidate nonlinear functions $\boldsymbol{\Theta}(\mathbf{X})$ may be constructed from the data in $\mathbf{X}$:

$$\boldsymbol{\Theta}(\mathbf{X}) = \begin{bmatrix}\mathbf{1} & \mathbf{X} & \mathbf{X}^2 & \cdots & \mathbf{X}^d & \cdots & \sin(\mathbf{X}) & \cdots\end{bmatrix}. \tag{12.60}$$

Here, the matrix $\mathbf{X}^d$ denotes a matrix with column vectors given by all possible time-series of dth degree polynomials in the state $\mathbf{x}$. In general, this library of candidate functions is only limited by one's imagination.

The dynamical system in (12.56) may now be represented in terms of the data matrices in (12.59) and (12.60) as

$$\dot{\mathbf{X}} = \boldsymbol{\Theta}(\mathbf{X})\boldsymbol{\Xi}. \tag{12.61}$$

Each column $\boldsymbol{\xi}_k$ in $\boldsymbol{\Xi}$ is a vector of coefficients determining the active terms in the kth row in (12.56). A parsimonious model will provide an accurate model fit in (12.61) with as few terms as possible in $\boldsymbol{\Xi}$. Such a model may be identified using a convex ℓ_1-regularized sparse regression:

$$\boldsymbol{\xi}_k = \operatorname{argmin}_{\boldsymbol{\xi}'_k} \|\dot{\mathbf{X}}_k - \boldsymbol{\Theta}(\mathbf{X})\boldsymbol{\xi}'_k\|_2 + \lambda\|\boldsymbol{\xi}'_k\|_1. \tag{12.62}$$

Here, $\dot{\mathbf{X}}_k$ is the kth column of $\dot{\mathbf{X}}$, and λ is a sparsity-promoting knob. Sparse regression, such as the LASSO (Tibshirani 1996) or the sequential thresholded least-squares (STLS) algorithm used in SINDy (Brunton, Proctor & Kutz 2016a), improves the numerical robustness of this identification for noisy overdetermined problems, in contrast to earlier methods (Wang et al. 2011) that used compressed sensing (Donoho 2006, Candès, 2006, Candès et al. 2006a, 2006b, Candès & Tao 2006, Baraniuk 2007, Tropp & Gilbert 2007). We advocate the STLS to select active terms.

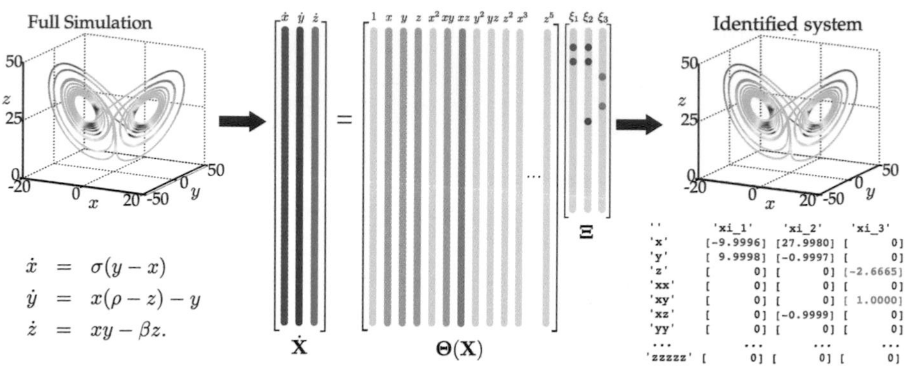

Figure 12.4 Schematic of the sparse identification of nonlinear dynamics (SINDy) algorithm (Brunton et al. 2016a). Parsimonious models are selected from a library of candidate nonlinear terms using sparse regression. This library $\mathbf{\Theta}(\mathbf{X})$ may be constructed purely from measurement data. Reproduced from Brunton et al. (2016a).

The sparse vectors $\boldsymbol{\xi}_k$ may be synthesized into a dynamical system:

$$\dot{x}_k = \mathbf{\Theta}(\mathbf{x})\boldsymbol{\xi}_k. \tag{12.63}$$

Note that x_k is the kth element of $\mathbf{x}$, and $\mathbf{\Theta}(\mathbf{x})$ is a row vector of symbolic functions of $\mathbf{x}$, as opposed to the data matrix $\mathbf{\Theta}(\mathbf{X})$. Figure 12.4 shows how SINDy may be used to discover the Lorenz equations from data.

The result of the SINDy regression is a parsimonious model that includes only the most important terms required to explain the observed behavior. The sparse regression procedure used to identify the most parsimonious nonlinear model is a convex procedure. The alternative approach, which involves regression onto every possible sparse nonlinear structure, constitutes an intractable brute-force search through the combinatorially many candidate model forms. SINDy bypasses this combinatorial search with modern convex optimization and machine learning. It is interesting to note that for discrete-time dynamics, if $\mathbf{\Theta}(\mathbf{X})$ consists only of linear terms, and if we remove the sparsity promoting term by setting $\lambda = 0$, then this algorithm reduces to the DMD (Rowley et al. 2009, Schmid 2010, Tu et al. 2014, Kutz et al. 2016). If a least-squares regression is used, as in DMD, then even a small amount of measurement error or numerical roundoff will lead to every term in the library being active in the dynamics, which is nonphysical. A major benefit of the SINDy architecture is the ability to identify parsimonious models that contain only the required nonlinear terms, resulting in interpretable models that avoid overfitting.

Applications, Extensions, and Connections to Fluid Dynamics

The SINDy algorithm has recently been applied to discover models in fluid dynamics (El Sayed M et al. 2018, Loiseau & Brunton 2018, Loiseau et al. 2018, Loiseau 2019, Guan et al. 2020, Deng et al. 2020). Because of the high-state dimension, SINDy models of fluids are typically based on POD coefficients or low-dimensional measurements such as the lift and drag coefficients (Brunton et al. 2016a, Loiseau

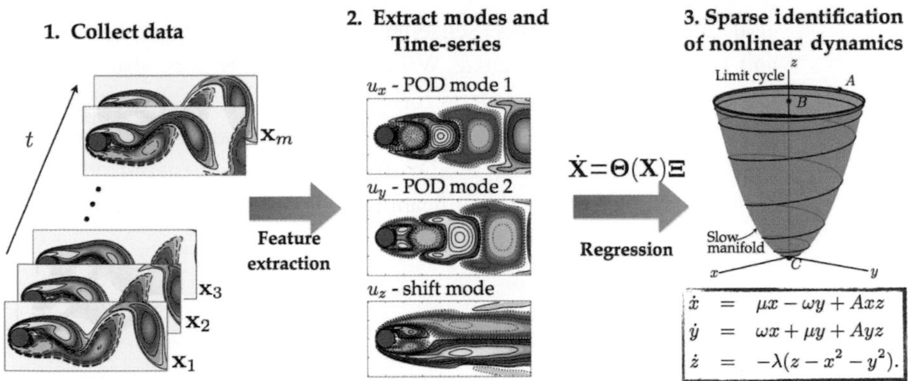

Figure 12.5 Schematic overview of nonlinear model identification from high-dimensional data using the sparse identification of nonlinear dynamics (SINDy) (Brunton et al. 2016a). This procedure is modular, so that different techniques can be used for the feature extraction and regression steps. In this example of flow past a cylinder, SINDy discovers the model of Noack et al. (2003). Reproduced from Brunton et al. (2016a).

& Brunton 2018, Loiseau et al. 2018). Recent studies have also leveraged SINDy for turbulence modeling (Beetham & Capecelatro 2020, Schmelzer et al. 2020). Figure 12.5 illustrates the application of SINDy to the flow past a cylinder, where the generalized mean-field model of Noack et al. (2003) was discovered from data. SINDy has also been applied to identify models in nonlinear optics (Sorokina et al. 2016), plasma physics (Dam et al. 2017), chemical reaction dynamics (Hoffmann et al. 2019), numerical algorithms (Thaler et al. 2019), and structural modeling (Lai & Nagarajaiah 2019), among others (Narasingam & Kwon 2018, de Silva et al. 2019, Pan et al. 2020).

Because SINDy is formulated in terms of linear regression in a nonlinear library, it is highly extensible. The SINDy framework has been recently generalized by Loiseau and Brunton (Loiseau & Brunton 2018) to incorporate known physical constraints and symmetries in the equations by implementing a constrained sequentially thresholded least-squares optimization. In particular, energy-preserving constraints on the quadratic nonlinearities in the Navier–Stokes equations were imposed to identify fluid systems (Loiseau & Brunton 2018), where it is known that these constraints promote stability (Majda & Harlim 2012, Balajewicz et al. 2013, Carlberg et al. 2017). This work also showed that polynomial libraries are particularly useful for building models of fluid flows in terms of POD coefficients, yielding interpretable models that are related to classical Galerkin projection (Brunton et al. 2016a, Loiseau & Brunton 2018). Loiseau et al. (2018) also demonstrated the ability of SINDy to identify dynamical systems models of high-dimensional systems, such as fluid flows, from a few physical sensor measurements, such as lift and drag measurements on the cylinder in Figure 12.5. For actuated systems, SINDy has been generalized to include inputs and control (Brunton et al. 2016b), and these models are highly effective for model predictive control (Kaiser et al. 2018). It is also possible to extend the SINDy algorithm to identify dynamics with rational function nonlinearities (Mangan et al. 2016), integral terms (Schaeffer & McCalla 2017), and based on highly corrupt and incomplete data (Tran & Ward 2016). SINDy was also recently extended to incorporate information criteria for objective model selection (Mangan et al. 2017), and to identify models with hidden variables using delay coordinates (Brunton et al. 2017). Finally, the SINDy framework was generalized to include partial derivatives, enabling the identification of partial differential equation models (Rudy et al. 2017, Schaeffer 2017).

13 Modern Tools for the Stability Analysis of Fluid Flows

P. J. Schmid

The response behavior of a fluid flow to small perturbations is a key measure of its dynamic properties and provides insight into the prevalence and dominance of coherent structures and their evolution in time or space. This chapter gives a brief introduction to the principal concepts of stability analysis of fluid flows, advocating a general mathematical framework based on finite-time or frequency gains, coupled to an optimization environment. This framework is particularly flexible in describing complex fluid behavior characterized by multi-physics and multiscale mechanisms interacting in a feedforward or feedback manner. Special cases of the same framework yield more familiar stability and receptivity concepts based on linear-algebra tools. Various implementation details will be discussed, and an example from computational aeroacoustics will be presented to showcase the mathematical techniques and illustrate the merits of the proposed approach.

13.1 Introduction

Stability analysis is a key discipline in fluid dynamics and is ubiquitous in the fluid dynamics literature, either as a way of analyzing complex fluid behavior or as a means to utilize it to manipulate intrinsic fluid motion. Instabilities are often postulated as the driving force for pattern formation, for the rise of specific scales, or for the bifurcation into a different flow regime.

Instabilities can be observed all around us. The flow patterns forming behind bluff objects, the breakup of water jets and drops, the clustering of stars in rotating galaxies, the formation of thermal plumes, the shaping of glaciers and stalactites/stalagmites in caves, the sand ripples in river beds, and desert dunes fall among applications where stability analysis of the governing equations has contributed to our understanding of the dominant physical processes at play (Figure 13.1).

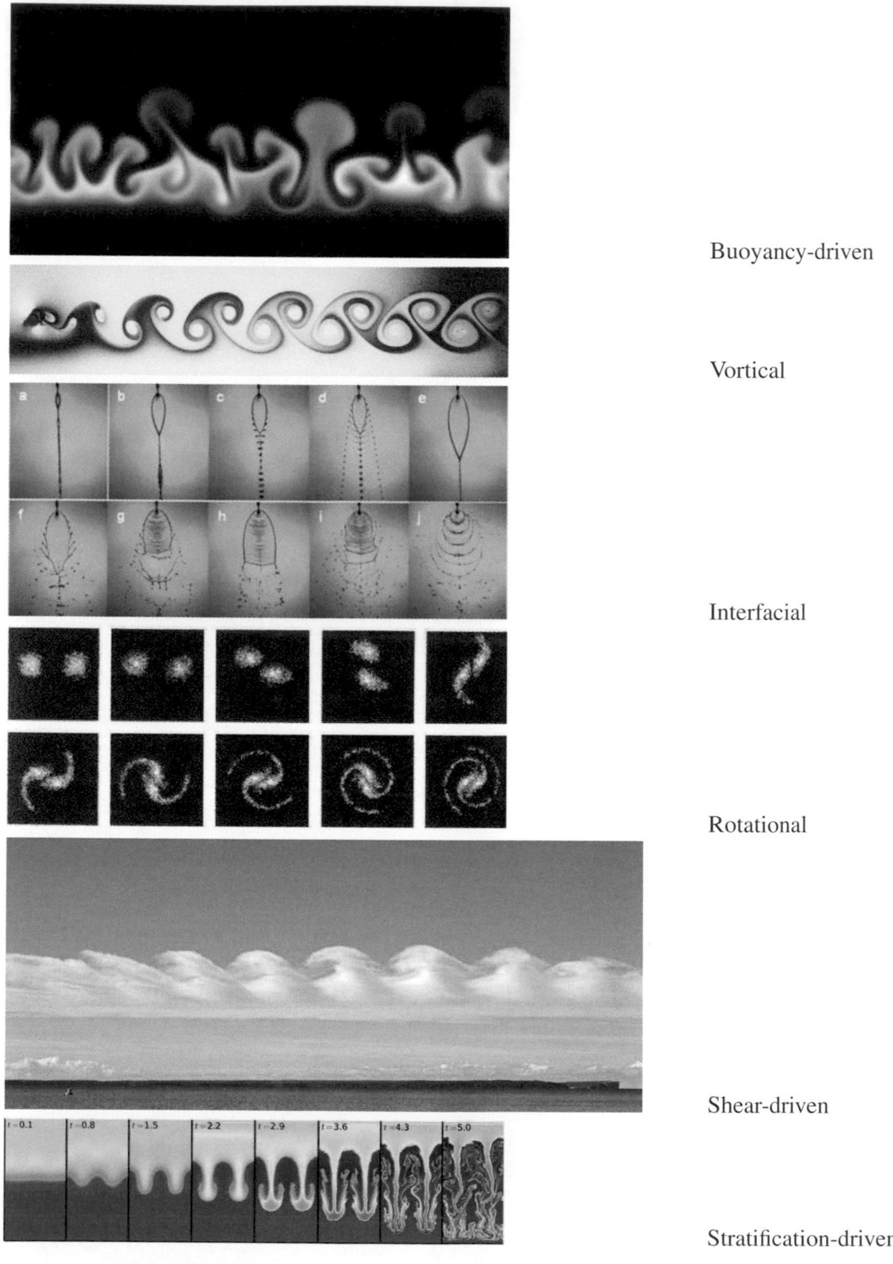

Figure 13.1 Examples of instabilities.

As much as instabilities are responsible for the observable features in deformable media around us, they are at the same time undesirable in many technological applications, or key players in natural phenomena. They often place limitations on the safe and efficient operation of many fluid devices; in flames, they lead to incomplete and nonstoichiometric combustion, increased soot formation and significant NO_x output;

instabilities in the inlet of a jet engine lead to non-smooth operation, while acoustic instabilities lead to premature material fatigue; and strong magnetic instabilities are one of the key reasons why nuclear fusion has not yet matured into a reliable technology. In natural settings, instabilities in the atmospheric boundary layer are responsible for weather abnormalities, and in lakes, rivers, and oceans they influence nutrient transport.

The underlying mechanisms for instabilities can be rather multifaceted. Yet, it is common to classify instabilities by their principal physical mechanism responsible for the growth (or decay) of disturbances: buoyancy-driven, shock-induced, rotational, magnetic, shear-driven, morphological, vortical, interfacial, thermal, reactive, acoustical, chemical, and so on. Instabilities have been studied intensively over the past decades. Along with this categorization, each effect is characterized and parameterized by a nondimensional number, such as the Reynolds number, Rayleigh number, Rossby number, Weber number, Mach number, Damkoehler number, and so on. These numbers quantify the relative importance of the involved processes and act as bifurcation parameters that, after passing a critical value, establish the presence of an instability.

Besides the key physical processes and their nondimensional number, the size of the disturbance background or of an initial condition, necessary to induce an instability, is important. Infinitesimal disturbances describe the early departure from an established equilibrium state. They have the added mathematical advantage that linearization is easily justified. The same is not true for finite-amplitude disturbances whose analysis requires more sophisticated mathematical techniques. In what follows, we will outline techniques for a general stability analysis, but will focus on common techniques for a linear analysis.

Hydrodynamic Stability Theory: A Brief Look Back

Owing to its central position within the broad field of fluid mechanics, hydrodynamic stability theory has a long history that is worth recalling.

While it is difficult to name a precise scientific event or discovery that acts as the defining moment of hydrodynamic stability theory, it is commonly acknowledged that the experiments of Osbourne Reynolds in the 1880s have, for the first time, solely focused on stability issues of shear flows and proposed a criterion (the Reynolds number) for the distinction of laminar, turbulent, and transitional flow regimes (Reynolds 1883). At the same time, Lord Rayleigh has given a formal mathematical framework for the evolution of infinitesimal perturbations in an inviscid fluid (Rayleigh 1887). Around the same time, the definition of stability has been influenced by the mathematical work of A. Lyapunov, who gave a precise meaning to the asymptotic fate of perturbations from an equilibrium point (Lyapunov 1892).

At the beginning of the twentieth century, the stability formulation by Orr and Sommerfeld, a viscous extension of the Rayleigh equation, has moved stability issues in the focus of fluid dynamicists and mathematicians and is still in widespread use today (Orr 1907a, Sommerfeld 1908). Many subsequent investigations focused on the associated Orr–Sommerfeld eigenvalue problem, and determined instability modes for a wide range of fluid configurations, but in particular for wall-bounded unidirectional shear flows. The incorporation of nonlinear saturation effects has further extended the range of applicability of stability results to finite-amplitude perturbations.

The advent of the computers and numerical algorithms has also greatly influenced hydrodynamic stability theory, and it did not take long before the central equations of stability theory have been solved numerically to high precision and for increasingly complex flows, such as high-speed, compressible flows.

At the same time, transition to turbulence and hydrodynamic stability have been linked in an effort to predict the onset of turbulent fluid motion using stability calculations. Secondary instability theory, a two-stage stability concept, has been introduced and proposed as the route of many flows to a highly disordered state (Orszag & Patera 1983, Herbert 1988, Koch et al. 2000).

A new development emerged in the 1990s whereby the original Lyapunov stability concept has been called into question, as it does not contain any notion of a time horizon. In this vein, eigenvalues have been increasingly superseded by more abstract operator concepts (Butler & Farrell 1992, Reddy & Henningson 1993, Trefethen et al. 1993, Schmid & Henningson 2001, Schmid 2007). This non-eigenvalue-based stability theory has succeeded in uncovering and explaining a great deal of experimentally observed phenomena and is often a key component of instability and amplification processes.

The past decade has continued to produce a great many tools related to stability theory: direct numerical simulations produce high-fidelity fluid solutions and give unprecedented insight into all aspects of transport processes and instabilities; the parabolized stability equations (PSE) provide a powerful and efficient tool for the calculation of stability characteristics in configurations that go far beyond the early simple geometries (Bertolotti et al. 1992). The stability of flow can also be determined globally using high-performance computers and iterative eigenvalue algorithms (Theofilis 2011). Following these techniques, stability properties have been computed for such complex flows as airfoil sections, turbomachinery stages, or wing-tip vortices. Moreover, variational techniques (Hill 1995), which recast stability concepts into the form of an optimization problem, have resulted in novel and powerful tools that extract stability information (even in the nonlinear case) directly from simulation software. It is these latter optimization techniques – in a rudimentary form – that will be the focus of this chapter.

A Broader Viewpoint on Instabilities

Besides the rise of sophisticated tools to address stability issues for complex flows, the definition of stability has also changed over the years. Starting from a rather binary definition (stable or unstable equilibrium), based on the existence of eigenvalues in a specific half-plane (Drazin & Reid 1981), we have brought back a timescale into the definition of stability, thus dismissing the asymptotic definition of Lyapunov, and reformulated stability-related issues under a more general sensitivity concept (sensitivity to initial conditions, to forcing, to internal changes, etc.). Section 13.2 gives a brief summary of stability notion, starting with the mathematical definition

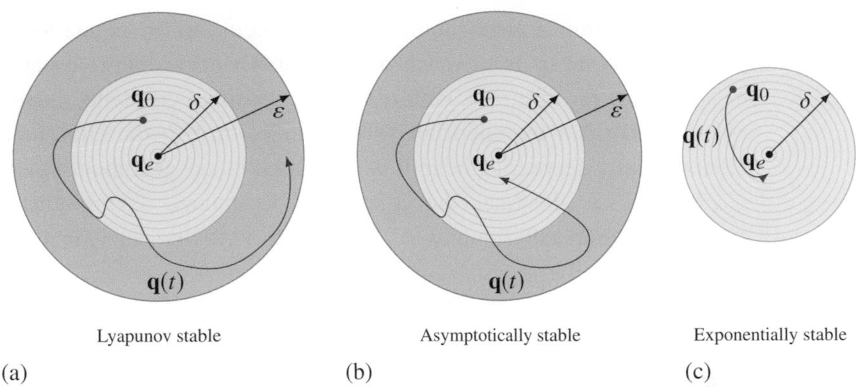

Figure 13.2 Stability concepts, illustrating (a) Lyapunov stability, (b) asymptotic stability, and (c) exponential stability.

of Lyapunov stability and progressing to more adapted concepts for hydrodynamic stability calculations (Schmid & Henningson 2001).

13.2 Definitions of Stability

The stability of a dynamical system has been formalized mathematically by Lyapunov in his landmark paper at the end of the nineteenth century (Lyapunov 1892). This definition has stood the test of time and is still principally applied in a wide range of situations from fluid dynamics to other dynamical systems in the engineering or life sciences.

The first definition of **Lyapunov stability** states that, given a general autonomous dynamical system $\mathbf{q} = \mathsf{N}(\mathbf{q})$ for the state vector $\mathbf{q}$ with $\mathbf{q}(0) = \mathbf{q}_0$, an equilibrium point $\mathbf{q}_e$ satisfying $\mathsf{N}(\mathbf{q}_e) = 0$ is Lyapunov stable, if for every $\varepsilon > 0$, there exists a $\delta > 0$ such that, if $\|\mathbf{q}_0 - \mathbf{q}_e\| < \delta$, then for every $t > 0$ we have $\|\mathbf{q}(t) - \mathbf{q}_e\| < \varepsilon$.

Lyapunov stability thus relates an initial condition to the long-term fate of the state vector following the governing equations. If the initial condition $\mathbf{q}_0$ is δ-close to the equilibrium point $\mathbf{q}_e$, a Lyapunov-stable system will remain in a neighborhood of size ε of this equilibrium for all times. This condition has to hold true for any ε one may choose. The previous definition ensures boundedness of the solution as long as the initial condition is sufficiently close to the equilibrium state.

A more stringent definition of stability, extensively used in fluid dynamics application, is **asymptotic stability**. It adds to the definition of Lyapunov stability the supplementary condition that, for an infinite time horizon, the equilibrium state $\mathbf{q}_e$ is approached. In other words, the solution to the dynamical system does not only stay bounded in a neighborhood of the equilibrium point $\mathbf{q}_e$, it approaches the equilibrium point as time tends to infinity.

A third definition of stability adds yet more restrictions to asymptotic stability by specifying the manner in which the equilibrium point is approached. For initial conditions within a δ-neighborhood of the equilibrium point, we require that the equilibrium point $\mathbf{q}_e$ is approached exponentially according to $\|\mathbf{q}(t) - \mathbf{q}_e\| < \alpha\|\mathbf{q}_0 - \mathbf{q}_e\| \exp(-\beta t)$ for all times, with α and β as positive constants. Systems of this type are referred to as **exponentially stable** and return to the equilibrium state within monotonically decreasing bounds when started in a δ-neighborhood of $\mathbf{q}_e$.

Figure 13.2 illustrates the concept of Lyapunov, asymptotic and exponential stability schematically for a dynamical system with two dependent variables.

For Hydrodynamic Applications

The definitions of Lyapunov have been adopted for hydrodynamic stability calculations; only minor modifications have been added to quantify concepts observed in fluid systems. We will follow Joseph (1976) and introduce four variants of hydrodynamic stability. These definitions require the introduction of a disturbance measure, which we take as the perturbation kinetc energy E.

The first definition (**asymptotic stability**) states that an equilibrium solution is asymptotically stable, if the perturbation energy satisfies

$$\lim_{t\to\infty} \frac{E(t)}{E(0)} \to 0. \tag{13.1}$$

This definition coincides with Lyapunov's asymptotic stability definition, where the distance to the equilibrium state is measured by the kinetic energy of the perturbation. This definition, however, does not specify an ε-neighborhood bounding the distance from equilibrium as the equilibrium state is approached.

The second definition (**conditional stability**) accounts for the fact that an initial energy threshold value has to be respected for asymptotic stability. It states that an equilibrium is conditionally stable, if there exists a positive energy level δ such that asymptotic stability is observed as long as the initial energy of the perturbation falls below δ, that is, $E(0) < \delta$.

A further restriction on this concept yields the third definition (**global stability**): for an infinite threshold value δ, the equilibrium is said to be globally stable. In other words, the equilibrium state is attracting trajectories, no matter how far we start from the equilibrium point.

Finally, even further constraints lead to the fourth definition (**monotonic stability**). We state that an equilibrium is monotonically stable, if the energy of the perturbation decreases monotonically for all times. Mathematically, this condition requires

$$\frac{dE}{dt} < 0 \qquad \text{for all} \quad t > 0. \tag{13.2}$$

The Role of Time

It is interesting to note that none of the previous definitions, whether set forth by Lyapunov or proposed by Joseph (1976), contains a characteristic timescale. Instead, time is either absent in the definition, or an infinite time horizon is envisioned. This lack of restriction on the time it takes to return to the equilibrium point and thus to confirm stability, suggests that such-defined stability measures (such as growth rates,

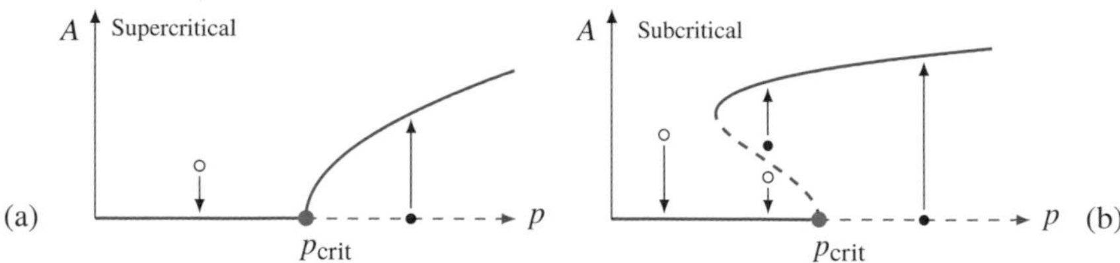

Figure 13.3 Bifurcation behavior, illustrated in the amplitude-parameter plane. In the supercritical bifurcation scenario (a), finite-amplitude states exist only for parameter values p larger than the critical one, p_{crit}. In the subcritical bifurcation scenario (b), finite-amplitude states exist at parameter values p where the base state has not yet gone unstable to infinitesimal perturbations, that is, for $p < p_{\text{crit}}$.

frequencies, modal shapes) may misrepresent the dynamic features of a flow that is characterized and dominated by fluid processes on finite timescales.

Critical Parameters

The equations governing the temporal evolution of the state variable, denoted symbolically by the function **f**, commonly contain a variety of physical parameters often expressed as nondimensional numbers (such as the Reynolds number, Rayleigh number, Mach number, etc.) or parameters linked to the shape of the disturbances (such as their wavenumbers in homogeneous coordinate directions). The stability properties thus depend on these parameters as well, and a specific perturbation may be stable at one parameter setting, but become unstable at another. The parameter value(s) at which this transition from stability to instability occurs are known as **critical parameters**. Different critical parameters can be computed, depending on the chosen definition of stability; for example, the critical Reynolds number for monotonic stability of plane channel flow is $Re_c = 49.6$, while the critical Reynolds number for asymptotic stability of the same flow is $Re_c = 5772.2$, where the Reynolds number is based on the centerline base velocity and the half-channel height (Joseph & Carmi 1969, Orszag 1971).

Bifurcation Behavior

Once we surpass the critical parameter, an instability ensues and the disturbance grows in amplitude/energy until nonlinearities saturate the growth and establish a new, nonlinear equilibrium state of finite amplitude. Finite-amplitude states, however, can also exist (due to a conditional stability; see the earlier definition) at parameter values below the critical one.

The existence of these finite-amplitude states determines the **bifurcation behavior** of the flow (see also Chapter 10 for a discussion of bifurcations in dynamical systems). Two cases have to be distinguished: supercritical and subcritical bifurcation behavior. In the supercritical case, finite-amplitude states exist only past the point (in parameter space) where the equilibrium state has gone asymptotically unstable to infinitesimal perturbations. This situation is illustrated in Figure 13.3(a). Infinitesimal perturbations below p_{crit} are asymptotically stable; for $p > p_{\text{crit}}$, finite-amplitude states exist owing to an asymptotic instability of the infinitesimal state (dashed line). In the subcritical case (see Figure 13.3(b)), we have a parameter regime where finite-amplitude states coexist with asymptotically stable infinitesimal states. In this regime, the infinitesimal state is conditionally stable: for an initial energy below a critical value, the perturbation returns to the infinitesimal state (open circle); while for an initial energy surpassing a threshold value (indicated by the dashed blue curve), a higher-energy state is approached at the same parameter value p (closed circle). After passing p_{crit}, this threshold value is zero, indicating that infinitesimal energy is necessary to approach the higher-energy state via an asymptotic instability.

Examples of supercritical bifurcation behavior in fluid systems include Raleig-Bénard convection and Taylor–Couette flow (flow between two differentially rotating coaxial cylinders) within certain parameter regimes, while wall-bounded shear flows such as plane Poiseuille, plane Couette, or pipe flow are governed by subcritical bifurcations.

Temporal, Spatial, Spatio-temporal Evolution

While many stability calculations use time as the evolution coordinate and treat disturbances as temporally evolving entities, there are configurations where a different description is more appropriate. For example, disturbance evolution in flow past a roughness element is more aptly described as a spatial problem. In this case, the proper mathematical formulation is a signaling problem, rather than a temporal initial-value problem. In a signaling problem, a harmonic or steady point source drives the flow perpetually, and the instability manifests itself as amplitude growth as we move downstream from the disturbance source. The evolution direction is the spatial direction given by a representative streamline of the base flow. In this case, the time coordinate is often taken as harmonic and can be Fourier transformed.

Besides the purely temporal and the purely spatial approach, we can also formulate a general spatio-temporal approach, where we postulate a spatially and temporally localized initial condition and track its development in space and time, without assumptions on either evolution direction. Figure 13.4 illustrates the three concepts of perturbation evolution: temporal, spatial, and spatio-temporal. Each concept has to be motivated by the specific flow configuration and requires corresponding mathematical tools for its analysis.

13.3 General Formulation

Most treatises on hydrodynamic stability theory start with many simplifying assumptions about the flow (steady, parallel, unidirectional, incompressible base flow of Newtonian fluid), guiding the reader toward a set of mathematical techniques for the solution of the resulting stability equations. Only in a second step are some of the initial assumptions relaxed or eliminated, leading to additional mathematical complications. This simple-to-complex route has also been the historic route: simplifications

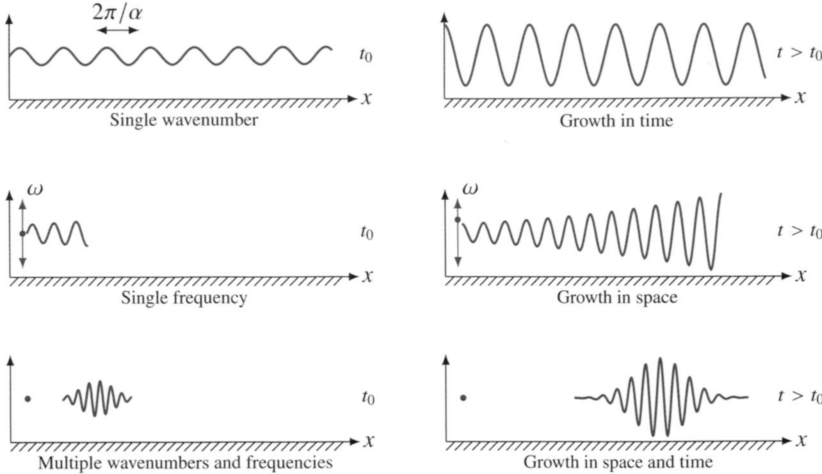

Figure 13.4 Stability formalisms for a temporally evolving perturbation (top), a spatially evolving disturbance (middle), and a general spatio-temporal evolution from a point source (bottom).

have been necessary, since the mathematical tools have been restricted to analytical, approximate, or asymptotic methodology, such as, for example, perturbation and asymptotic methods. With the advent of computational and data-driven techniques, we are now in a position to address directly – with few imposed restriction and forced assumptions – the hydrodynamic stability behavior of complex flows. The analysis of unsteady, separated, and multiscale flows are now within our range, as is the nonlinear evolution of disturbances in these configurations.

For this reason, we will advocate a reverse route of exposition: starting from the general, unrestricted stability problem to the more confined, but perhaps more familiar setup. We will describe a rather general computational framework for hydrodynamic stability problems based on the optimization of a cost functional subject to constraints, among them our nonlinear partial differential equation governing the fluid motion. This formalism is very flexible in describing and quantifying the perturbation dynamics under minimal restrictions and limitations. After having established this general framework, we will then make progressive assumptions – either about the base flow or the perturbation dynamics – to find reduced descriptions of the general formulation. This will naturally introduce familiar techniques for the analysis of a fluid system's stability characteristics.

The general framework for hydrodynamic stability analysis relies on a mathematical formalism that uses calculus of variations, optimization techniques, and concepts from linear algebra. A basic knowledge of these disciplines is helpful in understanding the derivations later; it can be gained from excellent textbooks, monographs, and review articles, such as Kot (2015) and Cassel (2013) for calculus of variations, Gunzburger (2003) and Magri (2019) for adjoint techniques, Kochenderfer and Wheeler (2019) and Nocedal and Wright (2006) for optimization, and Trefethen and Bau (1997) for linear algebra, among many other resources.

An Optimization Formulation for Hydrodynamic Stability

We assume a general form of the governing equations according to

$$\frac{d}{dt}\mathbf{q} = \mathsf{N}(\mathbf{q}, \mathbf{f}), \tag{13.3}$$

where the components of the state vector $\mathbf{q} \in \mathbb{C}^{n_q}$ represent all variables required to fully describe the flow state. The vector $\mathbf{f} \in \mathbb{C}^{n_f}$ stands for the control variable. The exact form of the governing equations is given by the operator N, where we additionally have assumed an appropriate spatial discretization of the state vector $\mathbf{q}$. We consider n_q degrees of freedom for the state vector, comprising the product of the total number of grid points (for example, $n_x \times n_y \times n_z$ in three Cartesian dimensions) and the number of tracked variables (for example, three velocities and pressure for three-dimensional incompressible flow). The control vector has n_f degrees of freedom, indicating the number of grid points and variables we have chosen to manipulate. Boundary conditions are enforced within the spatial discretization schemes, and the initial condition will be imposed via the control variable $\mathbf{f}$ (see below for details).

Next, we introduce a cost functional $\mathcal{J}(\mathbf{q}, \mathbf{f})$, consisting of a real scalar that depends, in a yet unspecified way, on the variables $\mathbf{q}$ and $\mathbf{f}$. We recall that a functional is a mapping of functions onto a (generally complex) scalar. Thus, $\mathcal{J}$ takes the time-dependent input functions $\mathbf{q}$ and $\mathbf{f}$ and assigns a scalar value to them. More specifically, this cost functional represents the objective of the optimization, for example, a maximization of energy amplification (for stability analyses) or a minimization of instabilities (for flow control applications; see also the control framework developed in Chapter 9). During the optimization, the governing equations (13.3) have to be satisfied, which renders our optimization setup as a constrained optimization problem. Invoking standard techniques, we convert this constrained optimization problem into an unconstrained equivalent by adding the constraints via Lagrange multipliers to the cost functional; this results in the augmented Lagrangian:

$$\mathcal{L}(\mathbf{q}, \mathbf{q}^\dagger, \mathbf{f}) = \mathcal{J}(\mathbf{q}, \mathbf{f}) - \left\langle \mathbf{q}^\dagger, \frac{d}{dt}\mathbf{q} - \mathsf{N}(\mathbf{q}, \mathbf{f}) \right\rangle. \tag{13.4}$$

This mathematical step required the introduction of Lagrange multipliers $\mathbf{q}^\dagger$ and the choice of a scalar product. It follows from expression (13.4) (more specifically, from term with the scalar product) that the Lagrange multipliers $\mathbf{q}^\dagger$ have the same dimensionality as $\mathbf{q}$, that is, $\mathbf{q}^\dagger \in \mathbb{C}^{n_q}$. We thus realize that the Lagrange multipliers represent time-dependent "flow fields." At this stage, however, we do not yet have an equation governing their temporal evolution. Proceeding, we define a scalar product as

$$\langle \mathbf{a}, \mathbf{b} \rangle \equiv \int_0^T \mathbf{a}^H \mathbf{b} \, dt. \tag{13.5}$$

The superscript ^{H} stands for the transpose conjugate operation. In the previous definition, we have introduced the time horizon τ, which has to be user-specified and adapted to the relevant timescales of the flow under consideration. In what follows, it will prove mathematically advantageous to recast the cost functional $\mathcal{J}$ in terms of the scalar product (13.5). We write

$$\mathcal{J}(\mathbf{q},\mathbf{f}) = \langle w(t), j(\mathbf{q},\mathbf{f})\rangle\,, \tag{13.6}$$

where we included a weight function $w(t)$ that is responsible for the enforcement of j within the time interval $[0,\ \tau]$. As an example, specifying a cost functional j over the entire time horizon, we choose $w = 1$; for focusing only on the value of j at the end of the time horizon, we choose $w = \delta(t - \tau)$.

With the augmented Lagrangian $\mathcal{L}$ fully specified, we seek an optimum of $\mathcal{L}$ with respect to all independent variables, that is, $\mathbf{q}, \mathbf{q}^\dagger, \mathbf{f}$. This mathematical problem requires calculus of variations: we determine three time-dependent functions $(\mathbf{q}, \mathbf{q}^\dagger, \mathbf{f})$ that optimize a scalar value (our augmented Lagrangian $\mathcal{L}$). Analogous to computing an optimum for a function of three independent variables $f(x, y, z)$ by setting the first-order partial derivatives with respect to the three variables to zero, $f_x = f_y = f_z = 0$, we have to take the **first-order variations** of the scalar $\mathcal{L}$ (see (13.4)) with respect to the three functions $\mathbf{q}, \mathbf{q}^\dagger, \mathbf{f}$ and set them to zero simultaneously (see, e.g., Cassel (2013) or Kot (2015)). The resulting three conditions are referred to as the Karush–Kuhn–Tucker (KKT) system. Formally, we have

$$\frac{\delta \mathcal{L}}{\delta \mathbf{q}^\dagger} = 0, \qquad \frac{\delta \mathcal{L}}{\delta \mathbf{q}} = 0, \qquad \frac{\delta \mathcal{L}}{\delta \mathbf{f}} = 0, \tag{13.7}$$

where each of the above conditions furnishes a governing equation.

The first condition reads, more explicitly,

$$\Big\langle \delta \mathbf{q}^\dagger, \underbrace{\frac{d}{dt}\mathbf{q} - \mathsf{N}(\mathbf{q},\mathbf{f})}_{=0} \Big\rangle = 0, \tag{13.8}$$

a condition that has to hold true for all variations $\delta \mathbf{q}^\dagger$. This expression thus recovers our original governing equation for $\mathbf{q}$.

The second condition is slightly more involved, as the state variable $\mathbf{q}$ appears in the governing equation as well as in the cost functional. We have

$$\Big\langle w, \frac{\partial j}{\partial \mathbf{q}}\delta \mathbf{q}\Big\rangle - \Big\langle \mathbf{q}^\dagger, \frac{d}{dt}\delta \mathbf{q} - \frac{\partial \mathsf{N}}{\partial \mathbf{q}}\delta \mathbf{q}\Big\rangle = \Big\langle \underbrace{\left(\frac{\partial j}{\partial \mathbf{q}}\right)^H w + \frac{d}{dt}\mathbf{q}^\dagger + \left(\frac{\partial \mathsf{N}}{\partial \mathbf{q}}\right)^H \mathbf{q}^\dagger}_{=0}, \delta \mathbf{q}\Big\rangle = 0, \tag{13.9}$$

where integration by parts (in time) is required to transfer the d/dt-operator onto the adjoint variable $\mathbf{q}^\dagger$. After isolating the first variation $\delta \mathbf{q}$ in the scalar products, we can extract an evolution equation for the adjoint variable $\mathbf{q}^\dagger$.

Finally, the third condition yields the following explicit expression:

$$\left\langle w, \frac{\partial j}{\partial \mathbf{f}} \delta \mathbf{f} \right\rangle - \left\langle \mathbf{q}^{\dagger}, -\frac{\partial \mathsf{N}}{\partial \mathbf{f}} \delta \mathbf{f} \right\rangle = \left\langle \underbrace{\left(\frac{\partial j}{\partial \mathbf{f}}\right)^{H} w + \left(\frac{\partial \mathsf{N}}{\partial \mathbf{f}}\right)^{H} \mathbf{q}^{\dagger}}_{=0}, \delta \mathbf{f} \right\rangle = 0, \tag{13.10}$$

from which we can extract an algebraic equation for the cost functional gradient with respect to the control vector $\mathbf{f}$.

In summary, we obtain from the three optimality conditions the following three equations:

$$\frac{\delta \mathcal{L}}{\delta \mathbf{q}^{\dagger}} = 0: \qquad \frac{d}{dt}\mathbf{q} = \mathsf{N}(\mathbf{q}, \mathbf{f}), \tag{13.11a}$$

$$\frac{\delta \mathcal{L}}{\delta \mathbf{q}} = 0: \qquad -\frac{d}{dt}\mathbf{q}^{\dagger} = \left(\frac{\partial \mathsf{N}}{\partial \mathbf{q}}\right)^{H} \mathbf{q}^{\dagger} + \left(\frac{\partial j}{\partial \mathbf{q}}\right)^{H} w, \tag{13.11b}$$

$$\frac{\delta \mathcal{L}}{\delta \mathbf{f}} = 0: \qquad \left(\frac{\partial j}{\partial \mathbf{f}}\right)^{H} w = -\left(\frac{\partial \mathsf{N}}{\partial \mathbf{f}}\right)^{H} \mathbf{q}^{\dagger}. \tag{13.11c}$$

By design, this set of equations provides an extremum $(\mathbf{q}, \mathbf{f})$ of the chosen cost functional $\mathcal{J}$. Analogously to finding the extremum of a multivariate function, the system (13.11) has to be solved simultaneously. In practice, however, we approach the solution in an iterative manner. We solve the two evolution equations (13.11a,b) exactly, while we use the algebraic optimality condition (13.11c) iteratively to converge toward an optimal control variable $\mathbf{f}$. The above optimization problem is the basis for our analysis of fluid problems as to their stability, receptivity, and sensitivity to internal or external changes. At the same time, the optimization framework (with only minor modifications) can also be used to compute optimal flow control strategies or optimal designs of fluid devices.

An Iterative Process: Direct-Adjoint Looping

The full iterative process unravels as follows: we commence with an initial control-variable estimate $\mathbf{f}^{(0)}$ and solve the governing equation (13.11a) over the chosen time interval $[0, \tau]$. This *direct* solution will then act as the forcing term for the adjoint equation (13.11b), which is solved in a time-reversed manner from $t = \tau$ to $t = 0$. The time-reversed integration of the adjoint stems from the integration by parts (in time) during the derivation of the adjoint equation; the change in sign for the first-order time derivative requires a reverse integration or, alternatively, a change of time-coordinate in the form $t' = \tau - t$. The resulting adjoint solution $\mathbf{q}^{\dagger}$ then is used in expression (13.11c) to determine the cost functional gradient with respect to the control variable $\mathbf{f}$. This gradient information is provided to a standard optimization routine, such as an algorithm from the conjugate gradient or quasi-Newton family, and a new control variable $\mathbf{f}^{(1)}$ is determined. The iterative process then repeats with this improved control strategy, until a user-specified convergence criterion is

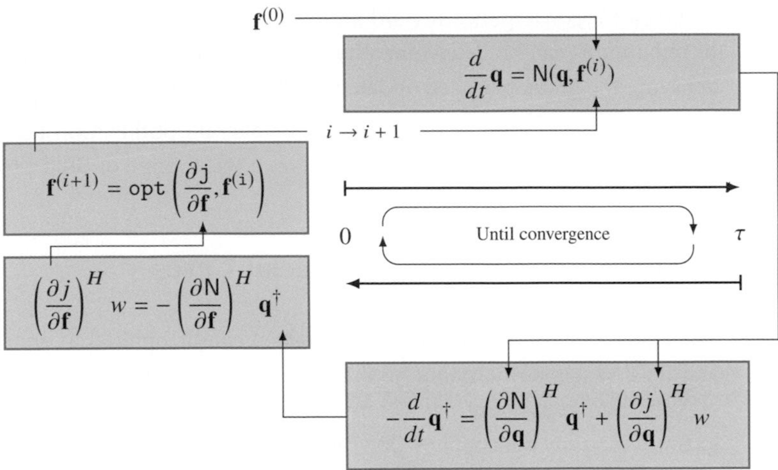

Figure 13.5 Schematic of adjoint looping. An initial guess $\mathbf{f}^{(0)}$ for the control variable is used to solve the governing equations over a time horizon $t \in [0, \tau]$. This is followed by a solution of the adjoint equations backward in time (from $t = \tau$ to $t = 0$), accounting for the driving by the forward solution $\mathbf{q}$. The adjoint solution $\mathbf{q}^\dagger$ is then substituted into the cost functional gradient $\nabla_{\mathbf{f}} j$, which in turn is passed to an optimization algorithm `opt`. A new and improved control variable $\mathbf{f}$ is determined, and another iteration starts. The iterations are continued until convergence is reached.

satisfied, or the computational resources are exhausted. Figure 13.5 gives a schematic representation of this iterative process, often referred to as adjoint looping.

A few observations about the full system (13.11) and the iterative scheme are worth pointing out.

(i) While the direct problem (13.11a) may be nonlinear, the adjoint equation (13.11b) is always linear in $\mathbf{q}^\dagger$. This fact results from the linear appearance of $\mathbf{q}^\dagger$ in the augmented Lagrangian. The linearity of the adjoint equation can be exploited in parallel-in-time algorithms for the adjoint part of the loop (Skene et al. 2020); increased performance and computational efficiency can be gained in this manner.

(ii) Expression $\partial \mathsf{N}/\partial \mathbf{q}$ is recognized as the Jacobian of the governing equation (13.3), and constitutes a matrix of size $n_q \times n_q$. For nonlinear N, the Jacobian appears in complex conjugate form in the adjoint equation and is evaluated at the time-local flow field $\mathbf{q}$. Details on how to compute this Jacobian – or more specifically its action on a given state vector – are given below.

(iii) The system matrix $(\partial \mathsf{N}/\partial \mathbf{q})^H$ for the adjoint equation (13.11b) does not dependent on the cost functional $\mathcal{J}$. The influence of $\mathcal{J}$ in the adjoint equation only stems from the external forcing term $(\partial j/\partial \mathbf{q})^H w$. Different cost functionals produce different adjoints via this forcing term, even though the inherent system dynamics is independent of our choice of cost functional.

(iv) Given a nonlinear N or a non-convex cost functional $\mathcal{J}$, any optimization scheme based on gradient information can only reach a local extremum. No guarantee can be given for convergence toward a global optimum. The type of optimum (minimum or maximum) can be determined by monitoring the cost functional; a sign change in applying the gradient-based update may be necessary.

Despite some obvious limitations, the general optimization formalism is versatile and flexible, and can be brought to bear on a wide range of fluid problems.

13.4 Reduced Formulation for Special Cases

The above system of equations (13.11a-b-c) can be applied to general configurations and fluid systems, including nonlinear governing equations. In this section, we will make common assumptions about the fluid system and link the general framework to more familiar concepts of flow analysis. To this end, we take the right-hand side of the governing equations in the special form $\mathsf{N}(\mathbf{q},\mathbf{f}) = \mathsf{L}\mathbf{q} + \mathbf{f}$, with L representing a linear(ized) set of equations (see also the material in Chapter 9). With this assumption, we can recover common stability and receptivity results (Schmid 2007).

Stability of Linear, Time-Invariant Systems

The linear stability of fluid flow is very common type of flow analysis. It rests on the assumption of a steady equilibrium state, which (i) can be computed exactly, such as, for example, in channel or pipe flow, or (ii) can be argued for as an approximation following a timescale separation. Linearization about this equilibrium state yields a linear, time-invariant (LTI) system, and the stability of the equilibrium state depends on the dynamics of infinitesimal perturbations about it.

Stability can be defined in a variety of ways. We will use a gain concept: the amplification of the size of a perturbation over its initial size, optimized over all initial conditions. This gain definition runs counter to Lyapunov's definition of stability, as it retains the timescale over which we consider the amplification. In addition, our gain definition requires a choice of measure of disturbance size. This choice is nontrivial in many cases and crucially depends on the flow configuration as well as the fluid properties. For incompressible flows, the choice is bit more straightforward: the kinetic perturbation energy is commonly used as a measure of disturbance size. If additional physical effects, such as, for example, buoyancy, surface tension, compressibility, more thought has to be put into an appropriate measure. Mathematically, we define the norm of the state vector according to

$$\|\mathbf{q}\|_{\mathsf{Q}}^2 = \mathbf{q}^H \mathsf{Q}\mathbf{q}, \tag{13.12}$$

in which Q is a symmetric, positive definite weight matrix. Factoring Q into $\mathsf{Q} = \mathsf{F}^H\mathsf{F}$ (using a Cholesky factorization), we can establish a link to the more standard L_2-norm as

$$\|\mathbf{q}\|_{\mathsf{Q}}^2 = \mathbf{q}^H \mathsf{F}^H \mathsf{F}\mathbf{q} = \|\mathsf{F}\mathbf{q}\|_2^2. \tag{13.13}$$

We state our cost objective as the quotient of the state-vector norm over the control-vector norm, evaluated over a time interval $[0,\ \tau]$. The norms for the state vector and the control vector do not necessarily have to coincide; for this reason, we choose two different weight matrices in the respective norms. Furthermore, we evaluate the gain at the end of the time horizon, at $t = \tau$, which coincides with more traditional notions of (in)stability. This choice of evaluation leads to $w = \delta(\tau - t)$ in (13.6). We have

$$j(\mathbf{q}, \mathbf{f}) = \frac{\|\mathbf{q}\|_{\mathsf{Q}}^2}{\|\mathbf{f}\|_{\mathsf{R}}^2} = \frac{\mathbf{q}^H \mathsf{Q} \mathbf{q}}{\mathbf{f}^H \mathsf{R} \mathbf{f}}, \tag{13.14}$$

with R denoting the weight matrix for the control variable, which in our case is taken as the initial condition, that is, $\mathbf{f} = \mathbf{q}(t = 0)$. Consequently, the evolution operator takes on the form $\mathsf{N}(\mathbf{q}, \mathbf{f}) = \mathsf{L}\mathbf{q} + \mathbf{f}\delta(t)$.

With this setup, we can reconsider the general iterative optimization framework and use the linearity of our governing equations to make simplifications. The first step in the direct-adjoint loop consists of solving the governing equations (13.11a) over the time interval $[0, \tau]$. The solution of this problem can formally be written as

$$\mathbf{q}(\tau) = \exp(\mathsf{L}\tau)\, \mathbf{f}, \tag{13.15}$$

introducing the **matrix exponential** that transforms the initial condition $\mathbf{f}$ into the output perturbation $\mathbf{q}(\tau)$. The second step, based on equation (13.11b), simplifies for our LTI-case[1]

$$-\frac{d}{dt}\mathbf{q}^{\dagger} = \mathsf{L}^H \mathbf{q}^{\dagger} + \frac{2\mathsf{Q}}{\|\mathbf{f}\|_{\mathsf{R}}^2} w \mathbf{q}, \tag{13.16}$$

which again can be solved formally according to

$$\mathbf{q}^{\dagger}(0) = \exp(\mathsf{L}^H \tau) \frac{2\mathsf{Q}}{\|\mathbf{f}\|_{\mathsf{R}}^2} \mathbf{q}(\tau). \tag{13.17}$$

The third and final step uses the cost functional gradient

$$\left(\frac{\partial j}{\partial \mathbf{f}}\right)^H w = -\delta(t)\mathbf{q}^{\dagger}, \tag{13.18}$$

where the right-hand side term, $\delta(t)$, is equivalent to the gradient $\partial \mathsf{N}/\partial \mathbf{f}$ for our special case.

At this stage, we would use the aforementioned gradient in a user-specified optimization routine (such as conjugate gradient, for example) to arrive at an improved initial condition $\mathbf{f}$. In this special case, however, we further simplify the direct-adjoint loop (Figure 13.5) by choosing a steepest-descent procedure as our gradient-based optimization routine. This results in an explicit expression for the new initial condition. We have

$$\left(\frac{\partial j}{\partial \mathbf{f}}\right)^H w = -\frac{\|\mathbf{q}\|_{\mathsf{Q}}^2}{\|\mathbf{f}\|_{\mathsf{R}}^4} 2\mathsf{R}\mathbf{f}\, w. \tag{13.19}$$

Combining (13.15), (13.17), (13.18), and (13.19), we obtain

$$\mathsf{R}^{-1} \exp(\mathsf{L}^H \tau)\, \mathsf{Q} \exp(\mathsf{L}\tau)\, \mathbf{f} = \frac{\|\mathbf{q}(\tau)\|_{\mathsf{Q}}^2}{\|\mathbf{f}\|_{\mathsf{R}}^2} \mathbf{f}. \tag{13.20}$$

[1] Recall that the derivative of the quadratic form $\mathbf{x}^H \mathsf{A} \mathbf{x}$ with respect to $\mathbf{x}$ is $2\mathsf{A}\mathbf{x}$ for symmetric A.

This last formula can be further manipulated into an expression for $\mathbf{f}$

$$\left[\exp(\bar{\mathsf{L}}\tau)\right]^H \exp(\bar{\mathsf{L}}\tau)\,\bar{\mathbf{f}} = \sigma^2\,\bar{\mathbf{f}}, \tag{13.21}$$

describing a full iteration of the direct-adjoint loop. We have introduced the energy gain $\sigma^2 = \|\mathbf{q}(\tau)\|^2_{\mathsf{Q}}/\|\mathbf{f}\|^2_{\mathsf{R}}$, together with the transformed matrix $\bar{\mathsf{L}} = \mathsf{FLG}^{-1}$ and the transformed initial condition $\bar{\mathbf{f}} = \mathsf{G}\mathbf{f}$. The matrix $\mathsf{R} = \mathsf{G}^H\mathsf{G}$ has been decomposed into its Cholesky factors G.

Equation (13.21) is an eigenvalue problem for the energy gain σ^2. It then follows that the largest eigenvalue produces the optimal energy growth, and the associated eigenvector $\bar{\mathbf{f}}$ gives the optimal initial condition $\mathbf{f}$. The two matrix exponentials in (13.21) are conjugate transpose to each other; the product yields real gains σ^2. This configuration also suggests that the eigenvalue problem can be transformed to a singular value problem for $\exp(\bar{\mathsf{L}}\tau)$: the largest singular value (i.e., the L_2-norm of the matrix exponential) produces σ, the principal left singular vector corresponds to $\mathbf{f}$ and the principal right singular vector gives the corresponding optimal output $\mathbf{q}(\tau)$. This latter relation is at the core of nonmodal stability theory: the norm of the matrix exponential measures the maximum transient energy amplification. The above demonstrates that it can now be thought of as a special case of the direct-adjoint looping procedure for LTI systems. The iterative optimization procedure is equivalent to a power iteration applied to the composite matrix in (13.21), and the optimization problem can be solved by straightforward linear algebra operations.

Receptivity of Linear, Time-Invariant Systems

Section 13.3 addressed the stability of an equilibrium state by computing the maximal gain of a perturbation over a specified time horizon. Alternatively, amplification of perturbations can also be accomplished by constant forcing. This alternative setup addresses the issue of receptivity, that is, the susceptibility of the flow to its disturbance environment (e.g., free-stream turbulence or wall roughness). Due to the linearity of the underlying equations, we can recast the general, forced problem into a frequency response analysis to harmonic forcing and use superposition to capture more general disturbance environments. An analysis of this type is referred to as a linear receptivity analysis.

As before, we consider an LTI system and simplify the right-hand side in (13.11) to $\mathsf{N}(\mathbf{q}, \mathbf{f}) = \mathsf{L}\mathbf{q} + \mathbf{f}\exp(i\omega t)$, where ω denotes the forcing frequency and $\mathbf{f}$ its spatial shape. We define a frequency response gain as the energy (norm) of the response divided by the energy (norm) of the forcing. Due to linearity, the response will adopt the frequency of the forcing; hence, only the spatial shapes of forcing and response are relevant to the analysis. We choose

$$j(\mathbf{q}, \mathbf{f}) = \frac{\|\mathbf{q}_w\|^2_{\mathsf{Q}}}{\|\mathbf{f}\|^2_{\mathsf{R}}}, \tag{13.22}$$

where $\mathbf{q}_w$ denotes the spatial shape of the forced response $\mathbf{q} = \mathbf{q}_w\exp(i\omega t)$. We note that $j(\mathbf{q}, \mathbf{f})$ is independent of time, and thus we do not need to specify w, but still assume it as a constant. The above analysis is linked to transfer functions, and the reader is urged to compare with related material in Chapters 4 and 9.

The direct problem (13.11a) reduces to

$$\mathbf{q}(t) = \int_0^t \exp(\mathsf{L}(t-t'))\,\mathbf{f}\exp(i\omega t')\,dt', \tag{13.23a}$$

$$= (i\omega - \mathsf{L})^{-1}\,[\exp(i\omega t) - \exp(\mathsf{L}t)]\,\mathbf{f}. \tag{13.23b}$$

We see that equation (13.23a) represents a convolution of the harmonic forcing with the input response of the linear system (see also Chapter 5 for additional material). Before moving forward, we stress that the above solution assumes an asymptotically stable system, with all eigenvalues of L contained in the stable half-plane. In essence, we only consider the long-term response of the system and ignore transient processes of establishing this long-term response. As a consequence, the term in (13.23b) containing the matrix exponential can be neglected. We have

$$\mathbf{q} = (i\omega - \mathsf{L})^{-1}\mathbf{f}\,\exp(i\omega t) = \mathbf{q}_w \exp(i\omega t). \tag{13.24}$$

Reformulating the adjoint equation in (13.11b) for our special case, we obtain

$$-\frac{d}{dt}\mathbf{q}^\dagger = \mathsf{L}^H\mathbf{q}^\dagger + \frac{2\mathsf{Q}}{\|\mathbf{f}\|_\mathsf{R}^2}\,\mathbf{q}_w \exp(i\omega t)\,w, \tag{13.25}$$

which has the formal solution

$$\mathbf{q}^\dagger = (-i\omega - \mathsf{L}^H)^{-1}\frac{2\mathsf{Q}}{\|\mathbf{f}\|_\mathsf{R}^2}\,\mathbf{q}_w \exp(i\omega t)\,w. \tag{13.26}$$

From the optimality condition (13.19) and the expression for $\mathbf{q}^\dagger$ we get

$$-\frac{\|\mathbf{q}_w\|_\mathsf{Q}^2}{\|\mathbf{f}\|_\mathsf{R}^4}\,2\mathsf{R}\,\mathbf{f}\,w = \left(\frac{\partial \mathsf{N}}{\partial \mathbf{f}}\right)^H \mathbf{q}^\dagger = \exp(-i\omega t)\,\mathbf{q}^\dagger. \tag{13.27}$$

Combining with (13.24) and (13.26) yields

$$\mathsf{R}^{-1}\,(-i\omega - \mathsf{L}^H)^{-1}\,\mathsf{Q}\,(i\omega - \mathsf{L})^{-1}\,\mathbf{f} = \sigma^2\,\mathbf{f}, \tag{13.28}$$

with σ^2 as the sought-after response gain, defined as $\|\mathbf{q}_w\|_\mathsf{Q}^2/\|\mathbf{f}\|_\mathsf{R}^2$. Using the Cholesky factorization of the weights Q and R (see above), we can write the above expression more compactly as

$$\left[(i\omega - \bar{\mathsf{L}})^{-1}\right]^H (i\omega - \bar{\mathsf{L}})^{-1}\,\bar{\mathbf{f}} = \sigma^2\,\bar{\mathbf{f}}, \tag{13.29}$$

with $\bar{\mathsf{L}} = \mathsf{FLG}^{-1}$ and $\bar{\mathbf{f}} = \mathsf{G}\mathbf{f}$.

As before, we obtain an eigenvalue problem for the gain σ^2 : the optimal frequency response gain for a given frequency ω is the largest eigenvalue of $\left[(i\omega - \bar{\mathsf{L}})^{-1}\right]^H (i\omega - \bar{\mathsf{L}})^{-1}$, or, equivalently, the first singular value of $(i\omega - \bar{\mathsf{L}})^{-1}$. This latter operator is referred to as the **resolvent**. The optimal forcing and output are given, respectively, by the principal right and left singular vectors (Schmid & Henningson 2001, Schmid 2007). The resolvent has found widespread application in the investigation of fluid behavior (see, e.g., McKeon & Sharma 2010, Moarref et al. 2014, Yeh & Taira 2019).

The Time-Asymptotic Limit

In some cases, we may be interested in the time-asymptotic limit $\tau \to \infty$. This may be justified by the presence of a very large timescale in the fluid system, or by the fact that the linear system matrix L is of normal type, that is, commutes with its adjoint. Fluid systems in that category are, for example, Rayleigh–Benard convection or Taylor–Couette flow in the small-gap limit. Other flows, while strictly nonnormal, may be nearly normal, and thus justify a simplified analysis. In the normal or nearly normal case, the direct-adjoint looping and the evaluation of the matrix functions $\exp(\tau\bar{\mathsf{L}})$ (for the initial-value problem) or $(i\omega - \bar{\mathsf{L}})^{-1}$ (for the harmonically driven problem) further reduced. The dynamics of a problem in this category is governed by the eigenvalues of $\bar{\mathsf{L}}$. For the initial-value problem, the time-asymptotic fate of perturbations is given by the least stable mode, that is, the eigenvector of $\bar{\mathsf{L}}$ corresponding to the eigenvalue with the largest growth rate. This dominant structure will grow exponentially in time, with the growth rate (and frequency) given by the least stable eigenvalue. For the harmonically driven case, the resolvent analysis reduces to a simple resonance argument: the response to harmonic forcing is inversely proportional to the distance of the forcing frequency ω to the spectrum (eigenvalues) of $\bar{\mathsf{L}}$. The closeness of the forcing frequency to the spectrum is the only mechanism to solicit a large response from the fluid system. In the general, non-normal case, we can also generate a substantial response, even though the forcing frequency is far from the spectrum of $\bar{\mathsf{L}}$ – a phenomenon referred to as pseudo-resonance.

A Link Between the Optimization and Linear-Algebra Approach

We have seen that for linear time-invariant systems, there is a link between the general direct-adjoint looping method, which finds optimal solutions to stability or receptivity problems iteratively, and the direct (non-iterative) evaluation of matrix function norms. These latter linear-algebra solutions are preferable when they apply, but the flexibility and generality of the direct-adjoint approach can hardly be overstated.

The connection between linear algebra and optimization, as a solution technique for hydrodynamic stability and receptivity problems, can be traced to the optimal character of the L_2-norm of a linear operator (matrix): the L_2-norm of a matrix is defined as the maximum ratio of input norm to output norm. If a mapping between the input and output is available and given by a function f of L, we have

$$||f(\mathsf{L})||_2 = \max_{\mathbf{q}\neq\mathbf{0}} \frac{||f(\mathsf{L})\mathbf{q}||_2}{||\mathbf{q}||_2}. \tag{13.30}$$

In the previous sections, we have seen that $f(z) = \exp(tz)$ for LTI-stability problems, and $f(z) = 1/(i\omega - z)$ for LTI-receptivity analyses. Other mappings, not covered here, include, for example, $f(z) = az^n$ for linear time-periodic (LTP) stability studies, also known as Floquet analysis (Schmid 2007).

The norm of the mapping $f(\mathsf{L})$ represents the maximum amplification over all input vectors $\mathbf{q}$. This latter statement is underlying all our gain optimizations of the previous sections: we wish to optimize an output-to-input ratio (gain) where the mapping between input and output is given by a linear operator (matrix function).

The optimal solution is then given by a singular value decomposition that results in the optimal gain (largest singular value), the optimal input (principal left singular vector), and the associated output (principal right singular vector); see Chapter 6 for additional material. The connection between optimization and linear algebra seems rather attractive, but does come with its limitations, particularly when we consider more complex flow configurations.

The foremost limitation is the restriction to linear and time-invariant systems. For nonlinear or nonautonomous linear systems, we have to revert to the iterative optimization techniques. Another limitation stems from the definition of the cost functional as a gain or a quadratic form. The stability/receptivity of some fluid problems require "exotic" norms that cannot be reduced to L_2-norms. In many other circumstances, we may wish to augment the cost functional by terms and norms that enforce additional constraints, for example, spatial sparsity. In other cases, the norm may not include all components of the state vector, resulting in the so-called semi-norm problem and requiring side constraints to ensure convergence of the iterative optimization scheme.

The most obvious advantage of the optimization approach lies in its versatility to adapt to/accommodate linear as well as nonlinear governing equations, to L_2-norms as well as other, more exotic norms, to gains as well as to more general functionals, and to additional side constraints.

13.5 Practical Issues and Implementation Details

The iterative optimization approach is a versatile concept to find optimal solutions or gradients of flow characteristics with respect to user-defined control variables, but an efficient implementation is crucial for making it a practical tool for quantitative flow analysis, particularly for large-scale applications, multi-physics problems, and nonlinear governing equations.

High Dimensionality

Whether using the iterative direct-adjoint looping or the special formulation as a linear-algebra problem, any effort should be taken to reduce the number of degrees of freedom and thus ensure a swift convergence to a solution. The most obvious reduction for linear problems is the use of transforms in the homogeneous directions, which is equivalent to a separation-of-variable approach. If our linear geoverning equations are of constant-coefficient type in one (or more) of the spatial coordinates, we can employ a Fourier transform in this directions and introduce an associated wavenumber. We then have to solve the iterative or linear-algebra system for each wavenumber (or wavenumber tuple) – a far more efficient undertaking than solving the global governing equations. Other techniques to reduce the number of degrees of freedom involve similarity transformations or the exploitation of symmetries.

Efficient Extraction of Linearized and Adjoint Information

The overall efficiency of the iterative optimization scheme (13.11) depends on efficiency in determining all its computational ingredients. In particular, we need a fast and convenient way to determine the key components of the direct-adjoint loop:

$$\left(\frac{\partial \mathsf{N}}{\partial \mathbf{q}}\right)^H, \qquad \left(\frac{\partial j}{\partial \mathbf{q}}\right)^H. \tag{13.31}$$

We wish to determine these expressions directly from the computational program used to solve the direct problem. To this end, we start with the numerical code that time steps the direct problem $d\mathbf{q}/dt = \mathsf{N}(\mathbf{q}, \mathbf{f})$ over a specified interval $[0, \tau]$, given an initial condition or an external driving term. The associated adjoint evolution operator $(\partial \mathsf{N}/\partial \mathbf{q})^H$ is required, but only necessary as a matrix-vector multiplication. Assuming that the (generally nonlinear) direct problem is given as a sequence of linear and nonlinear submodules that are encoded in our numerical program, we seek to extract the necessary linearized and adjoint information using ideas from reverse-mode automatic differentiation (Fosas de Pando et al. 2012).

Let us illustrate the key steps of this process on a simplified example. We choose

$$\frac{d}{dt}\mathbf{q} = \mathsf{N}(\mathbf{q}) = \mathsf{N}_2(\mathsf{D}_x\mathbf{q}, \mathsf{D}_y\mathsf{N}_1(\mathsf{D}_x\mathbf{q})), \tag{13.32}$$

for which we determine the adjoint evolution matrix for the second leg of the direct-adjoint iterative optimization. The functions $\mathsf{N}_1(\mathbf{q})$ and $\mathsf{N}_2(\mathbf{q}, \mathbf{q})$ are assumed nonlinear in their argument(s), but acting locally on the grid points. We next break the composite right-hand side of (13.32) into procedural steps (similar to the modules/subroutines of a computer program), introducing auxiliary variables for each step as needed. We obtain

$$\mathbf{q}_0 = \mathbf{q}, \tag{13.33a}$$

$$\mathbf{q}_1 = \mathsf{N}_1(\mathsf{D}_x\mathbf{q}_0), \tag{13.33b}$$

$$\mathbf{q}_2 = \mathsf{N}_2(\mathsf{D}_x\mathbf{q}_0, \mathsf{D}_y\mathbf{q}_1), \tag{13.33c}$$

$$\frac{d}{dt}\mathbf{q} = \mathbf{q}_2. \tag{13.33d}$$

The traversal from a given flow field $\mathbf{q}$ to its final time rate of change $d\mathbf{q}/dt$ is given by a directed, acyclic graph (DAG) connecting the various modules. For simplicity, we assume that numerical differentiation (indicated by multiplication by D_x or D_y) is a linear operation.

We first linearize the process of evaluating $\mathsf{N}(\mathbf{q})$. With differentiation assumed linear, we only have to linearize the modules N_1 and N_2. This is accomplished by a simple Taylor-series expansion about the linearization state (generally, the base flow denoted by $\bar{\mathbf{q}}$); we have

$$\left.\frac{d\mathsf{N}_1}{d\mathbf{q}}\right|_{\bar{\mathbf{q}}} \mathbf{q} \approx \frac{\mathsf{N}_1(\bar{\mathbf{q}} + \epsilon\mathbf{q}) - \mathsf{N}_1(\bar{\mathbf{q}})}{\epsilon} = \mathsf{A}_1\mathbf{q}, \tag{13.34a}$$

$$\frac{\mathsf{N}_2(\bar{\mathbf{q}} + \epsilon\mathbf{q}, \bar{\mathbf{q}}) - \mathsf{N}_2(\bar{\mathbf{q}}, \bar{\mathbf{q}})}{\epsilon} = \mathsf{A}_{2,0}\mathbf{q}, \tag{13.34b}$$

$$\frac{\mathsf{N}_2(\bar{\mathbf{q}}, \bar{\mathbf{q}} + \epsilon\mathbf{q}) - \mathsf{N}_2(\bar{\mathbf{q}}, \bar{\mathbf{q}})}{\epsilon} = \mathsf{A}_{2,1}\mathbf{q}. \tag{13.34c}$$

We see that the linearization of the two nonlinear modules is accomplished by two function calls. If the linearization point remains constant, it even reduces to one function call. The parameter ϵ in the above expression has to be chosen sufficiently small to ensure adequate accuracy of the linearization, but large enough to avoid effects due to round-off errors.

Replacing the nonlinear procedural steps by their linearized equivalents then yields

$$\mathbf{q}_0 = \mathbf{q}, \tag{13.35a}$$

$$\mathbf{q}_1 = \mathsf{A}_1\mathsf{D}_\mathsf{x}\mathbf{q}_0, \tag{13.35b}$$

$$\mathbf{q}_2 = \mathsf{A}_{2,0}\mathsf{D}_\mathsf{x}\mathbf{q}_0 + \mathsf{A}_{2,1}\mathsf{D}_\mathsf{y}\mathbf{q}_1, \tag{13.35c}$$

$$\frac{d}{dt}\mathbf{q} = \mathbf{q}_2. \tag{13.35d}$$

Combining all steps into one, results in the linearized governing equations. They read

$$\frac{d}{dt}\mathbf{q} = \underbrace{\left(\mathsf{A}_{2,0} + \mathsf{A}_{2,1}\mathsf{D}_\mathsf{y}\mathsf{A}_1\right)\mathsf{D}_\mathsf{x}}_{\frac{d\mathsf{N}}{d\mathbf{q}}}\mathbf{q}. \tag{13.36}$$

The evolution equation for the adjoint variables involves the complex transposition of this equation. We have to form

$$\frac{d}{dt}\mathbf{q}^\dagger = \underbrace{\mathsf{D}_\mathsf{x}{}^H\left(\mathsf{A}_{2,0}^H + \mathsf{A}_1^H\mathsf{D}_\mathsf{y}{}^H\mathsf{A}_{2,1}^H\right)}_{\left(\frac{d\mathsf{N}}{d\mathbf{q}}\right)^H}\mathbf{q}^\dagger, \tag{13.37}$$

or, broken down into procedural steps,

$$\mathbf{q}_0^\dagger = \mathbf{q}^\dagger, \tag{13.38a}$$

$$\mathbf{q}_1^\dagger = \mathsf{A}_{2,1}^H\mathbf{q}_0^\dagger, \tag{13.38b}$$

$$\mathbf{q}_2^\dagger = \mathsf{A}_1^H\mathsf{D}_\mathsf{y}{}^H\mathbf{q}_1^\dagger, \tag{13.38c}$$

$$\mathbf{q}_3^\dagger = \mathsf{A}_{2,0}^H\mathbf{q}_0^\dagger + \mathbf{q}_2^\dagger, \tag{13.38d}$$

$$\frac{d}{dt}\mathbf{q}^\dagger = \mathsf{D}_\mathsf{x}{}^H\mathbf{q}_3^\dagger. \tag{13.38e}$$

Comparing the linearized direct problem to the above adjoint problem, we notice that the involved operators have to be invoked in reverse order and as their conjugate transpose version. The involved operators are, however, readily available, in their original and conjugate transpose form.

The aforementioned computational technique presents an efficient way of extracting linearized and adjoint information required to perform the direct-adjoint optimization (see Figure 13.5).

Since it is not linked to a specific set of equations, but rather to a simulation code, it can easily be extended, without the need for further derivations. The gradient information is numerically accurate to the precision of the chosen numerical scheme, and requires a computational effort similar to the direct problem.

Krylov Time-Stepping

The previous sections have addressed the spatial discretization and the automated extraction of linearized/adjoint information. The temporal evolution of the governing equations also has to be considered carefully for an overall efficient optimization procedure.

We advocate the use of Krylov time-stepping to advance the direct and adjoint governing equations over the chosen time horizon $[0, \tau]$. To this end, we approximate the matrix exponential as the map over finite time interval, by forming a m-dimensional Krylov subspace based on the evolution matrix L according to

$$\mathcal{K}_m = \mathrm{span}\{\mathbf{v}, \mathsf{L}\mathbf{v}, \mathsf{L}^2\mathbf{v}, \ldots, \mathsf{L}^{m-1}\mathbf{v}\}, \tag{13.39}$$

where $\mathbf{v}$ represents a starting vector. We continue by projecting the matrix L onto this subspace to obtain the approximation

$$\mathsf{L} \approx \mathsf{Q}_m \mathsf{H}_m \mathsf{Q}_m^H, \tag{13.40}$$

where Q_m denotes an m-dimensional, orthonormal basis for the subspace $\mathcal{K}_m$. The $m \times m$ upper Hessenberg matrix H_m is a representation of the full matrix L. Any evaluation of any matrix function involving L can then be replaced by a more efficient evaluation using H_m instead. More specifically, for a matrix-vector operation involving the function $g(z)$, we have

$$g(\mathsf{L})\mathbf{b} \approx \mathsf{Q}_m\, g(\mathsf{H}_m)\, \mathsf{Q}_m^H \mathbf{b}, \tag{13.41}$$

requiring a far smaller effort, since the matrix function evaluation is performed on the reduced matrix H_m.

Still, even with these approximations, the repeated computation of the matrix-function-vector product constitutes the most costly part of our time-stepping method. For this reason, we choose a time-stepping scheme that uses a minimum number of Krylov projections for a desired performance (accuracy).

We then use this reduced representation of matrix functions to design an exponential time-stepping scheme and apply it to a nonlinear system of ordinary differential equations of the general form:

$$\frac{d}{dt}\mathbf{q} = \mathsf{N}(\mathbf{q}), \qquad \mathbf{q}(t_0) = \mathbf{q}_0. \tag{13.42}$$

We use a Taylor expansion about $\mathbf{q}_0$ to yield

$$\frac{d}{dt}\mathbf{q} = \mathsf{N}(\mathbf{q}_0) + \nabla\mathsf{N}|_0 (\mathbf{q} - \mathbf{q}_0) + \mathbf{r}(\mathbf{q}), \tag{13.43}$$

with $\mathbf{r}$ as the (nonlinear) remainder term. We apply an integrating factor based on the Jacobian $\nabla\mathsf{N}|_0$, together with a variable change $s = (t - t_0)/\Delta t$, to arrive at

$$\mathbf{q}(t_0 + \Delta t) = \mathbf{q}_0 + \Delta t \, \frac{\exp(\mathsf{L}_0\Delta t) - \mathsf{I}}{\mathsf{L}_0\Delta t} \, \mathsf{N}(\mathbf{q}_0) + \Delta t \int_0^1 \exp(s\mathsf{L}_0\Delta t)\, \mathbf{r}(\mathbf{q}(s)) \, ds, \tag{13.44}$$

where we used the notation $\mathsf{L}_0 = \nabla\mathsf{N}|_0$. This expression is the starting point of any exponential time-stepping method. Various manifestations of exponential time-stepping schemes distinguish themselves by how the integral term is treated and by how the second term on the right-hand side is approximated (Hochbruck & Ostermann 2010, Schulze et al. 2009). The following scheme (referred to as the exponential Rosenbrock-type scheme) blends many of the advantages of exponential time-stepping using Krylov techniques to approximate the various matrix functions arising in the formula. We have

$$\mathbf{k}_1 = \mathbf{q}_0 + \frac{\Delta t}{2}\, \varphi_1\left(\frac{\Delta t}{2}\mathsf{L}_0\right) \mathsf{N}(\mathbf{q}_0), \tag{13.45a}$$

$$\mathbf{k}_2 = \mathbf{q}_0 + \Delta t \, \varphi_1(\Delta t \mathsf{L}_0)\, \mathsf{N}(\mathbf{q}_0) + \Delta t \, \varphi_1(\Delta t\mathsf{L}_0)\, \mathbf{r}(\mathbf{k}_1), \tag{13.45b}$$

$$\begin{aligned}\mathbf{q}(t_0 + \Delta t) = {} & \mathbf{q}_0 + \Delta t \, \varphi_1(\Delta t\mathsf{L}_0)\, \mathsf{N}(\mathbf{q}_0) + \Delta t \, [16\varphi_3(\Delta t\mathsf{L}_0) - 48\varphi_4(\Delta t\mathsf{L}_0)]\, \mathbf{r}(\mathbf{k}_1) \\ & + \Delta t \, [-2\varphi_3(\Delta t\mathsf{L}_0) + 12\varphi_4(\Delta t\mathsf{L}_0)]\, \mathbf{r}(\mathbf{k}_2),\end{aligned} \tag{13.45c}$$

which requires three Krylov projections per time step. In the above time-stepping scheme, we encounter higher exponential functions φ_k that can be derived from the recurrence relation

$$\varphi_k(z) = \frac{1}{k!} + z\varphi_{k+1}(z), \qquad k = 0, 1, \ldots, \tag{13.46}$$

with $\varphi_k(0) = 1/k!$ and $\varphi_0(z) = \exp(z)$. They can be evaluated using the general matrix function expression for a Krylov subspace framework.

Checkpointing

Using the direct-adjoint optimization scheme with a nonlinear governing equation or a general form of the driving term $(\partial j/\partial \mathbf{q})^H$, the adjoint equation (13.11b) depends not only on $\mathbf{q}^\dagger$, but also on $\mathbf{q}$. This link between the direct and adjoint equation leads to additional computational challenges: we have to retain the flow fields $\mathbf{q}$ during the temporal evolution of the direct problem, and inject them, in reverse order, into the coefficients of the adjoint problem (13.11b). For small-size problems, we may have sufficient memory to store all generated flow fields $\mathbf{q}$ during the direct part of the loop. But even more moderately sized problems, we have to resort to a technique known as **checkpointing**.

During the computation of the direct solution, we store the flow fields $\mathbf{q}$ only at a given number of checkpoints $\{t_c\}$. These flow fields will subsequently be used as initial conditions to recompute the necessary flow fields between checkpoints as needed by the adjoint equation. Figure 13.6 demonstrates the two techniques: (i) storing all flow fields, and (ii) storing only at specific checkpoint and recovering the required flow fields by additional direct simulations. In the first case, we store all flow fields at all time steps within the interval $[0, \tau]$. These fields are then injected (in reverse order) into the coefficients of the adjoint equation as well as the driving term of the adjoint equation. No computational overhead arises. In the second case, we assume memory restrictions that allow us to only store a finite number of flow fields in the interval $[0, \tau]$ (in the illustration, seven checkpoints have been chosen). The checkpoints are distributed unevenly, with the final points stored for every time step. These last, densely stored fields are immediately used in the coefficients of the adjoint equation during the beginning adjoint loop. Once these stored fields have been used up, we recompute the direct flow fields starting from the next earliest checkpoint. Once these fields have been restored, we continue with the integration of the adjoint equation. This direct-recovery adjoint-integration procedure continues until we reach the initial time $t = 0$. It is obvious that we trade efficiency in using the available memory with inefficiency in simulation time, as most direct fields are computed twice: once during the direct loop and another time during the recovery part of the adjoint loop. The extra work is hence the cost of an additional direct loop. The distribution of the checkpoints is critical for the overall efficiency; libraries dealing with the storage and recovery management are readily available (Griewank & Walther 2000, Wang et al. 2009).

Miscellaneous

The previous sections outlined the most common computational tools to speed up the optimization by direct-adjoint looping. More recently, additional techniques have arisen and are currently validated and implemented.

For large-scale systems and moderate-to long-time horizons, a parallelization technique in time may be advantageous. These techniques, which break the full-time interval $[0, \tau]$ into p disjoint subintervals on which the governing equations are solved in parallel and subsequently adjusted and merged into a global solution, have been developed over the past five decades and have seen many applications in the computational sciences. Parallel-in-time techniques are particularly attractive for the solution of the adjoint equations, as the adjoint equation is linear by design, and can thus be solved efficiently by splitting the homogeneous and inhomogeneous part of the solution and treating them parallel in time.

Model reduction techniques to speed up the direct and adjoint loop of the iterative optimization scheme are also being developed for large-scale applications. These techniques provide approximate gradient and sensitivity information based on a reduced description of the full dynamics.

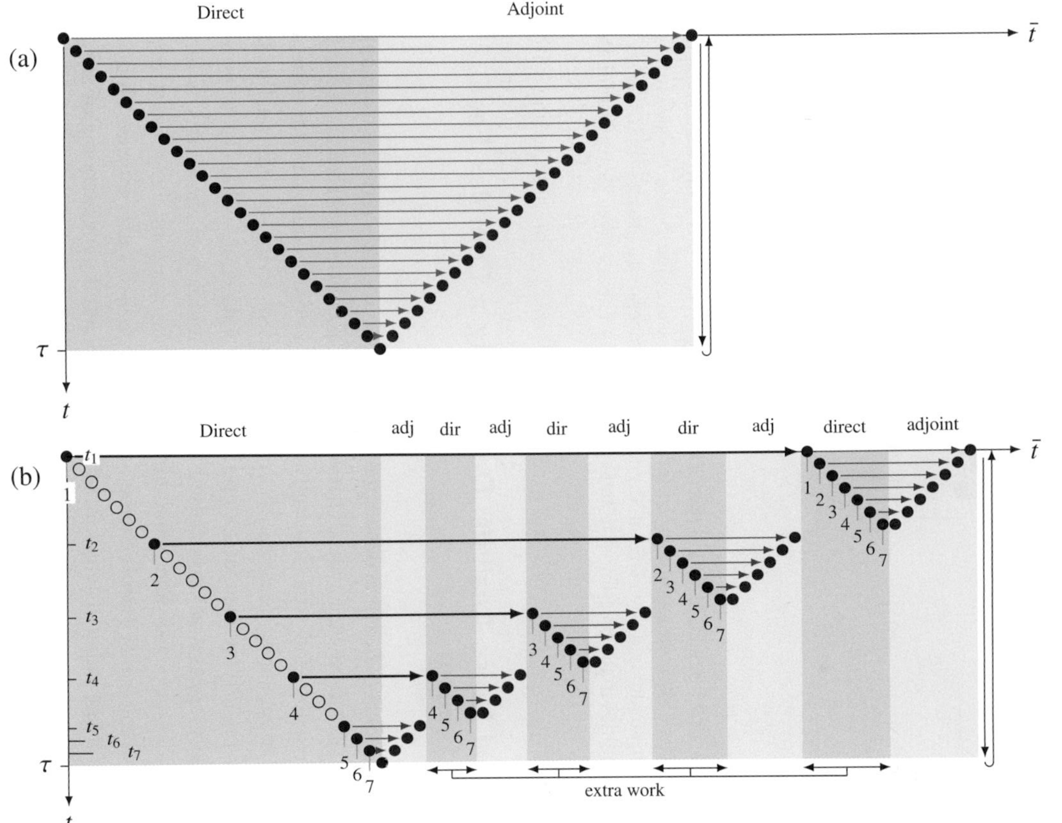

Figure 13.6 Schematic of checkpointing. (a) Checkpoints at all 25 time steps of the direct problem are stored and used as variable coefficients in the adjoint part of the iteration. (b) Checkpoint at 7 time steps (filled blue circles) are stored during the direct solution; they are used as starting points (initial conditions) for launching direct simulations to restore direct solutions for the entire interval $[0, \tau]$. We note that exactly seven checkpoints are stored at all stages of the process.

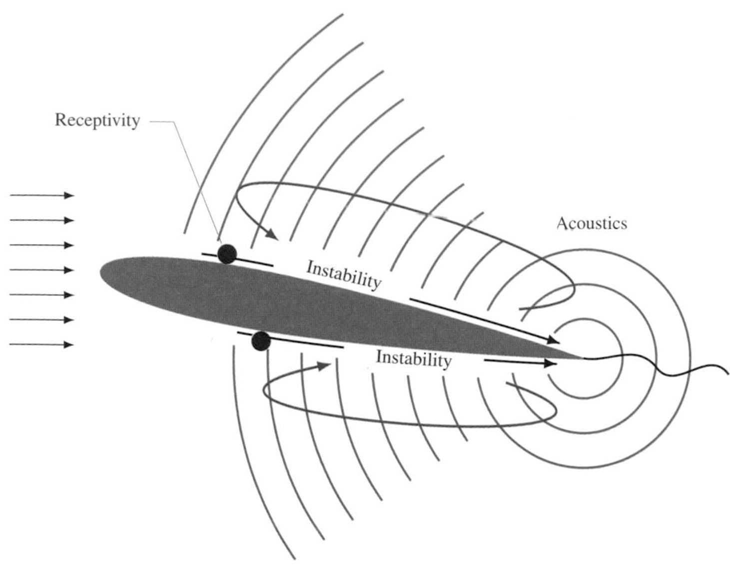

Figure 13.7 Sketch of tonal noise. Flow around a NACA-0012 airfoil at an angle of attack develops a shear instability on the suction and pressure side. Noise radiates from the trailing edge upstream and downstream, triggering via a receptivity process (indicated by red symbols) instabilities in the pressure-side and suction-side boundary layer, hence closing the feedback loop.

13.6 Tonal Noise Analysis: Quantifying a Complex Feedback Mechanism

Tonal noise is an aeroacoustic process by which sound is radiated from a moving body and maintained via a hydrodynamic-acoustic feedback mechanism. It has been studied theoretically and experimentally over many years, and recently direct numerical simulations have joined earlier investigations in trying to uncover the key elements of the underlying feedback loop (Fosas de Pando et al. 2017). Tonal noise on airfoils typically appears at chord-based Reynolds numbers of $Re \sim 10^5$, with applications to glider airplanes and wind turbines. Tonal noise in these latter applications is undesirable, and control schemes – active or passive – are being developed to suppress or eliminate radiated sound. In order to do so, a full understanding of the global tonal-noise instability is required after which we can attempt to break the feedback loop at the weakest link.

The optimization framework, outlined in this chapter, is particularly suited to analyze the key elements of the tonal noise instability; the same framework can also be used, in a second step, to design control schemes or to suggest geometric modifications to weaken or suppress the undesired tonal effects.

Our understanding of tonal noise is based on a sequence of physical processes as follows. The boundary layers on the suction and pressure side of the airfoil support instabilities on either side. Convected downstream toward the trailing edge, the

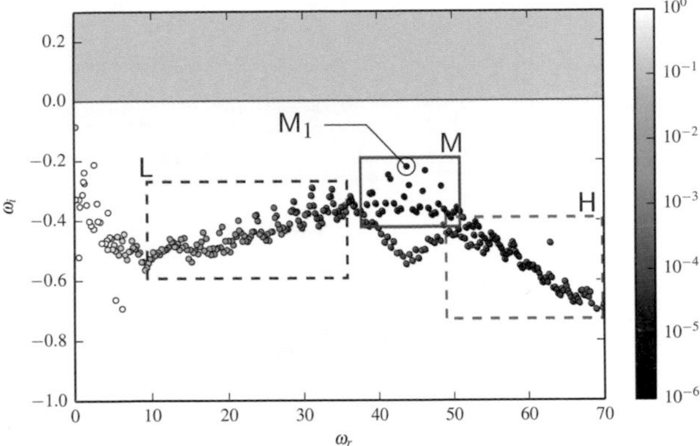

Figure 13.8 Global spectrum of the tonal noise problem in the complex frequency plane. The eigenvalues are colored according to the size of the relative residual with respect to the linearized operator. The spectrum is divided into least stable modes (labeled M), low-frequency modes (labeled L) and high-frequency modes (labeled H).

instability structures collide and generate a noticeable sound wave. This wave travels radially outward and reinitiates the instabilities in the pressure-side and suction-side boundary layers through a receptivity process. The feedback loop is thus closed. Figure 13.7 presents a sketch of this scenario. Two feedback loops – one based on the pressure-side boundary layer, one based on the suction-side boundary layer – linked at the trailing edge characterize the overall tonal noise process. It seems obvious that a local stability analysis, based on frozen boundary-layer velocity profile, cannot account for the complexity of the hydrodynamic-acoustic loop and cannot quantitatively capture the main features. Instead, a global perspective is necessary, and a DNS-based analysis beyond local instabilities should be attempted.

We consider a chord-based Reynolds number of $Re_c = 200{,}000$, a Mach number of $Ma = 0.4$, and describe the flow by the two-dimensional compressible Navier–Stokes equation. The airfoil has a NACA-0012 profile and a 2° angle of attack to the free-stream flow direction. The specific heat γ and Prandtl number Pr are taken as constant, with values 1.4 and 0.71, respectively. At these parameter values, the mean flow separates and reattaches on both sides.

We can gain insight into the tonal noise problem by first looking at the global spectrum of the governing equations, linearized about the mean state (Fosas de Pando et al. 2014). This is accomplished using an Arnoldi algorithm coupled to a linearized simulation code for the physical problem; alternatively, a dynamic mode decomposition (DMD) could have been used. As outlined earlier, we are less interested in the modal solutions that represent the time-asymptotic structures (possibly) encountered as $\tau \to \infty$, we compute the global spectrum (see Figure 13.8) to provide a first picture of preferred frequencies supported by the linear dynamics.

The global spectrum consists of various branches, showing equispaced, discrete eigenvalues. We classify these branches according to their frequency range or the

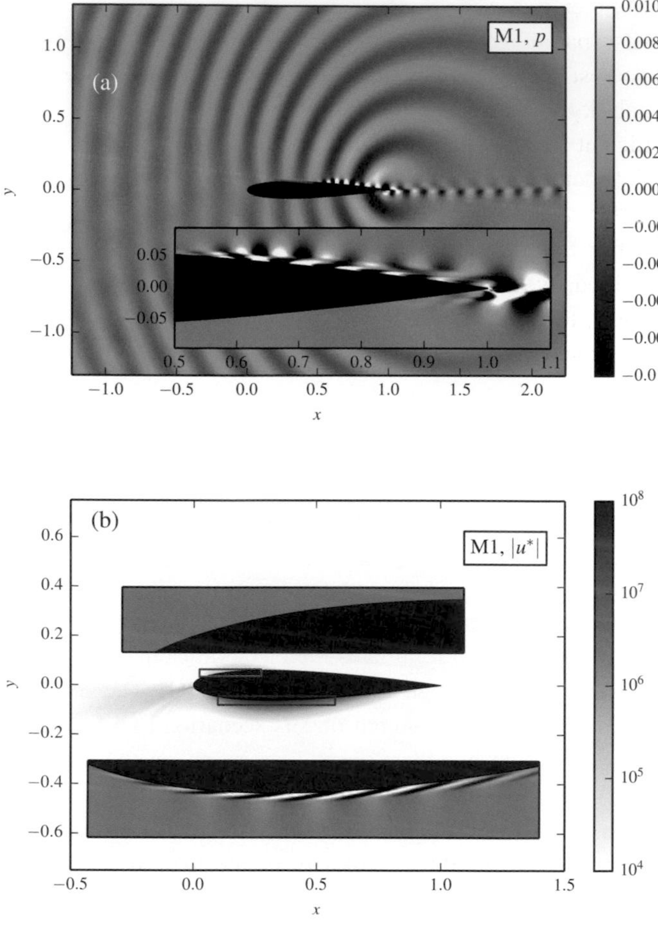

Figure 13.9 (a) Spatial structure of the global mode labeled M_1, visualized by the real part of the associated near-field pressure levels, and the real part of the streamwise velocity levels in the vicinity of the aerofoil surface (inset). The mode has been normalized by the maximum value of the velocity field in the near wake $1 < x < 1.2$. (b) Spatial structure of the associated adjoint global mode (labeled M_1 in Figure 13.8), visualized by the magnitude of the streamwise velocity levels, and the real part in the insets.

structure of the corresponding modes. They are outlined in Figure 13.8. Inspecting the modes across the full frequency range reveals that only the modes labeled M, which are also among the least damped modes, exhibit significant acoustic activity. A similar analysis shows that the low-frequency modes (designated L in the figure) represent the separation bubble dynamics and the reattachment dynamics. The high-frequency modes (marked by H in the figure) describe the Kelvin–Helmholtz-type shear layer instabilities on the suction side. The modes labeled H or L do not show a large acoustic component, and thus shall be ignored in our analysis of the tonal noise problem. Instead, we concentrate on the M-modes.

We then focus on the least stable mode, labeled as M_1 in the figure. A visualization of its spatial structure (see Figure 13.9(a)) demonstrates all components of the hypothesized feedback loop: the boundary layer on the suction side clearly shows an instability, and the collision of instabilities from the suction and pressure side creates sufficient shear to trigger an acoustic wave that emanates from the trailing edge and radiates omnidirectionally. Other modes from the M-branch show a similar structure, at different frequencies, but with only minor differences in their spatial composition.

After the dominant structure, with a significant acoustic component, has been identified, we need to find its structural sensitivity from the adjoint structure. In essence, this undertaking asks the question of how to trigger or influence the M_1-mode in an optimal manner. The answer to this question should point at regions of the flow where a minimal effort (input) is necessary to stimulate the occurrence of tonal noise. The flow field adjoint to the M_1-structure will give this information, since it is directly proportional to the amplitude of a general perturbation as it projects onto M_1. In other words, the adjoint mode associated with M_1 will identify the location, shape, and state-vector components where the feedback loop that sustains tonal noise (based on M_1) can most easily be broken. Figure 13.9(b) shows the adjoint M_1-mode; it has been visualized by the magnitude of the adjoint streamwise velocity. A localized support of this structure is found on the pressure side of the airfoil, between roughly 20% and 50% of chord. We can argue that the tonal noise structure associated with M_1 has its "origin" at this location, and it is at this location where tonal noise can most easily be manipulated. Early experiments and in more recent numerical simulations have proposed this position, although without an associated analysis.

The above analysis shows that complex stability/receptivity issues require mathematical and computational tools that transcend local analyses and time-asymptotic assumptions. An optimization-based framework using direct as well as adjoint information about the flow is sufficiently flexible and effective (once implemented in an efficient manner) to answer many of the questions of flow analysis and to provide a more complete and satisfactory picture of fluid flow behavior.

13.7 Further Extensions

We have demonstrated the optimization framework based on the direct governing equations and the adjoint equations (carrying sensitivity information) on a physically complex, but still comparatively simple, application: the aeroacoustic phenomenon of tonal noise. It is easy to recognize that by studying this example, we have only scratched the surface of what can be accomplished by this type of analysis. Many generalizations are within the range of an optimization-based computational framework; many of them have been realized over the past years. Minimal-seed computations for transition to turbulence, nonlinear harmonic analysis of driven fluid systems, the exploration of regularizing constraints on the optimization are but a few extensions that will bring new ideas to the quantitative analysis of fluid flows.

13.8 Summary and Conclusions

Hydrodynamic stability theory has played a central role in the development of theoretical fluid mechanics. It has motivated the development of a great many mathematical and computational tools to probe the behavior of perturbations about a steady or quasi-steady flow state and has formed the foundation for subsequent research fields, such as flow control. Starting in the beginning to determine the large-time stability of generic configurations, it has matured into a sophisticated set of tools applicable to large-scale systems and multiscale simulations. During this process, many fundamental assumptions and limitations of early stability theory have been (and will continue to be) challenged and abandoned. They will be replaced by a general and flexible numerical platform based on complex optimization, fast solvers, and parallel algorithms. In this manner and form, modern hydrodynamic stability theory will assert its central position in the quantitative analysis of complex fluid systems.

Part V

Applications

14 Machine Learning for Reduced-Order Modeling

B. R. Noack, D. Fernex, and R. Semaan

We describe data-driven reduced-order modeling (ROM) approaches using powerful methods of machine learning. The focus is gray-box models (GBMs) distilling coherent structure dynamics from snapshot data. We highlight two generally applicable methods: the proper orthogonal decomposition (POD) Galerkin method and cluster-based models. The Galerkin method has deep roots in the mathematical investigation of the Navier–Stokes equations and comes with illuminating physical insights. Yet, the accurate data-driven variant requires a rich set of tuning tools and tends to have a narrow range of validity. As an alternative, the recently developed cluster-based network model is inherently robust, can easily integrate numerous operating conditions, and its construction can fully be automated. However, we lose connection to first principles. Of course, there are many different shades of gray, which will briefly be reviewed. The chapter concludes with a tutorial of `xROM`, a freely available software for POD and cluster modeling for different type grids and a large volume of data.

14.1 Introduction

ROM has a long tradition in fluid mechanics. The focus of this chapter is *gray-box modeling*,[1] which resolves the coherent structure flow dynamics. Over 500 years ago, Leonardo da Vinci made the first paintings of wake vortices. von Helmholtz (1858) laid the foundation of their dynamic modeling with his famous vortex laws. In the years that followed, many vortex models were proposed. Examples are a pair of two equal vortices rotating about their axis, a pair of two equal but opposite vortices moving uniformly, leapfrogging ring vortices, the Föppl (1913) vortex model of the near wake, the von Kármán (1911) vortex model for vortex shedding, a shear-layer vortex model (Hama 1962), the recirculation zone model (Suh 1993), the vortex-model-based feedback control (Noack et al. 2004, Protas 2004), just to name a few. The reduced-order vortex models inspired high-dimensional simulations with vortex blobs (Leonard 1980), and vortex filaments (Ashurst & Meiburg 1988). Vortex models are closely tied to first principles and robustly model coherent structures and their convection.

[1] The term is adopted from Wiener (1948).

Another branch of reduced-order models can be traced back to stability analysis of Orr (1907) and Sommerfeld (1908) and the Galerkin (1915) method. These foundations corroborated the idea that fluid dynamics can be understood as a superposition of global physical modes. Landau (1944) and Hopf (1948) even considered turbulence as originating from a sequence of many Hopf bifurcations. The mean-field theory of Stuart (1958) beautifully explained the nonlinear saturation of instability modes. Subsequently, the Galerkin method was not only applied to instability modes but also to mathematical modes from Hilbert space considerations (Busse 1991, Noack & Eckelmann 1994). As with vortex models, the Galerkin method was increasingly generalized and became a foundation of computational fluid mechanics.

Data-driven reduced-order models have arguably originated from Galerkin methods with a POD expansion (Lorenz 1956, Lumley 1967). The POD of a given snapshot ensemble yields a Galerkin expansion with minimal representation error for given number of modes. The pioneering wall turbulence model of Aubry et al. (1988) demonstrated their use in explaining complicated dynamics. In the last three decades, a myriad of POD models have been proposed. The initial excitement of POD was mitigated by the increasing realization of the fragility of the resulting Galerkin models (Rempfer 2000) and their limited range of validity. Galerkin methods are detailed in Section 14.2.

A recent development of cluster-based reduced-order models (CROM) was pioneered by Burkardt et al. (2006) and Hof et al. (2004). Here, the dynamics are modeled as a stochastic transition between a small number of representative flow states. The k-means clustering of snapshots yields centroids as representative flow states with a minimal representation error. The dynamics can be framed in a Markov (Kaiser et al. 2014) and network model (Li et al. 2021). In Section 14.3, this branch of ROM is discussed including the latest innovations (Fernex et al. 2019, 2021).

Of course, there are different shades of gray with GBMs, which are outlined in Section 14.4. The chapter concludes with a tutorial of POD modeling (Section 14.5) with freely available software by the authors.

14.2 Proper Orthogonal Decomposition Galerkin Model

This section gives recipes and suggestions for the construction of POD Galerkin models. First, the considered configurations are specified in Section 14.2.1. Then, Sections 14.2.2 and 14.2.3 describe the kinematical and dynamical steps to a POD model. Sections 14.2.4 and 14.2.5 provide closures for an existing POD model and identification methods for experiment. Section 14.2.6 reviews common applications.

14.2.1 Flow Configuration

We assume an incompressible viscous flow in a steady domain Ω. A location in this domain is denoted by $\boldsymbol{x} = (x, y, z) = (x_1, x_2, x_3)$. The time is represented by t. The velocity and pressure fields read $\boldsymbol{u} = (u, v, w) = (u_1, u_2, u_3)$ and p, respectively. The Newtonian fluid is described by the density ρ and dynamic viscosity μ.

Let D and U be characteristic size and velocity scales of the configuration. The flow properties are characterized by the Reynolds number $Re = \rho U D/\mu$, or, equivalently by its reciprocal $\nu = 1/Re$. In the sequel, we assume that all quantities are nondimensionalized with D, U, ρ, and μ.

The continuity and Navier–Stokes equations read

$$\nabla \cdot \boldsymbol{u} = 0, \tag{14.1}$$

$$\partial_t \boldsymbol{u} + \nabla \cdot (\boldsymbol{u} \otimes \boldsymbol{u}) = -\nabla p + \nu \triangle \boldsymbol{u}. \tag{14.2}$$

The Dirichlet boundary conditions at stationary walls $\boldsymbol{x} \in \partial\Omega$ are

$$\boldsymbol{u}(\boldsymbol{x}, t) = \boldsymbol{0}. \tag{14.3}$$

In addition, a free-stream, stress-free, or von Neumann condition may be used for open flows.

The initial condition $\boldsymbol{u}(\boldsymbol{x}, 0)$ at time $t = 0$ may be a perturbed steady solution $\boldsymbol{u}_s(\boldsymbol{x})$ as in Chapter 1 of this book.

In what follows, we assume to have gathered M snapshots $\boldsymbol{u}^m$, $m = 1, \ldots, M$ from a post-transient flow solution. The snapshots are requested to be statistically representative for the computation of first and second moments.

14.2.2 Proper Orthogonal Decomposition

POD is a data-driven realization of a Galerkin approximation

$$\hat{\boldsymbol{u}}(\boldsymbol{x}, t) = \boldsymbol{u}_0(\boldsymbol{x}) + \sum_{i=1}^{N} a_i(t)\, \boldsymbol{u}_i(\boldsymbol{x}) \tag{14.4}$$

of a velocity field $\boldsymbol{u}(\boldsymbol{x}, t)$, as detailed in Chapter 6. Here, $\boldsymbol{u}_0$ is the basic mode, for example, the steady Navier–Stokes solution or the mean flow, and $\boldsymbol{u}_i$ are expansion modes with amplitudes a_i. These approximations rely on a square-integrable Hilbert space $\mathcal{L}^2(\Omega)$ equipped with an inner product between two elements $\boldsymbol{v}$ and $\boldsymbol{w}$,

$$(\boldsymbol{v}, \boldsymbol{w})_\Omega := \int_\Omega d\boldsymbol{x}\ \boldsymbol{v} \cdot \boldsymbol{w}. \tag{14.5}$$

The corresponding norm reads

$$\|\boldsymbol{u}\|_\Omega := \sqrt{(\boldsymbol{v}, \boldsymbol{w})_\Omega}. \tag{14.6}$$

POD modes are an orthonormal set of expansion modes that minimize the in-sample representation error

$$E = \frac{1}{M} \sum_{m=1}^{M} \|\boldsymbol{u}^m - \hat{\boldsymbol{u}}^m\|_\Omega^2, \tag{14.7}$$

with respect to the snapshot ensemble.

The computation of the POD modes is performed in the following steps:

Computation of the basic mode: The basic mode is a time-averaged flow and absorbs a potential inhomogeneity of the steady boundary condition,

$$\boldsymbol{u}_0(\boldsymbol{x}) := \frac{1}{M} \sum_{m=1}^{M} \boldsymbol{u}^m(\boldsymbol{x}). \tag{14.8}$$

One example is the uniform free-stream condition $\boldsymbol{u} = U\hat{\boldsymbol{e}}_x$ in the far-field, where $\hat{\boldsymbol{e}}_x$ is the unit vector in streamwise direction. The remaining fluctuation $\boldsymbol{u}' = \boldsymbol{u} - \boldsymbol{u}_0$ satisfies homogenized boundary conditions, for example, $\boldsymbol{u}' = 0$ in case of the uniform free-stream condition. This implies that any expansion mode $\boldsymbol{u}_i$ will satisfy these homogenized boundary conditions. In other words, the expansion (14.4) satisfies the full boundary conditions for arbitrary choices of amplitudes a_i, $i = 1, \ldots, N$.

Computation of the correlation matrix: The $M \times M$ correlation matrix $\boldsymbol{R}$ has the elements

$$\boldsymbol{R}_{mn} := \frac{1}{M} (\boldsymbol{u}^m - \boldsymbol{u}_0, \boldsymbol{u}^n - \boldsymbol{u}_0)_\Omega, \quad m, n = 1, \ldots, M. \tag{14.9}$$

Spectral analysis of the correlation matrix: This Gramian matrix $\boldsymbol{R}$ is symmetric and positive semi-definite. This implies real and nonnegative eigenvalues as well as orthogonal eigenvectors. Let

$$\boldsymbol{\alpha}_i := \begin{bmatrix} \alpha_i^1 \\ \vdots \\ \alpha_i^M \end{bmatrix}$$

be the ith eigenvector of the correlation matrix, that is,

$$\boldsymbol{R}\, \boldsymbol{\alpha}_i = \lambda_i\, \boldsymbol{\alpha}_i, \quad i = 1, \ldots, M. \tag{14.10}$$

Without loss of generality, we assume the real eigenvalues to be sorted in decreasing order

$$\lambda_1 \geq \lambda_2 \geq \cdots \geq \lambda_M = 0,$$

and that the eigenvectors are orthonormal

$$\boldsymbol{\alpha}_i \cdot \boldsymbol{\alpha}_j = \sum_{m=1}^{M} \alpha_i^m \alpha_j^m = \delta_{ij}, \quad i, j = 1, \ldots, M.$$

Note that the Mth eigenvalue must vanish, because M snapshots span at most an $M - 1$-dimensional hyperplane.

Computation of the POD modes: The POD modes are given by

$$\boldsymbol{u}_i(\boldsymbol{x}) := \frac{1}{\sqrt{M\,\lambda_i}} \sum_{m=1}^{M} \alpha_i^m\, (\boldsymbol{u}^m(\boldsymbol{x}) - \boldsymbol{u}_0(\boldsymbol{x})), \quad i = 1, \ldots, N. \tag{14.11}$$

Here, $N \leq M - 1$ is the number of expansion modes in (14.4). The POD modes are orthonormal by construction:

$$(\boldsymbol{u}_i, \boldsymbol{u}_j)_\Omega = \delta_{ij}, \quad i, j = 1, \ldots, N.$$

Computation of the amplitudes: The mode amplitudes scale – not accidentally – with the corresponding eigenmodes of the correlation matrix

$$a_i^m = \sqrt{\frac{\lambda_i}{M}} \alpha_i^m,$$

and satisfy

$$\overline{a_i} = \frac{1}{M} \sum_{m=1}^{M} a_i^m = 0, \quad i = 1, \ldots, N, \tag{14.12a}$$

$$\overline{a_i\, a_j} = \frac{1}{M} \sum_{m=1}^{M} a_i^m a_j^m = \lambda_i\, \delta_{ij}, \quad i, j = 1, \ldots, N. \tag{14.12b}$$

14.2.3 Galerkin Method

The projection of the Navier–Stokes equations (14.2) on the expansion (14.4) is performed in two steps. First, the inner product between the Navier–Stokes equations and the mode $\boldsymbol{u}_i$ is taken. Second, the resulting equations are integrated over the domain Ω. This Galerkin projection yields

$$\frac{da_i}{dt} = \nu \sum_{j=0}^{N} l_{ij}^{\nu}\, a_j + \sum_{j,k=0}^{N} q_{ijk}^{c}\, a_j\, a_k \quad i = 1, \ldots, N, \tag{14.13}$$

with the Galerkin system coefficients

$$l_{ij}^{\nu} = \left(\boldsymbol{u}_i, \triangle \boldsymbol{u}_j\right)_\Omega, \tag{14.14a}$$

$$q_{ijk}^{c} = \left(\boldsymbol{u}_i, \nabla \cdot \left(\boldsymbol{u}_j \otimes \boldsymbol{u}_k\right)\right)_\Omega. \tag{14.14b}$$

For analytical convenience, $a_0 = 1$ following Rempfer and Fasel (1994b). The pressure term vanishes for many boundary conditions. Otherwise, it leads to an additional quadratic term (Noack et al. 2005). Again, the reader is asked to refer to the original literature for the details.

Even for a 100% accurate Galerkin approximation with vanishing representation error, there is no guarantee that the stability properties of the Navier–Stokes equations are conserved in the Galerkin system. A stable Navier–Stokes solution may become an unstable Galerkin solution (Noack et al. 2003, Rempfer 2000). In fact, Galerkin systems are often observed to be fragile with unphysically long transient times or even finite-time divergence to infinity. Schlegel and Noack (2015) provide necessary and sufficient conditions for bounded solutions only based on the Galerkin system coefficients. Section 14.2.4 suggests cures.

14.2.4 Closures and Stabilizers

For turbulent flows, the representation error of the POD expansion may well be 50% or larger. This large neglected fluctuation needs to be accounted for in the Galerkin

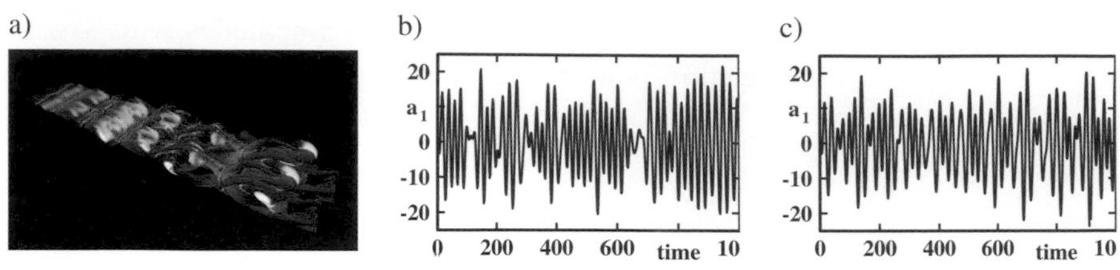

Figure 14.1 POD Galerkin model for the 3D mixing layer. For details see Noack et al. (2004). (a) Snapshot; (b) a_1 of LES; (c) a_1 of POD model.

system. This residual has two effects: a high-frequency noise excitation and an energy dissipation. Typically, only the energy dissipation is accounted for.

Mean-flow model: Aubry et al. (1988) account for the stabilizing coupling between fluctuations and mean-flow by replacing the constant basic mode with variable one computed from the Reynolds equation. The shift mode (Noack et al. 2003) has the same purpose for an oscillatory flow.

Global eddy viscosity: Aubry et al. (1988) employs a Boussinesq ansatz by replacing the kinematic viscosity ν with an effective one $\nu_{\text{eff}} = \nu + \nu_T$, where the global eddy viscosity ν_T is a tuning parameter. This ansatz will be dissipative but the Galerkin model inaccurately implies that a high-Reynolds number flow behaves like the laminar solution at a low-Reynolds number.

Modal eddy viscosities: Rempfer and Fasel (1994a) has refined this Boussinesq ansatz by a more realistic mode-dependent eddy viscosity – inspired by spectral turbulence theory.

Nonlinear eddy viscosities: Global and modal eddy viscosities imply that a nonlinear dynamics can be approximated by a linear Galerkin system term. Östh et al. (2014) have refined this ansatz by a fluctuation-dependent scaling. The scaling has been justified with a finite-time thermodynamics closure (Noack et al. 2008) and can be shown to guarantee boundedness (Cordier et al. 2013).

We keep the list of closures this short and simple. Figure 14.1 previews results of a modal eddy viscosity closure for a turbulent mixing layer.

14.2.5 Model Identification

A Galerkin projection is also possible on subdomains. However, the Navier–Stokes equations residual needs to be evaluated (Rempfer 1995). 2D particle image velocimetry (PIV) field measurements of 3D flows do not provide the plane normal velocity components, let alone their normal derivatives. In some cases, for example, non-Newtonian fluids, the evolution equation may not be known. These are motivations for model identification. We keep the ansatz of a constant-linear-quadratic dynamics,

$$\frac{da_i}{dt} = f_i(\boldsymbol{a})c_i + \sum_{j=1}^{N} l_{ij}\, a_j + \sum_{j=1}^{N}\sum_{k=j}^{N} q_{ijk}\, a_j\, a_k, \tag{14.15}$$

with the state $\boldsymbol{a} := [a_1, \ldots, a_N]^{\mathrm{T}}$.

A classical calibration method aims to minimize the state difference between the computed/measured POD mode amplitudes $\boldsymbol{a}^{\bullet}(t) = \left[a_1^{\bullet}(t), \ldots, a_N^{\bullet}(t)\right]^{\mathrm{T}}$ and the modeled one $\boldsymbol{a}^{\circ}(t)$,

$$E_1 := \frac{1}{t_1 - t_0} \int_{t_0}^{t_1} dt\ \|\boldsymbol{a}^{\bullet} - \boldsymbol{a}^{\circ}\|^2 = \min. \tag{14.16}$$

This formulation requires the integration of the model in the time interval $[t_0, t_1]$. The 4D Var method is a powerful method for minimizing the model error (Semaan et al. 2015). Yet, the challenge is that a short interval may not provide enough information about the dynamics while a long interval leads to inevitable "phase drifts" between the full plant and the model. If the full plant and the model are "out of sync," an overly dissipative dynamics with $\boldsymbol{a}^{\circ} \equiv 0$ will become the best model.

An alternative optimization avoids this phase-drift problem. Now, the dynamics residual is minimized along the full-plant trajectory $t \mapsto \boldsymbol{a}^{\bullet}$,

$$E_2 := \frac{1}{t_1 - t_0} \int_{t_0}^{t_1} dt\ \left\| \frac{d\boldsymbol{a}^{\bullet}}{dt} - \boldsymbol{f}\left(\boldsymbol{a}^{\bullet}\right) \right\|^2 = \min. \tag{14.17}$$

Here, a long time integral helps in the model accuracy as the whole attractor is incorporated in the calibration. This optimization requires the temporal state derivative and leads to an analytically solvable least-means-square problem. The challenge is that a long-term accumulation of errors, like amplitude errors, can easily occur, because they are weakly penalized by the error (14.17).

A large arsenal of methods have been developed for improving long-term asymptotics, for mitigating the effect of insufficient data (Cordier et al. 2010), and for sparsifying the dynamics, that is, make it more human interpretable (Brunton et al. 2016a). The reader is encouraged to visit the literature if needed.

14.2.6 Applications

Reduced-order models have many classical applications:

Understanding: A sufficiently low-dimensional sparse ROM may help to understand the dynamics by distilling the human-interpretable modes and the key linear and nonlinear terms.

Analysis: The Galerkin method provides powerful analytical frameworks for the energy-flow analysis and for explanation of aerodynamic forces and torques (Noack et al. 2011).

Estimation: A dynamic ROM may be the basis of a dynamic observer, that is, may help to predict flow states based on one or few sensor signals.

Prediction: A dynamic ROM may help to predict future states. One example is an undesirable flutter in an airplane experiment and an early shutdown of the wind-tunnel. A second example is a weather forecast.

Exploration: A ROM may help to explore unobserved behavior for new initial conditions or new parameters. Given the narrow range of validity of data-driven ROM, this new behavior needs to be validated in the full plant.

Control: ROM may help in the control design of first- and second-order dynamics.

Closures: ROM may guide closure terms for unsteady dynamics in the spirit of Liu (1989).

Response model: ROM may also provide a mapping from actuation or configuration parameters to the time-averaged flow (Chapter 16) or the performance (Albers et al. 2020).

14.3 Cluster-Based Reduced-Order Modeling

CROM can be considered as an antidote to POD Galerkin models. POD expansions are replaced by discrete representative flow states, centroids. And the analytically elegant POD Galerkin system is sacrificed for a dynamic transition model between the centroids, loosing the possibility to perform stability analyses, compute energy flows, and so on. The big advantage of CROM is the possibility to automate the complete model identification from snapshot data and to arrive at an inherently robust dynamical model. Unlike POD models, CROM do not require stabilizing auxiliary models, such as shift modes, calibrated cubic terms, or tuned eddy viscosity parameters. In contrast to POD, CROM requires *time-resolving* snapshots $\boldsymbol{u}^m(\boldsymbol{x}) = \boldsymbol{u}^m(\boldsymbol{x}, t^m)$, $t^m = m\Delta t$ with fixed time step Δt. In the following, we outline the kinematic compression with clustering (Section 14.3.1) and present two CROM, one focusing on statistics (Section 14.3.2) and one better resolving dynamics (Section 14.3.3). Section 14.3.4 outlines generalizations.

14.3.1 Clustering

Following Chapter 1, we employ clustering as an autoencoder. Let $\boldsymbol{c}_k(\boldsymbol{x})$, $k = 1, \ldots, K$ be the centroids, that is, representative flow states. For a given flow field $\boldsymbol{u}$, the encoder provides the index of the closest centroid

$$k(\boldsymbol{u}) = \arg \min_{i=1,\ldots,K} \|\boldsymbol{u} - \boldsymbol{c}_i\|_\Omega. \tag{14.18}$$

The decoder $\boldsymbol{H}$ returns the corresponding centroid

$$\hat{\boldsymbol{u}} = \boldsymbol{H}(\boldsymbol{u}) = \boldsymbol{c}_{k(\boldsymbol{u})}. \tag{14.19}$$

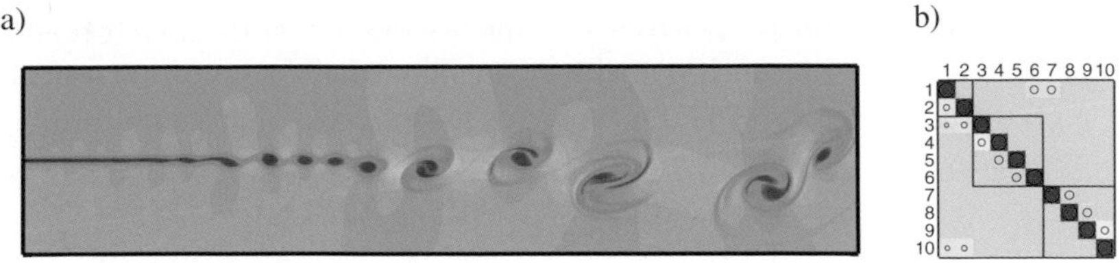

Figure 14.2 Cluster-based Markov model for the mixing layer. For the details, see Kaiser et al. (2014). (a) Vorticity snapshot; (b) Markov matrix $\boldsymbol{P}$.

In analogy to POD, the set of K centroids are chosen to minimize the in-sample representation error

$$E = \frac{1}{M} \sum_{m=1}^{M} \|\hat{\boldsymbol{u}}^m - \boldsymbol{u}^m\|_\Omega^2, \tag{14.20}$$

with respect to the snapshot ensemble. The number of centroids K is a design parameter. Burkov (2019) provides a useful criterion based on an out-of-sample error. Li et al. (2021) propose an alternative criterion based on the dynamic prediction error.

14.3.2 Cluster-Based Markov Models

The dynamics may be stroboscopically monitored at the sampling time step Δt. Let n_i be the number of snapshots in cluster i and n_{ij} be the number of observed transitions from cluster j to cluster i in the next time step. Then, $P_{ij} = n_{ij}/n_i$ estimates the probability to go from cluster j to cluster i. Let $\boldsymbol{p}^m = [p_1, \ldots, p_K]^{\mathrm{T}}$ comprise the probabilities to be in any of the clusters at time t^m. Then, the *Markov model* advances this probability vector according to

$$\boldsymbol{p}^{m+1} = \boldsymbol{P}\, \boldsymbol{p}^m, \quad \boldsymbol{P} = (P_{ij}). \tag{14.21}$$

Under reasonably generic conditions and sufficient data, the iteration (14.21) will converge against a fixed point approximating the cluster population of the snapshots $p_i^\infty \approx q_i := n_i/M$. In the sequel, the model will be referred to as a *cluster-based Markov model* (CMM).

Kaiser et al. (2014) provides the mixing layer (see Figure 14.2) and the Ahmed body wake as examples for CMM. The statistics of CMM is very accurate. On the downside, the temporal evolution is too diffusive for a small number of clusters.

14.3.3 Cluster-Based Network Models

The diffusion effect has been mitigated by a recent network model (Li et al. 2021). Let Q_{ij} be the data-inferred probability of a trajectory to go from cluster j directly to cluster i. Let T_{ij} be the corresponding averaged transition time. Both parameters are comprised in $K \times K$ matrices $\boldsymbol{Q} = \left[Q_{ij}\right]$ and $\boldsymbol{T} = \left[T_{ij}\right]$. And, for concreteness, let us assume we start at centroid $k = 1$ at time $t_0 = 0$. Now, the *cluster-based network model*

(CNM) stochastically generates a sequence of times $t_0, t_1, t_2, \ldots$ and cluster affiliations $k_0 = 1, k_1, k_2, \ldots$ consistent with the mentioned direct transition probabilities Q_{ij} and transition times T_{ij}. The trajectory for $t \in [t_m, t_{m+1}]$ moves uniformly from centroid $\boldsymbol{c}_{k_m}$ to $\boldsymbol{c}_{k_{m+1}}$,

$$\boldsymbol{u}^{\circ}(\boldsymbol{x},t) = \frac{t_{m+1} - t}{t_{m+1} - t_m} \boldsymbol{c}_{k_m}(\boldsymbol{x}) + \frac{t - t_m}{t_{m+1} - t_m} \boldsymbol{c}_{k_{m+1}}(\boldsymbol{x}). \tag{14.22}$$

In contrast to CMM, this model has a time-continuous flow representation $\boldsymbol{u}^{\circ}(\boldsymbol{x},t)$, that is, does not jump discretely between centroids. Moreover, the model-based autocorrelation function is found to be much more accurate. We refer to Li et al. (2021) for a detailed discussion of the model and the shear-flow example. The price for this increased dynamic resolution is a slightly larger error in the model-based cluster population.

14.3.4 Generalizations

CROM, in particular CNM, can be significantly generalized. One model may describe many operating conditions, if we enrich the set of centroid correspondingly and include the parameters $\boldsymbol{b}$ of the operating conditions in the model parameters, for example, $\boldsymbol{P}(\boldsymbol{b})$ for the CMM and $\boldsymbol{Q}(\boldsymbol{b})$, $\boldsymbol{T}(\boldsymbol{b})$ for the CNM (Fernex et al. 2021, Li et al. 2021). In addition, the assumed motion from centroid to centroid may be relaxed to a "flight" using the centroids and their transition characteristics as lighthouses. With these generalizations, a CROM has successfully approximated a dragreduction study of wall turbulence with dozens of spanwise traveling surface waves (to be published soon by the authors). Summarizing, CROM is based on a local interpolation between centroids as "collocation points" of the data set, and there is no limit to the structure of the data. The simple nature of clustering makes it a solid foundation for a very general modeling toolkit. The loss of analytical relationships to first principles is rewarded by significant gain in human-interpretable model complexity.

14.4 General Principles: Different Shades of Gray

This section outlines general principles and alternative realizations of data-driven GBMs. The notation is adopted from Wiener (1948) who distinguished between white-box models (WBMs) for the full plant, black-box models (BBMs) for input–output dynamics, and GBMs for low-dimensional state dynamics. Many GBMs follow the diagram (14.23).

$$\begin{array}{lccccc} & \text{WBM} & & \text{GBM} & & \text{state estimate} \\ \text{Kinematics} & \boldsymbol{u}(\boldsymbol{x}) & \longrightarrow & \boldsymbol{a} & \longrightarrow & \hat{\boldsymbol{u}}(\boldsymbol{x}) \\ & \downarrow & & \downarrow & & \\ \text{Dynamics} & \dfrac{\partial \boldsymbol{u}}{\partial t} = \boldsymbol{N}(\boldsymbol{u}) & \longleftarrow & \dfrac{d\boldsymbol{a}}{dt} = \boldsymbol{f}(\boldsymbol{a}) & & \end{array} \tag{14.23}$$

The first row defines an autoencoder, that is, a low-dimensional flow representation with an encoder to and decoder from the low-dimensional feature vector $\boldsymbol{a}$. The dynamics of this feature vector (bottom, middle) approximates the Navier–Stokes equations (bottom, left). The POD and cluster-based models follow this scheme. Also vortex models fit into this approach.

GBM models differ in the chosen autoencoder and the chosen model identification of the dynamics $\boldsymbol{f}(\boldsymbol{a})$. The autoencoder may be mathematical (Fourier modes, Tchebychev polynomials, etc.), physical (stability modes, Stokes modes, etc.), or data-driven (POD, DMD, centroids, etc.). The dynamics may be derived from first principles (Galerkin projection, etc.), or from data (model identification), or a mixture of both (model identification with a Thikonov penalization of the Navier–Stokes-based model, etc.). We shall not pause to declinate the different shades of gray, but refer to the excellent literature on the topic (Fletcher 1984, Holmes et al. 2012).

Instead, we mention two different approaches. *Feature-based manifold models* (FeMM) (Loiseau et al. 2018, 2021) assume a typical experimental situation. A sensor signal $\boldsymbol{s}$ is recorded in time, lifted to a dynamic feature $\boldsymbol{a}$ for which a dynamic model can be identified. The feature $\boldsymbol{a}$ is employed for an estimator of the flow using PIV and simultaneous sensor data. K-nearest neighbor interpolation is a simple yet effective method for state estimation. The following equation outlines the data structure:

$$\begin{array}{lccccc} \text{Kinematics} & \boldsymbol{s} & \longrightarrow & \boldsymbol{a} & \longrightarrow & \hat{\boldsymbol{u}}(\boldsymbol{x}) \\ & & & \downarrow & & \\ \text{Dynamics} & & & \dfrac{d\boldsymbol{a}}{dt} = \boldsymbol{f}(\boldsymbol{a}). & & \end{array} \tag{14.24}$$

Effectively, FeMM is a BBM with a sensor-based estimator of a manifold.

Another approach relies on brute-force data interpolation. Let us assume that the data contains the snapshots $\boldsymbol{u}^m$ and the time derivative $\partial_t \boldsymbol{u}^m$ at the same instants, for example, from double-shot PIV. Now, the flow field $\boldsymbol{u}^\circ$ can be integrated by an Euler scheme,

$$\boldsymbol{u}^\circ(t + \Delta t) = \boldsymbol{u}^\circ(t) + \Delta t \, \partial_t \boldsymbol{u}^\circ(t).$$

In the simplest 1-nearest neighbor realization, the time derivative is taken to be $\partial_t \boldsymbol{u}^m$, where $\boldsymbol{u}^m$ is the nearest neighbor to $\boldsymbol{u}^\circ$. Obvious refinements, like higher-order interpolation and smoothing are possible and probably advised.

We pause the discussion of the spectrum of GBM enabled by machine learning. Evidently, the choice of the GBM is guided by the available data and by the purpose of the model. It pays to start with a simple purely data-driven model before advancing to more refined and targeted ones. Moreover, every mapping can be identified with a rich set of tools – from Taylor's expansion to neural networks.

We conclude with a few words of warning. Data analysis is easy, insightful, and fun. Even with validation procedures, results are guaranteed. Dead data cannot complain anymore. Dynamic models attempt to predict the unknown future, which is easy for attractor dynamics, that is, transitions between similar recurrent states, but is next to impossible for extrapolation into terra incognita. Data-driven control-oriented dynamic models attempt to predict the unknown future with a rich set of possible actuations. The available data for control-oriented modes will typically be sparse, that is, data has to be replaced by good simplifying guesses.

14.5 Tutorial: `xROM`

This tutorial aims to give readers a hands-on experience with the learned Galerkin processes. This includes Galerkin expansion, Galerkin projection, and model calibration.

The tutorial shall solely rely on the numerical package `xROM` (Semaan et al. 2020), which is a freely available tool for ROM. One purpose of `xROM` is to perform a modal decomposition from snapshot data and to carry out the Galerkin projection using the Navier–Stokes equations. Moreover, `xROM`
(i) can analyze native CFD and PIV data;
(ii) can handle a large spectrum of 2D and 3D grids (Cartesian, structured, unstructured, etc.);
(iii) can create modal expansions and other modes (e.g., shift modes);
(iv) is user-friendly through a single configuration file;
(v) is parallelizable.
On the full capabilities of `xROM`, the reader is referred to the user manual, which is included in the package.

For this tutorial, we shall employ direct numerical simulations (DNS) of a cylinder flow at $Re = 100$ following Chapter 1 and duplicating the results presented in Noack et al. (2003).

The tutorial is organized as follows. In Section 14.5.1, we shall get acquainted with the numerical setup and the considered flow. General characteristics of `xROM` and how to use it are introduced in Section 14.5.2. We shall deploy `xROM` to perform three tasks in Section 14.5.3.

14.5.1 Numerical Setup

We consider the 2D cylinder wake at $Re = 100$. The DNS is performed with an FEM of third-order accuracy in space and time. The computational domain Ω_{DNS} is bounded by the rectangle $-5 < x < 15$ and $-5 < y < 5$, excluding the cylinder $\Omega_{\mathrm{cyl}} = \{(x, y) : x^2 + y^2 \leq 1/2\}$. Details can be inferred from Noack et al. (2003). Figures 14.3 and 14.4 provide an impression of the domain, the grid, and the periodic flow. In the ROM analysis, the observation domain $\Omega = \Omega_{\mathrm{DNS}}$ is identical to the computational domain.

14.5.2 General Characteristics of `xROM`

`xROM` capabilities

`xROM` is mainly designed to generate POD Galerkin models for a wide range of flow problems. To accomplish this, it has several additional preprocessing tools and options

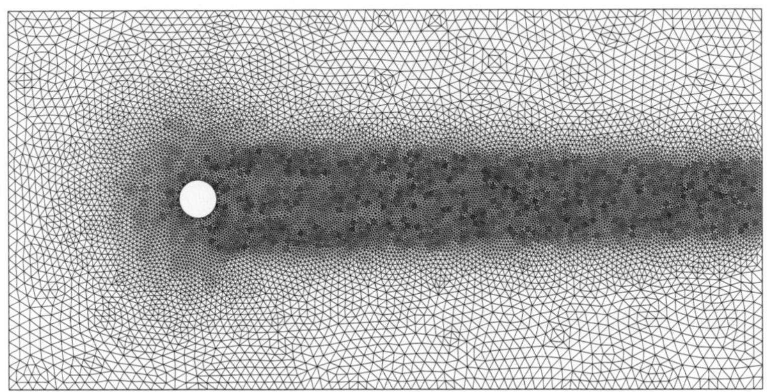

Figure 14.3 Unstructured grid for the FEM direct numerical simulation of cylinder wake.

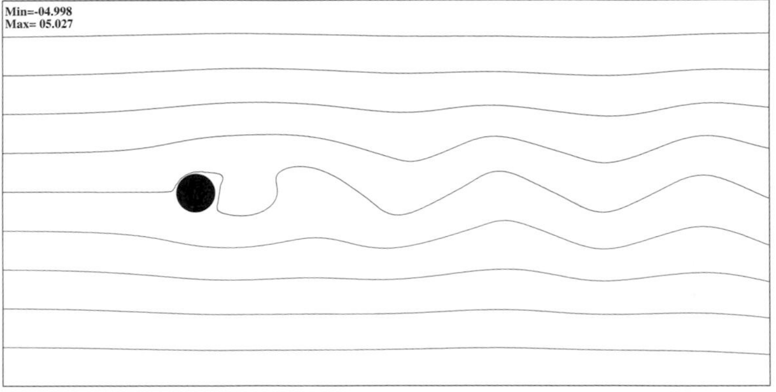

Figure 14.4 Streamlines of a snapshot.

that extend its capabilities and configurability. The main features that make the *full version* of `xROM` attractive are presented in the following.

Preprocessing: Two tools are available for the preprocessing phase. First, a PIV converter that directly converts PIV snapshots from their original `.txt` or `.dat` format to snapshots readable by `xROM`. Second, a tool to reduce the domain according to user-defined specifications, to compute the POD modes only in a specific region of interest.

Formats: `xROM` currently reads/writes two formats: CFD General Notation System CGNS, an efficient binary open-source format, and the ASCII format from Tecplot, `.dat`, which is widely used in the research community.

Case types: `xROM` supports a wide variety of mesh types: Cartesian equidistant, Cartesian, structured, and unstructured grids, in $2D$ and $3D$.

Parallelization: The software is fully parallelized using Massively Parallel Implementation (MPI), which offers several benefits.

Requirements and Download

xROM is distributed as a single binary executable file, which makes its usage very easy. It should work on any Linux distribution with a C library version 2.19 or newer. To check the C library version, run in a terminal:

```
$ ldd --version
```

For this tutorial, you can download a *lighter version* of the program that excludes CGNS capabilities (link).[2] The archive contains three items: the xROM software, the documentation, and the current sample case. Once the archive is extracted, open a terminal and go to where the binary is stored. Make sure that the binary is executable. If not, run:

```
$ chmod +x xROM
```

To run xROM, the program needs the path to the input folder as argument. There are two possibilities to define it: by specifying it directly as command line argument or by using the pop-up window. The two ways to define the input folder are shown in the following:

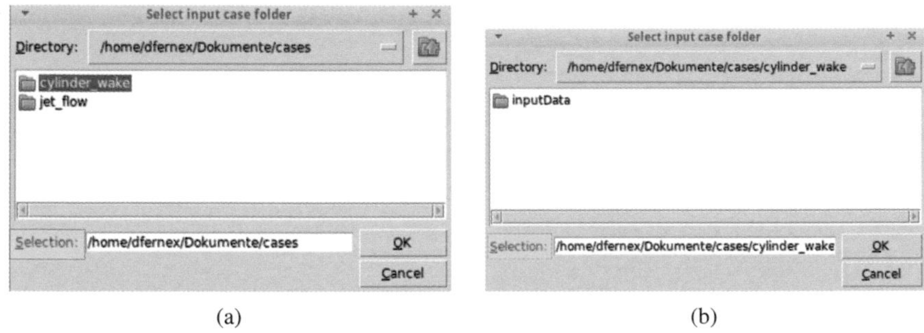

Figure 14.5 (a) Cylinder wake folder incorrectly selected. (b) Cylinder wake folder correctly selected.

Input folder from the command line: In that case, simply specify the correct path to the case folder when running the program, such as:

```
$ ./xROM /absolute/path/to/folder
```

Please note that it has to be the absolute path, which can be found using:

```
$ pwd
```

Input folder from the pop-up: If no path is defined in the command line, a pop-up window opens in which the user must select the right folder. In this case, the correct command is:

```
$ ./xROM
```

Be careful to correctly select the folder: *it is valid only when the field "Selection" of the window contains the full path, including the input folder.* In practice, the user has to click twice on the case folder (i.e., enter the

[2] https://cloudstorage.tu-braunschweig.de/getlink/fi9HQ68borJL8dJ9t4Yxv2au/

folder) to get the path, unlike other common GUIs where the user only have to click once. This is exemplified in Figure 14.5, which shows how the `cylinder_wake` folder is (a) wrongly and (b) correctly selected.

Configuration file and execution steps

The configuration file contains all the processes and their options to be executed: the tools that will be used (PIV conversion, domain reduction, ...), the steps that will be executed, and the parameters for each of these steps. It is the *only* required control input from the user. All capabilities of `xROM` can be set up in this configuration file.

The general form of an input is

```
<parameter> : <value>
```

The detailed specifications are the following:

- There is only one parameter and its value per line.
- The parameter and its value are separated by a colon ":."
- The spaces in the line are irrelevant.
- The position of a line in the file does not matter: it is for instance the same if a parameter is defined in the first line or the last line.
- To deactivate an option, the user must add the comment symbol "#" at the beginning of the line.

The standard process of `xROM` is divided into three main steps:

1. Galerkin expansion,
2. Galerkin projection,
3. dynamical system.

Each of these steps contains its owns options and tools in the configuration file. The user can selectively execute individual steps. It is possible, for instance, to compute only the Galerkin expansion or the Galerkin expansion *and* the projection without the dynamical system.

14.5.3 Exercises

As already mentioned in Section 14.5.1, we shall use DNS snapshots of a cylinder flow, which are included in the package under `CylinderWake/InputData/Snapshots`. Since all three exercises use the same data set, we shall use the same case folder and simply generate (automatically) a different run directory for each exercise. Before each run, we shall make the necessary modifications to the configuration file.

Exercise 1: Galerkin Expansion of the Cylinder flow

In this first exercise, we perform a POD on the computed snapshots. The configuration file for this exercise is located under `CylinderWake/ConfigFile`, which you can open with any text editor. Most of the options are self-explanatory. For more details, you are referred to the user manual.

As you can see in ConfigFile, only the first step GalerkinExpansion is enabled. The rest are disabled as these options are not relevant for this first exercise. Besides activating GalerkinExpansion, the only other option we need to concern ourselves with is the BaseFlow selection, which we choose as MeanFlow.

After executing the program using ./xROM, a new folder called OutputData will be generated. Inside, the POD results are found under GalerkinExpansion. Here you should find the output of the entire Galerkin expansion, which includes:

- The base flow as a Tecplot-compatible .dat file.
- The correlation matrix.
- The eigenvalues.
- The POD mode coefficients as an array, and as .png images under PODAmplitudesPlot.
- The POD modes as Tecplot-compatible .dat files under PODModes.
- The POD spectrum as a .png image.

If you wish to plot the eigenvalues and the mode coefficients yourself, you can simply disable the option plotFigures. All generated field data are saved in ASCII format and are Tecplot-compatible.

Examining the modes and the mode coefficients reveals the expected results for this shedding flow:

- The turbulent kinetic energy is highly concentrated in the first modes and drops quickly with higher mode numbers.
- A strong mode pairing between (at least) the first 8 modes.
- The mode pairs exhibit a phase shift between them.
- Modes and mode coefficients are each dominated by a single frequency that increases with higher mode pair.

Exercise 2: POD Galerkin Model without Calibration

Now, as we have computed the Galerkin expansion, we shall perform a Galerkin projection and generate our first reduced-order model. To that end, we shall activate the second (GalerkinProjection) and third (DynamicalSystem) steps in the ConfigFile as follows:

```
GalerkinProjection : yes          # (yes, no)}
   DynamicalSystem : yes          # (yes, no)
```

GalerkinProjection and DynamicalSystem perform a Galerkin projection on the POD modes and solve the dynamical system, respectively. You should leave all other parameters unchanged. We note here the deactivated model calibration option.

After executing the program, two new directories for the Galerkin projection and the dynamical system will be generated under outputData/Run2. Under GalerkiProjection, you should now have three data files: ConvectionTensor, ViscosityMatrix, and MassMatrix, which are the linear, quadratic, and mass matrix terms from the projection. Since there is no forcing and the POD modes are orthonormal, the resultant mass matrix is (within machine accuracy) a unity matrix.

These coefficients are then used to generate the dynamical system, which is integrated between time `tStartDS` and `tEndDS`. Inspecting the reference POD mode coefficient against those of the dynamical system informs us of the model accuracy. All eight reference and generated mode coefficients are found under `DynamicalSystem/DSAmplitudesPlot`. Comparing both mode coefficients shows a good match for the initial modes and a gradual degradation toward an unstable behavior for the higher modes. Without calibration, this behavior is expected.

Exercise 3: POD Galerkin Model with Calibration

In the second exercise, the Galerkin projection yielded an unstable model. This instability can be attributed to truncation errors, to structural instability of the Galerkin projection, or to insufficient numerical precision. There exists many techniques to calibrate a dynamical model (Östh et al. 2014). In this exercise, we shall rely on the easiest and very accurate *modal nonlinear eddy viscosity*. You can simply enable it in the `ConfigFile` as follows:

```
Calibration : ModalNonLinear     # (no,
    ↪ ModalNonLinear)
```

After rerunning the program, you can inspect the mode coefficients under `DynamicalSystem/DSAmplitudesPlot`. The improvement should be clear. The new model shows better accuracy and stability for all modes.

15 Advancing Reacting Flow Simulations with Data-Driven Models

K. Zdybal[1], G. D'Alessio[2], G. Aversano[3], M. R. Malik[4], A. Coussement, J. C. Sutherland, and A. Parente[5]

The use of machine learning algorithms to predict behaviors of complex systems is booming. However, the key to an effective use of machine learning tools in multi-physics problems, including combustion, is to couple them to physical and computer models. The performance of these tools is enhanced if all the prior knowledge and the physical constraints are embodied. In other words, the scientific method must be adapted to bring machine learning into the picture, and make the best use of the massive amount of data we have produced, thanks to the advances in numerical computing. The present chapter reviews some of the open opportunities for the application of data-driven reduced-order modeling of combustion systems. Examples of feature extraction in turbulent combustion data, empirical low-dimensional manifold (ELDM) identification, classification, regression, and reduced-order modeling are provided.

15.1 Introduction

The simulation of turbulent combustion is a very challenging task for a number of aspects beyond turbulence. Combustion is intrinsically multiscale and multi-physics. It is characterized by a variety of scales inherently coupled in space and time through

[1] Kamila Zdybał acknowledges the support of the Fonds National de la Recherche Scientifique (F.R.S.-FNRS) through the Aspirant Research Fellow grant.

[2] Giuseppe D'Alessio acknowledges the support of the Fonds National de la Recherche Scientifique (F.R.S.-FNRS) through the FRIA fellowship.

[3] Gianmarco Aversano acknowledges the funding from the European Research Council (ERC) under the European Union's Horizon 2020 research and innovation programme under grant agreement No. 714605.

[4] Mohammad Rafi Malik acknowledges the funding from the European Research Council (ERC) under the European Union's Horizon 2020 research and innovation programme under grant agreement No. 714605.

[5] Alessandro Parente acknowledges the funding from the European Research Council (ERC) under the European Union's Horizon 2020 research and innovation programme under grant agreement No. 714605.

thermochemical and fluid dynamic interactions (Pope 2013). Typical chemical mechanisms describing the evolution of fuels consist of hundreds of species involved in thousands of reactions, spanning 12 orders of magnitude of temporal scales (Frassoldati et al. 2003). The interaction of these scales with the fluid dynamic ones defines the nature of the combustion regime as well as the limiting process in determining the overall fuel oxidation rate (Kuo & Acharya 2012). When the characteristic chemical scales are much smaller than the fluid dynamic ones, the combustion problem becomes a mixing one (i.e., *mixed is burnt* (Magnussen 1981)): combustion and chemistry are decoupled, and the problem is highly simplified. Likewise, for chemical timescales much larger than the fluid dynamic ones, the system can be described taking into account chemistry only, neglecting the role of fluid dynamics altogether.

The intensity of interactions between turbulent mixing and chemistry is measured using the Damköhler number, defined as the ratio between the characteristic mixing, τ_m, and chemical, τ_c, timescales:

$$Da = \frac{\tau_m}{\tau_c}. \tag{15.1}$$

In terms of the Damköhler number, $Da \gg 1$ indicates a mixing-controlled, fast chemistry process. On the other hand, $Da \ll 1$ denotes a chemistry-controlled, slow chemistry process. Most practical combustion systems operate at conditions characterized by a non-negligible overlap between flow and chemical scales. This is particularly true for novel combustion technologies, where the use of diluted conditions and the enhanced mixing leads to a Da distribution close to unity. This grants some control on the combustion process, thanks to the increase of the characteristic chemical scales compared to the mixing ones. In particular, the operating conditions (temperature and compositions) can be adjusted in such a way that the emissions are kept below the required values (Wünning & Wünning 1997, Cavaliere & de Joannon 2004, Parente et al. 2011). The condition $Da \approx 1$ is generally referred to as finite-rate chemistry, to indicate that combustion is not infinitely fast but of finite speed.

Modeling finite-rate combustion regimes is very challenging because both fluid mechanics and chemistry effects must be accounted for accurately. In particular, chemistry cannot be described using simplified global mechanisms, and this results in a significant computational burden of combustion simulations. The resolution of turbulent combustion problems requires the resolution of hundreds of transport equations for (tightly coupled) chemical species on top of the conservation equations for mass, momentum, and energy. Beside the high dimensionality of the problem (Lu & Law 2009), the transport equations of reacting scalars require closure models, when the reacting structures are not fully resolved on the numerical grid.

The challenges associated with turbulent combustion modeling make the use of machine learning very attractive. While turbulent combustion models are spread across combustion industries, their current predictive capabilities fall well short of what would be needed in decision making for new design and regulation (Pope 2013). High-fidelity, direct numerical simulations (DNS) of combustion systems are still limited to isolated aspects of a turbulent combustion process and simple *building blocks*.

Still, these high-fidelity simulations are rich in information that could help decode the complexity of turbulence–chemistry interactions and guide the development of filtered and lower-fidelity modeling approaches for faster evaluations.

The objective of the present chapter is to demonstrate the potential of data-driven modeling in the context of combustion simulations. In particular, we present:

- The application of principal component analysis (PCA) and other linear and nonlinear techniques to identify low-dimensional manifolds in high-fidelity combustion data sets, and to reveal the key features of complex nonequilibrium phenomena. Different techniques are compared to PCA, including nonnegative matrix factorization (NMF), autoencoders, and local PCA in Section 15.3.
- The development of reduced-order models (ROMs), to be used in conjunction with, or to replace high-fidelity simulation tools, to reduce the burden associated with the large number of species in detailed chemical mechanisms. First, the use of transport models based on PCA is presented in Section 15.4. Finally, the application of the data-driven adaptive-chemistry approach based on the combination of classification and chemical mechanism reduction is discussed in Section 15.5.

15.2 Combustion Data Sets

The structure of combustion data sets differs from the one seen in pure fluid mechanics applications (presented in Chapter 6). The data set encountered in multicomponent reactive flows is stored in the form of a matrix $\mathbf{X} \in \mathbb{R}^{N \times Q}$. Each column of $\mathbf{X}$ is tied to one of the Q thermochemical state-space variables: temperature T, pressure p, and $N_s - 1$ chemical species[6] mass (or mole) fractions, denoted by Y_i for the ith species. For open flames and atmospheric burners the pressure variable can be omitted. Each of the N rows of $\mathbf{X}$ contains observations of all Q variables at a particular point in the physical space and/or time (and sometimes, a point in the space of other independent parameters, as briefly discussed later). This structure of the data matrix is presented below:

$$\mathbf{X} = \begin{bmatrix} \vdots & \vdots & \vdots & \vdots & & \vdots \\ T & p & Y_1 & Y_2 & \dots & Y_{N_s-1} \\ \vdots & \vdots & \vdots & \vdots & & \vdots \end{bmatrix}. \tag{15.2}$$

For such a data set, $Q = N_s + 1$. We denote the ith row (observation) in $\mathbf{X}$ as $\mathbf{x}_i \in \mathbb{R}^Q$ and the jth column (variable) in $\mathbf{X}$ as $\mathbf{X}_j \in \mathbb{R}^N$. When the data set is only resolved in space and not resolved in time, N represents the number of points on a spatial grid, and $\mathbf{X}$ can be thought of as a data *snapshot* (a notion much like the one discussed in Chapters 6–9). Typically, we can expect $N \gg Q$. However, the magnitude of Q will strongly depend on the number of species, N_s, involved in the chemical reactions, and can even reach the order of thousands for more complex fuels (Lu & Law 2009).

[6] Since mass (or mole) fractions sum up to unity for every observation, out of N_s species only $N_s - 1$ are independent.

Combustion data sets can be obtained from numerical solutions of combustion models, or from experimental measurements of laboratory flames. When generating a numerical data set, we incorporate information about the chemistry of the process by selecting a chemical mechanism for combustion of a given fuel. The mechanism determines which chemical species are involved in the specified chemical reactions, and it can vary in complexity.

For the discussion of data sets, it is useful to introduce the distinction between premixed and non-premixed combustion. In the premixed case, fuel and oxidizer are first mixed before they are burned. In the non-premixed combustion, fuel and oxidizer originate from separate streams and have to first mix for the combustion to occur. In this latter case, the mixture fraction variable is an important scalar quantity that specifies the local fuel to oxidizer ratio and is defined as

$$Z = \frac{\nu Y_F - Y_{O_2} + Y_{O_2,2}}{\nu Y_{F,1} + Y_{O_2,2}} \tag{15.3}$$

for any point in space, where Y_F is the local mass fraction of fuel, Y_{O_2} is the local mass fraction of oxidizer in the mixture, and ν is their stoichiometric ratio (Bilger et al. 1990). $Y_{F,1}$ is the mass fraction of fuel in the fuel stream and $Y_{O_2,2}$ is the mass fraction of oxidizer in the oxidizer stream. The complete combustion of fuel happens at the stoichiometric mixture fraction Z_{st}. If $Z < Z_{st}$, the mixture is called lean (fuel deficient) and if $Z > Z_{st}$, the mixture is called fuel rich.

Figure 15.1 presents an overview of frequently encountered numerical data sets, ordered schematically by the amount of information about the combustion process they contain. In particular:

- The zero-dimensional (0D) reactor model (e.g., perfectly stirred reactor [PSR]) assumes that combustion happens in a single point in space. This model can thus only include information about chemistry and thermodynamics of the combustion process evolving in time. It carries no information about spatial gradients of the thermochemical variables. The 0D reactor describes an ideal mixing process with $Da \ll 1$. When the data set is formed from the 0D reactor simulation, the N rows of $\mathbf{X}$ are linked to points in time only.
- The steady laminar flamelet (SLF) model (Peters 1984) assumes that fuel and oxidizer are two impinging streams, originating from the fuel and the oxidizer feed respectively. Combustion happens in infinitely thin sheets (flamelets) where fuel and oxidizer meet at varying stoichiometric proportions. The stoichiometry is specified by the mixture fraction variable, Z, and the thermochemical variables vary along the axis between pure oxidizer ($Z = 0$) and pure fuel ($Z = 1$). By varying the distance between the feeds (or the outlet velocities), a strain (specified by the strain rate χ) is imposed on the flamelet, which can displace it from its equilibrium. The SLF model describes a non-premixed system with $Da \gg 1$. This is thus a two-parameter model with each row of $\mathbf{X}$ linked to one pair (Z, χ).
- The counterflow diffusion flame (CFDF) model (Law 2010) is similar to the SLF model, as it assumes two separate streams of fuel and oxidizer diffusing into one another (non-premixed system). However, in contrast to SLF, the flow equations

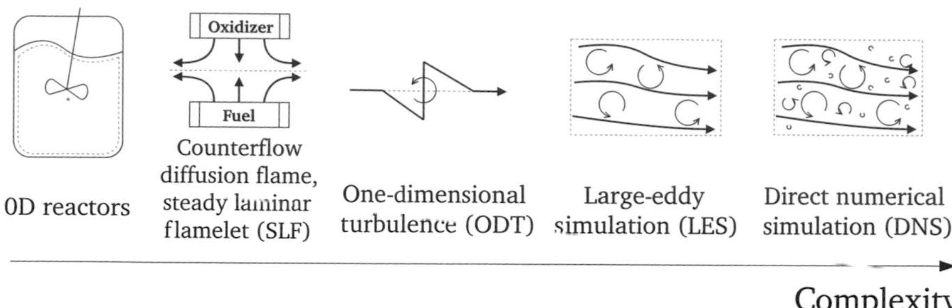

Figure 15.1 Examples of common numerical combustion data sets schematically presented on the axis of increasing complexity.

for CFDF are solved in the physical space. Flow is parameterized by the velocity gradient parameter a, which is an equivalent of the local strain rate. In the CFDF, each row of $\mathbf{X}$ is linked to a point in the physical space and time.

- The one-dimensional turbulence (ODT) model (Kerstein 1999, Echekki et al. 2011) can additionally incorporate the effects of turbulence by introducing eddy events on a one-dimensional domain. It allows for both spatial and temporal evolution of the flow. ODT can be used as a standalone model but can also serve as a subgrid model in large eddy simulations (LES).
- The Reynolds-averaged Navier–Stokes (RANS) models (Ferziger et al. 2002) solve the time-averaged equations describing fluid dynamics and hence any sources of unsteadiness coming from turbulence are averaged.
- The LES (Ferziger et al. 2002) resolves large scales associated with the fluid dynamics processes but a subgrid model is required to account for processes occurring at the smallest scales. The choice for the subgrid model becomes particularly important in reactive flows, since combustion is inherently tied to the smallest scales.
- The DNS (Ferziger et al. 2002) allows for the most accurate description of the coupled interaction between fluid dynamics and the thermochemistry since all subgrid processes are resolved directly. The scarcity of the DNS data sets is due to the large computational cost of performing DNS simulations, especially for more complex fuels.

More information on combustion theory and combustion models can be found in (Turns et al. 1996, Kee et al. 2005, Law 2010).

The selected data set is the starting point for applying a data science technique, and we often refer to those data sets as the *training* data. From the description of various data sets presented above, it is visible that the choice of the data set will be pertinent to the type and quality of the analysis performed. For instance, we can expect that features identified will depend on the amount of information about the coupled phenomena of turbulence and chemistry that was initially captured in the data set.

Scaling technique	Scaling factor d_j
Auto	σ_j
Pareto	$\sqrt{\sigma_j}$
Range	$max(\mathbf{X}_j) - min(\mathbf{X}_j)$
VAST	$\sigma_j^2/\bar{\mathbf{X}}_j$

Table 15.1 Few selected common data scaling criteria. σ_j is the standard deviation, $max(\mathbf{X}_j)$ is the maximum value, $min(\mathbf{X}_j)$ is the minimum value, and $\bar{\mathbf{X}}_j$ is the mean observation of the jth state-space variable $\mathbf{X}_j$.

15.2.1 Data Preprocessing

Combustion data sets typically contain variables of different numerical ranges (e.g., both temperature, which can range from hundreds up to thousands, and mass (or mole) fractions of chemical species, which take values in the range $Y_i \in \langle 0, 1 \rangle$). Scaling the data set is crucial in order to balance the importance of all state-space variables (Parente & Sutherland 2013). The data set $\mathbf{X}$ can be centered and scaled as follows:

$$\widetilde{\mathbf{X}} = (\mathbf{X} - \bar{\mathbf{X}})\mathbf{D}^{-1}, \tag{15.4}$$

where $\bar{\mathbf{X}}$ contains the mean observations of each variable, $\mathbf{D}$ is the diagonal matrix of scales, where the jth element d_j from the diagonal is the scaling factor corresponding to the jth state-space variable $\mathbf{X}_j$. A few of the common scaling criteria are collected in Table 15.1. The result is a centered and scaled data matrix $\widetilde{\mathbf{X}}$. Other preprocessing means can include outlier detection or data sampling. For the remainder of this chapter, we will assume that $\widetilde{\mathbf{X}}$ represents the matrix that has been adequately preprocessed.

15.3 Feature Extraction Using Dimensionality Reduction Techniques

Dimensionality reduction techniques offer a way to represent high-dimensional combustion data sets in a new, lower-dimensional basis. Such data representations are referred to as low-dimensional manifolds. Techniques such as PCA, NMF, or autoencoders can extract those manifolds in an empirical way from the training data sets. This approach belongs to the family of ELDMs (Pope 2013, Yang et al. 2013), and it is based on the idea that compositions occurring in combustion systems lie close to a low-dimensional manifold. In addition, these techniques offer a way to extract meaningful features by exploiting the fact that the new basis can be better suited to represent certain physical phenomena underlying the original data. In this section, we review a few popular dimensionality reduction techniques, and we present their potential to detect features in combustion data sets.

15.3.1 Description of the Analyzed Data Set

In this section, we apply various dimensionality reduction techniques on a numerical data set from a DNS simulation of a CO/H_2 turbulent jet (Sutherland et al. 2007) using the chemical mechanism consisting of 12 chemical species and 33 reactions (Yetter et al. 1991). Additional information regarding the numerical simulation can be found in Sutherland et al. (2007). The data set consists of a two-dimensional slice extracted from the 3D domain arranged in a matrix $\mathbf{X}$. Each row of $\mathbf{X}$ corresponds to a point on the two-dimensional grid, and therefore the matrix $\mathbf{X}$ can be thought of as a single-time snapshot. Several such snapshots from different times in the simulation are available. The matrix $\mathbf{X}$ has 13 columns corresponding to temperature and 12 species' mass fractions[7].

15.3.2 Extracting Features with Data Reduction Techniques

Principal Component Analysis

PCA (Jolliffe 2002) projects the preprocessed data set $\widetilde{\mathbf{X}}$ onto a new basis represented by the orthonormal matrix of modes $\mathbf{A} \in \mathbb{R}^{Q\times Q}$. The matrix $\mathbf{A}$ can be obtained as the eigenvectors resulting from the eigendecomposition of the data covariance matrix $\mathbf{S} \in \mathbb{R}^{Q\times Q}$,

$$\mathbf{S} = \frac{1}{N-1}\widetilde{\mathbf{X}}^\top\widetilde{\mathbf{X}}, \tag{15.5}$$

or from the singular value decomposition (SVD) of the data set,

$$\widetilde{\mathbf{X}} = \mathbf{U}\boldsymbol{\Sigma}\mathbf{A}^\top, \tag{15.6}$$

where $^\top$ denotes matrix transpose. This formulation makes PCA equivalent to proper orthogonal decomposition (POD) (see Chapter 6). The preprocessed data matrix $\widetilde{\mathbf{X}}$ can then be projected onto $\mathbf{A}$ to obtain the principal components (PCs) matrix $\mathbf{Z} \in \mathbb{R}^{N\times Q}$:

$$\mathbf{Z} = \widetilde{\mathbf{X}}\mathbf{A}. \tag{15.7}$$

PCs represent the original observation in the new coordinate system defined by $\mathbf{A}$. It is worth noting that the columns of $\mathbf{Z}$ are linear combinations of the original state-space variables. The linear coefficients of that combination are specified in the entries of $\mathbf{A}$, and we will refer to them as *weights*. Adopting the general notion of the data set as presented in (15.2), the jth PC, $\mathbf{Z}_j$ (the jth column of $\mathbf{Z}$), is computed as

$$\mathbf{Z}_j = \sum_{i=1}^{Q} a_{ij}\widetilde{\mathbf{X}}_i = a_{1,j}\cdot\widetilde{T} + a_{2,j}\cdot\widetilde{p} + a_{3,j}\cdot\widetilde{Y}_1 + \cdots + a_{Q,j}\cdot\widetilde{Y}_{N_s-1}, \tag{15.8}$$

where a_{ij} are the elements (weights) from the jth column of $\mathbf{A}$ and tildes represent the preprocessing applied to each state-space variable. Since the basis matrix is orthonormal, $a_{ij} \in [-1, 1]$ and $\mathbf{A}^{-1} = \mathbf{A}^\top$.

By keeping a reduced number of the $q < Q$ first PCs, we obtain the closest[8] rank-q approximation of the original matrix,

[7] For the purpose of the feature extraction analysis presented here, all N_s species were included in the data set.

[8] in terms of L_2, Frobenius, or trace norms, which follows from the Eckart–Young–Mirsky theorem (Eckart & Young 1936).

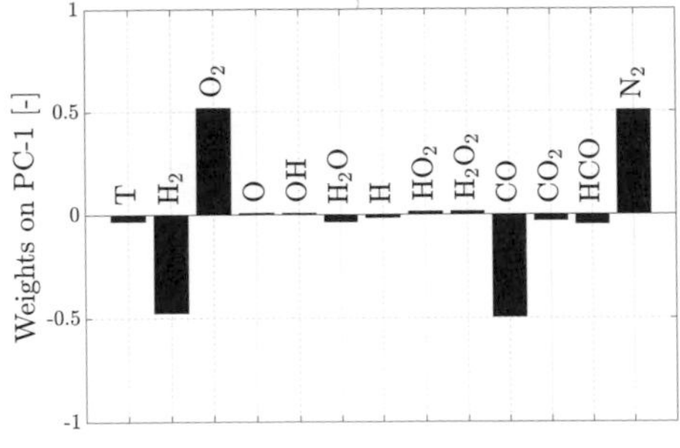

Figure 15.2 The eigenvector weights a_{ij} associated with the first PC found by PCA on a data set preprocessed with Range scaling.

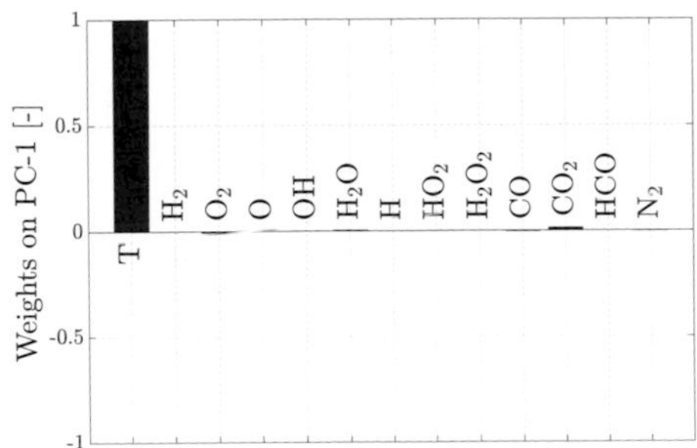

Figure 15.3 The eigenvector weights a_{ij} associated with the first PC found by PCA on a data set preprocessed with Pareto scaling.

$$\mathbf{X} \approx \mathbf{X_q} = \mathbf{Z_q}\mathbf{A_q}^{\top}\mathbf{D} + \bar{\mathbf{X}}, \tag{15.9}$$

where the index q denotes the truncation from Q to q components. Note that the reverse operation to the one defined in (15.4) has to be applied using matrices $\mathbf{D}$ and $\bar{\mathbf{X}}$.

We can assign physical meaning to the PCs by looking at the linear coefficients (weights) a_{ij} from the basis matrix $\mathbf{A}$ (Parente et al. 2011, Bellemans et al. 2018). High absolute weight for a particular variable means that this variable is identified by the PCA as important in the linear combination from (15.8). Moreover, since the vector $\mathbf{A}_j$ identifies the same span as $-\mathbf{A}_j$, only the relative sign of a particular weight with respect to the signs of other weights is important. In addition, PCs are ordered so that each PC captures more variance than the following one. Thus, we can expect the most important features identified by PCA to be visible in the first few PCs. This property of PCA can also guide the choice for the value q. Since PCs are decorrelated, increasing q in the PCA reconstruction from (15.9) guarantees an improvement in the reconstruction errors.

Preprocessing the data set, prior to applying a dimensionality reduction technique, can have a significant impact on the shape of the low-dimensional manifold and on the types of features retrieved from the data set (Parente & Sutherland 2013, Peerenboom et al. 2015). Figures 15.2 and 15.3 present the weights a_{ij} associated with the first PC resulting from Range and Pareto scaling (see Table 15.1) of the original data set. First, we observe that the structure of the PC can change significantly with the choice of the scaling technique. If we further consider the mixture fraction variable as defined in (15.3) as a linear combination of fuel Y_F and oxidizer Y_{O_2} mass fractions, it can be observed that the coefficients in front of Y_F and Y_{O_2} are of opposite signs in that

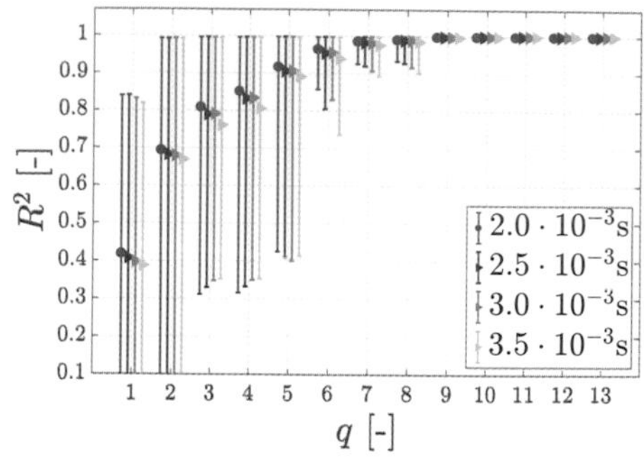

Figure 15.4 Mean, minimum (worst reconstruction), and maximum (best reconstruction) R^2 values for reconstructing three snapshots separated by $\Delta t = 5$ ms (blue triangles) using modes from the snapshot at time $t = 2.0 \cdot 10^{-3}$ s (red circles). Shown for increasing q in the PCA reconstruction.

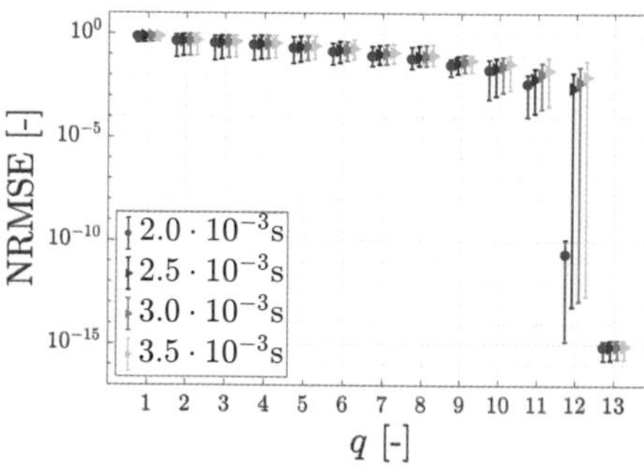

Figure 15.5 Mean, minimum (best reconstruction), and maximum (worst reconstruction) NRMSE values for reconstructing three snapshots separated by $\Delta t = 5$ ms (blue triangles) using modes from the snapshot at time $t = 2.0 \cdot 10^{-3}$ s (red circles). Shown for increasing q in the PCA reconstruction.

definition. Using Range scaling (Figure 15.2), the first PC can be attributed to the mixture fraction variable, where the only high weights are for the fuel (H_2 and CO) and oxidizer components (O_2 and N_2), and the two have opposite signs. The correlation between the mixture fraction variable and the first PC is in this case 99.96%. It is worth noting that the mixture fraction was not among the variables in the original data set $\mathbf{X}$ and PCA identifies it automatically. With Pareto scaling (Figure 15.3), the first PC is almost entirely aligned with the temperature variable and carries almost no information about the mass fractions of the chemical species.

In a previous study (Biglari & Sutherland 2012), the authors have demonstrated that PCA can identify PCs that are independent of the filter width on a fully resolved jet flame. PCA was performed on the state-space variables filtered using a top-hat filter of varying widths. To test the capability of PCA to extract time-invariant features of the data set, we can also use a fixed set of modes to reconstruct new, unseen data, such as data from future time steps of the same temporally evolving system. If the reconstruction process leads to errors that are comparable with the ones obtained for the training data, we can expect that PCA captures the *essence* of the physical processes underlying the system. We demonstrate this using 2D slices from the DNS data set from four time intervals separated by $\Delta t = 5$ ms. In Figures 15.4 and 15.5, three future data snapshots (blue triangles) are reconstructed using q first eigenvectors found on the initial snapshot (red circles). Figure 15.4 shows the coefficients of determination, which can be computed for the jth variable $\mathbf{X}_j$ in the data set as

$$R_j^2 = 1 - \frac{\sum_{i=1}^{N}(x_{ij} - \hat{x}_{ij})^2}{\sum_{i=1}^{N}(x_{ij} - \bar{x}_j)^2}, \tag{15.10}$$

where x_{ij} is the ith observation of the jth variable, $\hat{x}_{ij}$ is the PCA reconstruction of that observation, and $\bar{x}_j$ is the average observation of $\mathbf{X}_j$. The coefficient of determination R^2 measures the *goodness of the model fit* with respect to fitting the data with the mean value $\bar{x}_j$. Values $R^2 \in (-\infty, 1]$, where $R^2 = 1$ means a perfect fit. The smaller the R^2 value gets, the worse the model fit. Figure 15.5 shows the normalized root-mean-squared errors (NRMSE) on a logarithmic scale. NRMSE can be computed for the jth variable $\mathbf{X}_j$ in the data set as

$$\text{NRMSE}_j = \frac{1}{\bar{x}_j} \cdot \sqrt{\frac{\sum_{i=1}^{N}(x_{ij} - \hat{x}_{i,j})^2}{N}}. \tag{15.11}$$

In Figures 15.4 and 15.5, the markers represent the mean values of R^2 or NRMSE averaged over all variables in the data set. The bars range from the minimum and the maximum value achieved for any variable in the data set. It can be observed that for all the reconstructed snapshots, the error metrics show comparable values. This indicates the capability of PCA to capture generalized, time-invariant features of temporally evolving systems. It can also be observed that the mean errors grow for future snapshots, which indicates that there might be a limit on the time separation Δt for which we can extend the applicability of the features found.

Local Principal Component Analysis

PCA can also be applied locally (LPCA), on portions of the original data set. In LPCA, a clustering algorithm is first applied to the data set to partition observations into local clusters. Next, PCA is performed separately in each cluster. LPCA can not only allow for further reduction in dimensionality by adjusting to the potential nonlinearities of the data set, but it can also aid in detecting local features of the data. Local PCA can also guide the decision on how to partition the data into clusters, since it grants control of the data reconstruction errors. Figure 15.6 shows the difference between the global and the local approach on a synthetic data set that is visibly composed of two distinct clusters. In the global case, PCA is performed on the entire data set, ignoring the apparent clusters. When the data set is partitioned and PCA is performed locally in each cluster, a different set of eigenvectors and PCs is tied to each identified cluster of data. Those local PCs can better represent the physics or the underlying features of a particular cluster by capturing the local variance instead of the global one. In LPCA, the ith observation $\mathbf{x}_i$ is reconstructed from the local, q-dimensional manifold in an analogous way to (15.9):

$$\mathbf{x}_i \approx \mathbf{x}_{\mathbf{q},i} = \mathbf{z}_{\mathbf{q},i}(\mathbf{A}_{\mathbf{q}}^{(k)})^{\top}\mathbf{D} + \mathbf{c}^{(k)}, \tag{15.12}$$

where k is the index of the cluster to which the ith observation belongs, $\mathbf{z}_{\mathbf{q},i}$ is the ith observation represented in the local, truncated basis $\mathbf{A}_{\mathbf{q}}^{(k)}$ identified on the kth

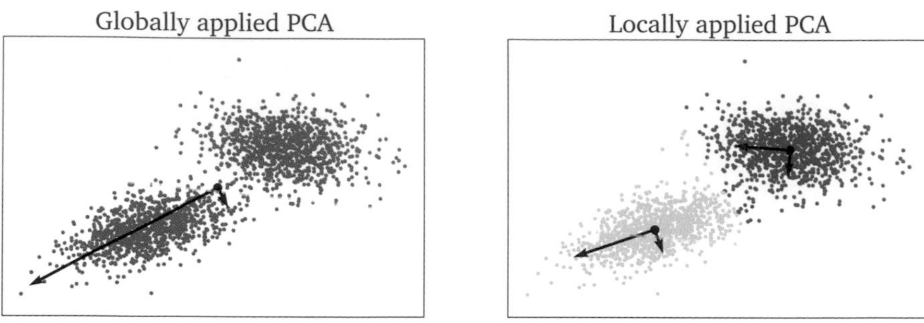

Figure 15.6 Schematic distinction between global and local PCA on a synthetic two-dimensional data set. The arrows represent the two global/local modes from the matrix **A**, defining the directions of the largest variance in the global/local data.

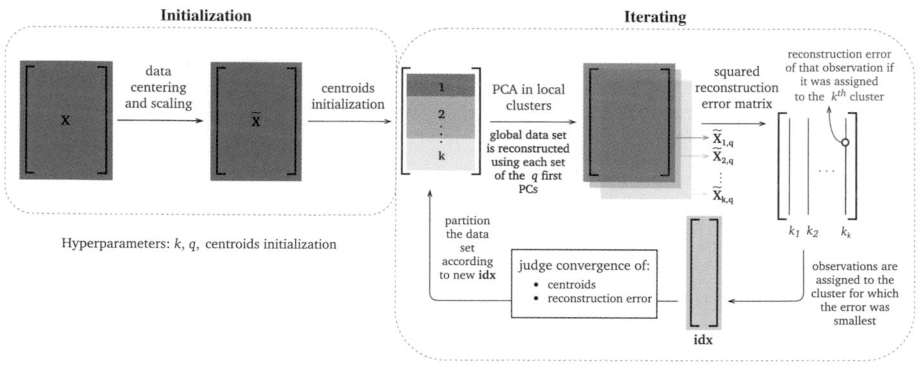

Figure 15.7 Diagram presenting schematically the VQPCA clustering algorithm.

cluster. Each cluster is centered separately using the centroid $\mathbf{c}^{(k)}$ of the kth cluster and typically the global diagonal matrix of scales $\mathbf{D}$ is applied in each cluster.

Data clustering, prior to applying local PCA, can be performed with any algorithm of choice. One of the techniques discussed in this chapter is the vector quantization PCA (VQPCA) algorithm (Kambhatla & Leen 1997, Parente et al. 2009), presented schematically in Figure 15.7. This is an iterative algorithm in which the observations are assigned to the cluster for which the local PCA reconstruction error of that observation is the smallest. The hyperparameters of the algorithm include the number of clusters k to partition the data set, the number of PCs q used to approximate the data set at each iteration, and the initial cluster partitioning. The latter is predefined by setting the initial cluster centroids. The most straightforward way is to initialize centroids randomly, but another viable option is to use partitioning resulting from a different clustering technique such as K-means (MacQueen et al. 1967). The VQPCA algorithm iterates until convergence of the centroids and of the reconstruction error is

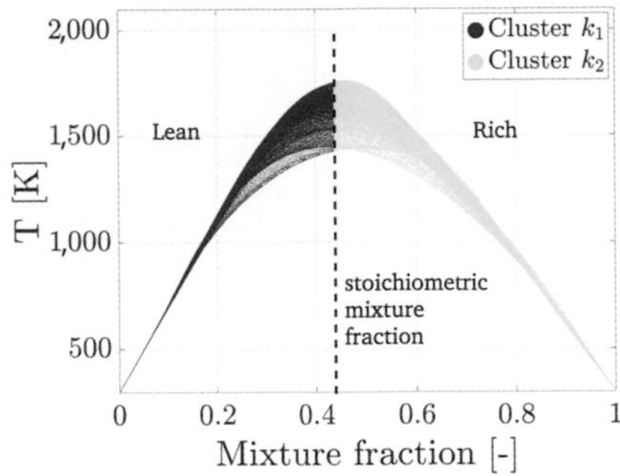

Figure 15.8 Conditioning variable partitioning into $k = 2$ clusters in the mixture fraction-temperature space.

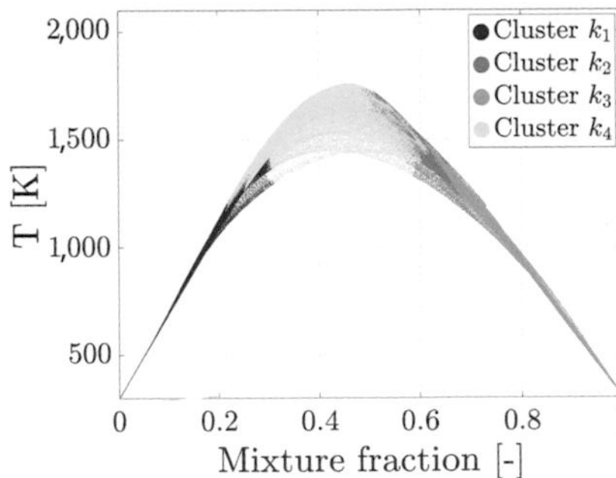

Figure 15.9 VQPCA partitioning into $k = 4$ clusters in the mixture fraction-temperature space.

reached. The reconstruction error is measured between the centered and scaled data set $\widetilde{\mathbf{X}}$ and the approximation $\widetilde{\mathbf{X}}_{\mathbf{i},\mathbf{q}}$ using the q first PCs computed from the eigenvectors found on the ith cluster. More details on the VQPCA algorithm can be found in Parente et al. (2009) and Parente et al. (2011).

Another possible way of clustering the combustion data sets is to use a conditioning variable, such as a mixture fraction, and partition the observations based on bins of that variable. If the mixture fraction is used, observations are first split into fuel-lean and fuel-rich parts at the stoichiometric mixture fraction Z_{st}. If more than two clusters are requested, lean and rich sides can then be further divided. The approach of performing local PCA on clusters identified through binning the mixture fraction vector is referred to as FPCA.

Local PCA was investigated on the benchmark DNS data set (Section 15.3.1) using two clustering algorithms. Figures 15.8 and 15.9 show a comparison between two clustering results in the space of temperature and the mixture fraction variable. Figure 15.8 presents clustering into $k = 2$ clusters using bins of mixture fraction as the conditioning variable. This partitioning can be thought of as *hardcoded* in a sense that the split into two clusters will always be performed at Z_{st}. The features retrieved on local portions of data can thus only be attributed to the lean and rich zones. In contrast, Figure 15.9 presents clustering into $k = 4$ clusters using the VQPCA algorithm. VQPCA could distinguish between the oxidizer (cluster k_1), the fuel (cluster k_3), and the region where the two meet close to stoichiometric conditions (clusters k_2 and k_4).

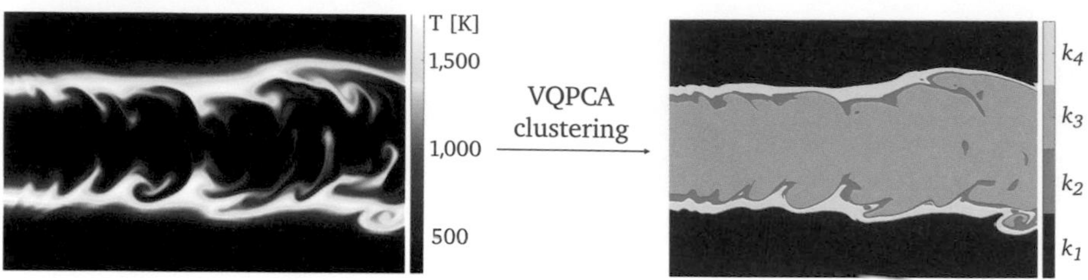

Figure 15.10 Temperature profile of the DNS data set (left) and the result of partitioning the data set into $k = 4$ clusters using the VQPCA algorithm (right).

This is even more apparent if we plot the result of the VQPCA clustering on a spatial grid in Figure 15.10. The space is clearly divided into the inner fuel jet (k_3), the outer oxidizer layer (k_1), and the two thin reactive layers (k_2, k_4) for which the temperature is the highest.

The success of local PCA in extracting features depends on the clustering technique used. In a previous study (D'Alessio et al. 2020), VQPCA has been compared to other clustering algorithms, and better results in terms of clustering quality and algorithm speed have been obtained. An unsupervised clustering algorithm based on the VQPCA partitioning has recently been proposed (D'Alessio et al. 2020b) to perform data mining on a high-dimensional DNS data set. If an algorithm such as VQPCA is used, the types of features found depend on data preprocessing and can additionally depend on the hyperparameters. By changing the number of clusters or the number of PCs in the approximation, the user can potentially retrieve different features. This has been also investigated in a previous study (D'Alessio et al. 2020a) for a more complex, high-dimensional DNS data set. This is in contrast with clustering strategies based on binning a single physical quantity, such as mixture fraction or heat release rate. For instance, taking the heat release rate as an example, we might anticipate that the cluster formed from the high heat release rate bin will identify the chemically reacting region.

Nonnegative Matrix Factorization

NMF (Paatero & Tapper 1994) is an algorithm for factorizing data matrices whose elements are nonnegative. This technique can thus be applied to the data set as defined in (15.2) since the thermochemical state-space variables are nonnegative, provided that at the preprocessing step the data set is scaled but not centered ($\widetilde{\mathbf{X}} = \mathbf{X}\mathbf{D}^{-1}$). This is to ensure that we are not introducing negative elements in the matrix $\widetilde{\mathbf{X}}$ through centering by mean values[9] as is done in (15.4). Given the nonnegative data set $\mathbf{X} \in \mathbb{R}^{N \times Q}$, the scaling matrix $\mathbf{D}$ and the value for q, NMF aims to find two nonnegative matrices $\mathbf{W} \in \mathbb{R}^{N \times q}$ and $\mathbf{F} \in \mathbb{R}^{q \times Q}$ such that

[9] Alternatively, one can also subtract minimum values from each variable, thus making the range of each state-space variable start at 0.

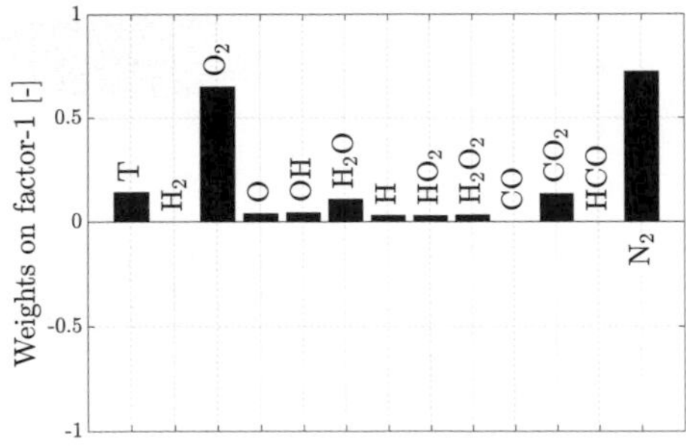

Figure 15.11 First nonnegative factor $\mathbf{f}_1$. NMF performed on Range-scaled data with $q = 2$.

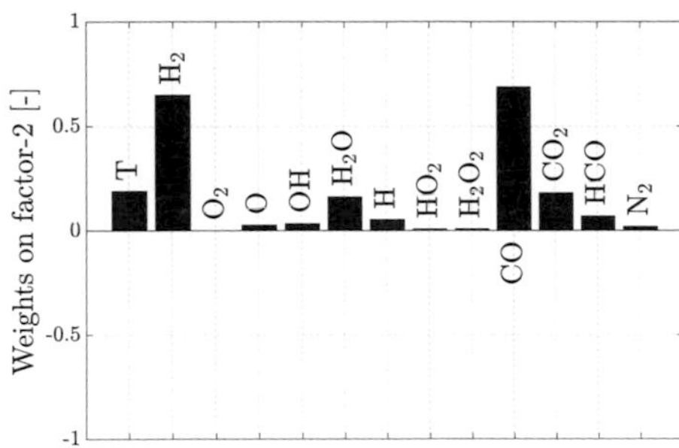

Figure 15.12 Second nonnegative factor $\mathbf{f}_2$. NMF performed on Range-scaled data with $q = 2$.

$$\mathbf{X} \approx \mathbf{X_q} = \mathbf{WFD}. \tag{15.13}$$

The matrix of nonnegative factors **F** can be regarded as the one containing a basis (analogous to the matrix **A** found by PCA). The matrix **W** represents the compressed data, namely the NMF scores and is thus analogous to the PCs matrix **Z**. The factorization to **W** and **F** is not unique, and various optimization algorithms exist (Berry et al. 2007, Lin 2007). In this chapter, we use the MATLAB® routine `nnmf`, which minimizes the root-mean-squared residual (Berry et al. 2007),

$$D = \frac{1}{\sqrt{N \cdot Q}} ||\mathbf{X} - \mathbf{WF}||_F, \tag{15.14}$$

starting with random initial values for **W** and **F**, where the subscript F denotes the Frobenius norm. This optimization might reach local minimum and thus repeating the algorithm can yield different factorizations.

Similarly to what was done in PCA, we can look at the nonnegative factor weights (the elements of **F**) to assign physical meaning to the factors. With the nonnegative constraint on **F**, only nonnegative weights can be found. Figures 15.11 and 15.12 show the first two nonnegative factors that together represent the mixture fraction variable (compare with Figure 15.2). The oxidizer components are included in the first factor and the fuel components in the second. If NMF is compared to PCA, the latter can be thought of as more robust since NMF required two modes to capture the same information (the mixture fraction variable) that was included in a single PCA mode.

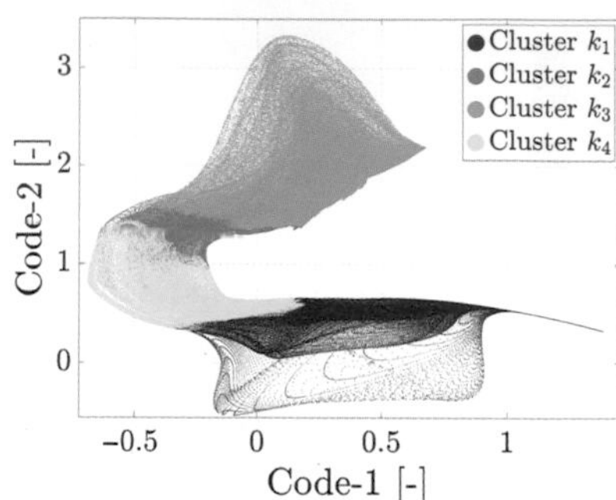

Figure 15.13 Two-dimensional manifold obtained using an Autoencoder with five hidden layers and SELU activation function. Colored by the result of VQPCA partitioning into $k = 4$ clusters.

Autoencoders

The autoencoder (Goodfellow et al. 2016, Wang et al. 2016) is a type of an unsupervised artificial neural network (ANN) whose aim is to learn the q-dimensional representation (*embedding*) of the Q-dimensional data set such that the reconstruction error between the input and the output layer is minimized. The standard form of an autoencoder is the feedforward neural network having an input layer and an output layer with the same number of neurons, and one or more hidden layers. Given one hidden layer, the *encoding* process takes as an input the preprocessed data matrix $\widetilde{\mathbf{X}} \in \mathbb{R}^{N \times Q}$ and maps it to $\mathbf{H} \in \mathbb{R}^{N \times q}$, with $q < Q$:

$$\mathbf{H} = f(\widetilde{\mathbf{X}}\mathbf{G} + \mathbf{B}), \tag{15.15}$$

where the columns of $\mathbf{H}$ are referred to as the *codes*, f is the activation function such as sigmoid, rectified linear unit (ReLU), or squared exponential linear unit (SELU), $\mathbf{G} \in \mathbb{R}^{Q \times q}$ is the matrix of weights and $\mathbf{B} \in \mathbb{R}^{N \times q}$ is the matrix of biases. At the *decoding* stage, $\mathbf{H}$ is mapped to the reconstruction $\widetilde{\mathbf{X}}_\mathbf{q}$,

$$\widetilde{\mathbf{X}} \approx \widetilde{\mathbf{X}}_\mathbf{q} = f'(\mathbf{H}\mathbf{G}' + \mathbf{B}'), \tag{15.16}$$

where f', $\mathbf{G}'$, and $\mathbf{B}'$ may be unrelated to f, $\mathbf{G}$, and $\mathbf{B}$. The encoding/decoding process can thus be summarized as

$$\widetilde{\mathbf{X}} \in \mathbb{R}^{N \times Q} \xrightarrow{\text{encoding}} \mathbf{H} \in \mathbb{R}^{N \times q} \xrightarrow{\text{decoding}} \widetilde{\mathbf{X}}_\mathbf{q} \in \mathbb{R}^{N \times Q}. \tag{15.17}$$

In this section, we use an autoencoder with five hidden layers and SELU activation function and generate a two-dimensional embedding ($q = 2$) of the original data set. Figure 15.13 shows the two-dimensional manifold obtained after the autoencoder compression (represented by the matrix $\mathbf{H}$). The manifold is colored by the previously obtained result of partitioning via the VQPCA algorithm. From Figure 15.13, it can be observed that the result of VQPCA partitioning still uniformly divides the autoencoder manifold. Clusters k_1 and k_3, representing the oxidizer and fuel respectively, are located at the opposing ends of the manifold. Thus, it is possible to think of that manifold as describing the progress of the combustion process, with the fuel and oxidizer meeting in the center (k_2 and k_4) where they finally react.

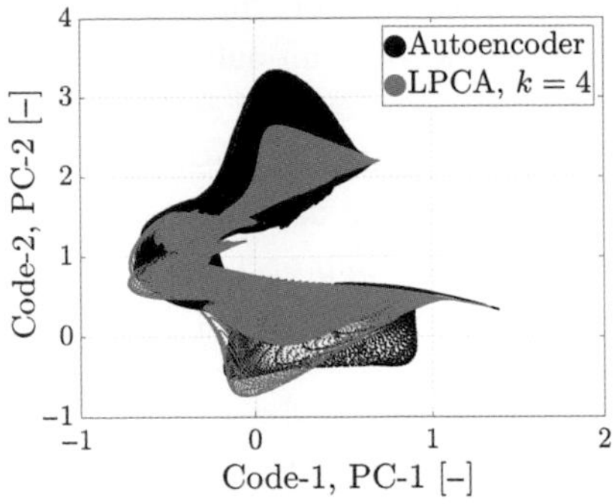

Figure 15.14 The two-dimensional manifold obtained from local PCA using VQPCA clustering algorithm and increasing the number of clusters, overlayed with the Autoencoder manifold using the Procrustes analysis. VQPCA performed with $k = 4$ clusters.

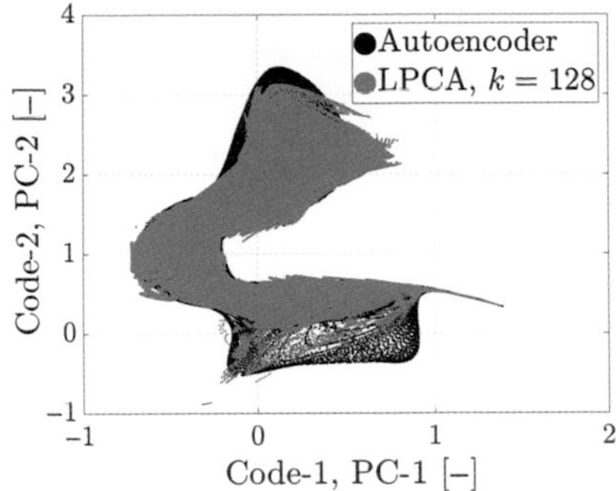

Figure 15.15 The two-dimensional manifold obtained from local PCA using VQPCA clustering algorithm and increasing the number of clusters, overlayed with the Autoencoder manifold using the Procrustes analysis. VQPCA performed with $k = 128$ clusters.

Linear Operations That Aid in Feature Interpretation

Several linear operations can aid the interpretation of PCs or low-dimensional manifolds. One such technique is the rotation of modes/factors (Abdi 2003, Bellemans et al. 2018) with the varimax orthogonal rotation (Kaiser 1958) used most commonly. The varimax rotation first selects which of the n factors should be rotated together. It then maximizes the sum of variances of the squared weights within each of the n rotated factors

$$\mathcal{V} = \sum (w_{ij}^2 - \bar{w}_{ij}^2)^2, \tag{15.18}$$

where w_{ij} is the ith weight on the jth factor and $\bar{w}_{ij}^2$ is the mean of the squared weights (Abdi 2003). After rotation, those n factors explain the same total amount of variance as they did together before the rotation, but variance is now redistributed differently among the selected factors. Varimax-rotated factors typically have high weights on fewer variables than the original factors that can aid in their physical interpretation. The rotated factors can then be used as the new basis to represent the original data set. Varimax and several other rotation methods are available within the MATLAB® routine `rotatefactors`.

Another interesting technique is the Procrustes analysis (Seber 2009), which is a series of linear operations that allow translation, rotation, or scaling of the low-dimensional manifold. This can be particularly useful when manifolds obtained from two different dimensionality reduction techniques should be compared.

Figures 15.14 and 15.15 present the Procrustes transformation (using the MATLAB® routine `procrustes`) of the LPCA manifold onto an autoencoder manifold. With the series of linear transformations the manifolds overlay each other, and the match between them becomes closer to exact as the number of clusters with which VQPCA was performed is increased. This suggests that the data reconstruction error for the two-dimensional approximation from LPCA will approach the one from autoencoder decoding when the number of clusters is high enough.

15.4 Transport of Principal Components

In Section 15.2, we have seen that the number of thermochemical state-space variables Q determines the original dimensionality of the data set. This number also reflects how many transport equations for the state-space variables should be solved in a numerical simulation. The general transport equation for the set $\mathbf{\Phi} = \mathbf{X}^\top = [T, p, Y_1, Y_2, \ldots, Y_{N_s-1}]^\top$ of state-space variables is

$$\rho \frac{D\mathbf{\Phi}}{Dt} = -\nabla \cdot (\mathbf{j_\Phi}) + \mathbf{S_\Phi}, \tag{15.19}$$

where $\mathbf{j_\Phi}$ is the mass-diffusive flux of $\mathbf{\Phi}$ relative to the mass-averaged velocity, and $\mathbf{S_\Phi}$ is the volumetric rate of production of $\mathbf{\Phi}$ (also referred to as the source of $\mathbf{\Phi}$). Performing detailed simulations with large chemical mechanisms with significant number of chemical species is still computationally prohibitive. Sutherland and Parente (2009) proposed to use PCA to reduce the number of transport equations that solve a combustion process. Instead of solving the original set of Q transport equations, the original variables are first transformed to the new basis identified by PCA on the training data set $\mathbf{X}$. Next, the truncation from Q to q first PCs is performed. Transport equations for the q first PCs can be formulated from (15.19) using the truncated basis matrix $\mathbf{A_q}$,

$$\rho \frac{D\mathbf{z}}{Dt} = -\nabla \cdot (\mathbf{j_z}) + \mathbf{S_z}, \tag{15.20}$$

where $\mathbf{z} = \mathbf{Z_q}^\top$ (with $\mathbf{Z_q} = (\mathbf{X} - \bar{\mathbf{X}})\mathbf{D}^{-1}\mathbf{A_q}$), $\mathbf{j_z} = \mathbf{A_q j_\Phi}$, and $\mathbf{S_z} = \mathbf{A_q S_\Phi D}^{-1}$ (also referred to as the PC-sources). This is the discrete analogous of the Galerkin projection methods described in Chapters 6 and 14. Note that the source terms of the PCs, $\mathbf{S_z}$, are scaled (but not centered) using the same scaling matrix $\mathbf{D}$ as applied on the data set $\mathbf{X}$.

The challenge associated with the resolution of the PC-transport equation is related to the PC-source terms. The latter is highly nonlinear functions (based on Arrhenius expressions) of the state-space variables. The nonlinearity of the chemical source terms strongly impacts the degree of reduction attainable using the projection of the species transport equations onto the PCA basis (Isaac et al. 2014, 2015). A solution to this problem is to use PCA to identify the most appropriate basis to parameterize the ELDM and then, both the thermochemical state-space variables and the PC-source terms can be nonlinearly regressed onto the new basis. This allows us to overcome the shortcomings associated with the multilinear nature of PCA and to reduce the number of components required for an accurate description of the state-space.

The construction of an appropriate low-dimensional manifold requires training data. This might be seen as a limitation of the approach since all system states are required before applying model reduction. Although initial studies on PCA models involved DNS data of turbulent combustion (Sutherland & Parente 2009, Pope 2013), recent studies have demonstrated (Biglari & Sutherland 2015, Echekki & Mirgolbabaei 2015, Coussement et al. 2016, Malik et al. 2018, Dalakoti et al. 2020) that PCA-based models can be trained on simple and inexpensive systems, such as zero-dimensional reactors and one-dimensional flames (see Section 15.2), and then applied to model complex systems, such as the flame–vortex interaction (Coussement et al. 2013), the flame–turbulence interaction (Isaac et al. 2015) as well as the turbulent premixed (Coussement et al. 2016) and non-premixed flames (Malik et al. 2020). Up-to-date, the PC-transport approach has been demonstrated for a wide range of problems involving ideal reactors, including batch and PSR reactors.

15.4.1 Nonlinear Regression Models

The thermochemical state-space variables $\mathbf{\Phi}$ and the PC-source terms $\mathbf{S_{Z_q}}$ can be mapped to the q-dimensional PC space using a nonlinear regression function $\mathscr{F}$, such that

$$\phi \approx \mathscr{F}(\mathbf{Z_q}),$$

where ϕ represents any of the dependent variables ($T, p, Y_1, Y_2, \ldots, Y_{N_s-1}$, or $\mathbf{S_{Z_q}}$). The models used in the literature to generate the function $\mathscr{F}$ include, but are not limited to

- multivariate adaptive regression splines (MARS) (Friedman 1991);
- artificial neural networks (ANNs) (Pao 2008);
- Gaussian process regression (GPR) (Rasmussen 2006).

In a previous study (Isaac et al. 2015), the authors compared different regression models in their ability to accurately map highly nonlinear functions (such as the chemical source terms) to the PCA manifold. In this section, we focus on the use of GPR for state-space and source term parameterization. The choice of GPR is due to its semi-parametric nature that increases the generality of the approach. GPR employs Gaussian mixtures to capture information about the relation between data and input parameters, making predictions of non-observed system states more reliable than in fully parametric approaches.

15.4.2 Validation of the PC-Transport Approach in Large Eddy Simulation Simulations

To demonstrate the application of the PC-transport approach, we show the results of the LES of a multidimensional flame (Malik et al. 2020). A piloted methane–air diffusion flame (referred to as Flame D) (Barlow & Frank 1998) for which high-fidelity experimental data is available is selected. Flame D is fueled by a mixture of CH_4 and air (25% / 75% by volume), at $Re = 22\,400$ and 294K, respectively. The fuel jet is surrounded by pilot flames that stabilize the flame, which is surrounded by a co-flow.

The development of a reduced-order model for subsequent application in LES requires the availability of training data. The most critical aspect when generating a training data set is to make sure that the generated state-space includes all the possible states accessed during the actual simulation. At the same time, that data cannot come from expensive simulations. Parametric calculations on inexpensive canonical configurations can provide an effective solution in this context, although the nature of the flame archetype chosen is an open subject of discussion. Considering the non-premixed nature of Flame D, it was decided to rely on unsteady one-dimensional laminar counter diffusion flames. The use of unsteady inlet velocity profiles allows us to explore transient flame behavior, including extension and reignition phenomena. The calculations were performed using the OpenSMOKE++ suite developed in Politecnico di Milano (Cuoci et al. 2015). The GRI 3.0 (Gardiner et al. 2012) mechanism, involving 35 species and 253 reactions (excluding NO_x), was used. The inlet conditions were set equal to the experimental ones (Barlow & Frank 1998). Multiple simulations were carried out by varying the strain rate, from equilibrium to complete extinction. The unsteady solutions were saved on a uniform grid of 400 points over a 0.15 m domain. All of the unsteady data from the various simulations was used collectively for the PCA analysis. The final data set consisted of $\sim 80,000$ observations for each of the state-space variables.

Table 15.2 shows the basis matrix weights obtained from the PCA analysis on the five major chemical species only. It can be seen that $\mathbf{Z}_1$ has a large positive weight for CH_4 and a large negative value for the oxidizer (O_2 and N_2). This can be linked to the definition of mixture fraction from (15.3). Figure 15.16 shows a plot of the first PC highly correlated with the mixture fraction (correlation factor exceeding 99%). Therefore, in the numerical simulation, the first PC is directly replaced by the mixture fraction, to avoid transporting a reactive scalar. The weights for $\mathbf{Z}_2$ also show an interesting pattern: a positive correlation for H_2O and CO_2, and a negative correlation for CH_4, O_2 and N_2. This can be linked to a progress variable, where products have positive stoichiometric coefficients and reactants negative ones. Figure 15.17 shows a plot of the second PC highly correlated with the mass fraction of CO_2, a variable that can be attributed to the progress of reaction. It is worth pointing out that PCA identifies these patterns without any prior assumptions or knowledge of the system of interest.

LES simulations were then performed in OpenFOAM using the PC-transport approach. The independent variables (PCs) are transported, and the state-space is recovered from the nonlinear regression using GPR. The low-Mach Navier–Stokes equations are solved on an unstructured grid, together with the PC-transport equations from (15.20). Since the state-space was accurately regressed using $q = 2$ PCs, the simulation was carried out using only $\mathbf{Z}_1$ and $\mathbf{Z}_2$ as transported PCs. More details about the numerical setting are reported in Malik et al. (2020). As part of the validation

Species	$\mathbf{A}_1$	$\mathbf{A}_2$	$\mathbf{A}_3$	$\mathbf{A}_4$	$\mathbf{A}_5$
H_2O	−0.02	0.51	0.45	−0.73	0.02
O_2	−0.18	−0.67	−0.22	−0.60	0.30
N_2	−0.64	−0.01	−0.15	−0.09	−0.74
CH_4	0.73	−0.14	−0.20	−0.27	−0.56
CO_2	−0.02	0.50	−0.82	−0.14	0.21

Table 15.2 The elements a_{ij} of the basis matrix $\mathbf{A}$ identified by PCA when the data set was composed of five major chemical species only.

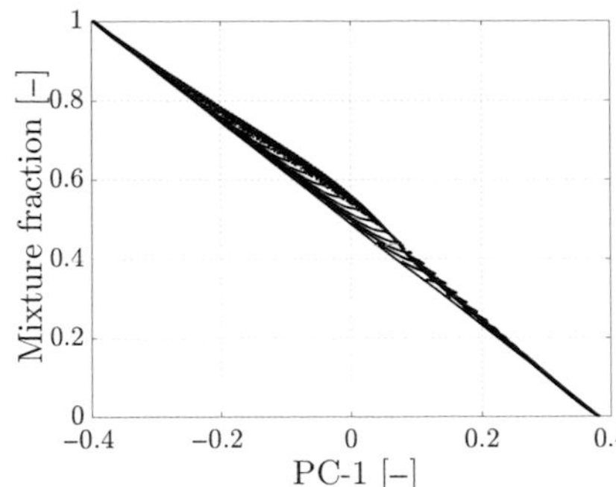

Figure 15.16 Scatter plot of the first PC (PC-1) versus the mixture fraction variable. From Malik et al. (2020), copyright 2020, with permission from Elsevier.

of the method, we look at the temperature profiles conditioned on mixture fraction at axial location $x/D = 60$ ($x = 432$ mm), shown in Figure 15.18. It can be observed that the predicted temperature lies well inside the single shot experimental data points.

Figure 15.19 shows the original manifold obtained from the training data set, while Figure 15.20 shows the manifold accessed during the simulation with major species at $t = 1$ s. It can be observed that the simulation does not leave the training manifold, which bounds all the points. It is also apparent that most of the solution data is clustered toward the equilibrium solution, showing that the Flame D does not experience significant extinction and reignition.

The PC–GPR model can accurately simulate a complex multidimensional flame, using only two PCs instead of the 35 chemical species. The model can be built using inexpensive one-dimensional simulations, as long as the training data cover the potential state-space accessed during the actual simulation. The PCs stay bounded to

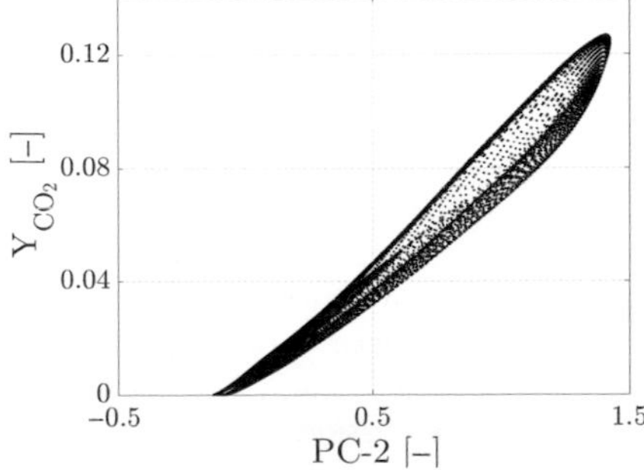

Figure 15.17 Scatter plot of the second PC (PC-2) versus the mass fraction of CO_2 species, Y_{CO_2}. From Malik et al. (2020), copyright 2020, with permission from Elsevier.

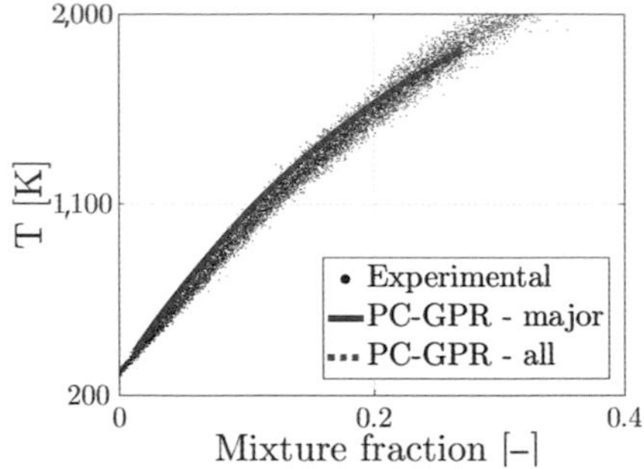

Figure 15.18 Conditional average of the temperature at $x/D = 60$. From Malik et al. (2020), copyright 2020, with permission from Elsevier.

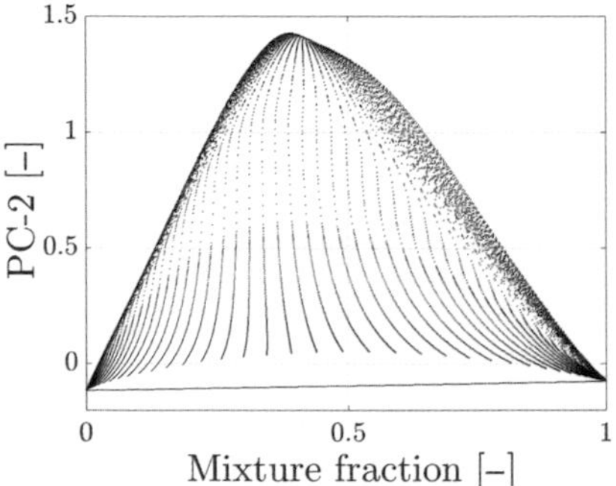

Figure 15.19 Scatter plot of the low-dimensional combustion manifold in the space of the second PC (PC-2), and the mixture fraction variable. Manifold obtained from the training data set. From Malik et al. (2020), copyright 2020, with permission from Elsevier.

the training manifold during the simulation, indicating that the choice of an unsteady canonical reactor ensures to span all the potential chemical states accessed during the simulation. The strength of the method resides in the fact that it does not require any prior selection of variables. Instead, it automatically extracts the most relevant variables to describe the system. From this perspective, the PC–GPR method can be regarded as a generalization of tabulated chemistry approaches (Pope 2013), particularly for complex systems requiring the definition of a larger number of progress variables.

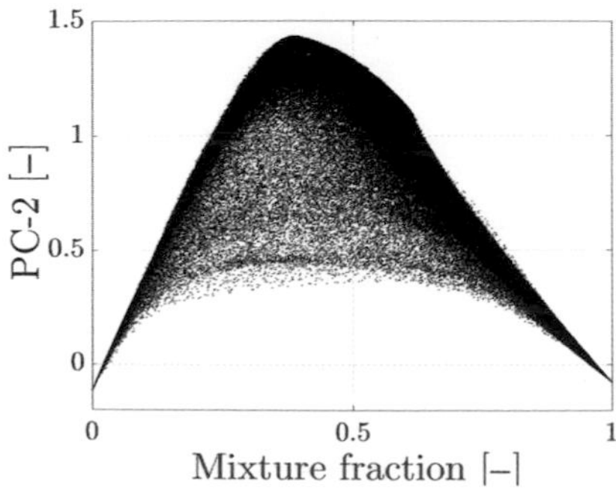

Figure 15.20 Scatter plot of the low-dimensional combustion manifold in the space of the second PC (PC-2), and the mixture fraction variable. Manifold accessed during the simulation. From Malik et al. (2020), copyright 2020, with permission from Elsevier.

15.5 Chemistry Acceleration via Adaptive-Chemistry

In Section 15.4, PCA was used to derive the reduced number of transport equations for the new set of variables, PCs. Regression was introduced to handle the nonlinearity of chemical source terms. In this section, we investigate the potential of local PCA (Section 15.3.2) to classify the thermochemical state-space into locally homogeneous regions and apply chemical mechanism reduction locally.

15.5.1 Description of the Approach

Ren and Pope (2008) proposed the operator-splitting strategy for application in numerical algorithms used to solve the multicomponent reactive flows. With this approach, chemistry acceleration techniques can be implemented to reduce the computational cost related to the inclusion of detailed kinetic mechanisms. If we consider $\mathbf{\Psi}$ as a vector representing the temperature and mass fractions of chemical species, the transport equation for $\mathbf{\Psi}$ can be written as

$$\frac{d\mathbf{\Psi}}{dt} = \mathbf{C}(\mathbf{\Psi},t) + \mathbf{D}(\mathbf{\Psi},t) + \mathbf{S}(\mathbf{\Psi}), \tag{15.21}$$

where $\mathbf{C}(\mathbf{\Psi},t)$ and $\mathbf{D}(\mathbf{\Psi},t)$ are the convective and diffusive transport terms, respectively, and $\mathbf{S}(\mathbf{\Psi})$ is the source term vector (representing the rates of change of $\mathbf{\Psi}$ due to chemical reactions) (Cuoci et al. 2013). For N grid points in the simulation, we obtain a system of N ordinary differential equations (ODEs). According to the Strang splitting scheme (Strang 1968), (15.21) can be solved by grouping the contributions of convection and diffusion and integrating them separately from the chemical source term

$$\begin{cases} \frac{d\mathbf{\Psi}^a}{dt} = \mathbf{S}(\mathbf{\Psi}^a), \\ \frac{d\mathbf{\Psi}^b}{dt} = \mathbf{C}(\mathbf{\Psi}^b,t) + \mathbf{D}(\mathbf{\Psi}^b,t), \end{cases} \tag{15.22}$$

where the indices a and b denote the separate contributions to $\mathbf{\Psi}$. In particular, three sub-steps are adopted for the numerical integration:

1. *Reaction step:* The ODE system corresponding to the source term $\mathbf{S}(\mathbf{\Psi}^a)$ is integrated over $\frac{\Delta t}{2}$.
2. *Transport step:* The ODE system accounting for the convection and diffusion terms $\mathbf{C}(\mathbf{\Psi}^b,t)$ and $\mathbf{D}(\mathbf{\Psi}^b,t)$ is integrated over Δt.
3. *Reaction step:* The source term $\mathbf{S}(\mathbf{\Psi}^a)$ is again integrated over $\frac{\Delta t}{2}$.

Unlike the transport step, the two reaction steps do not require boundary conditions and they do not have spatial dependence. The system of N ODEs from steps 1 and 3 can be solved independently from the system of N ODEs from step 2: this makes the adaptive-chemistry techniques very effective and easy to implement. The idea behind the adaptive-chemistry approach is that it is possible to locally consider only a subset of the chemical species implemented in the detailed mechanism, while the remaining subset consists of the species that locally have zero concentration, or result to be not chemically active. It is thus possible to build a library of reduced mechanisms in the preprocessing step, covering the composition space, which is expected to be visited during simulation of the flame of interest (Schwer et al. 2003, Banerjee & Ierapetritou 2006). To build the library of reduced mechanisms, the state-space must be first partitioned into a prescribed number of clusters, and a reduced kinetic mechanism is then created separately in each cluster. The high-dimensional space can be partitioned via the VQPCA algorithm (Section 15.3.2), and for each cluster a reduced kinetic mechanism is generated via a directed relation graph with error propagation (DRGEP) (Pepiot-Desjardins and Pitsch 2008). At each time step of the CFD simulation, the whole set of species in the detailed mechanism is transported. Before the chemical step, the grid points are classified *on-the-fly* using the local PCA reconstruction error metrics (Kambhatla & Leen 1997), and for each of them the appropriate reduced mechanism is selected from the library. Although a direct relation between the CPU time and the number of species implemented in the mechanism is found for both the transport and the reaction sub-steps described in the (15.22), alleviating the costs related to the reaction step appears to be more important. The reaction step, in fact, results to be the most consuming part of the computation requiring up to 80–85% of the CPU time (Cuoci et al. 2013). The full details of the overall procedure, called SPARC, as well as its validation for steady and unsteady laminar flames, can be found in D'Alessio et al. (2020).

15.5.2 Application of the Approach

The application of the adaptive-chemistry approach is briefly presented here for the simulation of an axisymmetric, non-premixed laminar nitrogen-diluted methane co-flow flame (Mohammed et al. 1998). The detailed simulation was first carried out using the `POLIMI_C1C3_HT_1412` kinetic mechanism (84 species and 1698 reactions) (Ranzi et al. 2014), and the adaptive simulations were then performed using four degrees of chemical reduction, ranging from $\epsilon_{DRGEP} = 0.03$ to $\epsilon_{DRGEP} = 0.005$. The average and the maximum number of species included in the reduced mechanisms depending on the reduction tolerance ϵ_{DRGEP} are reported in Table 15.3. An excellent agreement was observed for all the cases, although for $\epsilon_{DRGEP} = 0.03$ and 0.02 a negligible discrepancy was observed for the lift-off height. The inverse proportionality between the simulation accuracy and the degree of reduction, that is, the tolerance used for DRGEP, is due to a different number of species included in the reduced mechanisms. Figure 15.21 shows a comparison between the temperature and species profiles obtained from the detailed and the adaptive simulation carried out with $\epsilon_{DRGEP} = 0.005$.

The inclusion of a higher number of chemical species in the reduced mechanisms obviously entails an increase of the mean CPU time required to carry out the chemical step ($\bar{\tau}_{chem}$). Nevertheless, the adaptive simulation speedup with respect to the detailed simulation (S_{chem}) is large ($S_{chem} \sim 4$) even when the lowest degree of chemical reduction is adopted, as reported in Table 15.4.

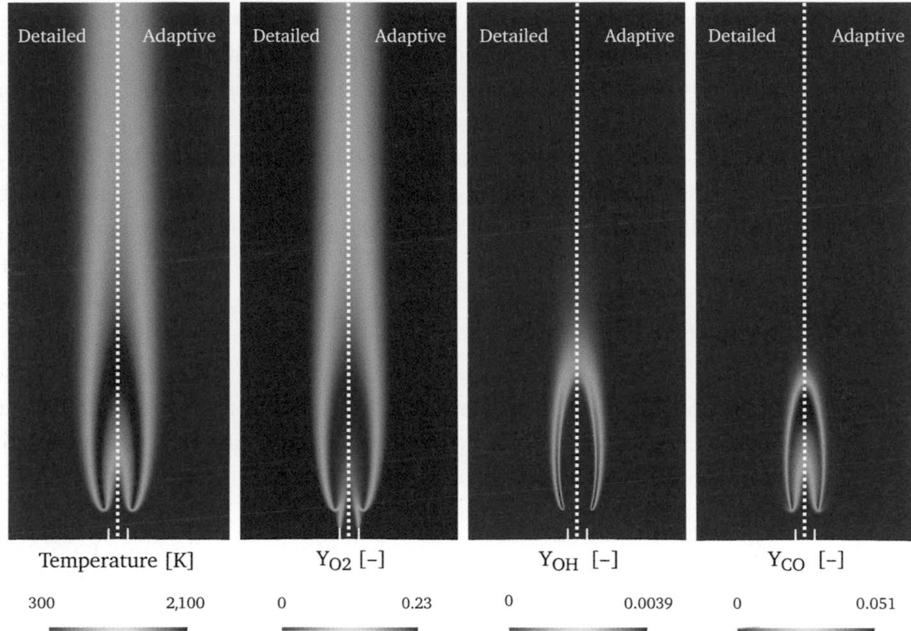

Figure 15.21 Temperature, O_2, OH and CO profiles obtained from the detailed and the adaptive-chemistry simulation of an axisymmetric, non-premixed laminar nitrogen-diluted methane coflow flame.

ϵ_{DRGEP}	n_{sp}^{mean}	n_{sp}^{max}
0.03	31	38
0.02	34	42
0.01	39	44
0.005	43	50

Table 15.3 The mean and maximum number of chemical species versus the uniformity coefficients for prescribed tolerances ϵ in DRGEP. The composition space was partitioned via the VQPCA algorithm.

If multidimensional simulations of the same system are not available, the SPARC model can be trained using lower-dimensional (0D or 1D) detailed simulations of the same chemical system. Reduced mechanisms can then be generated based on such training data sets and applied in the multidimensional (2D or 3D) adaptive simulations. D'Alessio, Parente, Stagni and Cuoci (2020) also considered a data set consisting

ϵ_{DRGEP}	$\bar{\tau}_{chem}$	S_{chem}
0.03	2.78	5.39
0.02	3.03	4.94
0.01	3.25	4.61
0.005	3.78	3.97
detailed	15.02	-

Table 15.4 Performances of adaptive-chemistry algorithm: average CPU-time per cell (in ms) for chemical step integration ($\bar{\tau}_{chem}$) and relative mean speed-up factor (S_{chem}) for the steady-state laminar methane flame with respect to the degree of chemistry reduction (ϵ_{DRGEP}).

of observations generated through steady-state CFDF simulations (see Section 15.2), evaluated at different strain rates from $10s^{-1}$ to $330s^{-1}$ (the latter corresponding to extinction conditions). This new data set consisted of $\sim$ 11 000 points, corresponding to 30 different strain rates randomly chosen in the considered interval. Also in this case, the accuracy using mechanisms obtained from a lower dimensional training data set was high, with the speedup with respect to the detailed simulation between 4.5 and 6. More details can be found in D'Alessio, Parente, Stagni and Cuoci (2020).

15.6 Available Software

Apart from the many available MATLAB® routines, some of which were mentioned in this chapter, two recently developed Python libraries can be used to perform dimensionality reduction using PCA and other techniques:

- `OpenMORe` (D'Alessio et al. 2020) is a collection of Python modules for reduced-order modeling (ROM), clustering, and classification. It incorporates many ROM techniques such as PCA, local PCA, kernel PCA, and NMF along with the varimax rotation for factor analysis. It can also be used for data clustering with VQPCA or FPCA algorithms, and it introduces utilities for evaluating the clustering solutions. The software, along with detailed documentation and several examples, is available at `https://github.com/burn-research/OpenMORe`.
- `PCAfold` (Zdybał et al. 2020) is a Python library that can be used to generate, improve, and analyze low-dimensional manifolds obtained via PCA. Several novel functionalities to perform PCA on sampled data sets and analyze the improved representation of the thermochemical state-space variables (Zdybał et al. 2021) were introduced. The effect of the training data preprocessing on the topology of the low-dimensional manifolds can be investigated with the novel approach to assess the quality of manifolds (Armstrong & Sutherland 2021, Zdybal et al. 2022a, Zdybal et al. 2022b). `PCAfold` also accommodates for treatment of the thermochemical source terms for use in PC-transport approaches (Section 15.4). The user can find numerous illustrative examples and the associated Jupyter notebooks in the software documentation (available at `https://pcafold.readthedocs.io/`) under **Tutorials & Demos**. The software is available at `https://gitlab.multiscale.utah.edu/common/PCAfold`.

15.7 Summary

In this chapter, several examples of the application of data-driven techniques to multi-component reactive flow systems have been shown. Data reduction techniques, such as PCA, NMF, or autoencoders also offer a way to explore hidden features of data sets (Section 15.3). These methods are capable of detecting information in an unsupervised way and can thus aid in the process of selecting the best set of variables to effectively parameterize complex systems using fewer dimensions. Moreover, it has been shown how the aforementioned algorithms can be effectively coupled with algorithms that are widely used in the combustion community, such as DRGEP, to obtain physics-informed reduced-order models. Two applications of reduced-order modeling were presented in this chapter: reduction of the number of transport equations (Section 15.4) and reduction of large chemical mechanisms (Section 15.5). The main power of data-driven techniques is that the modeling can be informed by applying the technique on simple systems that are computationally cheap to obtain. Further improvement can be achieved based on the feedback coming from the validation experiments.

16 Reduced-Order Modeling for Aerodynamic Applications and Multidisciplinary Design Optimization

S. Görtz[1], P. Bekemeyer, M. Abu-Zurayk, T. Franz, and M. Ripepi

Reduced-order models (ROMs) for both steady and unsteady aerodynamic applications are presented. The focus is on compressible, turbulent flows with shocks. We consider ROMs combining proper orthogonal decomposition (POD), Isomap, which is a nonlinear manifold learning method, and autoencoder networks with interpolation methods. Physics-based ROMs, where an approximate solution is found in the POD-subspace by minimizing the corresponding steady or unsteady flow-solver residual, are also being discussed. The ROMs are used to predict unsteady gust loads for rigid aircraft as well as static aeroelastic loads in the context of multidisciplinary design optimization (MDO) where the structural model is to be sized for the (aerodynamic) loads. They are also used in a process where an a priori identification of the critical load cases is of interest and the sheer number of load cases to be considered does not allow for using high-fidelity numerical simulations. The different ROM methods are applied to 2D and 3D test cases at transonic flow conditions where shock waves occur and in particular to a commercial full aircraft configuration.

16.1 Introduction and Motivation

The multidisciplinary design of a civil transport aircraft is a highly iterative optimization process, each design cycle requiring a large volume of computations to analyze the current performance, handling qualities and loads. For conventional aircraft, a load envelope may require on the order of 100 000 simulations to find all critical load cases, comprising gust encounters, maneneuvers, and failure scenarios at various, mostly off-design Mach numbers, flight levels, and trim states within the aircraft flight envelope. For unconventional aircraft, where little or no engineering experience is available, up to 10 million computations may be required. The use of high-fidelity

[1] The author would like to thank Airbus for providing the XRF-1 testcase as a mechanism for demonstration of the approaches presented in this chapter.

computational fluid dynamics (CFD) in this context is at the horizon, but still too costly and time-consuming to provide all the required aerodynamic data, that is, steady and unsteady pressure and shear stress distributions on the aircraft surface, at any point within this envelope. This motivates procedures and techniques aimed at reducing the computational cost and complexity of high-fidelity simulations to provide accurate but fast computations of, for example, the aerodynamic loads and aircraft performance.

A classical approach to reduce the numerical complexity is to simplify the physics. An example of this is the common use of linear potential flow equations during loads analysis. However, such physical model simplifications have the disadvantage of neglecting significant effects such as transonic flow, stall, and friction drag in the case of aerodynamics. This may be acceptable early on in the design process, while more detailed analysis may be applied at a later stage when the design space has been narrowed down sufficiently.

As an alternative to simplifying the physical model, reduced-order modeling (ROM) provides another approach to reduce numerical complexity and computational cost while providing answers accurate enough to support design decisions and to perform quick trade studies. In general, the various ROM methods realize such a goal by identifying a low-dimensional subspace or manifold based on an ensemble of high-fidelity solutions "snapshot," which sample a certain parametric domain of interest. The number of degrees of freedom (DoF) is then reduced while retaining the problem's physical fidelity, thus allowing predictions of the required aerodynamic data with lower evaluation time and storage than the original CFD model. ROMs are anticipated to enable incorporating high-fidelity-based aerodynamic data earlier into the design process. Methods for minimizing the number of expensive high-fidelity simulations needed to extract a reduced-order model from simulation data are sought after, including adaptive sampling strategies. Data classification methods are of interest to gain more physical insight, for example, to identify (aerodynamic) nonlinearities and to track how they evolve over the design space and flight envelope. Finally, data-driven methods, including machine learning methods stemming from other fields of research, that are able to extract relevant information from high-fidelity numerical and experimental data and for merging heterogeneous data can help to arrive at a more consistent, homogeneous digital aircraft model. This chapter focuses on reduced-order modeling approaches based on high-fidelity CFD in the context of aerodynamic applications and multidisciplinary analysis and optimization. The emphasis is on the basic ideas behind the methods, their accuracy and computational efficiency, and their application to industrial aircraft configurations. The mathematical formulation of the methods can be found in the corresponding references.

The basic ideas behind reduced-order modeling based in POD and Isomap are explained briefly in the next section, followed by a description of their implementation. The last section is concerned with their application to different two- and three-dimensional use cases.

16.2 Reduced-Order Modeling

The goal of a ROM is to emulate an expensive process like a full-order CFD simulation. Based on a given set of parameter-dependent CFD solutions, called snapshots, a ROM can be built and exploited to predict solutions at untried parameter combinations. For highly nonlinear functional dependencies of the underlying process, a proper set of snapshots is necessary to obtain a sufficient accuracy of the predictions (Franz et al. 2014). ROMs can be divided into intrusive and nonintrusive methods. Intrusive methods make use of the underlying equations, whereas nonintrusive ROMs are completely decoupled from the underlying equations, but coupled with an interpolation method, which inherits the parameter dependency of the solutions.

ROMs for aerodynamic applications operate on parametrically generated data represented by either surface quantities (e.g., surface pressure and shear stress) or volume quantities (e.g., the primitive variables). Here, the focus is on compressible, turbulent flows with shocks as described by the Navier–Stokes equations. The DLR TAU code (Kroll et al. 2014) is utilized as a CFD solver, employing hybrid unstructured grids, to obtain the aerodynamic data snapshots by solving the Reynolds-Averaged Navier–Stokes (RANS) equations. The parameters can be related to the flow (e.g., the angle of attack, the Mach number), the geometry (e.g., wing span, taper ratio, and sweep angle), the structure (e.g., Young's and shear modulus of the beam representation of the wing box), and the flight condition (e.g., load factor, altitude). The model order reduction techniques used are hereafter briefly described, and their application to aerodynamic problems and Multidisciplinary Design Analysis and Optimization (MDAO) is shown in the following.

16.2.1 Proper Orthogonal Decomposition-Based ROMs

A widely used tool for reduced-order modeling is POD, also known as principal component analysis (PCA) and Karhunen–Loéve expansion (KLE). In fluid dynamics, POD is applied to steady problems and unsteady problems in the time as well as frequency domain as illustrated in Chapter 6. The POD method generates a sequence of orthogonal basis functions, called POD modes, through modal analysis of an ensemble of snapshot flow solutions, where every snapshot corresponds to a different combination of parameters, for example, different combinations of Mach number and angle of attack. These solutions to the governing equations, here the RANS equations, form the solution space. The POD modes span an optimal linear subspace for the corresponding solution space. By choosing a subset of modes, the method seeks to isolate the few main structures whose linear combination represents the system in an optimal way. Several variants of POD-based ROMs have been developed that primarily differ in the way they connect the retained modes to the parameter space of the snapshot distribution. The POD may be embedded in a Galerkin projection framework; it may be combined with a CFD flux residual minimization scheme; or it may be coupled to an interpolation method (POD+I). Galerkin projection is an example of an intrusive method. It projects the underlying spatially discretized partial differential equations (PDEs) onto the POD subspace to obtain a system of ordinary differential equations (ODEs).

POD+I is a nonintrusive method as the interpolation technique does not require any details on the underlying governing equations. ROM-predicted solutions are determined by directly interpolating the coefficients of the POD modes, without the need to solve any ODE system. It generally establishes a multidimensional relationship between the modal coefficients or amplitudes and the parameter space, for example, by fitting a radial basis function in the modal space to the set of snapshot points in the parameter space. This requires simple interpolation of scalar-valued POD basis coefficients to get a surrogate model of these coefficients as a function of the parameters. To predict a flow solution at an untried combination of parameters, the surrogate is evaluated for these parameters, and the predicted POD coefficients are multiplied with the corresponding precomputed, invariant POD modes. This has the advantage of simplicity of implementation and independence of the complexity of the system and source of the modes being processed, which allows for application to multidisciplinary problems and the combination of different data sources such as CFD and experimental test results. The main disadvantage of nonintrusive POD methods stems from their reliance on interpolation techniques to accurately reproduce the possibly very nonlinear response surfaces of the modal coefficients.

Intrusive POD-based methods do better in this respect. This is done by solving a nonlinear least-squares (LSQ) optimization problem for the modal coefficients minimizing the steady (or unsteady) flow-solver residual of the governing equations in POD subspace (Zimmermann & Görtz 2010). For the semi-discrete Navier–Stokes equations,

$$\hat{\mathbf{R}} := \mathbf{R}\left(\mathbf{w}\left(t\right)\right) + \frac{\partial \mathbf{w}\left(t\right)}{\partial t} = \mathbf{0} \in \mathbb{R}^{N}, \tag{16.1}$$

with $\mathbf{w} = [\rho, \rho\mathbf{v}, \rho E^t, \nu_t] \in \mathbb{R}^N$ and N corresponding to the order of CFD model (number of variables times number of grid nodes), we search for an approximated solution corresponding

$$\mathbf{w} \approx \sum_{i=1}^{r} a_i \mathbf{U}^i + \overline{\mathbf{w}} = \mathbf{U}_r \mathbf{a} + \overline{\mathbf{w}} \tag{16.2}$$

in the POD subspace $\mathbf{U}_r \in \mathbb{R}^{N\ x\ r}$, $r \ll N$, minimizing a subset of the unsteady residual in the L_2 norm,

$$\min_{\mathbf{a}} \parallel \hat{\mathbf{R}}^* \left(\mathbf{U}_r \mathbf{a} + \overline{\mathbf{w}}\right) \parallel_{L_2}^2, \tag{16.3}$$

with $\mathbf{a}$ being the vector of the unknown POD coefficients a_i and $\overline{\mathbf{w}}$ the mean of the snapshots set. A greedy missing point estimation is used to select the subset. This approach is an alternative to POD-based ROMs based on Galekin–projection. In the following, such approaches will be referred to as POD+LSQ or LSQ-ROM.

16.2.2 Reduced-Order Models Based on Isomap

The linear nature of POD makes the method attractive but also is a source of its limitations. Highly nonlinear flow phenomena, such as shocks, are often insufficiently reproduced, because of the underlying assumption that the full-order CFD flow solutions lie in a low-dimensional linear subspace. Although the coefficients of the POD basis can handle nonlinearities, the nonlinearity might not be sufficiently accounted for in the POD basis due to insufficient sampling in the solution space.

One approach to improve the fidelity of linear ROMs is to substitute the POD with a nonlinear manifold learning, or, more generally, dimensionality reduction (DR) technique, which assumes that full-order data lie on a nonlinear manifold of low-dimension. The manifold can be approximated by sampling the full-order model. IsomapTenenbaum (2000), which is a nonlinear DR method based on multidimensional scaling (MDS), can be employed to extract low-dimensional structures hidden in a given high-dimensional data set. The Isomap method only provides a mapping from the high-dimensional input space onto a lower-dimensional embedding space for a fixed finite set of given snapshots. For any ROM of the Navier–Stokes equations, however, it is an essential requirement that the approximate reduced-order flow solutions are of the same type and dimension N as the CFD snapshots. Hence, once the set of low-dimensional vectors is obtained, a back-mapping from the reduced-order embedding to the high-dimensional solution space is mandatory. Coupled with an interpolation model formulated between the parameter space and the low-dimensional space, a ROM is obtained, which is capable of predicting full-order solutions at untried parameter combinations. This method will be referred to as Isomap+I. Furthermore, another back-mapping from the low-dimensional space to the high-dimensional space may be performed based on residual optimization. Its objective is to obtain a CFD-enhanced prediction by minimizing the discretized flux residual of the interpolated solution. This method will be referred to as Isomap+LSQ.

A schematic representation of the procedure for creating parametric ROMs for steady flow problems based on POD and Isomap in combination with interpolation or residual minimization is shown in Figure 16.1.

16.3 Implementation

The methods described herein are implemented in DLR's Surrogate Modeling for AeRo Data Toolbox (SMARTy) [Bekemeyer et al, 2022]. SMARTy is a modular, object-oriented Python package for data-driven modeling, simulation, and optimization. It is designed as an application programming interface (API) that provides a set of techniques that can be used out of the box, utilizing the provided application scripts. For more advanced scenarios, different methods can easily be combined in "user-defined" scripts and workflows.

Key features include design of experiments methods, adaptive sampling strategies, dimensionality reduction, and data fusion techniques (Mifsud et al. 2019), as well as methods for surrogate modeling, variable-fidelity modeling (Han & Görtz 2012a),

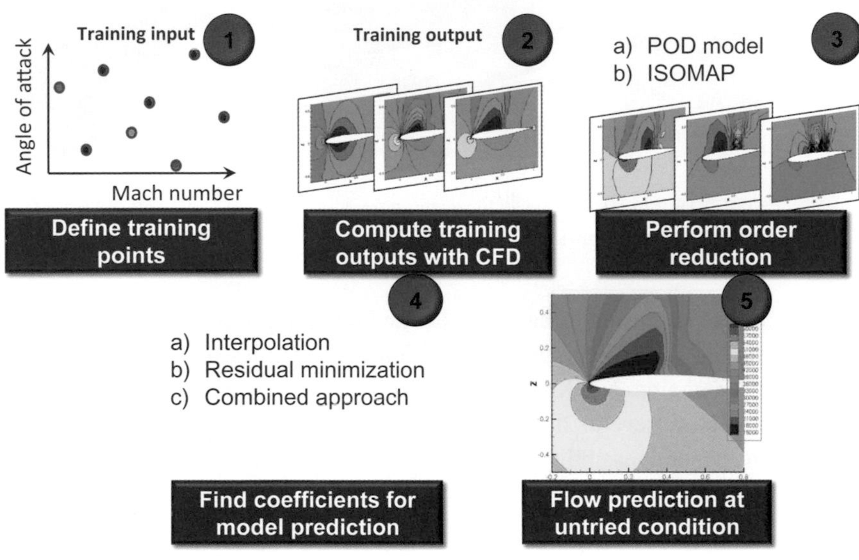

Figure 16.1 Steady ROMs in a nutshell.

and intrusive and nonintrusive reduced-order modeling (Zimmermann et al. 2014). In terms of interpolation, thin-plate spline (TPS) interpolation, Kriging, gradient-extended Kriging (Han et al. 2013), and hierarchical Kriging (Han & Görtz 2012b) are available. Various kernels (or correlation functions) are implemented and differentiated and can be used with these interpolation methods. Computationally demanding tasks of the interpolation module are implemented in Cython. Moreover, it has an interface to the FlowSimulator (Reimer et al. 2019) that enables easy access to the DLR CFD code TAU as well as allows using SMARTy as a plugin in complex workflows such as surrogate-based optimization. SMARTy is parallelized in order to handle distributed CFD data on multicore architectures. SMARTy can be used for rapidly predicting aerodynamic data based on high-fidelity CFD (Mifsud et al. 2014), fusing numerical and experimental data (Franz & Held 2017), surrogate-based optimization (Han et al. 2018), uncertainty quantification (Kroll et al. 2015), and robust design (Sabater et al. 2020).

16.4 Applications

The applications demonstrate how reduced-order modeling and dimensionality reduction methods can be employed to provide a digital description of an aircraft in terms

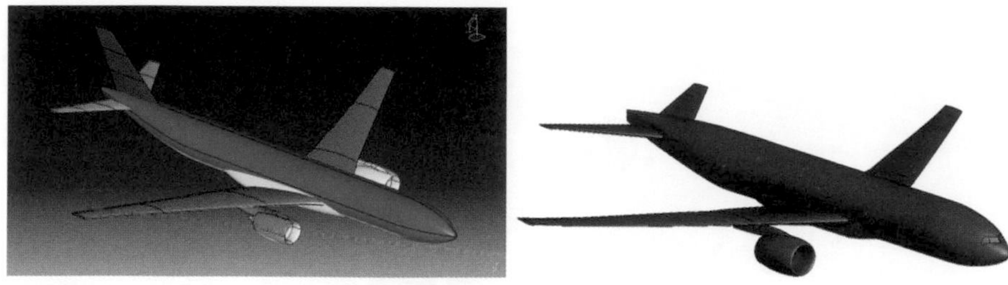

Figure 16.2 XRF-1 generic long-range transport aircraft (left) and NASA's Common Research Model (CRM) (right).

of its aerodynamic characteristics and properties based on a limited set of high-fidelity simulations. This digital description is sometimes referred to as "virtual aircraft" or "digital aircraft" (Kroll et al. 2015).

Below, several 2D and 3D test cases are used. The 2D test cases are used to validate the general feasibility of the methods. The 3D use cases demonstrate their applicability to problems of interest to industry. Three 3D use cases are employed, an isolated wing and two complete aircraft configurations. The first one is the XRF-1 long-range transport aircraft configuration. The XRF-1 is an Airbus provided industrial standard multidisciplinary research test case representing a typical configuration for a long-range wide body aircraft. The XRF-1 research test case is used by Airbus to engage with external partners on development and demonstration of relevant capabilities/technologies. The second one is the NASA Common Research Model (CRM) (Vassberg et al. 2008). The CRM consists of a contemporary supercritical transonic wing and a fuselage that is representative of a wide-body commercial transport aircraft. The geometry of both aircraft configurations is shown in Figure 16.1.

16.4.1 Parametric Reduced-Order Models Based on Proper Orthogonal Decomposition and Isomap for LANN Wing

A parametric ROM for steady CFD data was developed based on Isomap, a nonlinear "manifold learning" method. It is assumed that the space of all CFD solutions forms a nonlinear manifold, which in turn constitutes a sub-manifold of $\mathbb{R}^n$ of lower intrinsic dimensionality. This is a more elaborate approach compared to POD method in which a linear subspace of $\mathbb{R}^n$ is assumed. Isomap is comparable to the kernel-PCA, which uses kernels derived from simulation data. ML methods apply various approaches to identify the manifold geometry and to represent it in a low-dimensional Euclidean space. For this purpose, Isomap uses the pairwise geodesic distances between previously generated CFD solutions at selected parameter combinations. After approximating these distances, Isomap computes a data set of low-dimensional vectors in the so-called embedding space, whose pairwise Euclidean distances correspond to the approximated geodesic distances. A mapping from the embedding space back onto the manifold in the high-dimensional CFD solution space was developed to make predictions of CFD solutions at any point in the parameter

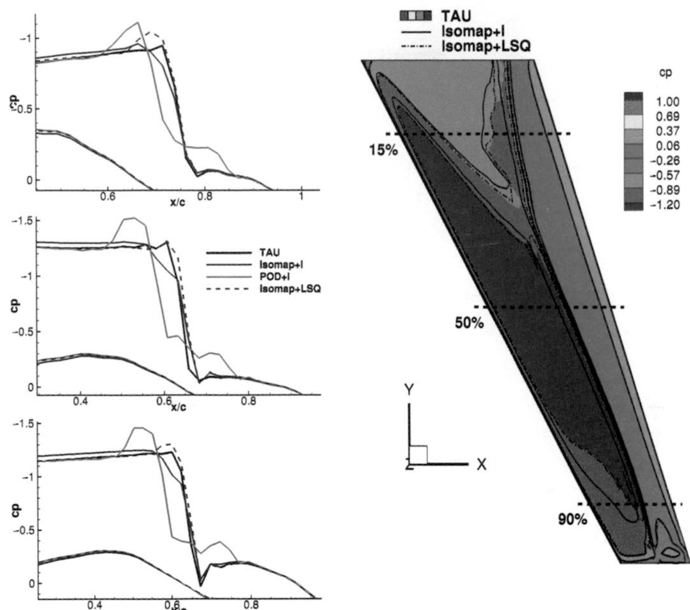

Figure 16.3 Steady Euler computation for the LANN wing compared to intrusive and nonintrusive ROM predictions based on Isomap and POD, $M = 0.81$, $\alpha = 2.6°$.

space. When combined with an interpolation model between the parameter space and the embedding space, similar to POD with interpolation (POD+I), an Isomap-based ROM (Franz et al. 2014), called Isomap+I, is obtained.

Parametric ROMs for the steady transonic flow around the LANN wing were generated. The LANN Wing is a well-known supercritical research wing model, built by the Lockheed-Georgia Company for the U.S. Air Force for testing at NLR and NASA Langley in 1983; the best reference is: LANN (Lockheed, AFWAL, NASA-Langley, and NLR) Wing Test Program: Acquisition and Application of Unsteady Transonic Data for Evaluation of Three-Dimensional Computational Methods (dtic.mil). Here, the ROMs were parametric with respect to the flow conditions. The parameter space, defined by variations of the angle of attack, α and the Mach number, Ma, was previously sampled in the range $[\alpha \times Ma] = [1°, 5°] \times [0.76, 0.82]$ using a "design of experiment" (DoE) based on a Latin-Hypercube sampling approach with 25 different α-Mach-combinations. At these parameter combinations, the corresponding CFD solutions were computed and used as input data for both ROMs. Figure 16.3 provides a comparison of the predictions of Isomap+I and POD+I with the computed TAU reference solution for the pressure coefficient at three wing sections for the LANN wing in inviscid flow. The two different curves for the pressure coefficient distributions correspond to the pressure and suction sides. The required parameters for Isomap+I were determined automatically, yielding a two-dimensional embedding space and seven nearest neighbors to be ideal for the embedding. Isomap+I yields better predictions than a POD-based interpolation method, in particular for the location and magnitude of the shock wave at transonic flow conditions. Isomap+LSQ is able to further improve the prediction around the shock wave, where Isomap+LSQ refers to the intrusive ROM approach based on residual minimization.

16.4.2 Parametric Reduced-Order Model Based on Proper Orthogonal Decomposition and Isomap for XRF-1

The validation of the POD and Isomap-based ROMs was performed using the XRF-1 wing fuselage configuration. Here, the focus was on how accurate the different ROMs predict shocks in the transonic flow regime and on evaluating the impact of the accuracy of the prediction on downstream analyses. In this specific case, the ROMs were used to predict the surface pressure distribution acting on the wing at a particular flow condition, which in turn was used to size the wing structural model of the (rigid) XRF-1 subject to this one load case. The wing of the XRF-1 was parameterized with five free-form deformation parameters in order to vary the twist of the wing. About 100 TAU simulations were performed for 100 differently twisted wings at a Mach number of 0.83, a Reynolds number of $43.3 \cdot 10^6$ and a target lift coefficient of 0.5. An Isomap-based adaptive sampling algorithm was used to select combinations of parameters in the five-dimensional parameter space where to compute snapshots with TAU (Franz et al. 2017).

The fully turbulent calculations were performed with TAU using the Spalart–Allmaras negative turbulence model. The hybrid unstructured numerical grid had 784,384 grid points. Using the 100 snapshots, ROMs for the surface pressure distribution as a function of the geometry variations were constructed. For this purpose, POD and Isomap were combined with an interpolation scheme to make predictions at 10 untried parameter combinations in the five-dimensional parameter space. These 10 parameter combinations were selected at the center of the 10 simplexes with the largest volume of the Delaunay triangulation of the parameter space, that is, where the largest "gap" between snapshots occurs in the five-dimensional parameter space. Predicting the pressure distribution at these points is most challenging for the ROMs, because the prediction points are the farthest away from the snapshots.

The prediction of the surface pressure distribution at one of the 10 untried parameter combinations, or test points, is shown in Figure 16.4. The twisting of the wing using the free-form deformation is shown in the upper right corner of the figure. There are slight differences in the contour distribution between the two ROM predictions, but the shock position and strength are sufficiently well reproduced compared to the TAU reference solution.

These aerodynamic loads were used to size the structural model of the XRF-1 wing, which consists of 2,167 elements. Normally, several critical load cases are used to size a structure; however, it should first be investigated how the accuracy of a ROM prediction of the aerodynamic load distribution affects the sizing of the structure. The result can be seen in Figure 16.5. Here only results for POD-based predictions are compared with TAU, but they are also available for Isomap. The picture shows the thickness distribution of the structural elements of 350 optimization regions (opt region), where one optimization region consists of several finite elements.

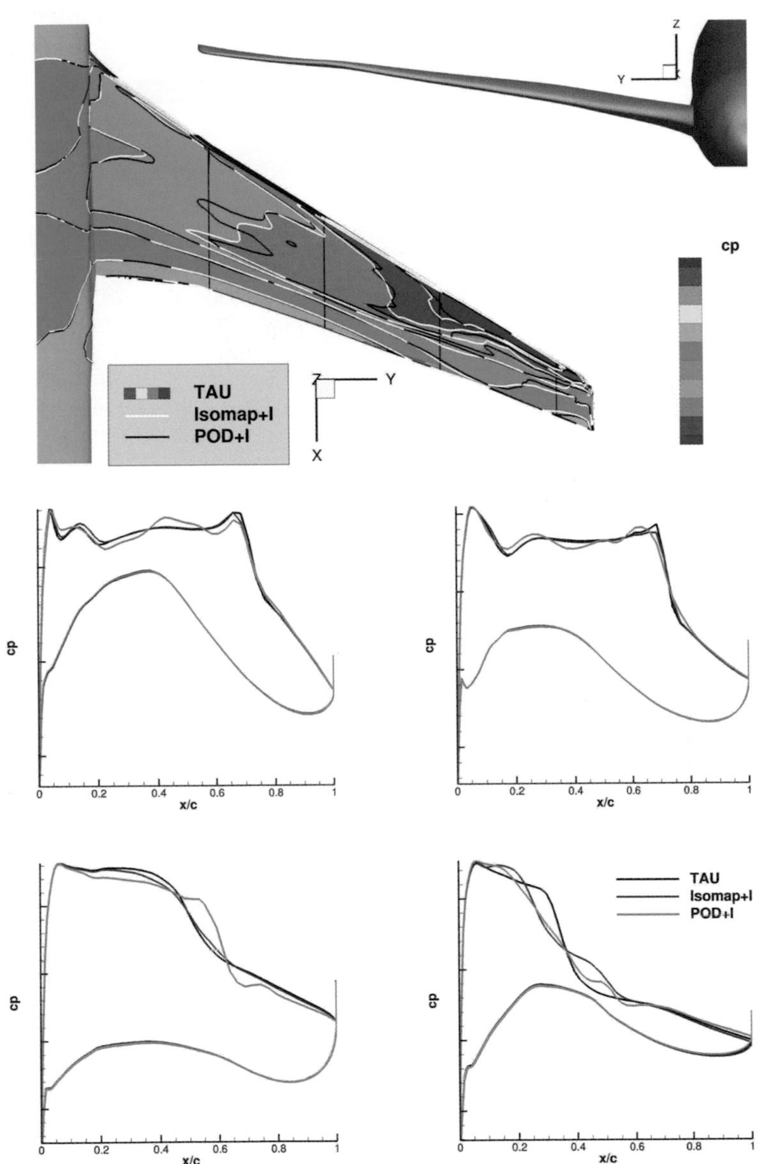

Figure 16.4 Comparison of the surface pressure distribution predicted by Isomap+interpolation and POD+interpolation for the rigid XRF-1 with TAU reference solution for an untried combination of geometry parameters at four different spanwise wing sections. Reprinted from Ripepi et al. (2018) by permission from Springer Nature.

The structural sizing leads to a wing mass of 3 761 kg when sizing with TAU reference loads, 3 776 kg when sizing with the POD-based predictions, and 3,755 kg when sizing the wing with Isomap-based predictions. The difference in wing mass is therefore only 0.4% and 0.1%, showing that ROM-based predictions of aerodynamic loads are of sufficient accuracy in the context of structural sizing.

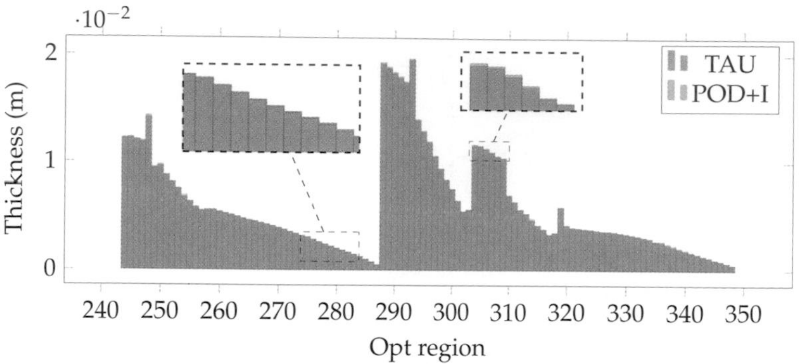

Figure 16.5 Comparison of the results optimizing the different regions of the wing structure of the rigid XRF-1 with the aerodynamic loads predicted by the POD-based ROM (red bars) and with the reference loads computed with the CFD code TAU (blue bars).

16.4.3 Reduced–Order Modeling for Static Aeroelastic Problems

Steady ROMs were used to predict the static aeroelastic loads for structural sizing within a multidisciplinary design and optimization context (Ripepi et al. 2018). They were also used in a process where an a priori identification of the critical load cases is of interest and the sheer number of load cases to be considered cannot be computed with high-fidelity CFD.

Within the framework of MDO, the CFD-code TAU and a computational structural mechanics (CSM) solver are used to perform coupled fluid–structure interaction (CFD–CSM) simulations to compute aerodynamic loads acting on the wing, which are in turn used to design or optimize the structure (structural sizing). Since TAU is repeatedly called for different structural models and different wing geometries to evaluate the load cases during the optimization, the flow solver should be replaced by a steady ROM for the elastic aircraft. Replacing the flow solver with a ROM has the advantage that the snapshots necessary to build the ROM can be calculated offline, that is, before the actual optimization begins. Once the snapshots have been computed, the ROM itself can also be generated offline. During the optimization, the much-faster ROM replaces the flow solver. This will help to speed up the overall design and optimization process and allow more load cases to be considered in the sizing of the structure. To set up the ROMs, snapshots (TAU solutions) for different aerodynamic shapes and structural modes are required. Here, a ROM for the static aeroelastic load prediction of the XRF-1 wing-fuselage configuration was created, that is, the shape was not varied.

For this purpose, 21 coupled computations with a structured mesh suitable for RANS simulations with one million mesh points (Figure 16.6, left) and an ANSYS finite element model with 2,167 elements (Figure 16.6, right) for different altitudes and Mach numbers at a load factor of 2.5g were performed to create snapshots for the wing deformation and the aerodynamic load in terms of the pressure distribution. The 21 points in the parameter space are shown in Figure 16.7 as blue symbols.

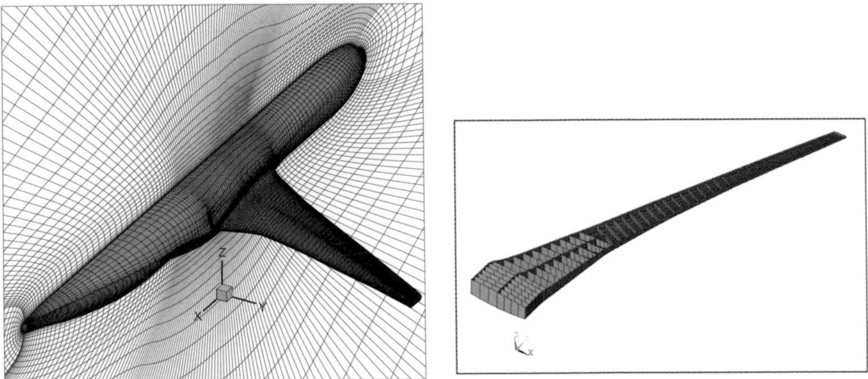

Figure 16.6 Structured CFD mesh for XRF-1 (left), finite element model of the wing (right). Reprinted from Ripepi et al. (2018) by permission from Springer Nature.

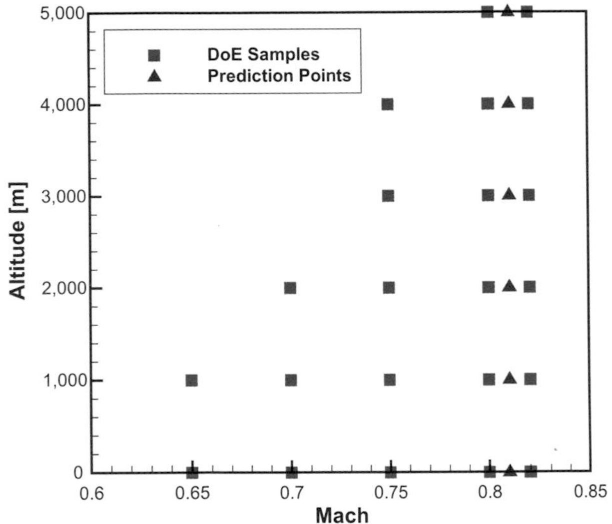

Figure 16.7 Sample locations for ROM generation in the parameter space (blue) spanned by Mach number and altitude and prediction points (red). Reprinted from Ripepi et al. (2018) by permission from Springer Nature.

The coupled calculations were carried out in such a way that the lift counteracts the weight and inertial forces due to the load factor, which corresponds to the ratio of the lift of an aircraft to its weight, and represents a global measure of the stress or "load" to which the structure of the aircraft is subjected. The snapshots were used to set up a ROM based on POD or Isomap and thin-plate spline (TPS) interpolation. The calculation of a snapshot with TAU and ANSYS took 60 minutes on 48 cores, while the generation of the ROM took one minute on one core and the ROM prediction of the surface pressure distribution and deformation were in real time on one core.

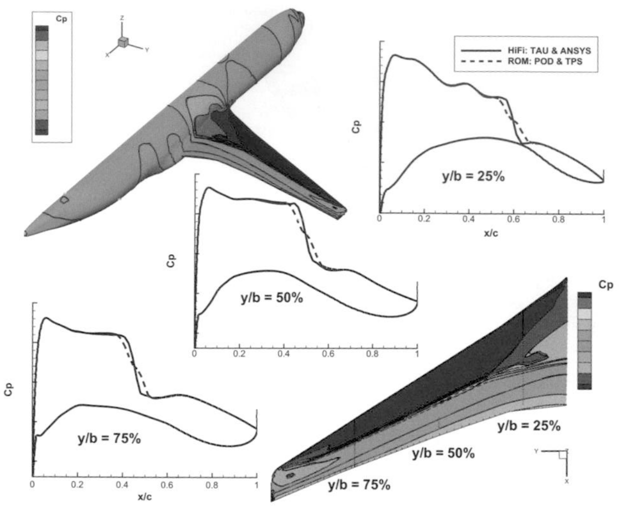

Figure 16.8 Comparison of the pressure distributions for the flexible XRF-1 at different wing sections for a Mach number of 0.81 and 4 000 m altitude (visualization on jig-shape). Reprinted from Ripepi et al. (2018) by permission from Springer Nature.

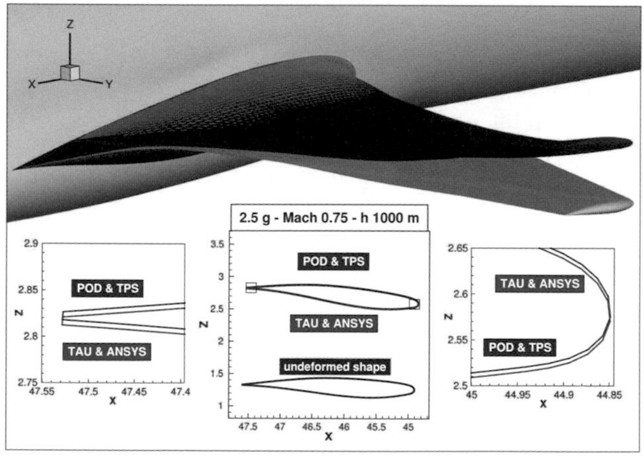

Figure 16.9 Comparison of steady ROM with coupled CFD/CSM simulation. Reprinted from Ripepi et al. (2018) by permission from Springer Nature.

In Figure 16.8, the ROM prediction of the surface pressure distribution (red) for a Mach number of 0.81 and 4 000 m altitude is compared with a TAU-ANSYS reference solution (black). The agreement is very good; only the gradient of the pressure distribution at the shock deviates slightly. For all prediction points, the relative error in the aerodynamic coefficients is below 0.5%. An excellent agreement between the elastic wing deformation predicted by the ROM and the coupled high-fidelity reference TAU-ANSYS computation is observed in Figure 16.9.

16.4.4 Reduced-Order Models in the Context of Multidisciplinary Design Optimization

A gradient-based optimization algorithm can efficiently spot local optima in high-dimensional design problems. A gradient-based approach, which can tackle hundreds of design parameters controlling the wing profile shapes and the structural thicknesses with a single optimizer, was used for aero-structural optimization of the XRF-1 (Görtz et al. 2016). Such optimization problems comprise the computation of disciplinary as well as interdisciplinary gradients, that is, gradients of, for example, aerodynamic quantities with respect to shape parameters as well as gradients of aerodynamic quantities with respect to structural parameters. The adjoint approach was deemed suitable for computing the aerodynamic gradients since the number of design variables is considerably higher than the number of constraints (Abu-Zurayk et al. 2017). On the structural side, a parallelized finite differences (FD) approach was used to compute the gradients. FD was feasible because the structural problem is linear, and its solution costs much less than that of the aerodynamic problem. Disciplinary gradients were compared with interdisciplinary gradients, and only the most significant gradients were employed during the optimization.

Aiming to an aeroelastic structural optimization of the generic long-range transport aircraft able to withstand the critical loads, the Mach-altitude envelope for five design mass cases has been computed for the baseline XRF-1 configuration using high-fidelity coupled CFD–CSM methods, and the relative design aerodynamic load cases have been determined. The Mach-altitude envelope for the five mass cases was computed in intervals of 0.02 in Mach number, and of 1 000 m in altitude for two load factors (−1.0, 2.5 g). After computing the aerodynamic load envelope of the flexible aircraft (i.e., the aerodynamic pressure distribution of the static aeroelastic solution), all the loads were passed to a sizing tool. This tool provides as an output to the designer the critical loads and the sized structure able to sustain such loads. The process is iterated, until convergence, by recomputing the aerodynamic loads associated to the static aeroelastic solution with the currently sized structure.

Since this process is computationally very expensive, POD-based ROMs were employed using the DLR's SMARTy toolbox. Here ROMs were used to explore the parameter space with a finer sampling at in-between altitudes. The ROM-predicted aeroelastic loads are sent to the sizing tool, which determines if the loads are critical. Whenever a newly predicted aeroelastic load is found to be potentially critical, the corresponding load case is recomputed with the high-fidelity coupled CFD–CSM methods and checked with the sizing tool if it is really critical or not.

As an example, two of the five critical mass cases were used to generate 400 sized high-fidelity aeroelastic snapshots. A parametric reduced-order model has been generated using such snapshots, and then used to compute 360 additional load predictions. Three of these 360 predictions were found to be additional candidates for design load cases, and by computing them with the high-fidelity tools, one case was found to be actually critical.

After identifying the sizing load cases for the baseline configuration, their definition was frozen, assuming that a gradient-based optimization only results in small changes

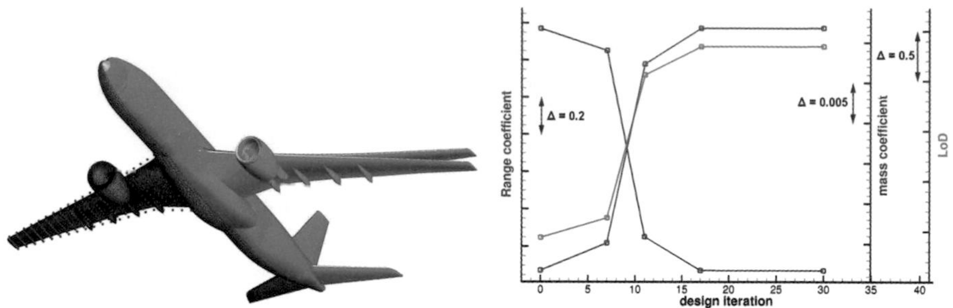

Figure 16.10 Gradient-based, high-fidelity aero/structure optimization; baseline aircraft configuration (left), course of optimization in terms of objective function and quantities of interest as a function of number of iterations after ROM-based selection of sizing load cases.

to the wing shape and the structure. The objective of the gradient-based MDO was to maximize the range factor for the same transport aircraft at the design cruise point as well as at 4 off-design points. The optimization resulted in an increase in the range by a factor of 6%, see Figure 16.10.

16.4.5 Aeroelastic Reduced-Order Model Based on Synthetic Modes and Gappy Proper Orthogonal Decomposition

Recently, a derivative of the constraint gappy POD method (Franz & Held 2017) has been developed and applied for the rapid prediction of aerodynamic loads while taking elastic deformations into account. This is of interest in steady fluid–structure interaction analysis. First, a set of synthetic surface deformations around an estimated elastic axis is determined, the so-called synthetic modes, which are an established concept for unsteady aeroelastic investigations including low-fidelity aerodynamic models (Voss et al. 2011).

Second, samples are computed with TAU for the different synthetic mode shapes based on a Halton sampling strategy (Kuipers & Niederreiter 2005), and a POD basis is computed, which contains the deformations corresponding to the mode shapes as well as the surface forces computed with TAU. Finally, surface forces for a given deformation of interest can then rapidly be obtained by solving a least-squares problem, minimizing the difference between the target surface deformation given by the current structural model of interest and the deformation entries within the POD modes to find a set of POD coefficients. The resulting POD coefficients can then be used to compute the surface force by a simple matrix-vector product. A more in-depth theoretical introduction, including synthetic mode generation and a formulation of the constraint least-squares optimization problem, can be found in Bekemeyer et al. (2019). Here results are presented for a long-range passenger aircraft with a semi-wingspan of approximately 29.0 m shown in Figure 16.11. During each analysis, an iterative fluid–structure coupling process is performed including a structural optimiza-

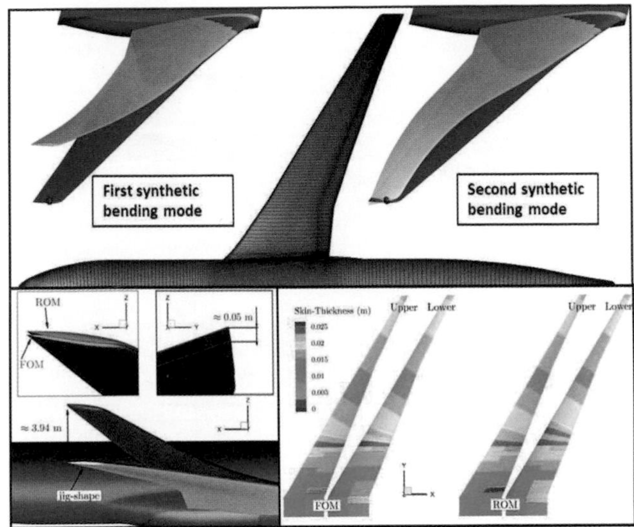

Figure 16.11 Assessment of an aeroelastic reduced-order model based on synthetic mode shapes; mode shapes and computational grid (top), global deformation behavior (bottom left) and optimized skin thicknesses (bottom right).

tion in each iteration step to minimize weight while respecting stress constraints. The final deformation behavior of the wing for a transonic load at M = 0.784 is compared between the proposed reduced-order model approach and a fully coupled CFD–CSM analysis with excellent agreement. Also final skin thicknesses show only minor deviations around the engine pylon wing junction. However, the online cost for performing the ROM-based analysis is a factor of nearly 2 000 faster than its full-order equivalent.

16.4.6 Nonlinear Dimensionality Reduction by Autoencoder Networks

The POD-based ansatz is still considered as the state-of-the-art method in reduced-order modeling, but also nonlinear dimensionality reduction methods like Isomap and autoencoders have been used to build ROMs. An autoencoder (Goodfellow et al. 2016) is a neural network used for unsupervised learning of efficient codings. The aim of an autoencoder is to learn a representation (encoding) for a set of data, typically for the purpose of dimensionality reduction or representing data in a lower-dimensional space. The decoder is used for mapping back to the original dimension of the problem (reconstruction), see Figure 16.12. There exist several modifications of the autoencoder; however, only the standard method was employed here (Hinton 2006).

The method was used in the context of reduced-order modeling of distributed aerodynamic data. Instead of computing a reduced-order representation of the data via POD, the autoencoder was employed to compute the reduced-order representation. The method directly yields a mapping from the reduced-order representation to the high-dimensional space; however, for huge input data (large dimension) it is difficult

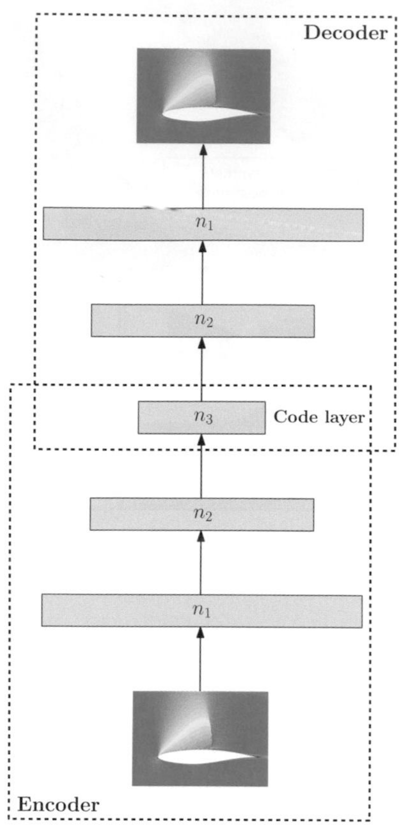

Figure 16.12 Schematics of an autoencoder network.

and time-consuming to train a deep network. Training times can be significantly reduced on GPUs when compared with CPUs.

The potential of autoencoder networks for nonlinear dimensional reduction was evaluated using a two-dimensional test case from the "virtual aircraft" application area. Thirty steady flow solutions were computed for the NACA 64A010 airfoil at transonic flow conditions with the DLR flow solver TAU by varying the angle of attack and the Mach number as shown by the black symbols in Figure 16.13. Predictions were made for three different flow conditions (red crosses in Figure 16.13) and compared with TAU reference solutions in terms of the surface pressure distribution. The autoencoder predictions were also compared with ROM predictions based on Isomap, POD and cluster POD (C-POD). K-means clustering was used to cluster a given set of parameter-dependent flow solutions in k clusters. The clustering was performed based on either the flow parameters or the reduced representation of the solutions given by Isomap. On each cluster, a local POD-based reduced-order model was built, which is similar to local principal component analysis (LPCA) described in

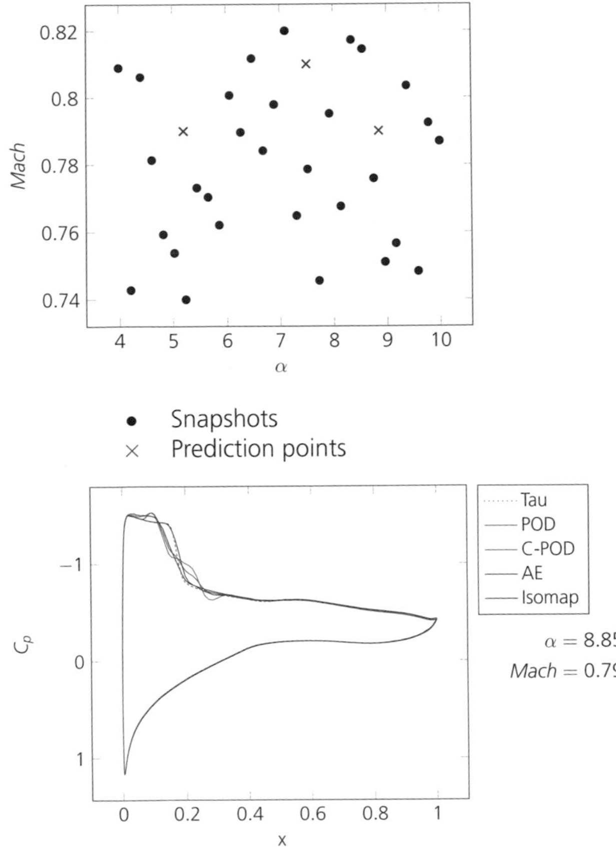

Figure 16.13 Autoencoder network for reduced-order modeling of distributed aerodynamic data: sample locations (black) and trial locations (red) in parameter space spanned by angle of attack and Mach number (top) and surface pressure predictions (bottom) with different interpolation-based ROMs at an untried parameter combination compared to TAU reference solution.

Chapter 15, where PCA is applied separately on clusters of observations. To predict a solution at an untried parameter combination, the model belonging to the cluster with the nearest mean was exploited.

Figure 16.13 compares the different ROM predictions with the reference solution at one of the trial points. The quality of the autoencoder (AE) prediction exceeds that of the established linear dimension reduction method (POD), which is the case for all three trial flow conditions; however, Isomap is seen to be superior to the autoencoder in terms of predicting the shock location and shock strength, while C-POD turns out to be only slightly better than POD. Note that in this study, convolutional autoencoder networks in particular were found to reproduce the strong nonlinearities in the quantity of interest very well, even for a small number of snapshots.

First, training data is generated by simulating a training maneuver (e.g., the 1-cos gust response of an aircraft), which is step 4 in Figure 16.14. This data is then used

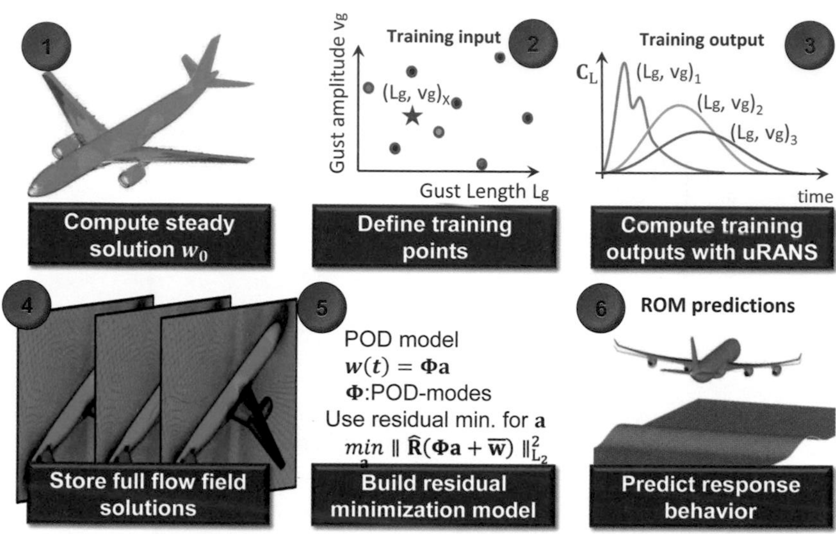

Figure 16.14 Unsteady ROMs in a nutshell.

to obtain a subspace representation using POD. Second, a subset of the grid points is computed by combining the classical Discrete Empirical Interpolation Method method and a greedy missing point estimation. This subset of grid points is used to restrict the evaluation of the unsteady flow solver residual to a small number of points during the online prediction phase for reasons of computational efficiency (Bekemeyer et al. 2019) (step 5 in Figure 16.14). The resulting model can then be used to simulate gust response behavior at varying parameter combinations such as gust length and amplitude by minimizing the unsteady residual on the subspace at each time step. ROM results are compared to a time-marching reference solution for a 1-cos gust with length of 213 m and amplitude as defined by international certification authorities at cruise conditions for NASA's CRM, compare Figure 16.2. In addition, the linearized frequency domain solver (LFD) based on the TAU code (Thormann & Widhalm 2013) was adopted to efficiently predict the dynamic behavior of the CRM assuming transonic and mildly separated flow and dynamically linear response behavior. The LFD was originally developed, offering a significant reduction in computational effort for small-perturbation problems while retaining the fidelity of the RANS flow solutions. The lift coefficient, pitching moment coefficient, and the differences in surface pressure distribution at the time step corresponding to the maximum lift coefficient are shown in Figure 16.15. Throughout, the unsteady ROM accurately predicts the full-order model (FOM) behavior, whereas the linearized approach deviates once flow separation occurs. Significant discrepancies between the three approaches are observed in the surface pressure distribution.

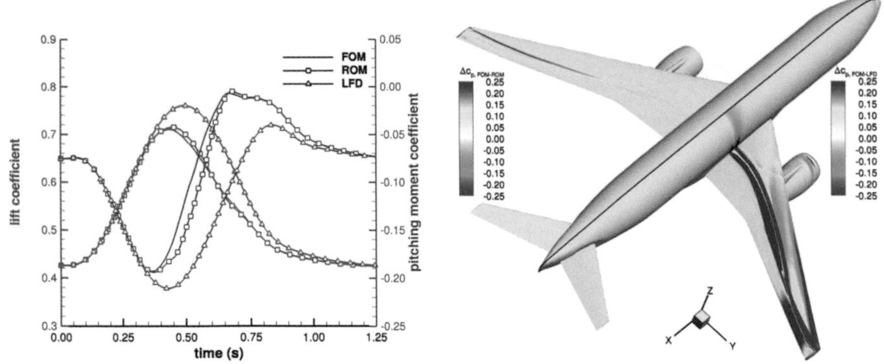

Figure 16.15 Comparison of gust response behavior of the CRM for the full-order model (FOM), unsteady nonlinear ROM and linear frequency domain (LFD) solver. Lift and pitching moment coefficient (left), surface pressure differences at maximum lift coefficient (right).

16.4.7 Unsteady Physics-Based Nonlinear Reduced-Order Modeling

The investigation of unsteady, gust induced air loads is a crucial part of aircraft design, and time-marching CFD simulations are infeasible due to their high computational cost. However, nonlinear aerodynamic effects should not be neglected as is the case for lower fidelity methods. Thus, based on previous work on steady transonic flows using a combination of POD and nonlinear least-squares-based residual minimization techniques (Zimmermann & Görtz 2010, 2012), an unsteady nonlinear reduced-order model (LSQ-ROM) was developed to predict dynamically nonlinear maneuver and gust response behavior in the transonic flow regime (Bekemeyer et al. 2019). The schematics of the unsteady physics-based nonlinear reduced-order modeling procedure are sketched in Figure 16.14.

16.5 Conclusions

Parametric ROMs based on both POD, cluster POD, and Isomap were developed and successfully applied to a variety of two- and three-dimensional test cases, including two industrial aircraft configurations, to predict aerodynamic loads over a range of flow or operating conditions as well as for changing geometry parameters. Isomap is a dimensionality reduction technique, which is based on a nonlinear "manifold learning" method. In this method, it is assumed that the CFD solutions form a nonlinear manifold, which in turn constitutes a sub-manifold of lower intrinsic dimensionality. Isomap yields better predictions than a POD-based ROM method, in particular for the location and magnitude of the shock wave at transonic flow conditions. This is due to the fact that Isomap makes predictions based on a local subspace, similar to cluster POD, while POD-based predictions make use of a global subspace.

17 Machine Learning for Turbulence Control

B. R. Noack, G. Y. Cornejo Maceda, and F. Lusseyran

Closed-loop turbulence control has current and future engineering applications of truly epic proportions, including cars, trains, airplanes, jet noise, air-conditioning, medical applications, wind turbines, combustors, and energy systems. A key feature, opportunity and technical challenge, is the inherent nonlinearity of the actuation response.

This chapter outlines model-based and model-free control taming the nonlinear dynamics. Model-based control employs the POD Galerkin method (see Chapter 14: Noack et al.: Machine learning for Reduced-Order Modeling). Many flow control results will be explained with a set of few different sparse human-interpretable models. Artificial intelligence (AI)/machine learning (ML) has opened another game-changing new avenue: the automated model-free discovery and exploitation of unknown nonlinear actuation mechanisms directly in the plant. Variants of this machine learning control (MLC) will be discussed. The chapter concludes with a tutorial of MLC on a simple dynamical system with a freely available MATLAB/Octave code.

17.1 Introduction

Controlling turbulence for engineering goals is one of the oldest and most fruitful academic and technological challenges. Engineering applications have epic proportion. The infrastructure of any industrial nation requires flawless pipe systems with turbulent flows of drinking water, gas, oil, and sewage. Drag reduction in these pipe flows spells energy saving for operation. Transport-related examples include drag reduction of road vehicles, airborne transport, ships, and submarines, lift increase of airfoils, and associated noise reduction. Energy and production-related examples include efficiency increase of harnessing wind and water energy, heat transfer, chemical, and combustion processes.

Closed-loop active control is increasingly investigated for performance improvements in industrial applications. Several trends support this development. First, aerodynamic shapes and passive control have been optimized for over 100 years. Thus, further optimization efforts may yield diminishing returns. Second, in many cases the engineering challenge is not the cruise or normal operating condition but short-term and rare events. For wind turbines, the effect of gusts needs to be mitigated. Aeroengines need to be safe during 60 to 90 seconds of takeoff, where they can produce six times the thrust required during cruise. The large nacelle is only required for this takeoff and may be replaced by a smaller version with active control. Third, actuators and sensors become increasingly cheaper, more powerful, more reliable, and thus more economical. Finally, modern control methods can harness increasingly more complex dynamics, last but not least through the powerful augmentations by ML. Goals, tools, and principles of turbulence control are reviewed in Section 17.2.

This chapter focuses on the control logic. The classical paradigm is from understanding to dynamic modeling to model-based control design to controller tuning in the full plant. Key applications are first (relaxational) and second-order (oscillatory) dynamics. The nonlinear dynamics need to be linearized or at least sufficiently understood. In Section 17.3, we present simple human-interpretable POD-based models showing the spectrum from linear to strongly nonlinear dynamics explaining a large range of current experiments.

Building a control-oriented model is a challenge for shear turbulence with a large range of temporal and spatial scales and their nonlinear interactions. The challenge becomes even larger for spatially distributed control with many actuators and many sensors. Here, powerful methods of ML may fully invert the classical control paradigm. First, the near-optimal control law is automatically learned in the full plant without any model. Then, simpler control laws with similar performance are identified. The performance-winning mechanism is then dynamically modeled, leading to a deeper understanding of the flow. In Section 17.4, different realizations of this MLC are described. The chapter concludes with a MLC tutorial for the control of nonlinearly coupled oscillators (Section 17.5). For more information, we refer to our recent reviews (Brunton & Noack 2015, Brunton et al. 2020) and our textbooks on model-based control (Noack et al. 2011) and MLC (Duriez et al. 2017) and the literature cited therein.

17.2 Goals, Tools, and Principles

This section provides a high-level perspective on turbulence control, reviewing typical goals (Section 17.2.1), hardware tools (Section 17.2.2), and control principles (Section 17.2.3). Subsequent sections assume that the configuration with goals, actuators, and sensors is given and the control logic needs to be developed.

17.2.1 Goals

Turbulent flows around cars, trucks, trains, airplanes, helicopters, ships, and submarines affect their performance. An important engineering goal is *drag reduction* or, more precisely, the reduction of net propulsion power and thus energy saving. Airplanes during takeoff and landing require high lift at low velocities. Here, a goal is to achieve *lift increase* with minimal means, for example, low extra weight and small material fatigue. During cruise, the increase of lift-to-drag ratio leads to reduced fuel consumption, and hence reduced environmental impact and increased profitability. Ground and airborne traffic produce undesirable noise. Here, *noise reduction* is another important goal. Chemical production, combustion processes, and heat exchangers profit from *mixing increase*. Another important goal is the *stabilization of flow oscillations*. Examples include flutter and buffeting of airfoils, cavity noise (Rossiter modes) associated with landing gears of airplanes and detrimental vortex shedding around buildings, chimneys, bridges, the pillars of oil-producing platforms, underwater pipelines, and heat-exchanging pipes in nuclear plants.

The goals of drag reduction, lift increase, mixing increase, noise reduction, and oscillation mitigation may be served by separation prevention, skin-friction reduction, and manipulation of free shear flows. As a rule of thumb, in most flow control strategies the transverse mixing of free and wall-bounded shear flows is increased (drag reduction, mixing increase) or decreased (skin friction reduction, noise reduction, oscillation prevention). We refer to the excellent discussion of underlying flow physics by Gad-el Hak (2007).

17.2.2 Tools

Flow control performance during cruise or standard operating conditions is well served by shape optimization under given constraints. An ubiquitous example is cars, which have become increasingly more aerodynamic over the past hundred years.

The performance may be further improved by *passive devices*, that is, small changes that require no energy input. Feathers at the end of arrows stabilize the flight and increase the range – perhaps the first example of man-made passive control. Another example is vortex generators on the wing of a Boeing 737 to prevent early separation with associated lift decrease and drag increase. The efficiency of wind turbines may also profit by up to 7% from SmartBlade vortex generators. Riblets reduce skin-friction drag by up to 11%. The chevron-shaped engine nozzles reduce jet noise by 1–3 dB. Spoilers at the trailing edge of a car stabilize aerodynamic forces and reduce drag. Vanes in ducts and at airplanes redirect the flows in a performance increasing way.

Passive devices may come with parasitic drag during cruise or standard operating conditions. In many cases, passive devices, like vortex generators, may be emulated by active (energy-consuming) devices, like fluidic jets. Active control has the advantage that it can be turned on or off and even operate at a large range of amplitudes and frequencies. This dynamic range often yields performance benefits in contrast to steady operation. Active control may be operated open-loop, that is, be blind to the flow state. Yet there may be significant potential in combining unsteady active control with sensor-based flow monitoring. In the sequel, we will focus on this closed-loop control.

The choice of the actuators, their kind, amplitude, frequency range, and placement has a large effect on control efficiency. Actuators are often placed at high-receptivity locations, for example, at the trailing edge of a bluff body, where the separating shear layer can be easily manipulated. The coherent structures associated with boundary-layer transition and skin friction can be directly opposed with wall motion or wall actuation. The actuation may be based on blowing, suction, and zero-net mass flux, each having its distinct advantages and disadvantages.

Yet, this choice is largely based on engineering experience with little guidance from mathematical and physical methods. The closest to a first-principle based method is the volume force optimization of linearized Navier–Stokes dynamics. Making a good choice of actuation is a challenge. The scaling properties to full-scale engineering configurations is the next task, again with no or little guidance from available theories.

Optimizing sensor placement is easier and may be guided by a direct numerical simulation or particle image velocimetry measurements. Yet, the choice of the optimization criterion is far from trivial. The chosen sensors may, for instance, be placed at locations to optimize (1) linear flow estimation, (2) linear-quadratic flow estimation, (3) Kalman filter-based flow estimation, (4) a dynamic observer based on a reduced-order model, (5) the estimate of the observable subspace (balanced truncation), or, in the most general case, (6) the feedback control performance for all considered control laws. In the following, we assume a specified control goal and a given actuator and sensor configuration.

17.2.3 Principles

The choice of the actuators and sensors is – lacking a computationally accessible mathematical method – always guided by a conceptual idea of the control mechanism. A long list of hypotheses can be produced. Skin-friction reduction relies on the suppression of sweeps and ejections (opposition control). Separation can be delayed by increasing the turbulence level (destabilizing control). Heat exchangers also rely on mixing increase. Vortex shedding behind bluff bodies can be mitigated with phasor control, or exciting larger or smaller frequencies and blowing the wake away (Coanda blowing, base bleeding). A few general dynamics principles can be distilled.

Kill the monster while it is little! The ideal scenario of stabilizing control is that the instability is successfully fought in statu nascendi. This requires vanishing energy in a noise-free environment and a noise-dependent level otherwise. Examples are mitigation of a Tollmien–Schlichting instability with local opposition control, the reduction of skin-friction drag by opposing the wall-normal velocity of sweeps and ejections and Coanda blowing / flow vectoring to symmetrize the wake.

Support the enemy of your enemy! In some cases, the actuation may be too weak or distant for a direct stabilization. We might exploit that it is much easier to excite a coherent structure than to prevent an instability. Vortex shedding behind a cylinder may be mitigated by exciting shear-layer vortices at high frequency. In this case, the shear-layer structures decrease the gradient of the mean flow and hence the growth rate of vortex shedding. The energized shear layer is the enemy of the enemy (vortex shedding).

Blow the instability away! A not-so-subtle control is based on blowing the instability away. The vortex shedding behind a D-shaped cylinder is successfully reduced by Coanda forcing, that is, redirecting the flow into the near wake.

Support your friend! In some cases, naturally occurring coherent structures serve the control goal, that is, they are your friends. Kelvin–Helmholtz vortices over a backward-facing step reduce the recirculation zone. An amplification of these vortices reduces the length of this zone even further.

Manipulate the sociology of coherent structures! Many aerodynamic goals have a well-founded conceptual picture of the beneficial and detrimental coherent structures to be augmented or mitigated. For mixing and noise of broadband turbulence, such a picture is largely missing. Mixing and noise is based on a longer-term integration of many different structures. This is the classic case for model-free exploration and exploitation of control.

To illustrate these principles, let us consider following dynamical system coupling a self-amplified amplitude limited oscillator (a_1, a_2) and a stable linear oscillator at 10-fold frequency (a_3, a_4) inspired by several wake and airfoil models (Luchtenburg et al. 2009, Semaan et al. 2015):

$$\begin{aligned} da_1/dt &= \sigma a_1 - a_2, & da_3/dt &= -0.1 a_3 - 10 a_4, \\ da_2/dt &= \sigma a_2 + a_1 + b_1, & da_4/dt &= -0.1 a_4 + 10 a_3 + b_2, \\ \sigma &= 0.1 - a_1^2 - a_2^2 - a_3^2 - a_4^2 - b_3. \end{aligned} \tag{17.1}$$

Without control, $b_1 = b_2 = b_3 = 0$, the first oscillator converges to the limit cycle $a_1^2 + a_2^2 = 0.1$ with unit frequency, while the second one vanishes, $a_3 = a_4 = 0$. We will ignore the stable oscillator (a_3, a_4) unless it is excited with b_2. The first oscillator can be stabilized with phasor control $b_1 = -0.4\, a_2$. This oscillator can also be indirectly stabilized by exciting the second oscillator $b_2 = k a_4$ to fluctuation level above 0.1 so that $\sigma < 0$. The constant blowing is mimicked by $b_3 = 0.2$, leading to $\sigma \leq -0.1$, that is, stabilization.

17.3 Nonlinear Low-Dimensional Models for Control

Despite the powerful tools for linear model reduction and control, the assumption of linearity is often overly restrictive for real-world flows. Turbulent fluctuations are inherently nonlinear, and often the goal is not to stabilize an unstable fixed point but rather to change the nature of a turbulent attractor. Moreover, it may be the case that the control input is either a bifurcation parameter itself, or closely related to one, such as the control surfaces on an aircraft.

The degree of nonlinearity is most easily characterized in a Galerkin modeling framework discussed in Chapters 1 and 15. We assume a Galerkin model with the steady solution $\boldsymbol{u}_s$ as basic mode and N expansion modes $\boldsymbol{u}_i$, $i = 1, \ldots, N$,

$$\boldsymbol{u}(\boldsymbol{x}, t) \approx \hat{\boldsymbol{u}}(\boldsymbol{x}, t) = \boldsymbol{u}_s(\boldsymbol{x}) + \sum_{i=1}^{N_a} a_i(t)\, \boldsymbol{u}_i(\boldsymbol{x}). \tag{17.2}$$

Thus, the state is approximated by the vector $\boldsymbol{a} = [a_1, \ldots, a_N]^{\mathrm{T}}$ comprising the modal amplitudes. The Galerkin system reads

$$\frac{da_i}{dt} = f_i(\boldsymbol{a}, b) = \sum_{j=1}^{N} l_{ij}\, a_j + \sum_{j,k=1}^{N} q_{ijk}\, a_j\, a_k + g_i b. \tag{17.3}$$

Here, the constant term vanishes identically, because the steady Navier–stokes solution $\boldsymbol{u}_s$ corresponds to the fixed point $\boldsymbol{a}_s = \boldsymbol{0}$ in the Galerkin framework. Following the example of Chapter 1, a single linear control input $g_i b$ is added.

In the following sections, three prototypic examples are discussed: (1) an oscillation around the fixed point (Section 17.3.1); (2) a self-excited amplitude-limited oscillation (Section 17.3.2); (3) frequency cross-talk with two different frequencies (Section 17.3.3); over the base-flow deformation. More details and the remaining irreducible cases are elaborated in Brunton and Noack (2015).

17.3.1 Linear Dynamics

First, a small oscillatory fluctuation around a steady solution is considered. Examples include the flow over a backward-facing step (Hervé et al. 2012) at subcritical Reynolds number with noise, transition delay of a boundary layer (Semeraro et al. 2013), or stabilization of a cylinder wake (Roussopoulos 1993, Weller et al. 2009). In the following, a control-oriented oscillator model is presented as a least-order description. Then energy-based control design is exemplified as a powerful method of linear and nonlinear models.

The considered flows can be described by

$$\boldsymbol{u}(\boldsymbol{x},t) = \boldsymbol{u}_s(\boldsymbol{x}) + \boldsymbol{u}^u(\boldsymbol{x},t), \tag{17.4a}$$

$$\boldsymbol{u}^u(\boldsymbol{x},t) = a_1(t)\, \boldsymbol{u}_1(\boldsymbol{x}) + a_2(t)\, \boldsymbol{u}_2(\boldsymbol{x}). \tag{17.4b}$$

Here, $\boldsymbol{u}_i$, $i = 1,2$ may correspond to the real and imaginary part of the unstable complex eigenmode or are distilled from the oscillatory data. Higher harmonics are neglected. By construction, the stable or unstable fixed point is $\boldsymbol{a}_s = \boldsymbol{0}$.

The linearized version of the dynamics (17.3) reads

$$\frac{d}{dt}\boldsymbol{a} = \boldsymbol{A}_0\, \boldsymbol{a} + \boldsymbol{B}\, b, \quad \text{where} \quad \boldsymbol{A}_0 = \begin{bmatrix} \sigma^u & -\omega^u \\ \omega^u & \sigma^u \end{bmatrix}, \quad \boldsymbol{B} = \begin{bmatrix} 0 \\ g \end{bmatrix}, \tag{17.5}$$

where $\lambda = \sigma^u \pm \imath\, \omega^u$ denotes the complex conjugated eigenvalue pair of the linear system. Without loss of generality, the modes can be rotated so that the gain in the first component vanishes. A similar equation holds for the measurement equation. The matrices $\boldsymbol{A}_0$ and $\boldsymbol{B}$ depend on the fixed point $\boldsymbol{a}_s$. The linear dynamics (17.5) may also be an acceptable approximation for turbulent flows with dominant oscillatory behavior with small modifications (Brunton & Noack 2015)

Control design of the linear system (17.5) can be performed with one of many control theory methods (Aström & Murray 2010). Here, we illustrate the idea of energy-based control, which is particularly suited for nonlinear dynamics. Moreover, energy-based control has a kinematic relation to phasor control and an energetic relation to opposition control.

The growth rate σ^u is assumed positive and small enough so that the timescale of amplitude growth is small compared to the timescale of oscillation. In this case, the state can be approximated by the polar coordinates r, θ with $a_1 = r \cos\theta$, $a_2 = r \sin\theta$. Here, r and $\omega = d\theta/dt$ are slowly varying functions of time. The energy equation is given by

$$\frac{dr^2}{dt} = 2a_1 \frac{da_1}{dt} + 2a_2 \frac{da_2}{dt} = 2\sigma^u\, r^2 + 2g\, a_2\, b.$$

The control goal is an exponential decay of the amplitude with $\sigma^c < 0$, that is,

$$\frac{dr^2}{dt} \stackrel{!}{=} 2\sigma^c r^2.$$

Eliminating the time derivative in both equation yields

$$\frac{dr^2}{dt} = 2\sigma^c r^2 = 2\sigma^u\, r^2 + 2g\, a_2\, b.$$

The control command b increases (decreases) the energy $r^2/2$ if it has the same (different) sign as a_2. To prevent wasting actuation energy with the wrong phase, the linear ansatz $b = -k\, a_2$ is made. The gain $k > 0$ is determined by substituting $a_2 = r \cos\theta$ in the energy equation and averaging over one period. The resulting control law reads

$$b = 2\frac{\sigma^c - \sigma^u}{g} a_2. \tag{17.6}$$

The gain increases with the difference between the natural and design growth rate $\sigma^c - \sigma^u$ and decreases with the forcing constant g in the linear dynamics. The factor 2 arises from the fact that the actuation is only effective in the $[0, 1]^{\mathrm{T}}$ direction. The control design is very simple and immediately reveals the physical mechanism. The method is easy to generalize for nonlinear systems, particularly if the fluctuation is composed of clean frequency components. Related approaches are the Lyapunov control design (Khalil 2002) and harmonic balance (Jordan & Smith 1988).

17.3.2 Weakly Nonlinear Dynamics

As a refinement to the linearization, mean-field theory is recapitulated (Stuart 1958, Stuart 1971), providing an important nonlinear amplitude selection mechanism. The onset of vortex shedding behind a cylinder is one well-investigated example fitting this description (Schumm et al. 1994). This section and Section 17.3.2 outline the dynamical model and the corresponding control design, respectively.

Qualitatively, mean-field theory describes the feedback mechanism between the fluctuations and the base flow. The fluctuation gives rise to a Reynolds stress, which changes the base flow. The base-flow deformation generally reduces the production of fluctuation energy with increasing fluctuation level until an equilibrium is reached. The resulting evolution equations are also referred to as *weakly nonlinear dynamics*, as they describe a mild form of nonlinearity.

The fluctuation has the same representation as in Section 17.3.1. However, the base flow is allowed to vary by another mode $\boldsymbol{u}_3$, called the 0th, base deformation, or shift mode (Noack et al. 2003). This mode can be derived from the (linearized) Reynolds equation and is assumed to be slaved to the fluctuation level. The resulting velocity field ansatz reads

$$\boldsymbol{u}(\boldsymbol{x},t) = \boldsymbol{u}_s(\boldsymbol{x}) + \boldsymbol{u}^u(\boldsymbol{x},t) + \boldsymbol{u}^\Delta(\boldsymbol{x},t), \tag{17.7a}$$

$$\boldsymbol{u}^u(\boldsymbol{x},t) = a_1(t)\,\boldsymbol{u}_1(\boldsymbol{x}) + a_2(t)\,\boldsymbol{u}_2(\boldsymbol{x}), \tag{17.7b}$$

$$\boldsymbol{u}^\Delta(\boldsymbol{x},t) = a_3(t)\,\boldsymbol{u}_3(\boldsymbol{x}). \tag{17.7c}$$

The evolution equation is given by

$$\frac{d}{dt}\begin{bmatrix} a_1 \\ a_2 \end{bmatrix} = \boldsymbol{A}(a_3)\begin{bmatrix} a_1 \\ a_2 \end{bmatrix} + \boldsymbol{B}\,b, \quad \boldsymbol{A}(a_3) = \boldsymbol{A}_0 + a_3\boldsymbol{A}_3, \tag{17.8a}$$

$$a_3 = \alpha^u\left(a_1^2 + a_2^2\right), \tag{17.8b}$$

where

$$\boldsymbol{A}_0 = \begin{bmatrix} \sigma^u & -\omega^u \\ \omega^u & \sigma^u \end{bmatrix}, \quad \boldsymbol{A}_3 = \begin{bmatrix} -\beta^u & -\gamma^u \\ \gamma^u & -\beta^u \end{bmatrix}, \quad \boldsymbol{B} = \begin{bmatrix} 0 \\ g \end{bmatrix}. \tag{17.9}$$

Without loss of generality, $\alpha^u > 0$. Otherwise, the sign of the mode $\boldsymbol{u}_3$ must be changed. A nonlinear amplitude saturation requires the constant $\beta^u > 0$ to be positive.

For $a_3 \equiv 0$, (17.8) is equivalent to (17.5). However, (17.8) has a globally stable limit cycle with radius $r^\infty = \sqrt{\sigma^u/\alpha^u\beta^u}$ in the plane $a_3 = a_3^\infty = \sigma^u/\beta^u$, and with center $\left[0, 0, a_3^\infty\right]$.

In the framework of weakly nonlinear stability theory, the growth rate is considered a linear function of the Reynolds number $\sigma_u = \kappa(Re - Re_c)$. Here, Re_c corresponds to its critical value where the steady solution with damped oscillations ($\sigma_u < 0$) becomes unstable ($\sigma_u > 0$) in a Hopf bifurcation, giving rise to self-amplified oscillatory fluctuations. The other parameters are considered to be constant. This yields the famous Landau equation for the amplitude $dr/dt = \sigma^u r - \beta r^3$, $\beta = \alpha^u\beta^u$ and a corresponding equation for the frequency. The Landau equation explains the famous square root amplitude law $r \propto \sqrt{Re - Re_c}$ for supercritical Reynolds numbers assuming a soft bifurcation. We refer to the literature for a discussion of the hard subcritical bifurcation with quintic nonlinearity (Stuart 1971). Intriguingly, even turbulent flows with dominant periodic coherent structures may be described by (17.8) (Bourgeois et al. 2013).

The mean-field model and variants thereof have been successfully used for the model-based stabilization of the cylinder wake at $Re = 100$ (Gerhard et al. 2003, Tadmor et al. 2011). In these studies, an energy-based control design as discussed in Section 17.3.1 has been used to prescribe a fixed decay rate of the model.

17.3.3 Moderately Nonlinear Dynamics

Some oscillatory flows may be tamed by direct mitigation with models and methods described in Sections 17.3.1 and 17.3.2. Not all plants, particularly turbulent flows, have the actuation authority for such a stabilization. However, periodic forcing at high frequency has mitigated periodic oscillations in a number of experiments. Examples are the wake stabilization with a oscillatory cylinder rotation (Thiria et al. 2006), suppression of Kelvin–Helmholtz vortices in transitional shear layers (Parezanovic et al. 2016), reduction of the separation zone in a high-lift configuration (Luchtenburg et al. 2009), and elongation of the dead-water region behind a backward facing step (Vukasonivic et al. 2010). Also a periodic frequency at 60–70% of the dominant shedding frequency may substantially delay the vortex formation in wall-bounded shear-layers[1] and D-shaped cylinders (Pastoor et al. 2008). Sections 17.3.1 and 17.3.2 outline a corresponding modeling and control strategy.

Here, a generalized mean-field model for such frequency cross-talk phenomena is reviewed from Luchtenburg et al. (2009). Let $\boldsymbol{u}^u$ denote the natural self-amplified oscillation represented by two oscillatory modes $\boldsymbol{u}_1$, $\boldsymbol{u}_2$. Analogously, the actuated oscillatory fluctuation $\boldsymbol{u}^a$ is described by two modes $\boldsymbol{u}_3$, $\boldsymbol{u}_4$. The base-flow deformation due to the unstable natural frequency ω^u and stable actuation frequency ω^a are described by the shift-modes $\boldsymbol{u}_5$ and $\boldsymbol{u}_6$, respectively. The resulting velocity decomposition reads

$$\boldsymbol{u}(\boldsymbol{x},t) = \boldsymbol{u}_s(\boldsymbol{x}) + \boldsymbol{u}^{\Delta}(\boldsymbol{x},t) + \boldsymbol{u}^u(\boldsymbol{x},t) + \boldsymbol{u}^a(\boldsymbol{x},t), \tag{17.10a}$$

$$\boldsymbol{u}^u(\boldsymbol{x},t) = a_1(t)\, \boldsymbol{u}_1(\boldsymbol{x}) + a_2(t)\, \boldsymbol{u}_2(\boldsymbol{x}), \tag{17.10b}$$

$$\boldsymbol{u}^a(\boldsymbol{x},t) = a_3(t)\, \boldsymbol{u}_3(\boldsymbol{x}) + a_4(t)\, \boldsymbol{u}_4(\boldsymbol{x}), \tag{17.10c}$$

$$\boldsymbol{u}^{\Delta}(\boldsymbol{x},t) = a_5(t)\, \boldsymbol{u}_5(\boldsymbol{x}) + a_6(t)\, \boldsymbol{u}_6(\boldsymbol{x}). \tag{17.10d}$$

Generalized mean-field arguments yield the following evolution equation:

$$\frac{d}{dt}\begin{bmatrix} a_1 \\ a_2 \\ a_3 \\ a_4 \end{bmatrix} = \boldsymbol{A}(a_5,a_6)\begin{bmatrix} a_1 \\ a_2 \\ a_3 \\ a_4 \end{bmatrix} + \boldsymbol{B}\, b, \tag{17.11a}$$

$$\boldsymbol{A}(a_5,a_6) = \boldsymbol{A}_0 + a_5\boldsymbol{A}_5 + a_6\boldsymbol{A}_6, \tag{17.11b}$$

$$a_5 = \alpha^u \left(a_1^2 + a_2^2\right), \qquad a_6 = \alpha^a \left(a_3^2 + a_4^2\right), \tag{17.11c}$$

where $\boldsymbol{A}_0 = \begin{bmatrix} \sigma^u & -\omega^u & 0 & 0 \\ \omega^u & \sigma^u & 0 & 0 \\ 0 & 0 & \sigma^a & -\omega^a \\ 0 & 0 & \omega^u & \sigma^u \end{bmatrix}$, $\boldsymbol{A}_5 = \begin{bmatrix} -\beta^{uu} & -\gamma^{uu} & 0 & 0 \\ \gamma^{uu} & -\beta^{uu} & 0 & 0 \\ 0 & 0 & -\beta^{au} & -\gamma^{au} \\ 0 & 0 & \gamma^{au} & -\beta^{au} \end{bmatrix}$,

$\boldsymbol{B} = \begin{bmatrix} 0 \\ 0 \\ 0 \\ g \end{bmatrix}$, $\boldsymbol{A}_6 = \begin{bmatrix} -\beta^{ua} & -\gamma^{ua} & 0 & 0 \\ \gamma^{ua} & -\beta^{ua} & 0 & 0 \\ 0 & 0 & -\beta^{aa} & -\gamma^{aa} \\ 0 & 0 & \gamma^{aa} & -\beta^{aa} \end{bmatrix}$.

[1] J.-L. Aider, Private communication.

It should be noted that a vanishing actuation implies $a_3 = a_4 = a_6 = 0$ and yields the mean-field model of (17.8) modulo index numbering. In the sequel, we assume that the fixed point is unstable ($\sigma^u > 0$) and the limit cycle is stable ($\beta^{uu} > 0$). Similarly, a vanishing natural fluctuation $a_1 = a_2 = a_5$ yields another mean-field model, again modulo index numbering. In the sequel, we assume that the actuated structures vanish after the end of actuation, implying $\sigma^{aa} < 0$ and $\beta^{aa} > 0$ in the model.

An interesting aspect of (17.11) is the frequency cross-talk. The effective growth rate for the natural oscillation reads

$$A_{11} = \sigma^u - \beta^{uu}\alpha^u \left(a_1^2 + a_2^2\right) - \beta^{ua}\alpha^a \left(a_3^2 + a_4^2\right). \qquad (17.12)$$

The forcing stabilizes the natural instability if and only if $\beta^{ua} > 0$. Complete stabilization implies $A_{11} \leq 0$. From (17.12), such complete stabilization is achieved with a threshold fluctuation level at the forcing frequency

$$a_3^2 + a_4^2 \geq \frac{\sigma^u}{\alpha^a \beta^{ua}}.$$

Thus, increasing the forcing at higher or lower frequency can decrease the natural frequency.

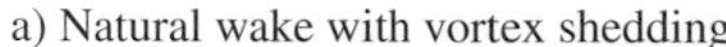

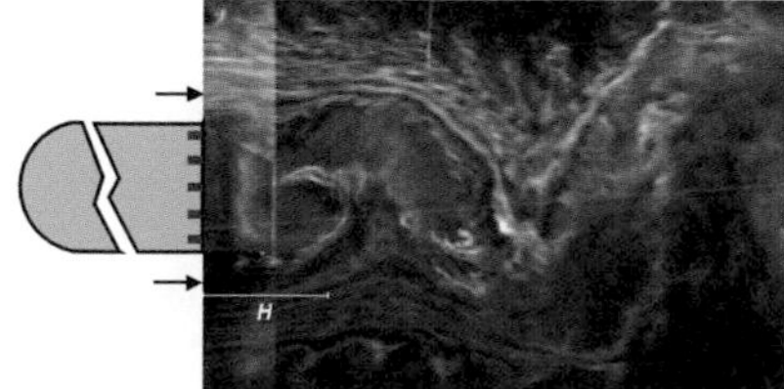

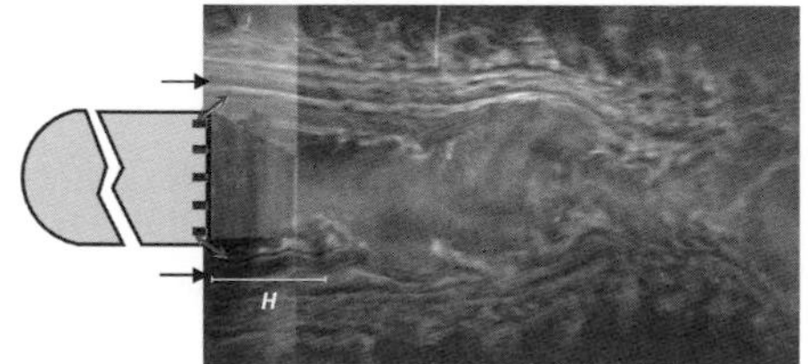

Figure 17.1 Flow visualization of the experimental wake behind a D-shaped body without (a) and with symmetric low-frequency actuation (b), reproduced with permission from Pastoor (2008). The D-shaped body is indicated in gray; the red squares mark the location of the pressure sensors, and the blue arrows indicated the employed ZNMF actuators.

The generalized mean-field model has been fitted to numerical URANS simulation data of a high-lift configuration (Luchtenburg et al. 2009) with high-frequency forcing. This model also accurately describes the experimental turbulent wake data with a stabilizing low-frequency forcing (Aleksic et al. 2010) as shown in Figure 17.1. The model the generalized mean-field model as expressed by (17.10) and (17.11) may guide in-time control (Duriez et al. 2017) and adaptive control design, providing the minimum effective actuation energy (Luchtenburg et al. 2010). In principle, (17.11) can be generalized for an arbitrary number of frequencies.

17.4 Model-Free Machine Learning Control

Model-based control has success stories in turbulence experiments for first- and second-order dynamics as outlined in the previous section. Control-oriented models for broadband turbulence are quite a challenge. Hence, most experimental turbulence control approaches are model-free, albeit typically restricted to tuning one or few actuation parameters. This section outlines the recently discovered MLC approaches, which solve a much more general problem: the optimization of general nonlinear multiple-input multiple-output (MIMO) control laws in an automated manner in the full plant without a model. First (Section 17.4.1), *cluster-based control* (CBC) is described allowing the rapid learning of simple smooth nonlinear control laws. Second (Section 17.4.2), *linear genetic programming control* is presented, allowing to optimize rather complex control laws with more testing.

17.4.1 Cluster-Based Control

Cluster-based control (Nair et al. 2019) is based on a low-dimensional empirical parameterization of MIMO control laws. The parameters are optimized with a gradient-based search. Let $\boldsymbol{b}$ comprise N_b actuation commands. Let N_s instantaneous sensor signals represent the output $\boldsymbol{s}$. The goal is to minimize a cost J, for example, drag or noise, with associated actuation penalty. The corresponding regression problem for the sensor-based control law

$$\boldsymbol{b} = \boldsymbol{K}(\boldsymbol{s}) \tag{17.13}$$

reads

$$\boldsymbol{K}^\star = \arg\min_{\boldsymbol{K}} J(\boldsymbol{K}). \tag{17.14}$$

Typically, the optimization (17.14) leads to a practically unsolvable variational problem in experiment and in simulations because of limited testing and simulation time.

The trick is to employ sensor data $\boldsymbol{s}^m$, $m = 1, \ldots, M$ for unforced and several forced flows. The sensor data is clustered to K centroids $\boldsymbol{c}_k$, $k = 1, \ldots, K$. The actuation vector at each centroid, $\boldsymbol{b}_k = \boldsymbol{K}(\boldsymbol{c}_k)$, $k = 1, \ldots, K$, is considered as tunable parameter. The CBC law (17.13) uses the actuation parameters at the centroids to interpolate between them, for example, with K-nearest neighbors. The optimization of this CBC law is performed with a robust downhill simplex method for the $K \times N_b$ actuation parameters. The centroids may be improved with added sensor data from the CBC law.

Nair et al. (2019) employs this approach for optimizing stall prevention over an actuated airfoil with large-eddy simulations in few dozen runs. Several other CBC variants have been proposed from brute-force testing (Kaiser et al. 2017b) to model-based control (Fernex et al. 2019). No experimental applications have been reported yet.

17.4.2 Linear Genetic Programming Control

Another approach to solving the regression problem (17.14) is facilitated by an evolutionary algorithm, more precisely *linear genetic programming* (LGP). In contrast to CBC, no advance data are assumed, and the control laws can, in principle, be rather complex.

Figure 1 of Li et al. (2018) illustrates the approach. A new MIMO control law (17.13) is tested for a given time in a fast control loop. The typical timescale from sensing to actuation is of the order of milliseconds in experiments. The performance J of the control is optimized in a slow outer loop. In aerodynamic experiments, the testing of one control law typically takes 5 to 10 seconds. After a few hundred up to one thousand control tests, the optimization is typically converged for $O(10)$ sensors and up to $O(10)$ actuators.

The optimization is based on LGP starting with the first generation of $N_i = 100$ random control laws $\boldsymbol{K}_i^1$, $i = 1, \ldots, N_i$. All these laws are tested in the experiment J_i^1 and sorted according to performance $J_1^1 \leq J_2^1 \leq \cdots \leq J_{N_i}^1$. Then, the $N_e = 1$ best control law K_1^1 is directly copied into the new generation. The other laws are obtained from genetic operations that preserve memory (replication), tend to breed better laws (cross-over), and explore new minima (mutation). The new generation is again tested and sorted, The iteration stops after a prescribed convergence criterion is reached. Typically up to $N_g = 10$ generations are required. We refer to the exquisite detailed description of LGP in Wahde (2008) and for control applications to Duriez et al. (2017).

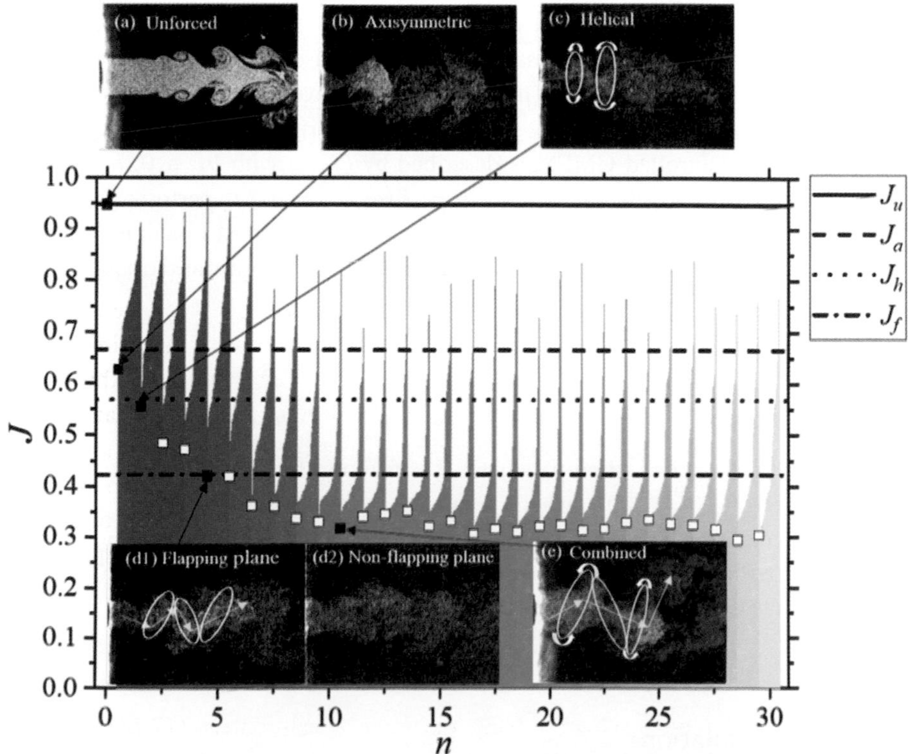

Figure 17.2 LGPC learning curve of jet mixing optimization. For details see Zhou et al. (2020).

Figure 17.2 shows the learning of a distributed actuation at a nozzle exit for jet mixing optimization. In 11 generations, the automated LGPC learning has discovered axisymmetric forcing, helical forging, and flapping forcing before discovering a much better performing hitherto unknown combined actuation. For details, the reader is referred to Zhou et al. (2020).

17.5 Tutorial: `xMLC` for Nonlinear Dynamics

In this section, MLC is exemplified for a dynamical system using the `xMLC` software. The tutorial begins with a short description of the *generalized mean-field model* (GMFM), a benchmark dynamical system of Duriez et al. (2017). Then, the download and the installation of the `xMLC` software are explained. `xMLC` capabilities and functions are illustrated by control design for the GMFM. Section 17.4 gives advice for user-defined problems.

17.5.1 The Generalized Mean-Field Model: A Nonlinear Benchmark Problem

The GMFM is a four-dimensional mean-field Galerkin model that describes the nonlinear interaction between two oscillatory fluctuations (see Section 17.3.3 and Chapter 5 of Duriez et al. (2017)). Here, for simplicity, specific values for all coefficients are assumed. It consists of two oscillators, (a_1, a_2) (unstable) and (a_3, a_4) (stable), coupled by a nonlinear growth rate σ of the (a_1, a_2) oscillator. The control objective is to stabilize the first oscillator. This can only be achieved via excitation of the second oscillator. Let $\boldsymbol{a} = [a_1, a_2, a_3, a_4]^{\mathrm{T}} \in \mathbb{R}^4$ be the state variable and $b \in \mathbb{R}$ the actuation command. The controlled GMFM is described by (17.1) with $b_1 \equiv 0$, $b_2 = b(a_1, \ldots, a_4)$ and $b_3 \equiv 0$. A full-state control $b = K(a_1, a_2, a_3, a_4)$ is performed by adding a forcing term b in the fourth equation of the system. The cost function J comprises the mean distance to the fixed point J_a and a penalization term for the actuation J_b:

$$
\begin{aligned}
J &= J_a + \gamma J_b, \\
J_a &= \frac{1}{T_{\max}} \int_0^{T_{\max}} (a_1^2 + a_2^2)\, dt, \qquad (17.15)\\
J_b &= \frac{1}{T_{\max}} \int_0^{T_{\max}} b^2\, dt.
\end{aligned}
$$

The penalization parameter γ is taken to be 0.01. The cost function is computed for $T_{\max} = 100T_1$ where $T_1 = 2\pi$ is the period of the first oscillator. The initial condition is set to $[a_1, a_2, a_3, a_4]^{\mathrm{T}} = [\sqrt{0.1}, 0, 0, 0]^{\mathrm{T}}$, that is, on the unforced limit cycle (radius $\sqrt{0.1}$) and the second oscillator is at the fixed point $[0, 0]^{\mathrm{T}}$. The dynamics of the GMFM without control is illustrated in Figure 17.3.

17.5.2 Installation

In this section, the necessary steps to install the `xMLC` software are presented.

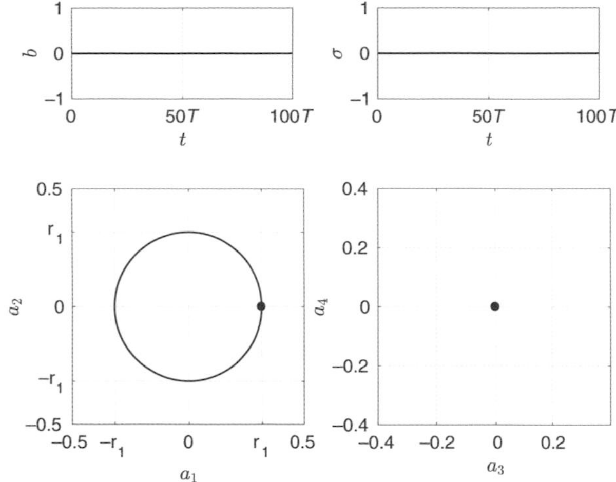

Figure 17.3 Unforced dynamics of the GMFM: actuation command b (top left), growth rate σ (top right), phase space for the unstable oscillator (bottom left), and phase space for stable oscillator (bottom right). The last period (T_1) is depicted in red. The circle (○) and the dot (•) represent, respectively, the initial condition and the final state for the oscillators. For the non controlled system, they overlap. This figure has been plotted with MATLAB with 100 points per period and 100 periods.

Requirements

The `xMLC` software has been coded in MATLAB version 9.2.0.556344 (R2017a). Natick, Massachusetts: The MathWorks Inc., 2017 and Octave version 4.2.2 (2018). John W. Eaton and others. Thus, any later version should be compatible with the software. No particular MATLAB package is needed for the functioning of the software. For Octave, the package `communications` is needed. It is freely available on the official webpage *https://octave.sourceforge.io*. It can also be downloaded directly from the graphic user interface with the `pkg` function.

Installation

The `xMLC` software can be download from the following link:

https://BerndNoack.com/programs/xMLC_v0.9.2.tar.gz

Once downloaded, untar the `xMLC_v0.9.2.tar.gz` file and copy the `xMLC_v0.9.2 folder` where it is needed. Installation is then complete.

To assure the compatibility with MATLAB or Octave, some specific files must be used. Those files are stored in the *Compatibility/* folder. To change from MATLAB to Octave and conversely, execute the adequate bash file `MatlabCompat.sh` or `OctaveCompat.sh` in the *Compatibility/* folders. If no Linux system is used, copy the files from *Compatibility/MATLAB/* or *Compatibility/Octave/* to the corresponding folders.

For further information about the content of the xMLC_v0.9.2 folder please consult the README.md file.

17.5.3 Execution

To use the xMLC software, launch a MATLAB or Octave session in the xMLC_v0.9.2 folder. Then run the Initialization.m script having replaced the files for compatibility as outlined in Section 17.5.2. This script loads all the necessary paths and creates a xMLC class object with the GMFM benchmark problem. This object is stored under the variable mlc by default. To create a new xMLC class object with the default parameters, use the following command: MyMLC = MLC;.

The following section describes some of the basic xMLC parameters and functions. All the commands are compatible with MATLAB and Octave except the saving and loading functions that are specific for each software.

xMLC **parameters**

Once the Initialization.m script is launched, a small description of the problem to be solved is printed. It contains information about the number of inputs (controllers), the number of outputs (sensors), the population size, and the strategy (genetic operators probabilities). This is shown each time a new instance of a xMLC class object is created. To show it again use the show_problem method by using the command mlc.show_problem;. The mlc object has four properties:

- population: contains all the information and database references for each generation of individuals;
- parameters: where all the parameters are defined, be it for the problem, the control law description or the xMLC parameters;
- table: is the database, it contains all the individuals explored, not all of them are evaluated thanks to screening options;
- generation: integer referring to the actual generation. It is set to 0 by default when the population is empty.

You can access and modify xMLC parameters with the following commands:

```
% To display all the parameters
mlc.parameters % MATLAB only, for Octave see default_parameters.m
% To display parameters related to the problem definition
mlc.parameters.ProblemParameters % MATLAB only
% To display parameters related to the control law definition
mlc.parameters.ControlLaw % MATLAB only
% To display parameters related to the MLC algorithm
mlc.parameters.CrossoverProb
% To modify some MLC parameters
mlc.parameters.Elitism = 0;
mlc.parameters.CrossoverProb = 0.5;
```

```
mlc.parameters.MutationProb = 0.5;
mlc.parameters.ReplicationProb = 0;
mlc.parameters.PopulationSize = 50;
```

Problem parameters and control law parameters should not be modified before starting a run; only xMLC parameters should be considered. To change them, see Section 17.5.4.

My first run

Once the xMLC parameters are appropriately set, the optimization process can be launched. The evolution of the population of individuals is done with the go method. A generation can be given as an argument to perform several evolution iterations. With no arguments one evolution step is performed leading to the evaluation of the generated individuals. Once the process is over, the xMLC object can be saved to be used later. All runs are saved in the *save_runs/* folder with their name and with their compatibility (MATLAB or Octave). The folders are created automatically.

```
% To compute the next generation and evaluate the new individuals.
mlc.go;
% To compute the next 5 generations :
mlc.go(5);
% To continue with the following 5 next generations :
mlc.go(10);
% Change the name of my run and save
mlc.parameters.Name = 'MyRun';
mlc.save_matlab; % for MATLAB
mlc.save_octave; % for Octave
% Load my run
mlc.load_matlab('MyRun'); % for MATLAB
mlc.load_octave('MyRun'); % for Octave
```

for loops in Octave are not as optimized as in MATLAB. Hence, the evaluation of the individuals may take much more time with Octave as compared to MATLAB. The reader is invited to optimize the evaluation of the individuals following ones needs by adjusting the evaluation parameters in mlc.parameters.ProblemParameters or in the GMFM_problem.m file, see Section 17.5.4.

Post-Processing and Analysis

Once the optimization process is done, the best individual can be accessed with the following method: mlc.best_individual;. This command plots information concerning the best individual. The ID number represents its index in the database. For the GMFM problem, a visualization of the best control law is plotted, see Figure 17.4. The first oscillator has been successfully stabilized whereas the second oscillator has been destabilized. Moreover the fixed point has been reached only after three periods. The growth rate σ oscillates around 0, and the second oscillator is on the limit cycle of radius $\sqrt{0.1}$.

For more information about the best control law, use the command `mlc.table.individuals(IDN);` where `IDN` is the ID number of the best individual. Other features can be extracted such as the learning process or the Pareto front, thanks to the command `mlc.convergence;`, see Figure 17.5. Figure 17.5(a) shows that the performance of the best control increases throughout the generations as well as the overall performance of the population. From first to last generation the cost function is dropped by one order of magnitude. In Figure 17.5(b) the latest generations are the ones pushing forward the Pareto front optimizing J_a and J_b even though the penalization parameter is only $\gamma = 0.01$.

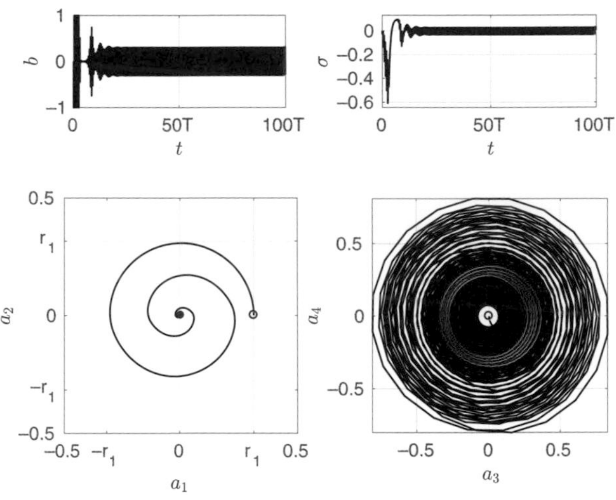

Figure 17.4 Same as Figure 17.3 for the best control law computed after the evolution of 100 individuals through 10 generations.

17.5.4 My MLC Problem

The definition of an MLC problem and its parameters is done by two files: `MyProblem_problem.m` and `MyProblem_parameters.m`, respectively. An example is provided in the *Plant\MyProblem* folder. The problem parameters and control law parameters should be modified only by creating an appropriate `MyProblem_parameters.m` file. Those parameters include the penalization parameter γ (gamma), the actuation limitation, the number of constants, the operators $(+, -, \times, \div, \ldots)$, the register initialization, and so on. The user is free to do any modification in the *MyProblem* files.

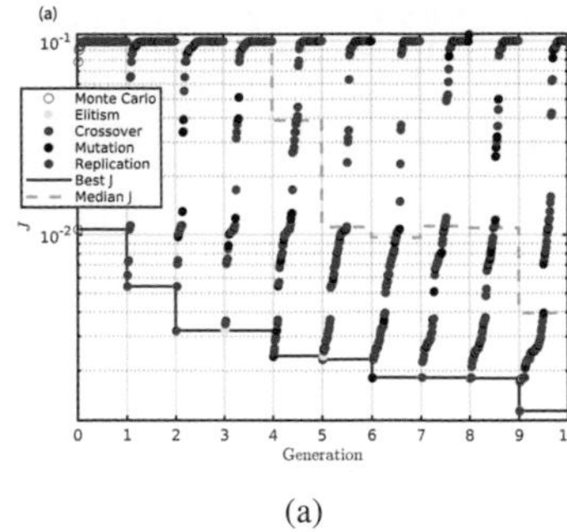

(a)

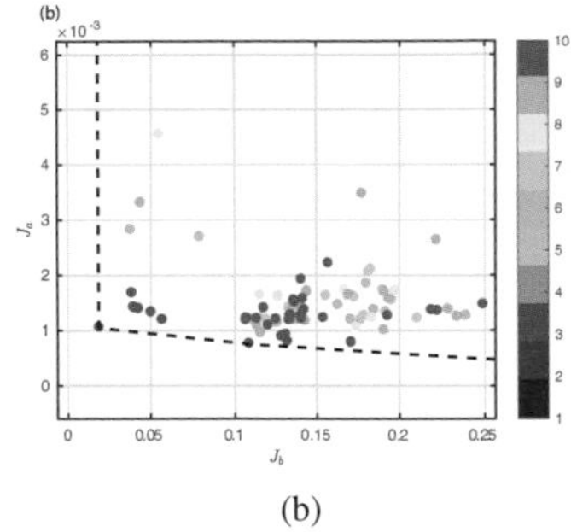

(b)

Figure 17.5 Two figures obtained thanks to the `convergence` method for the 100×10 GMFM runs. Those plots show the distributions of the individuals following their performance and cost throughout the generations. Plotted with MATLAB. (a) Cost distribution for each generation. The color represents how each individual has been created. The red solid line represents the best individual for each generation, and the green dashed line follows the median performance. (b) Zoom-in on the Pareto front (black dashed line) for the 1000 individuals. Each dot represents one individual, the color codes the generation.

Any malfunction in the evolution process should be reported to Guy Y. Cornejo Maceda (`gy.cornejo.maceda@gmail.com`) with the error message.

A cheat sheet containing more information about other features is also provided in the `xMLC_v0.9.2` folder.

18 Deep Reinforcement Learning Applied to Active Flow Control

J. Rabault[1] and A. Kuhnle

This chapter reviews the recent applications of deep reinforcement learning (DRL) to the control of fluid mechanics systems. DRL algorithms attempt to solve complex nonlinear, high-dimensionality control, and optimization problems using a direct exploration approach, that is, trial and error. Here, we start with an in-depth review of current state-of-the-art DRL algorithms, and we provide the background necessary to understand how these are able to perform learning through direct close-loop interaction with the system to control. Then, we offer an overview of the applications of DRL to active flow control (AFC) that can be found in the literature, and we use this review to draw concrete guidelines for the application of DRL to fluid mechanics. Finally, we discuss both outstanding challenges and directions for further improvement in the use of DRL methods for AFC. The chapter is supported by several examples of DRL application codes, and offers both a theoretical and a practical guide to new practitioners.

18.1 The Promise of Deep Reinforcement Learning in Fluid Mechanics

Reinforcement learning (RL) – often referred to as DRL when it is combined with deep neural networks (DNNs) as function approximators – has seen rapid progress in recent years. As discussed briefly in Chapter 3, while supervised learning is focused on learning a function such as a classification or regression task from observed input–output data points, and unsupervised learning is focused on identifying patterns without explicit target labels, RL is concerned with optimizing decision-making in interactive problems for which no solution is known a priori. Instead of target labels, a reward signal, which may be sparse and/or time-delayed, acts as feedback to reflect overall performance and to consequently guide the learning process. This implies that RL training takes place through direct interaction with the system to control in a trial-and-error, closed-loop fashion.

The general RL framework presents many similarities with the control framework presented in Chapter 10. However, while traditional proportional integral differential

[1] Jean Rabault was funded through the Research Council of Norway, by the Petromaks II project DOFI, grant number 280625.

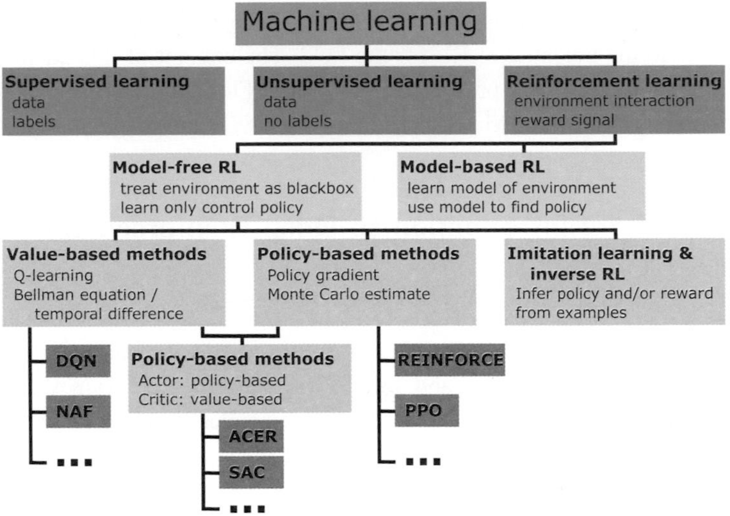

Figure 18.1 Overview of the different machine learning approaches related to reinforcement learning. The focus of this chapter is on model-free RL algorithms or, more specifically, on value-based / Q-learning algorithms and policy-based / policy gradient methods, which are discussed in Sections 18.2.2 and 18.2.3, respectively. Popular recent example algorithms like deep Q-learning (DQN) or proximal policy optimization (PPO) are introduced in the corresponding section.

(PID) controllers or linear quadratic regulators (LQR) rely on a local analysis of the system and assume properties such as linearity, RL aims at performing control without imposing requirements on the underlying system. In particular, the RL algorithms discussed in this chapter are able to effectively map nonlinear systems through dircct exploration, and to find complex nonlinear control strategies adapted to each system.

As a consequence of its generality, applications of the RL paradigm are many, ranging from playing a wide range of games, such as Atari, Chess, Go, or Poker (Silver et al. 2017, 2018), to providing high levels of dexterity to robots (Hwangbo et al. 2019), to controlling complex industrial systems in order to optimize, for example, power consumption (Knight 2018). All these applications illustrate that properly designed RL systems are able to perform a wide range of decision-making and control tasks involving complex systems. This makes RL methods a promising candidate for solving AFC problems in the context of fluid mechanics. AFC is well known to be very difficult due to the many challenges presented by the underlying Navier–Stokes equations, and the inherent nonlinearity and high dimensionality of their solutions. In this chapter, AFC strategies under consideration are of a closed-loop nature, as an observation of the state of the flow is provided to the RL agent.

This chapter provides an overview of RL applications for fluid mechanics problems. First, we introduce basic concepts of RL and briefly describe the most common classes of modern RL algorithms. Next, we focus on how RL can be applied to fluid mechanics and review recent work on this topic. Finally, we discuss best practices and caveats to consider when applying RL to fluid mechanics problems, as well as future prospects and the potential of RL for the field. The content of this chapter closely follows the material that was presented at the 2019 "Flow/Interface School on Machine Learning and Data Driven Methods (Rabault 2019)."

18.2 A Brief Introduction to Deep Reinforcement Learning

18.2.1 Learning through Trial and Error

RL is the area of machine learning concerned with learning optimal behavior through interaction and trial and error. The standard RL framework conceptualizes this problem as a sequence of interaction episodes in which an agent repeatedly observes the state of its environment, decides what action to perform next, and receives a reward signal as performance feedback. Typically, one distinguishes between model-based and model-free RL: **model-based RL** methods incorporate knowledge about the problem domain to learn a functional model of the environment simultaneously to optimizing its decision-making, whereas **model-free RL** considers the environment to be a black-box and instead focuses entirely on learning to maximize the attained reward through trial and error. Since these classes are very different in practice, and since the large majority of deep-learning-based RL methods are model-free, we will focus exclusively on model-free RL for the purpose of this chapter. Figure 18.1 presents an overview of the different classes of RL methods together with some key characteristics and popular example algorithms.

More formally, the agent is usually represented as a stochastic **decision policy** π that, given an observation of the environment state s, outputs an action distribution $\pi(a \mid s)$, from which a decision is sampled on which action a to take. The general framework is summarized in Figure 18.2. Similarly, the environment is modeled as an (unknown) stochastic process $W(r, s_{\text{next}} \mid s, a)$ that provides the likelihood of next state s_{next} and reward r, given that the system is currently in state s and action a has been performed. An interaction **episode** – also referred to as trajectory or rollout – can thus be represented as a sequence of state–action–reward triples $\tau = \big((s_0, a_0, r_0), (s_1, a_1, r_1), \ldots\big)$. In addition to this general framework, a few important points will be assumed in all of the following:

- Strictly speaking, the environment **state** s is a discrete, potentially noisy and incomplete observation of the true internal (and continuous) state of the environment. In some cases, the state can be captured perfectly as, for instance, the position of stones in the game of Go, whereas in most cases the observation comes from noisy sensors, providing partial information about the environment, like pressure values at a range of sensor locations on the surface of a cylinder immersed in a flow. The literature nonetheless generally refers to it as the environment "state" s, instead of a more appropriate term like "observation," which we will do here as well.
- The **action** a may be either a decision from a fixed finite set of choices or a continuous control signal. Taking the previous examples, the action space in

Go is a set of positions where the next stone can be placed at each time step, while an action in the cylinder control case could correspond to the discretized opening of valves, that is, the mass flow rate ejected, at some actuation slots on the surface of the cylinder.

- The **reward** r loosely reflects the performance of the agent and enables the agent to learn an effective policy from trial-and-error interaction with the environment. Importantly, the impact of a (sequence of) action(s) may not become apparent immediately, resulting in delayed and/or sparse rewards, which is why they cannot be used as independent supervised signal per time step. For instance, in the previous examples, an agent may only receive a final reward at the end of a round of Go depending on its success, and mass flow rate control decisions have longer-term impact on stability and drag of the cylinder system. Therefore, instead of optimizing the instantaneous reward directly, RL algorithms optimize the cumulative reward over an episode trajectory τ, $R(\tau) = \sum_{t=0}^{\infty} \gamma^t \cdot r_t$, discounted by a factor $0 \leq \gamma \leq 1$ (usually $\gamma \in [0.95, 0.999]$). On the one hand, the discount factor encourages shorter-term over longer-term gains while on the other hand, it ensures that the cumulative reward converges in the case of very long or unbounded episodes.
- The following reward-related value abstractions are used by different algorithms to make sense of the environment and its reward structure, and to provide guidance for the learning process:
  1. the **state-value** $V^\pi(s) = \mathbb{E}_{\tau\sim\pi}[R(\tau) \mid s_0 = s]$ represents the "expected value" of being in state s, that is, the expected return obtained through a sequence of interactions following the policy π when starting in environment state s;
  2. the **state-action-** or **Q-value**[2] $Q^\pi(s,a) = \mathbb{E}_{\tau\sim\pi}[R(\tau) \mid s_0 = s, a_0 = a]$ that, similarly, captures the "expected value" of taking action a in state s, that is, the expected value of taking action a when starting in state s and continuing to act according to π afterward;
  3. the **advantage** $A^\pi(s,a) = Q^\pi(s,a) - V^\pi(s)$ that indicates how much "better" (in terms of expected value) it is to take action a when in state s, compared to the expected return when randomly sampling an action from π.

RL is not a new field, and the framework represented in Figure 18.2 can be traced back to at least the 1980s. However, the learning algorithms used for training, and in particular the function approximators being trained, have undergone rapid development since the popularization of deep learning. The key idea behind DRL is to leverage the universal approximator properties of DNNs (Hornik et al. 1989), together with effective training methods such as stochastic gradient descent and backpropagation of errors (see Chapter 3), in order to make RL algorithms able to cope with complex systems. In the following sections, we will introduce the two main classes of RL algorithms – Q-learning and policy gradient methods – together with recent examples of DRL algorithms for each class.

[2] Q originally comes from *"quality"*, but nowadays is just known as Q-value and not used as an abbreviation anymore.

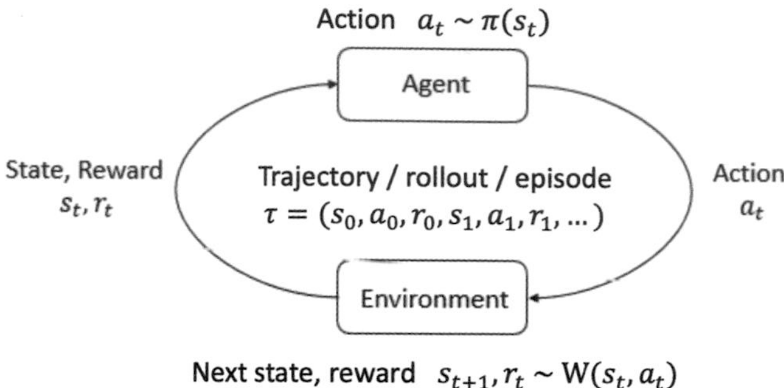

Figure 18.2 Generic reinforcement learning setup. Iteratively, an agent reacts to the observed environment state with an action, which in turn influences how the environment state evolves. Moreover, the agent receives a reward signal, which it aims to optimize over the course of multiple training episodes, that is, rollouts of agent–environment interactions.

18.2.2 Q-Learning

Q-learning comprises a range of RL algorithms, dating back to at least the work of Watkins (1989), though it mainly relies on the Bellman equation, which was formulated much earlier (Bellman 1957). A method belonging to this category, Deep Q-Network (DQN) (Mnih et al. 2015), was among the early breakthroughs of DRL. DQN combines DNNs and Q-learning, and its success in learning to play Atari games solely from vision and game score feedback triggered a renewal of interest in DRL. This success is made possible by the combination of RL for learning through trial and error, with DNNs for their ability to extract features and identify patterns in data.

Q-learning is a **value-based method**, meaning that the algorithm focuses on learning a (state-action) value function, $Q(s, a)$, which represents the overall reward obtained by taking action a in state s. Given this Q-function, Q-learning defines the resulting policy as $\pi(s) = \text{argmax}_a\, Q(s, a)$. Note that this definition crucially relies on the fact that the action space is finite, as otherwise the computation of argmax is not straightforwardly possible. The Q-value function is optimized via temporal difference (TD) learning in which its own estimates are used as optimization targets. More precisely, the TD target corresponding to an experience time step s_t, a_t, r_t, s_{t+1} can be recursively estimated as $Q^{\text{target}} = r_t + \gamma \cdot \max_{a'} Q(s_{t+1}, a')$. Note that the value of the next state is $\max_{a'} Q(s_{t+1}, a')$ since the policy always picks the max action. The Q-function is then updated by incrementally shifting its predictions in the direction of the TD residual $Q^{\text{target}} - Q$, with the update step size being controlled by a learning rate $\eta > 0$ (n indicates the current optimization step):

$$Q_{n+1}(s_t, a_t) := Q_n(s_t, a_t) + \eta \cdot \left[\left(r_t + \gamma \cdot \max_{a'} Q_n(s_{t+1}, a') \right) - Q_n(s_t, a_t) \right]. \tag{18.1}$$

In the context of deep RL, the value function Q is parametrized by an artificial neural network Q_θ with learned weights θ, which is optimized via gradient descent. By minimizing the quadratic loss function $\mathcal{L}(Q^{\text{target}}, Q_\theta) = 0.5 \cdot (Q^{\text{target}} - Q_\theta(s, a))^2$

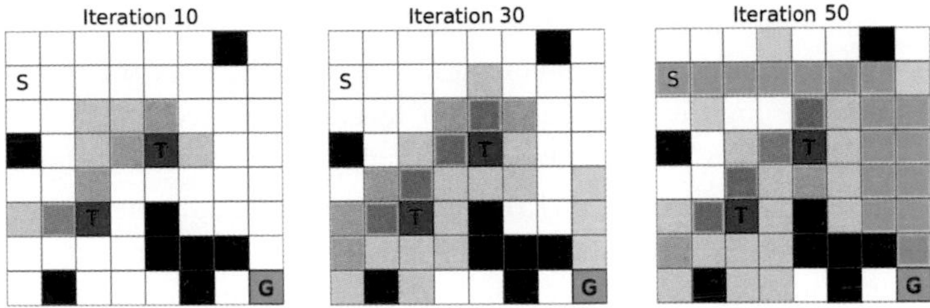

Figure 18.3 Illustration of the learning process of an RL agent in a grid-world with one positive-reward goal state G and two negative-reward failure states T. Starting with no prior knowledge, the agent learns to estimate the value of states more accurately over the course of training solely via interaction. Initially, the negative reward of the failure states propagates to neighboring states, so the agent learns to avoid moving there. As a consequence, it subsequently comes across the goal state whose positive reward propagates back to the starting point over time, yielding a first successful strategy. Ultimately, the agent learns to trade off avoiding failure states with quicker strategies of getting to the goal state.

between Q^{target} and the Q_θ to be optimized, we arrive at the equivalent update rule for DNN-parametrized Q-functions (note that Q^{target} is fixed with respect to the differentiation):

$$\begin{aligned}\theta_{n+1} &= \theta_n + \eta \cdot \nabla_\theta \mathcal{L}\left(Q_{\theta_n}^{\text{target}}, Q_\theta(s_t, a_t)\right) \\ &= \theta_n + \eta \cdot \nabla_\theta \left(0.5 \cdot \left[(r_t + \gamma \cdot \max_{a'} Q_{\theta_n}(s_{t+1}, a')) - Q_\theta(s_t, a_t)\right]^2\right) \\ &= \theta_n + \eta \cdot \left[(r_t + \gamma \cdot \max_{a'} Q_{\theta_n}(s_{t+1}, a')) - \nabla_\theta Q_\theta(s_t, a_t)\right]. \end{aligned} \tag{18.2}$$

Q-learning is typically illustrated at the example of a grid world, as shown in Figure 18.3. Given a finite state and action space, tabular Q-learning simply represents the Q-value function as an array with states on one axis and actions on the other. This scenario is useful for grasping how the algorithm learns to improve its behavior: during the process of repeatedly interacting with the world in a series of episodes, the update in (18.1) iteratively transmits the reward information of a state to adjacent states, ultimately communicating the signal from goal to start state(s) and thus arriving at a successful policy. The approach follows the "dynamic programming" pattern to make the optimization tractable, as is familiar from optimization methods in other domains.

A fundamental problem with the Q-learning formulation is that the policy derived from the learned value function as $\pi(s) = \text{argmax}_a Q(s, a)$ is deterministic, and thus can easily get stuck in local maxima during the training process, even if their corresponding Q-value is only marginally higher. A common solution to this is to introduce an exploration rate $0 \leq \epsilon \leq 1$, which determines how frequently the argmax policy is ignored and a random action is sampled instead. Q-learning can still learn from such **off-policy data** – that is, actions which do not follow the policy – since it can still extract useful information to update its Q-value estimates. How best to trade off exploration and exploitation is an important and problem-dependent hyperparameter decision, and it is common to use decay schedules which over time decrease the initially high exploration rate to leverage the accumulated knowledge. Note that exploration is only useful during training, and is turned off for deployment.

DQN (Mnih et al. 2015) is considered a breakthrough, being the first Q-learning algorithm to successfully leverage a DNN as a feature extractor instead of a table. This allowed researchers to move beyond low-dimensional finite states and to handle the complex visual state-space of Atari games. DQN introduced a range of technical improvements, which were required to stabilize the learning optimization, most notably: (a) an experience replay memory that stores a large amount of experience time steps and from which random time steps are sampled for an update, thus relaxing otherwise detrimental temporal correlation between update time steps; and (b) the use of a target network to compute the TD target in (18.2), which is an infrequently synced copy of the main network and thus reduces the problem of potentially detrimental feedback loops in the recursive updates.

Since this milestone achievement, more improvements and novel techniques have been introduced in follow-up papers: for instance, double DQN (Hasselt et al. 2016) improves a bias in the original DQN formulation that tends to overestimate Q-values. Dueling DQN (Wang et al. 2016) factorizes Q-values into state-value $V(s)$ and advantage $A(s, a)$, which helps the network to learn coherent Q-values per state. Categorical DQN (Bellemare et al. 2017) extends the Q-learning framework to work with (quantized) reward/value distributions, which enables the agent to capture the reward structure and uncertainty more faithfully. Normalized advantage function (NAF) (Gu et al. 2016) derives a continuous variant of the Q-learning algorithm, to be able to work with non-categorical action spaces. These are just a few examples of extensions, and many more can be found in the literature, see for example Recht (2019).

18.2.3 Policy Gradient Methods

Policy gradient algorithms are another popular class of RL approaches, dating back to at least the work of Williams (1992). These are **policy-based methods**, meaning that they attempt to optimize a parametrized (stochastic) policy π_θ directly with respect to its expected return $\mathbb{E}_{\tau \sim \pi_\theta}[R(\tau)]$. This is in contrast to Q-learning, where Q is optimized to approximate the state-action-value function and the (deterministic) policy is inferred via $\text{argmax}_a\, Q(s, a)$. In deep RL, the policy is parametrized by a DNN (the policy network) that outputs relevant distribution parameters, for instance, the probability per action for discrete actions, or the mean and standard deviation for normally distributed continuous actions.

Fundamental to this optimization is the policy gradient theorem, which uses the "log-derivative trick" to establish the following equality for the derivative of the expected return:

$$\begin{aligned}
\nabla_\theta \left(\mathbb{E}_{\tau\sim\pi_\theta}[R(\tau)]\right) &= \int R(\tau)\cdot\nabla_\theta\pi_\theta(\tau)\,\mathrm{d}\tau \\
&= \int R(\tau)\cdot\pi_\theta(\tau)\cdot\nabla_\theta\log\pi_\theta(\tau)\,\mathrm{d}\tau \\
&= \mathbb{E}_{\tau\sim\pi_\theta}\left[R(\tau)\cdot\nabla_\theta\log\pi_\theta(\tau)\right].
\end{aligned} \tag{18.3}$$

Using the conditional (in-)dependence between time steps, the fact that the logarithm of the product probability $\pi_\theta(\tau)$ can be turned into a sum, and the fact that the transition probabilities of the environment do not depend on the policy, a simple Monte Carlo gradient estimator $\hat{g}$ of this expectation can be derived as:

$$\begin{aligned}
\hat{g} &= \frac{1}{n}\cdot\sum_{i=0}^{n}\left(R(\tau_i)\cdot\nabla_\theta\log\pi_\theta(\tau_i)\right) \\
&= \frac{1}{n}\cdot\sum_{i=0}^{n}\left(R(\tau_i)\cdot\nabla_\theta\log\left[\prod_{t=0}^{\infty}\pi_\theta(a_t^{(i)}|s_t^{(i)})\cdot W(s_{t+1}^{(i)}|s_t^{(i)},a_t^{(i)})\right]\right) \\
&= \frac{1}{n}\cdot\sum_{i=0}^{n}\left(R(\tau_i)\cdot\nabla_\theta\left[\sum_{t=0}^{\infty}\log\pi_\theta(a_t^{(i)}|s_t^{(i)})+\sum_{t=0}^{\infty}\log W(s_{t+1}^{(i)}|s_t^{(i)},a_t^{(i)})\right]\right) \\
&= \frac{1}{n}\cdot\sum_{i=0}^{n}\left(R(\tau_i)\cdot\sum_{t=0}^{\infty}\nabla_\theta\log\pi_\theta(a_t^{(i)}\mid s_t^{(i)})\right),
\end{aligned} \tag{18.4}$$

where $W(s_{t+1}|s_t, a_t)$ is the probability that, starting from state s_t and taking action a_t, the environment evolves into state s_{t+1}. To sum it up, policy gradient methods work by sampling a set of trajectories at each training step, and use the collected **on-policy data** to shift the weights of the policy network in order to maximize its expected return, that is, to increase the likelihood of trajectories with high return and discourage decisions that have led to low episode returns. Note that, in contrast to Q-learning, exploration is intrinsic to policy gradient methods since the latter is based on a stochastic policy. Consequently, balancing exploration and exploitation is part of the training objective, which means that only after repeated positive feedback the action distribution narrows to the point where it is virtually deterministic. When deploying the trained algorithm, sampling is turned off and replaced by deterministically picking the maximum likelihood action.

The gradient estimator in (18.4) is also known as the REINFORCE algorithm (Williams 1992). However, it is a high-variance estimate since it scales all time steps of an episode by the same episode return $R(\tau)$. This in turn may impact convergence speed of the gradient descent optimization. Lower-variance estimators, which instead weight time steps according to the "reward to go", that is, the future return $R_t(\tau) = R(\tau[t, t+1, \dots])$ following the current timestep t can be derived similarly. Variance can be further reduced by observing that a state-dependent baseline function $b(s)$ can be subtracted from the return without affecting the expectation value in (18.3). It can be shown that $b(s) \approx V^\pi(s)$ is the optimal choice with respect to variance reduction for a state-dependent baseline. The following thus yields a lower-variance gradient estimator for the policy gradient objective:

$$\hat{g} = \frac{1}{nt} \cdot \sum_{i=0}^{n} \sum_{t=0}^{\infty} \left(R_t(\tau^{(i)}) - b(s_t^{(i)})\right) \cdot \nabla_\theta \log \pi_\theta(a_t^{(i)} \mid s_t^{(i)}). \tag{18.5}$$

Algorithms that focus on learning a baseline value function in addition to the actual policy are also called **actor-critic methods**. In this case, the value function is typically a second "critic" DNN besides the first "actor" policy DNN, trained for a regression objective similarly to Q-learning in (18.2). The purpose of the critic is to modify the return $R(\tau)$ in the original policy gradient equation (18.3) in order to improve algorithm convergence. This can be performed in several ways, for example, by subtracting a baseline as is done through using $R_t(\tau) - b(s_t)$ in (18.5), or by replacing the return altogether by the value output of a network trained through Q-learning, $V(s_t) \approx R_t(\tau)$.

A range of extensions and modifications of this approach to policy optimization have been explored. ACER (Wang et al. 2017) is an off-policy actor-critic algorithm that uses experience replay similar to DQN (see Section 18.2.2). Soft actor-critic (SAC) (Haarnoja et al. 2018) incorporates an entropy-based loss component into the policy optimization to encourage exploration. Deterministic policy gradient (DPG) (Silver et al. 2014) derives a deterministic variant of the policy gradient theorem. Trust-region policy optimization (TRPO) (Schulman et al. 2015) is a related class of algorithms that additionally constrain the update steps to a trust region, preventing the updated policy distribution from diverging too far from the old policy, thus limiting the confidence in each single, potentially detrimental update. Proximal policy optimization (PPO) (Schulman et al. 2017) is inspired by such trust-region methods and loosely enforces conservative updates by clipping the objective accordingly.

18.2.4 Current Research Directions in Deep Reinforcement Learning

Since its early breakthroughs, DRL has evolved into a mature research field with many research directions and open problems actively being worked on. The following paragraphs briefly introduce some interesting new lines of research to improve DRL algorithms and make them more practically applicable. Importantly, this review emphasises that DRL is a fast-moving field with lots of yet untapped potential.

Parallelized Data Collection and Distributed Learning

DRL algorithms can be very data-hungry and, different from supervised machine learning, do not learn from static data sets, but have to collect data via interaction. Consequently, methods have been developed to parallelize data generation across 100–1000s of "worker" agents, each of which interacts with a copy of the simulated environment. The collected experience is aggregated in either one central or multiple distributed parameter servers that perform the actual RL optimization algorithm, and the updated weights are synchronized with the workers and other servers. For instance, Asynchronous Advantage Actor-Critics (A3C) (Mnih et al. 2016) used around 16 workers, Importance-weighted actor-learner architectures (IMPALA)

(Espeholt et al. 2018) pushed the number to around 200, and the evolution-strategy-based system of Salimans et al. (2017) leveraged more than 1 000 cores in parallel. Such a parallelization approach is particularly interesting for the application of DRL to flow control because of its time-consuming simulations, as illustrated by Rabault and Kuhnle (2019).

Inverse Reinforcement Learning

Many environments cannot easily be parallelized, particularly if they are not simulation-based. In such cases it can be infeasible to collect enough interaction data to train an RL agent from scratch. Inverse reinforcement learning comprises techniques to instead make use of existing data that were not produced by the agent. By inferring a suitable reward function from such trajectories, the agent is provided a learning signal to bootstrap its own policy without the need for trial and error. Two popular approaches are generative adversarial imitation learning (GAIL) (Ho & Ermon 2016) and adverserial inverse reinforcement learning (AIRL) (Fu et al. 2018).

Model-Based Influences

The successes of DRL have mostly used model-free RL algorithms, which attempt to solve a task solely based on interaction and reward. However, trying to understand how the environment works can provide an additional learning signal that ultimately also benefits the agent's task-solving ability, and there has been work to augment model-free approaches with model-based features. For instance, Jaderberg et al. (2017) used auxiliary tasks such as next-frame prediction to encourage the network to learn representations that capture basic world dynamics. More recently, Ha and Schmidhuber (2018) presented an approach in which a full world model is learned from (random) interaction data and subsequently used by the agent for training, instead of interacting with the real environment.

Exploration and Curiosity

RL agents trained from scratch may either never observe any useful reward signal due to sparse rewards, or get stuck in local optima early on in the learning process and exploit this suboptimal solution instead of continuing to search for better strategies. While random exploration is sufficient for simple problems, bigger and more complex environments require more principled exploration approaches. A range of recent work has attempted to incorporate a sense of "intrinsic curiosity" into the agent's objective, by comparing the agent's expectation with the actual outcome and rewarding surprisal (Pathak et al. 2017, Savinov et al. 2018, Burda et al. 2019). This encourages the agent to explore uncertain areas of the problem space until they are sufficiently explored. In addition, exploration problems can be alleviated by incrementally making the task more difficult, for example, by starting episodes from a point closer to obtaining a reward than the actual initial state (Salimans & Chen 2018).

From Simulation to Real World

Simulations address the problem of DRL methods being data-hungry by providing a safe and infinite source of training data. However, the learned behavioral policy ultimately has to be transferred from an imperfect simulation to the real-world application. Some recent work has focused on techniques to bridge the simulation reality gap, as in Hwangbo et al. (2019).

18.3 Recent Applications of Deep Reinforcement Learning to Fluid Mechanics Problems

In this section, we focus on recent applications of DRL in the fluid mechanics literature. DRL can be applied to several classes of problems in fluid mechanics. First, it can be used to find high-level control strategies of complex systems, in a similar way to how robots are controlled. This class of applications includes, for example, swimming of fish schoolings, which we will review first. Second, we will discuss applications of DRL to a more detailed control of flow instabilities and detachment at small scales. Several such applications have been presented recently, such as the control of vortex shedding control or of Rayleigh–Benard instabilities.

In addition to these applications, there is a range of work that in other ways relates to fluid mechanics problems, or to the interaction between fluid mechanics and other engineering challenges, such as controlling unmanned aerial vehicles (UAVs) (Bohn et al. 2019), or optimizing the trajectory of gliders taking ascendences (Reddy et al. 2016). However, these are out of scope of the focus of this chapter, and we refer the reader directly to these papers for further information. Finally, another application of DRL within fluid mechanics is shape optimization (Yan et al. 2019, Viquerat et al. 2019). This last category is, so far, less mature than the use of DRL for control, in particular since shape optimization problems are less naturally expressed as a sequence of interactions with an environment. Therefore, we will not discuss this application in more detail either, and the reader is again referred to the most recent publications on the topic (Garnier et al. 2019, Viquerat et al. 2019, Yan et al. 2019).

18.3.1 Swimming and Locomotion

To the best of our knowledge, the first applications of DRL in fluid mechanics have been to study the collective behavior of swimmers. This is due to the fact that such behavior is relatively easy to simulate, especially if reduced-order models are used (Gazzola et al. 2014, 2016), while the complexity of the system composed of several swimmers makes it a challenging setup to investigate with other methods than via empirical simulation. In this context, DRL made it possible to both understand and reproduce a variety of schooling behaviors. In such studies, the environment state is a representation of the flow and the relative effect of swimmers on each other, while the actions describe the swimming effort of individual fishes, and reward is based on either attaining maximum speed or reducing energy consumption.

The most recent studies in this field solve the full Navier–Stokes equations, either in two or three dimensions (Novati et al. 2017, Verma et al. 2018). This is particularly impressive, given both the computational cost of such simulations, and the large state and action spaces associated with multiple 3D-simulated fishes swimming together as a group. In these studies, DRL is able to discover strategies that provide clear energy benefits compared to fishes swimming individually, and complex strategies are found that allow follower fishes to effectively extract energy from the vortex shedding alley behind leading fishes. Such results open the way to experimental understanding of fish schoolings and associated energy gains, as well as to applications in robotics. In a similar fashion, applications to gliding have been proposed (Novati et al. 2019). This category of applications can be seen as an effective way of finding optimal locomotion strategies for active swimmers or gliders, in a direct analogy to the use of DRL for legged robots locomotion.

These examples illustrate how DRL is used as a tool to experimentally study complex phenomena where a purely theoretical approach faces challenges, since describing a school of swimming fish is analytically intractable, and other experimental or numerical approaches do not cope well with the high dimensionality and analytical intractability of the problem.

18.3.2 Feedback Active Flow Control

The second category of fluid mechanics applications where DRL has attracted a lot of attention recently is feedback flow control (Rabault et al. 2020, Ren et al. 2020). Similar to the examples of the previous paragraph, feedback flow control is a natural fit for the closed-loop framework of DRL. Properties of the flow such as pressure or velocity maps make up the environment state, while the actions represent some form of forcing on the flow. Depending on the use case, the objective defined by the reward is either to reduce drag, or to damp unstable waves or modes.

The first application of DRL for such problems is arguably the control of the Karman vortex street presented in Rabault et al. (2019).[3] While this problem has been considered at several occasions using traditional control theory techniques (Roussopoulos 1993, Min & Choi 1999, Thiria et al. 2006), the wake dynamics are increasingly challenging to control as the Reynolds number – that is, flow nonlinearity – is increased. Here, DRL was used to control the separation and vortex shedding behind a 2D cylinder, as illustrated in Figure 18.4, and is able to reduce drag significantly. While the general situation is similar to what was presented in Chapter 10, the DRL approach is able to control the wake also in cases when a linear approximation does not suffice to capture the dynamics of the system. In a series of follow-up experiments, speeding up the learning process via parallelization as well as choosing an optimal reward function and DRL hyperparameters were investigated (Rabault & Kuhnle 2019).[4] The results in particular emphasized the sensitivity of

[2] The full code is released open-source, and the results can be reproduced following the instructions available at `https://github.com/jerabaul29/Cylinder2DFlowControlDRL`.

[3] The code is available here: `https://github.com/jerabaul29/Cylinder2DFlowControlDRL`

[4] The code is available here:
`https://github.com/jerabaul29/Cylinder2DFlowControlDRLParallel`

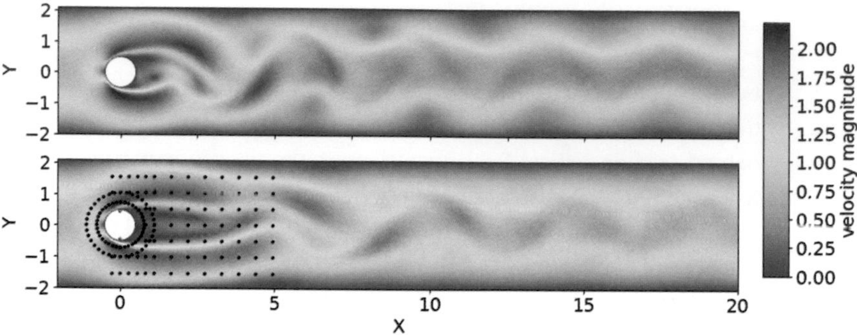

Figure 18.4 Illustration of active flow control of the von Karman alley by a deep reinforcement learning Agent (PPO algorithm). This figure is a slightly edited reproduction of the results presented in Rabault et al. (2019), reproduced with permission. The top panel shows a representative snapshot of the 2D flow around a cylinder, in a setting similar to the benchmark of Schäfer et al. (1996). The bottom panel shows the typical flow field obtained when a trained DRL agent is used to control the vortex shedding to reduce drag. The black dots indicate the location of a series of pressure probes, which form the state observation. The red dots at the top and bottom of the cylinder indicate the position of two small synthetic jets, whose mass flow rates are decided by the actions of the DRL agent. Here, the reward function encourages to minimize drag while keeping lift fluctuation low. A drag reduction of about 8% is obtained, close to the optimal drag reduction value predicted by the symmetric baseline flow analysis of Bergmann et al. (2005).

the resulting strategy to the exact choice of reward function, and more generally the importance of using physical insights when defining the reward.

Having established the potential of DRL for AFC, several questions remained to enable the application of DRL to more realistic and sophisticated systems. First, DRL needs to be able to cope with the large action space dimensionality of real-world systems. This is challenging since the size of the control space to explore is much larger when the number of actuation signals is increased (typically, the naive direct exploration cost is a power function of the action space dimensionality, as it scales with the combinatorial cost of exploring the effect of each combination of actuations). Fortunately, most realistic systems exhibit invariants of some form, which can be exploited to reuse the same policy across different parts of the system and, consequently, reduce exploration and learning costs. This was demonstrated empirically through the analysis of thin fluid film instability control (Belus et al. 2019),[5] and will be discussed in more details in Section 18.4.3, where we provide guidelines for practitioners.

Second, robustness of the final control strategy is critical for real-world applications employing DRL. Here again, recent results have been promising. Several robust applications of DRL, even in complex environments, have been presented by the

[5] The code is available here: `https://github.com/vbelus/falling-liquid-film-drl`

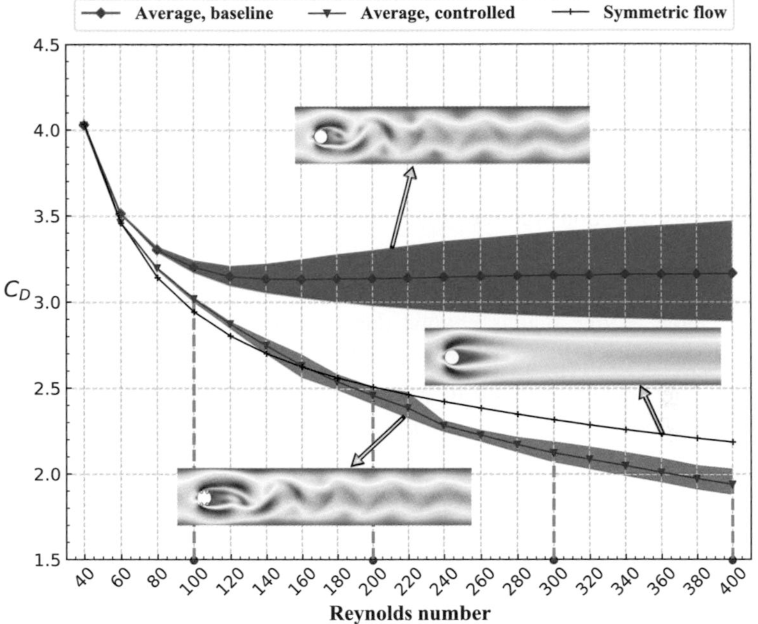

Figure 18.5 Demonstration of the robust closed-loop AFC strategy found by a PPO agent, as described in the work of Tang et al. (2020) (the figure is a slightly edited version of the results presented there, used with the permission of AIP Publishing). A single PPO agent is trained on simulations similar to the ones presented in Figure 18.4. However, in the present case, the Reynolds number used for each episode is chosen at random, in the discrete set {100, 200, 300, 400}. After training, the PPO agent is found both to exhibit highly effective control (comparable with the optimal control expected from the symmetric flow analysis suggested in Bergmann et al. (2005)), and to be robust enough to control the flow at any Reynolds number in the range [60, 400], including values that had not been seen during training. This illustrates both the robustness and generalization ability of AFC strategies discovered through DRL.

ML community, for instance, for playing games or for robotic control (Hwangbo et al. 2019). In the fluid mechanics community, the first investigation of the ability of DRL to produce robust control strategies was presented by Tang et al. (2020).[6] In this work, the DRL agent was trained on a range of Reynolds numbers rather than a single fixed number, as was done in previous studies. This means that the agent is trained on a range of flow conditions that, while broadly exhibiting the same patterns, still differ in the precise behavior and responsiveness of the wake. Experiments showed that the DRL agent is able to learn a robust global control strategy, and to perform control over a wide range of Reynolds numbers, including ones not seen during training. The main findings of this work are visualized in Figure 18.5. Such findings are very promising for applications in the real world, where one would expect a similar variation of flow properties.

[6] The code is available here: `https://github.com/thw1021/Cylinder2DFlowControlGeneral`

The examples presented so far still work with relatively low Reynolds numbers, that is, moderate nonlinearity and complexity, owing to the computational cost associated with increasing the number. More critically, all the AFC examples have been in the laminar (though, chaotic) regime. Therefore, a final milestone to achieve in order to truly qualify DRL as a possible AFC method for realistic applications is to show successful application in turbulent flow conditions. An important first step in this direction was recently achieved in 2D simulations of the Karman wake behind a cylinder, at an intermediate Reynolds number of 1000 (Ren et al. 2020). The setup is very similar to the one introduced by (Rabault et al. 2019), except for the flow regime that is chaotic due to a much higher Reynolds number. The drag in these conditions can be reduced by up to around 30% on average, and a transient drag reduction of up to around 50% can be observed, which clearly demonstrates the potential of DRL control even in the case of more complex, higher Reynolds number flows.

In addition to the application of DRL for controlling instabilities in the wake of a cylinder or at the surface of thin water films, at least three other applications were presented recently. The first is concerned with the control of the Kuramoto–Sivashinsky (KS) equation (Bucci et al. 2019), which is an example of a system exhibiting spatiotemporal chaos. The authors find that a simple Q-learning method is able to capture the dynamics of the system and control its behavior with only limited actuation. The second example is focused on controlling spatio-temporal structures happening during Rayleigh–Benard convection (Beintema et al. 2020). Again, the authors find that DRL is able to drastically reduce the development of instabilities in the system by controlling the distribution of heating power at the lower side of a convection cell. The strategy obtained through DRL is compared with classical control algorithms and found to significantly outperform them. Finally, Xu et al. (2020) studied the control of the wake behind a cylinder using small contra-rotative cylinders.

The work of Beintema et al. (2020) clearly illustrates the limitations of traditional "optimal control" algorithms. While these methods are very attractive both from a theoretical and mathematical point of view, in practice they rely on strong assumptions about the underlying system for both convergence and stability. These assumptions – for instance, some form of Lipschitz-continuity, convexity, or the existence of extrema/poles – are often not satisfied by real-world, nonlinear, high-dimensional systems. In such cases, DRL, which is based on learning to control through trial and error and leverages the approximation capabilities of deep learning, is not only more robust but it may be the only feasible approach to control systems that exhibit strong nonlinearities. Indeed, if a system is highly nonlinear, any method based on linearization or local-order reduction techniques will always have a very limited domain in the phase space within which it is valid. Consequently, one needs to explore the space and exploit the gathered knowledge, which is precisely the learning process formalized by the DRL framework, while relying on DNNs as general-purpose function approximators.

18.4 A Guide to Deploying Deep Reinforcement Learning in Fluid Mechanics

In this section, we present a range of technical points to keep in mind when designing a new application of DRL in fluid mechanics. This section is a synthesis of beneficial practices found in papers referred to in the Sections 18.2 and 18.3. While the discussion is focused on fluid mechanics problems, the techniques discussed here are generally applicable to many DRL applications.

18.4.1 Problem Identification and Coupling to Deep Reinforcement Learning Framework

The first critical requirement before attempting to leverage DRL in an application is to make sure that the problem and its characteristics fit the DRL framework. First, the problem needs to be expressible as a closed-loop interaction, as described in Section 18.2.1. Second, DRL is of particular interest since it imposes relatively few assumptions on the underlying system. so it is especially suitable to tackling problems with high nonlinearity and few mathematical properties such as convexity or Lipschitz-continuity, which would be required for analytical tools and traditional applied mathematical methods. This applies to most non-simplistic fluid mechanics applications.

Once these high-level requirements are met, the problem needs to be wrapped up as an interactable "environment." This is usually done by implementing a small set of predefined interface methods. Code examples here are based on the applied DRL library Tensorforce (Kuhnle et al. 2017) but similar patterns can be found in other DRL frameworks as well. To use the environment interface of Tensorforce, the user needs to provide a specification of the state and action space (`states()` and `actions()`), a method to start a new interaction episode (`reset()`) and a function that, given the agent actions, advances the environment by one time step (`execute(...)`):

```
# Inherit from the Environment base class
class MyEnvironment(Environment):

    def states(self):
        """State space specification"""
        return dict(...)

    def actions(self):
        """Action space specification"""
        return dict(...)

    def reset(self):
        """Resets the environment to start a new episode"""
        # ... prepare environment ...
        return initial_state

    def execute(self, actions):
        """Executes the given action(s) and advances the
            ↪ environment"""
        # ... advance environment ...
        return next_state, is_terminal, reward
```

The flow of interaction between the agent and the environment closely follows that of Figure 18.2. First, the environment is prepared for a new episode of interaction with the agent and returns the initial state (`reset()`). Starting from here, alternately, the agent reacts to the observed environment state, and the environment is advanced by one time step based on the agent's decision (`execute(...)`). In all this, the state and action space specify the communication format, that is, what observations of the environment look like (for instance, image size or number of sensor values) and what type of actions the environment expects in return from the agent (for instance, discrete or continuous decisions). Finally, the interaction either terminates once a failure or success state is reached, or is aborted after a predefined number of time steps, particularly if there are no natural terminal states (as is often the case for flow control problems). Subsequently, a new training episode is initiated. This is illustrated by the following code snippet:

```
# Train for 200 episodes
for _ in range(200):

    # Initialize environment
    states = environment.reset()
    terminal = False

    # Episode of agent-environment interaction
    while not terminal:
        actions = agent.act(states)
        states, terminal, reward = environment.execute(
            ↪ actions)
        agent.observe(terminal, reward)
```

While it may seem more flexible to reimplement a DRL algorithm from scratch instead of having to interface with an existing framework, we discourage fluid mechanics practitioners, particularly if they are new to DRL, from doing so, and highly recommend relying on well-established open-source implementations such as Tensorforce, Stable-Baselines or RLlib instead. The main reason is that it is a challenging task to make sure that a DRL implementation actually works correctly, and debugging it if it does not. This is due to a combination of factors, including their intrinsic non-determinism on various levels (action sampling, random weight initialization, etc) and the fact that internal representations are non-interpretable. Moreover, one frequently finds that DRL algorithms learn to "compensate" for implementation errors, meaning that the agent may be able to optimize the reward function reasonably well even in the context of critical bugs since, to the learning algorithm, an erroneous implementation is just yet another complex optimization problem to solve.

18.4.2 Problem Modeling: Invariances, Reward Shaping, and Parallelization

When modeling the problem environment, important practical considerations need to be taken. While they may seem obvious at first, challenges when deploying DRL, in the experience of the authors, often stem from inconsistent or lacking attention to these points.

First, one needs to make sure that the choice of the state input to the agent is consistent with the problem to solve, and provides a sufficient amount of information for the agent to identify potential causality relations between the state of the system, the action taken, the evolution in time of the system, and the reward observed. At this stage, physical understanding of the controlled system is very beneficial. In addition, the general setup of the control must take advantage of the structure of the problem in order to perform learning effectively. In particular, properties of invariance by translation, rotation, or any other form of symmetry should be exploited. Rabault et al. (2019) and Belus et al. (2019) investigated several possible methodologies for exploiting system invariants when designing DRL applications. Using a simple example application (controlling the development of instabilities in a thin fluid film over an extensive region in space using multiple jets), they found that failure to take into account invariants can lead to prohibitive exploration costs which make learning impossible when several control signals are needed. In contrast, successful exploitation of these invariants made it possible to control an arbitrary number of actuators. The best DRL deployment technique, shown in Figure 18.6, relies on applying the same DRL agent to control different parts of the system, which are modeled as separate environments but reuse the same agent DNN weights – and thus the distilled knowledge about the system. The same approach can be adopted to enforce other 2D, 3D, or axis-symmetric policy invariants, depending on the problem under consideration. A challenge with this approach is to formulate a suitable reward function. The reward function of each environment should, on the one hand, be as "local" as possible, so that each agent instance gets direct feedback on the consequence of its actions, while on the other hand, the overall optimization objective is usually a global problem that does not have a simple "local" solution. One approach to address this dilemma is to formulate the reward function as a weighted sum of two components, one limited to the local effects of each agent instance and the other reflecting the global state of the system.

Second, the choice, shaping, and physical meaning of the reward function is critical for obtaining efficient control strategies. This is well known both in RL where it is often referred to as "reward shaping" (Ng et al. 1999), as well as in a variety of applications within mechanics (Allaire & Schoenauer 2007). Indeed, a bad (or naive) choice of reward function may lead to unexpected and detrimental behavior despite the fact that an agent manages to achieve consistently high rewards. An example of this problem is described in Rabault et al. (2019) and Rabault and Kuhnle (2019): when using DRL to control the wake of a cylinder with the aim of reducing drag, a reward function focusing purely on drag reduction leads to degenerate strategies with a strong biased blowing. While this effectively reduces drag, it also creates a large lift bias. To prevent this problem, one has to modify the reward function so as to both encourage drag-reducing strategies while simultaneously discourage the use of detrimental tricks to get there. As is the case with specifying good state and action spaces, designing appropriate reward functions benefits greatly from engineering knowledge about the system to control.

Another important point to consider when deploying DRL in fluid mechanics applications is the parallelization of learning. Since most fluid mechanics simulations

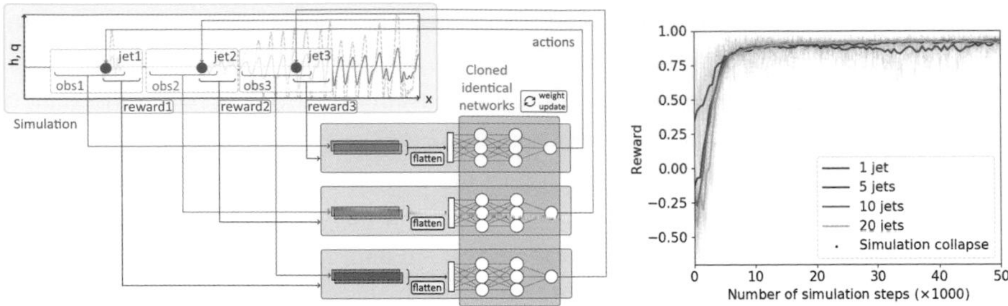

Figure 18.6 Illustration of the environment-splitting method for taking advantage of physical system invariants when performing AFC with DRL. The figures are reproduced from Belus et al. (2019). Left: the general setup of the environment-splitting method. The neighborhood of each actuator constitutes an individual environment, and all the environments are controlled by clones of the same DRL agent. This way, invariance of the policy is enforced. Right: applying this method translates in a training cost that is independent of the number of jets. This is in stark contrast to methods that do not take advantage of these invariance properties. Indeed, as described in Belus et al. (2019), failing to represent invariants inside the structure of the DRL problem formulation leads to prohibitive exploration cost as the control space dimension increases, and training fails completely in a setting without such invariant setup when as little as five jets are involved.

are expensive, they are frequently the time bottleneck in the overall training process. Fortunately, the DRL side of the training process can be parallelized simply by running several copies of the environment simulation (e.g., on different cores or machines) and enabling the agent's learning algorithm to incorporate data from parallel experience streams. Consequently, one can leverage parallelized CFD simulations – plus, where applicable, multiple experience streams per simulation when implementing invariant control, as described earlier – to effectively sample a large amount of training data, given enough computational resources.

18.4.3 Other Techniques: Normalization, Actuation Frequency, and Transfer Learning

So far, we have mostly discussed high-level considerations around problem modeling when applying DRL to AFC. In this section, we focus on concrete technical points that can have a large impact on the learning process and thus should be considered "best practice".

A first important detail for effective learning is to normalize state, action, and reward values. Generally, DNNs perform best when their inputs and outputs are, roughly, centered around zero and normally distributed or within the range $[-1; 1]$. This is due to several reasons, for instance, most typical nonlinear activation functions change their "firing" state within that range, or most weight initialization schemes base their effectiveness on assuming such input values. Therefore, both the input state and the action output should be scaled accordingly to fall within this range. In addition, the signals should be normalized in such a way that meaningful differences for the application are reflected in corresponding variation of the input/output values and hence are "visible" to the DNN within that normalized range. For instance, if DRL is supposed to avoid instabilities in a dynamic system, it may not be appropriate to scale a signal linearly between -1 and 1 if a residual value of, say, 10^{-10} is to be obtained. In such cases, it is probably better to first apply a logarithmic transformation before scaling the results to $[-1; 1]$. As a "rule of thumb," the signals for which a DNN is supposed to react and behave differently should be visible to the naked eye in a plot displaying the range from -1 to 1.

Similar reasoning applies to the reward function. However, since different methods use the reward signal in different ways, this requires some knowledge about the type of algorithm used (see the overview of different classes of algorithms in Section 18.2). In particular, the normalization needs to consider the effect of the discount factor, the episode length, and the fact that the actual training signal is not the per time step reward, but is based on the discounted cumulative sum of rewards. Empirically, zero-centered values not exceeding the range from -10 to 10 work well. Again, it is important that meaningful performance differences for the application should be reflected in the reward signal accordingly.

Another critical point for successful and stable learning is to choose an appropriate action "frequency" or "granularity" (Rabault & Kuhnle 2019). Whereas other applications like games often naturally define the moments when an agent needs to act, this is not obvious for physical control problems. The dynamics of a physical system usually take place at a characteristic timescale, and the relation between this natural frequency of the system, f_s, and the frequency of agent decision time steps, f_a, is important. If $f_a < f_s$, the system is evolving too fast for the agent, meaning that, on the one hand, it may miss important details between the unobserved action intervals and, on the other hand, it is not able to react to system developments fast enough. However, $f_a \gg f_s$ is also problematic, which is probably best illustrated at the beginning of training (but applies throughout training), when the agent is mainly exploring by randomly picking actions. Each individual random action on its own will have little impact on the system, and it becomes increasingly unlikely with growing f_a/f_s that the agent randomly picks a long-enough sequence of coherent actions that results in meaningful positive feedback. Consequently, the action frequency f_a should be chosen such that it is granular enough to control the system, but coarse enough to observe moments of successful control accidentally during random exploration. In practice, after having identified the approximate timescale f_s for the system to control, $f_a/f_s \in [5; 10]$ is a reasonable initial choice, which can subsequently be tuned. This is illustrated in Figure 18.7. A related technique in the RL literature is "frame skip," which refers to skipping a fraction of screen frames if the game simulation is rendered at a too high frequency.

In some situations, the appropriate action frequency may be lower than either the numerical time step in a simulation, or the maximum response frequency of sensors and actuators in a real-world application. If such constraints are encountered, an interpolation law should be used to obtain a smooth control law at the high frequency corresponding to the numerical time step or capability of the hardware. The choice of

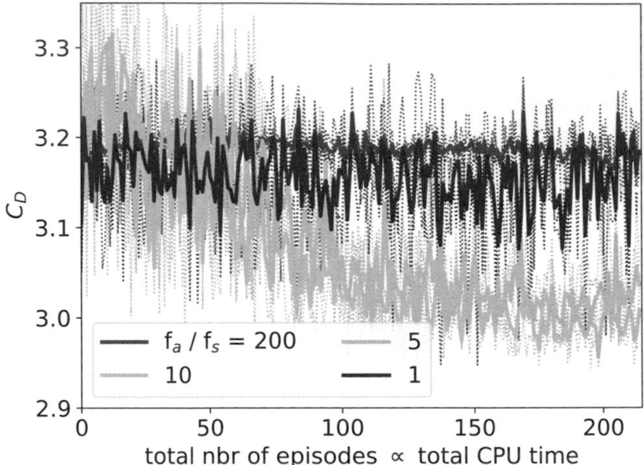

Figure 18.7 Illustration of the importance of choosing consistent timescales across the DRL setup. The figure is reproduced with minor editing from Rabault and Kuhnle (2019), with the permission of AIP Publishing. Depending on the choice of action frequency f_a relative to the typical timescale of the underlying system f_s, the learning curves of a PPO agent attempting to reduce drag show substantial variation. On the one hand, if the action frequency f_a is too low ($f_a / f_s = 1$), the DRL agents cannot control the system as it is unable to react fast enough to affect the vortex shedding. On the other hand, if the action frequency f_a is too high ($f_a / f_s = 100$), learning is not successful either, since the agent's random behavior at the beginning of training simply adds high-frequency white noise to the system, which is effectively smoothed out at the larger timescale of vortex shedding. However, between these two extremes, there is a range of f_a / f_s ratios for which the agent successfully learns to control the system.

the interpolation function may be important for the quality of control, as discussed in Tang et al. (2020).

Another technique that can improve learning quality and stability is transfer learning, as illustrated by recent work. First, Ren et al. (2020) initially trained a DRL agent in a situation of reduced complexity – in this case, a flow at moderate Reynolds number – before moving to the full problem complexity, a flow at a higher Reynolds number. This approach stabilized learning and yielded better final results. Second, Tang et al. (2020) trained a DRL agent to control a range of Reynolds numbers at once, where the Reynolds number was randomly chosen at the beginning of each episode. Here again, the learning process improved and the final strategies were more efficient than comparable strategies trained to control flow just at a single Reynolds number. This indicates that training agents on a broader range of similar control problems and environment conditions help to find better and more robust strategies.

18.5 Perspectives and Remaining Questions

DRL is a promising methodology for solving closed-loop control problems involving high-dimensional nonlinear systems. In such conditions, traditional optimization and control techniques often fail, whereas DRL with its approach of exploring the dynamics of the system to control offers a solution to the challenges encountered by traditional methods. Moreover, DRL is attractive for real-world applications because of its computational requirements. While training an RL control policy can be expensive in terms of the number of training episodes/interactions required, applying a trained agent is not, as it simply requires a neural network inference step, whose computational cost is usually linear in the input size. This is in stark contrast to, for example, adjoint-based methods.

This explains why the fluid mechanics community has been investigating the use of DRL to control increasingly complex systems. Recent work has evidenced the fact that DRL can control complex flow dynamics in 2D configurations, even for intermediate Reynolds number and chaotic conditions. Moreover, it has been shown that the learned control strategies can be made robust to a range of flow regimes, which is important for the prospect of real-world application. Consequently, next steps in that direction involve, on the one hand, showcasing the ability of DRL to handle even more challenging configurations and, on the other hand, investigating the application of DRL methods to the control of 3D flows. The latter is a critical milestone for fully qualifying DRL as an effective approach to AFC.

A more fundamental question is whether DRL can enable us to gain a better understanding of the physical properties of the systems being controlled. No results have been presented on this aspect yet, however, there is hope that the actuation strategies discovered by DRL through trial and error can be combined with more traditional analysis to gain insights into physical mechanisms. In this context, DRL would be used as a tool to experimentally investigate the global nonlinear response of a system, while more traditional techniques relying on, for example, linearization or modal analysis, are employed to quantify the stability or robustness of the newly discovered configurations.

Another promising direction for future research is to attempt to combine classical control methods with strategies found via DRL. While traditional control methods are often better understood and more robust, DRL by contrast is often able to find more efficient control strategies, but with the downside that it is very challenging to substantiate them with theoretical proofs of their stability and safety. The promise of a combined approach is to "get the best of both worlds," which would greatly facilitate their adoption for practical and industrial applications. There are several possibilities. One may, for instance, define a domain in the phase space where traditional controllers are able to guarantee safe operation, and hence allow DRL algorithms to further optimize the control while the system remains in this safe domain. If the system gets close to the boundaries of guaranteed safety, DRL can be turned off and the more interpretable traditional algorithm takes over control. Alternatively, DRL can be applied on top of traditional control strategies, as illustrated by the residual learning approach (Johannink et al. 2019), meaning that DRL learns how best to modify (within a safe margin) the control given by the underlying system for further gains.

As a final thought, we want to emphasize the importance of reproducibility of experimental results obtained by applying DRL, in particular with respect to open-sourcing code and providing containerized environments. While it may be sufficient to describe the algorithm and theoretical foundation in the case of traditional approaches, the control strategies found by DRL cannot be described in the same way, since they are encoded in the precise weights of the neural network. Consequently, it is important to publish environment and training code together with the trained DRL agent to be able to reproduce, analyze, and leverage the reported results. Moreover, it is well known in the DRL community that reproducibility is an important problem (Henderson et al. 2018), and that even seemingly irrelevant technical implementation choices can play a significant role for the quality of the learned behavior (Engstrom et al. 2020). We hope that the fluid mechanics community will follow the example set by the ML community in this aspect, and will increasingly adopt the policy of releasing code open source and enabling reproducibility by providing containerized environments.

Part VI

Perspectives

19 The Computer as Scientist

J. Jiménez

Using the identification of causally significant flow structures in two-dimensional turbulence as an example, this chapter explores how far the usual procedure of planning experiments to test hypotheses can be substituted by "blind" randomized trials, and notes that the increased efficiency of computers is beginning to make such a "Monte Carlo" approach practical in fluid mechanics. The processes of data generation, model creation, validation, and verification are described in some detail. Although the purpose of the chapter is to explore the procedure, rather than to develop new models for two-dimensional turbulence, it is intriguing that the Monte Carlo process naturally led to the consideration of vortex dipoles as building blocks of the flow, on a par with the more classical individual vortex cores. Although not completely novel, this "spontaneous" discovery supports the claim that an important advantage of randomized experiments is to bypass researcher prejudice and paradigm lock. The method can be extended to three-dimensional flows in practical times.

19.1 Introduction

We discussed in Chapter 2 the different meanings of the word "Science," and how they are related to the scientific method and to the idea of causality. We also mentioned two approaches to determining causes: the *observational* one, which uses correlations between past and future events to deduce rules that describe the future behavior of the system, and the *interventional* one, in which the researcher monitors the consequences of changing the initial conditions in the (often aspirational) hope of eventually controlling it.

Most of the discussion in Chapter 2 was observational, because experiments in our chosen field of turbulence are expensive. In this chapter, we explore the interventional determination of causality in physical systems, and discuss whether recent technological developments provide us with new ways of doing it. To fix ideas, we illustrate our argument with the study of a particular flow, two-dimensional decaying homogeneous turbulence, about which the general feeling is that most things are understood. As such, we will be able to test our conclusions against accepted wisdom, although this

This work was supported by the European Research Council under the Coturb grant ERC-2014.AdG-669505.

may lead some readers to think that the discussion is uninteresting, because nothing new is likely to be discovered. We will see that this is not quite true. Some new aspects can still be unearthed, or at least reemphasized, and the method for doing so is interesting because it would have been impractical a decade ago.

We mentioned in Chapter 2 that fluid mechanics has only recently accumulated enough data to even consider a data driven approach, but the same has not been true in other fields. Most medieval science was data-driven. Kepler's laws (1618) were the empirical results of observations, with no plausible explanation until the publication of Newton's *Principia*, 70 years later (Newton 1687). Engineering and medicine have always been at least partly empirical and data-driven, and biochemistry has probably been the first subject to reach the data-driven level in modern times, made possible by the fast methods of DNA synthesis that became available in the 1980s. Much of the modern discussion on raw empiricism relates to biochemistry (Voit 2019), and so did the first article claiming to describe a "robot scientist" (King et al. 2009). In fluid mechanics, the availability of enough data was made possible by the numerical simulations of the 1990s, which appeared to promise that any question that could be posed to a computer would eventually be answered (Brenner et al. 2019, Jiménez 2020a), but this promise was tempered by the limitation that the questions first had to be put to the computer by a researcher.

In this chapter, we discuss to what extent this last roadblock can be removed. The new enabling technology is the increased speed and memory of computers, which can now do in minutes what used to take days a decade ago. We will argue that this enables a new way of asking questions, not based on a plausible hypothesis, but randomly, in the hope that some of them might turn out to be interesting. This procedure does not avoid the necessity of answering the questions, although this can presumably be done quickly by the computer, nor of evaluating how "interesting" the answers are, which computers most probably cannot yet do. This "Monte Carlo" procedure does not point towards a future of "human-less" research, but to a new level of partnership between humans and computers. Similar steps have been taken before: we no longer dig canals or throw spears by hand, except as a sport, nor do we integrate ordinary differential equations using special functions. Most of these human–machine symbioses are usually considered beneficial, although many created their own disruptions when they were introduced. There is no reason to believe that this time will be different, but it is fair to question what the new advantages and difficulties will be.

Relinquishing control of the questions to be asked is not an altogether new experience to anybody who has trained graduate students or mentored postdocs, not to speak of managers of large industrial or academic research groups. Any such person knows the feeling that, at some point, the research is no longer theirs, although most of us console ourselves by arguing that we are training our peers and that, in the end, the overall direction is set by us. Monte Carlo science lacks both of these (probably spurious) consolations. This may be its main advantage. It is inevitable that our students or subordinates share some of our ideas and, most probably, some of our prejudices. It is often argued that researcher prejudice is the main roadblock to qualitative scientific advances, and that "paradigm shifts," always hard to come by, are delayed because of it (Kuhn 1970). The main advantage to be gained from a Monte Carlo questioning algorithm is probably its lack of prejudice. If we can avoid transferring our biases to it, such an algorithm would act as an *unprejudiced*, although probably not yet very smart, scientific assistant.

The rest of this chapter is structured as follows. Section 19.2 describes how the Monte Carlo ideas just described can be applied to the particular problem of two-dimensional turbulence, and Section 19.3 briefly discusses the physics that can be learned from them. Section 19.4 closes with a discussion of the connections that can be established between the scientific method and modern data analysis or artificial intelligence, in the light of the experience of Sections 19.2 and 19.3.

19.2 2-D Turbulence: A Case Study

Two-dimensional turbulence is an old problem in fluid mechanics, of importance in geophysical flows (Maltrud & Vallis 1991), highly stratified situations (Voropayev et al. 1991), and Bose–Einstein condensates (Neely et al. 2010). It differs from three-dimensional turbulence in that the absence of vortex stretching inhibits the amplification of velocity gradients. In the inviscid limit, energy and enstrophy are thus conserved, and the usual energy cascade is disrupted (Betchov 1956). Early papers centered on the statistical consequences of these conservation laws (Kraichnan 1967, 1971, Batchelor 1969), but it was soon realized that the formation of long-lasting coherent vortices (Onsager 1949, Fornberg 1977, McWilliams 1984) interferes with the regular statistical theory (McWilliams 1990a). Since then, two-dimensional turbulence has mostly been analyzed in terms of the behavior of these vortices (McWilliams 1990b, Carnevale et al. 1991). General reviews can be found in Kraichnan and Montgomery (1980) and Boffetta and Ecke (2012).

Because two-dimensional turbulence is relatively cheap to simulate, it serves as a good system in which to test how Monte Carlo science can be applied to a real flow and, since a lot is already known about it, any conclusions drawn in this way can be checked against received wisdom, both for accuracy and for originality. In this section, we apply the ideas in Section 19.1 to test whether the vortex model mentioned earlier is the only possible one for causality in two-dimensional decaying turbulence. The problem, the general procedure, and some of the physical conclusions derived from it are described in Jiménez (2018b, 2020a, 2020b). Here we center on the data-processing aspects, and on how they can be related to the current interest in big data analysis and artificial intelligence.

19.2.1 Data Generation and Analysis

The problem to be addressed can be summarized as follows: given a two-dimensional field of decaying turbulence, can we identify which properties of a flow neighborhood make it more or less likely to result in widespread flow changes when it is perturbed? The underlying theme is turbulence control. For example, given a turbulent

Case	Perturbation to cell
A	$\omega \Rightarrow -\omega$
B	$\boldsymbol{u} \Rightarrow -\boldsymbol{u}$

Table 19.1 Initial perturbations for the two experiments discussed in the text. In both cases, the mean velocity and vorticity are zeroed over the full computational box after the perturbation is applied. An extra "pressure" step is applied to enforce continuity after modifying $\boldsymbol{u}$ in case B, and may substantially modify the intended perturbation.

atmosphere in which simulations tell us that a hurricane is likely to form after a certain time, and assuming that we have no limitations on what can be done, what should we do? Target existing storms? Target the jet stream?

Thus posed, the problem is to clarify causality, and our method is interventional: that is, we change the initial conditions and classify the results (see the discussion in the Section 2.1). A number (N_{exp}) of experimental flow fields ("flows" from now on) are prepared, and each of them is perturbed to create a variety of initial conditions ("experiments" from now on). Each experiment is run for a prescribed time, T, of the order of $\omega_0' T \approx 10$ turnovers, where $\omega_0' = \langle \omega^2 \rangle^{1/2}$ is the root-mean-square (r.m.s.) vorticity of the initial unperturbed flow, and the time-dependent average $\langle \cdot \rangle$ is taken over the full computational box. As the evolution of the perturbed flow diverges from that of the unperturbed initial condition, the magnitude of the evolving perturbation is defined as the norm, $\epsilon(t)$, of the difference between the perturbed and unperturbed flow fields, evaluated over the entire computational box. The experiments for which this magnitude is largest at some predetermined time, T_{ref}, are defined as most "significant." For each experiment and test time, the n_{keep} perturbations with the largest and smallest deviations $\epsilon(T_{ref})$ are classified as most or least significant, respectively, because, in common with many complex systems, it is empirically found that the first few most- and least-significant experiments result in fairly similar perturbation intensities (LeCun et al. 2015). Because significance is defined as the effect on the flow at some future time, the most significant experiments are also defined as being most "causally important" for the flow evolution. In our study, the initial perturbation is applied by dividing each flow into a regular grid of $N_c \times N_c$ square cells, each of which is in turn modified in a number of different ways. The two experiments discussed here are listed in Table 19.1; several others are presented in Jiménez (2020b). In most cases, the results are averaged over $N_{exp} = 768$ flows, with $n_{keep} = 5$, on a grid with $N_c = 10$.

Simulations are performed in a doubly periodic square box of side $L = 2\pi$, using a standard spectral Fourier code dealiased by the 2/3 rule. Time advance is third-order Runge–Kutta. The flow is defined by its velocity field $\boldsymbol{u} = (u, v)$ as a function of the spatial coordinates $\boldsymbol{x} = (x, y)$, and time. The scalar vorticity is $\omega = \partial_x v - \partial_y u$, and the rate-of-strain tensor is $s_{ij} = (\partial_i u_j + \partial_j u_i)/2$, where the subindices of the partial derivatives range over (x, y), and those of the velocity components over (u, v). The rate-of-strain magnitude is $S^2 = 2 s_{ij} s_{ij}$, where repeated indices imply summation.

Time and velocity are respectively scaled with ω_0', and with $q_0' = (u'^2 + v'^2)^{1/2}$, both measured at the unperturbed initial time, $t = 0$. All the cases discussed here have Fourier resolution 256^2, with $Re = q_0' L/\nu = 2500$, where ν is the kinematic viscosity. Further details can be found in Jiménez (2018b, 2020b).

Figure 19.1(a,b) shows the typical initial vorticity and kinetic energy fields, with the 10×10 grid overlaid, and Figure 19.1(c) shows the corresponding enstrophy and energy spectra. This figure also includes the transfer function for a window corresponding to the isolation of an individual cell. Because such a window acts as a multiplicative factor in physical space, it acts as a convolution kernel for the spectra, smoothing details narrower in wavenumber (or wider in wavelength) than the width of the transfer function. The cell outlined in red in Figure 19.1(a,b) is one of the most significant ones, and the one outlined in black is one of the less significant. Both are modified as in case A in Table 19.1 (i.e., inverting the vorticity in the cell). These two cells have been chosen so that they have similar initial perturbation intensities, but different intensities at the classification time, $\omega_0' T_{ref} = 4.5$. The evolution in physical space of their perturbations is shown in Figure 19.1(d,e). It is clear that the more significant perturbation rearranges the flow in its vicinity, and its effect eventually spreads to the full field, while the less significant one stays localized and soon decays.

The geometric size of the perturbations can be estimated by the ratio between their integral quadratic and point-wise maximum. It is shown in Jiménez (2020b) that all perturbations start with sizes of the order of the cell size, L_c, and either decay or spread to the size of the computational box, $10L_c$, over times of the order of $\omega_0' t \approx 5$ turnovers. This can be taken as the time over which perturbations lose their individuality, and over which causality can be usefully studied.

The purpose of our study is to determine which properties of the unperturbed initial cells best correlate with the effect of modifying them. Several factors are important, such as the modification method (Table 19.1), the norm used to measure the perturbation intensity, and the time T_{ref} at which the classification is done. The following analysis uses the kinetic energy norm $\|\boldsymbol{u}\|_2$ as a measure of intensity, and a reference time $\omega_0' T_{ref} = 4.5$ for the classification, but similar experiments were repeated using $\|\boldsymbol{u}\|_\infty$, $\|\omega\|_2$, and $\|\omega\|_\infty$, with little difference in the results. The choice of T_{ref} is discussed in Jiménez (2020b), and its effect on the results in Section 19.2.2. The cell properties used as diagnostic variables are the average cell enstrophy, $\omega_c^2 = \langle \omega^2 \rangle_c$, defined over individual cells, the averaged kinetic energy, $q_c^2 = \langle u^2 + v^2 \rangle_c$, and the average magnitude of the rate-of-strain tensor, S_c^2. Jiménez (2018b, 2020a) diagnosed significance using an optimal threshold computed for each of these variables in isolation. Each modification of an individual cell was classified as significant or not according to the flow behavior at T_{ref}, and the resulting labeled set was used to train the threshold in such a way that the number of misclassified cells was minimized. Jiménez (2020b) and the present notes use a version of the same idea with the capacity of multivariable classification: support vector machines (Cristianini & Shawe-Taylor 2000), implemented by the `fitcsvm` MATLAB routine, which determines a separating hyperplane instead of a scalar threshold (see Figure 19.2).

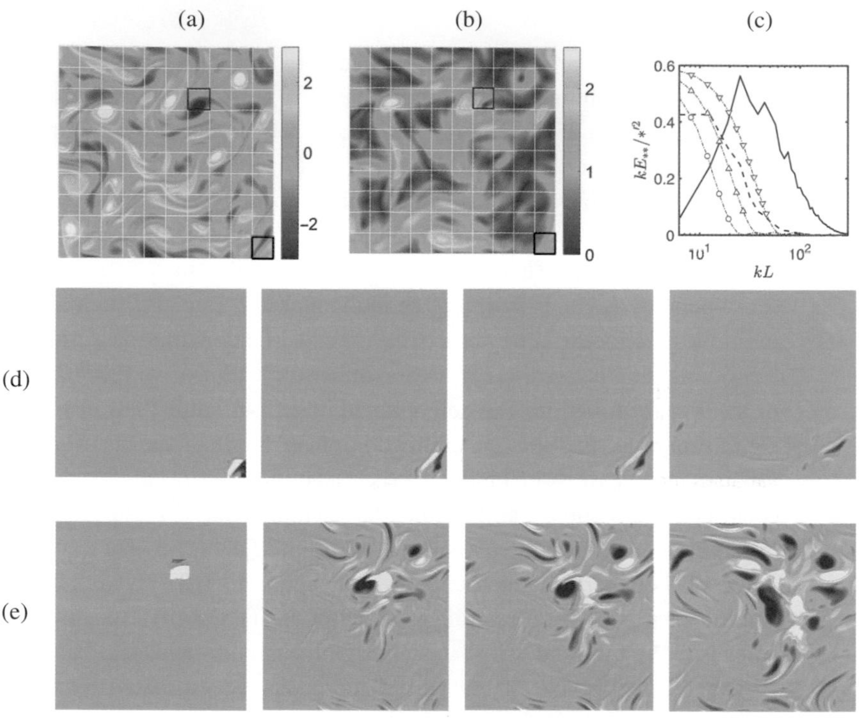

Figure 19.1 (a) Initial vorticity field, ω/ω'_0, used for the evolutions in (d,e). Case A in Table 19.1. The cells outlined in black (less significant) and in red (more significant) have relatively similar initial perturbation intensities, but very different later evolutions. (b) As in (a), for the velocity magnitude, $|\boldsymbol{u}|/q'_0$. (c) Premultiplied spectra of the initial flow fields used in the experiments, in terms of the wavenumber magnitude, k. ———, Enstrophy spectrum; — — —, energy; —·—·—, transfer functions of a box window corresponding to the cell size of the experiments, vertically scaled to fit the plot. ∘, $N_c = 4$; △, $N_c = 6$; ▿, $N_c = 10$. (d) Evolution of the perturbation vorticity for the less significant cell, marked in black in (a,b) 19.1. From left to right: $\omega'_0 t = 0$, 1.3, 2.6, 4.5. (e) As in (d) for the more significant cell.

The efficiency of the classifier, defined as the fraction of correctly classified events, is collected in the tables in Figure 19.3 for various experimental perturbations and combinations of diagnostic variables. It is clear that some linear combination of enstrophy and kinetic energy is always able to separate the data almost perfectly, but that S_c rarely helps. The best combination of vorticity and velocity used by the classifier depends on the initial perturbation method, as shown in Figure 19.2. In general, the enstrophy is the best diagnostic variable for perturbations that manipulate the vorticity (case A in Figure 19.2(a)), while the kinetic energy is the dominant variable for the cases that manipulate the velocity (case B in Figure 19.2(b–d)). The fifth column in Figure 19.3 shows that some improvement can be achieved by including contributions from both the enstrophy and the energy, but that the effect is marginal.

The classification efficiency depends on the size of the experimental cells. In general, the efficiency degrades for larger cells, as shown in Figure 19.3 for $N_c = 4 - 10$. The details of the degradation in case B can be followed by the increasing overlap of the joint p.d.f.s in Figure 19.2(b–d). This dependence of the effectiveness of the

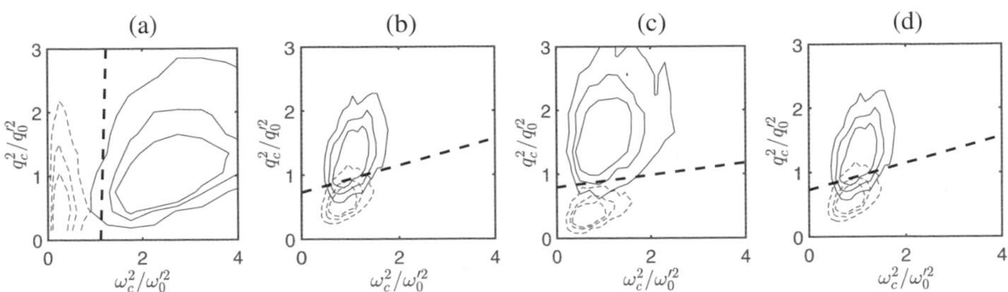

Figure 19.2 Optimum classification lines for different initial perturbations, in terms of the kinetic energy and of the enstrophy. (a) Case A, $N_c = 10$ (b) Case B, $N_c = 10$. (c) Case B, $N_c = 6$. (d) Case B, $N_c = 4$. In all cases, 768 flows and $n_{keep} = 5$. The contour lines contain 50%, 70%, and 90% of the joint probability density functions (p.d.f.) of the diagnostic variables for: ———, most significant cases; − − −, least significant cases.

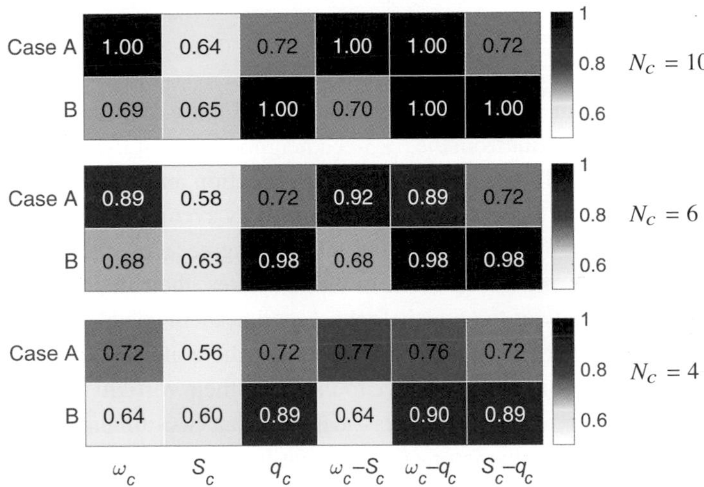

Figure 19.3 Efficiency of support vector machine classifiers. Unit efficiency is perfect classification, and 0.5 is random choice. Rows are the cases in Table 19.1, and columns are the different combinations of diagnostic variables.

enstrophy is consistent with the spectrum in Figure 19.1(c), which shows that the cell size when $N_c = 10$ preserves details of the spectrum of the order of the vortex size, but it is little surprising that small cells work so well for the kinetic energy, whose spectrum peaks at the scale of the box, suggesting that, at least at this low Reynolds number, the causality of the kinetic energy remains concentrated at the scale of individual vortex pairs. In fact, the effectiveness of the enstrophy and of the energy behave differently with the cell size. While the mean effectiveness of ω_c^2 as a diagnostic variable in case A decays from 1 to 0.72 as $N_c = 10 \rightarrow 4$, that of q_c^2 in case B only decays from 1 to 0.89 in the same range.

The difference among perturbation methods is best displayed by constructing for each case a conditional "template" for the immediate neighborhood of the significant

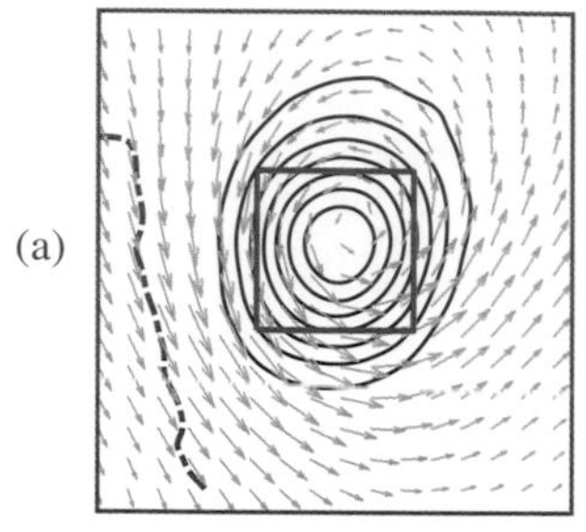

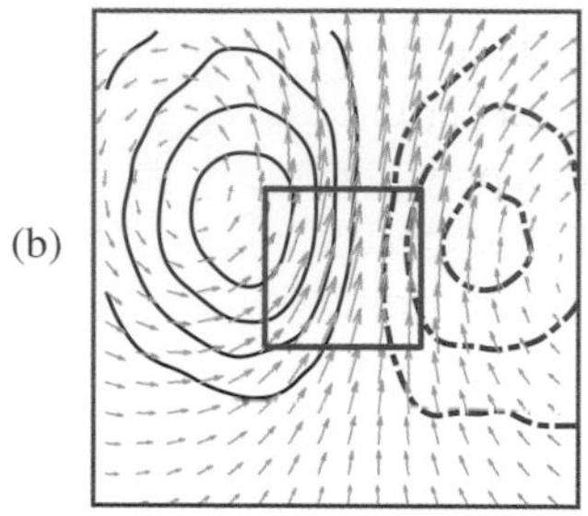

Figure 19.4 Conditional vorticity and velocity distributions, normalized with their unconditional r.m.s., in the neighborhood of the most significant cells. The test cell is outlined in blue and, because of the invariances of the problem, the orientation and the vorticity sign are immaterial. The vorticity contours are spaced by $\omega/\omega_0' = 0.3$. Solid lines and positive, and dashed ones are negative. (a) Case A in Table 19.1. (b) Case B. Compiled from 768 flows, $n_{keep} = 1$.

cells of the initial unperturbed flow. Figure 19.4(a,b) includes templates for cases A and B, built from the 3×3-cell neighborhood of the most significant cell in each experiment. To take into account the reflection and rotational symmetries of the equations of motion, the template is computed by averaging these flow patches after rotating and reflecting them so that they mutually agree as much as possible. To compensate for the effect of the magnitude of the templates, their intensity is scaled to match the global intensity of the flow before comparing them to individual neighborhoods.

Figure 19.4(a), which represents the conditional structure in the vicinity of cells which are most sensitive to inverting their vorticity, is an isolated vortex. This would appear to support the classical view that two-dimensional turbulence is a system controlled by the interactions among individual vortices (McWilliams 1990b, Carnevale et al. 1991). But the template in Figure 19.4(b), which corresponds to cells that have been perturbed by inverting their velocity and that are thus best diagnosed by the magnitude of their kinetic energy, is a vortex dipole. This is a less expected result, but a reasonable one, because the velocity field of a dipole contains a local jet, and it makes sense that blocking it has a strong effect. In general, the templates for the most significant structures in experiments that manipulate the vorticity are isolated vortices, while the manipulation of the velocity results in dipoles (Jiménez 2020b). The templates obtained from coarser experimental grids are similar to those for $N_c = 10$, but they become progressively less well defined as N_c decreases.

19.2.2 Verification and Validation

Any result based on the exploitation of large data sets is statistical and, as such, needs to be verified and validated. Verification refers to whether the experimental procedure can be trusted, while validation tests whether the results are relevant to the problem at hand. Further details for most of this section can be found in Jiménez (2020b).

Verification typically boils down to testing statistical significance, but it may include other things. The simplest question is whether enough experiments have been performed to justify trusting the averages obtained, but it is not always enough to apply the theory for sums of independent variables. For example, consider the templates in Figure 19.4. When they are computed using either 768 or 128 experiments, they differ by approximately $\|\boldsymbol{u}_{T(128)} - \boldsymbol{u}_{T(768)}\|_2/\|\boldsymbol{u}_{T(768)}\|_2 \approx 0.2$. This appears high, because if we assume that most of the error is due to the case with the fewest experiments, and treat the templates as simple averages, we would not expect a variability with respect to the average higher than $1/\sqrt{128} \approx 0.09$. However, a conditional average is more complicated than a simple sum of independent variables. Templates are reoriented and rescaled during the conditioning, and a relatively small misalignment of the final average can result in a large L_2 difference even for essentially equivalent templates. A more relevant question is whether our templates are more or less stable than those computed for randomly chosen cells. Jiménez (2020b) repeated the template calculation for random cells and for the least significant ones, and found differences of 40% between 128 and 768 random experiments, and 40–60% for the templates of the neighborhoods of the least significant cells, compared with the 20% found above for the most significant templates cells. This suggests that the latter represents physical features, while the other two do not.

Another important test is whether the distribution of relative approximation errors between the flow and the template is similar when the template is tested against the same set of experiments used to create it, or against an independent data set. This is tested in Figure 19.5 in which the experiments have been divided into halves. The templates are computed using the first half, and tested against either that half or against the second one. The template-flow approximation error is displayed in the ordinates, and both sets of results agree well, reducing the probability of template overfitting. On the other hand, the average approximation errors are substantial, of the order of $\|\boldsymbol{u} - \boldsymbol{u}_T\|_2/\|\boldsymbol{u}\|_2 \approx 0.7$. This is lower than the expected value of 1.4 between unrelated flows of the same intensity, or than the experimental values of order unity obtained by Jiménez (2020b) when repeating the experiments with randomly chosen cells, but it underscores that, although templates contain information about the structure of the flow in the neighborhood of significant cells, we should not expect to find many "pure" circular vortices or dipoles in the flow (see Figure 19.1(a)).

Although the verification tests show that the templates in Figure 19.4 can be statistically trusted, it remains to be shown that they represent flow properties rather than experimental artifacts.

Consider first the sensitivity of the results to experimental conditions. We have already seen that different experiments produce different templates, but this should be considered a desirable property of the Monte Carlo exploration of possibilities. More disturbing is the dependence on the testing time. The experiments above produce a classification of flow features at time $t = 0$, based on the intensity of the perturbations at a later time, T_{ref}, and this classification is considered to be a property of the flow at $t = 0$. Whether this is true or not needs to be tested, starting with whether cells

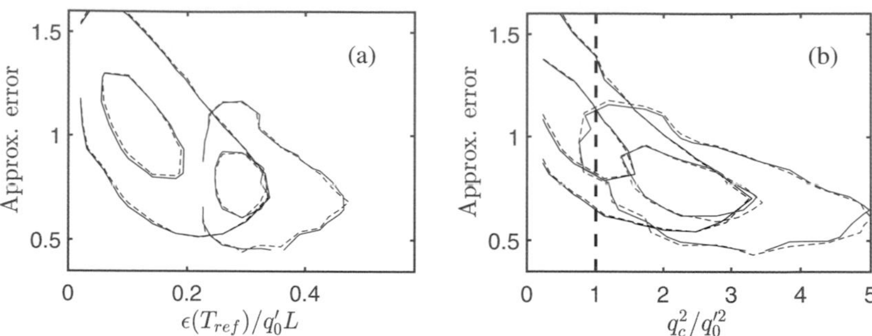

Figure 19.5 (a) Joint p.d.f. of the relative approximation error, defined as the difference between the templates and the unperturbed initial flow, $\|\boldsymbol{u} - \boldsymbol{u}_T\|_2/\|\boldsymbol{u}\|_2$, versus the measured divergence at $\omega_0' T_{ref} = 4.5$. Case B in Table 19.1, plotted against the dipole template in Figure 19.4(b). The black lines use all the cells in 384 experiments, and the red ones only use the most significant ones in each experiment. $n_{keep} = 5$ and $N_c = 10$. Contours contain 50% and 95% of the probability. ———, Tested using the training set; – – –, using an independent test set. (b) As in (a), but plotted versus q_c^2. The vertical dashed line is the discrimination threshold in Figure 19.2(b).

classified as significant for a given T_{ref}, remain significant when classified at a different time.

This was checked by Jiménez (2020b), who showed that the fraction of cells classified as significant at $T_{ref} = T_1$ and at T_2 decays approximately as $1 - 0.04\omega_0'(T_2 - T_1)$. Since we have seen that perturbations lose their individuality in $\omega_0' T \approx 5$, this means that approximately 80% of the cells classified as significant remain so during the time of the experiment. This is substantially higher than random persistence. If two independent sets of $n_{keep} = 5$ cells in a 10×10 grid are compared, their average intersection is only expected to be about 0.25%.

More interesting is the question of whether the relation between templates and significance can be reversed, so that not only can we say that significant cells look like vortices or dipoles, but that vortices and dipoles are predominantly significant. This is tested in Figure 19.5(a), which shows joint p.d.f.s of the flow-template approximation error versus the perturbation magnitude at T_{ref}. Contrary to Figure 19.2, the p.d.f.s drawn in black in the figure are compiled over all the cells of the original flow, without classification, and it is clear that there is an inverse relation between the two quantities: Cells whose flow neighborhood is better approximated by a significant template are also more causally significant, and vice versa. The p.d.f.s drawn in the figure with red lines only include cells previously classified as most significant. They are the right-most end of the unconditional distributions. As already mentioned, the figure displays separately the result of using test data from the set used to train the template, and data from a separate set of experiments, not used in the training. They agree well.

Figure 19.5(b) further tests the validity of the template by displaying the fitting error against the kinetic energy of the central cell, which is the diagnostic variable of

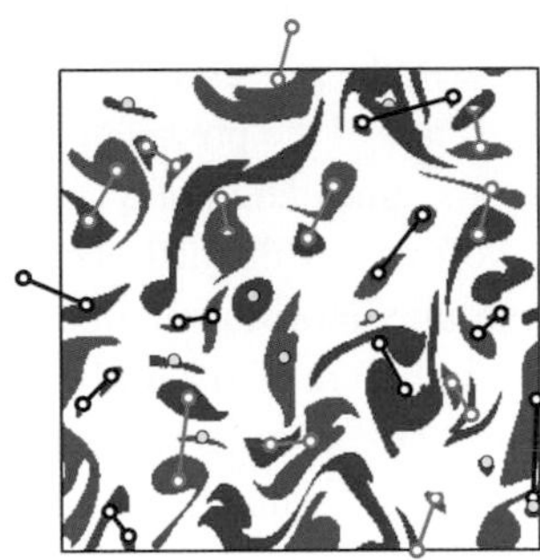

Figure 19.6 Sample segmented vorticity field with pair identification. Blue vortices are counterclockwise, and red ones are clockwise. Black connectors are corrotating pairs, magenta are counterrotating dipoles, and yellow markers are unpaired vortices. Note that, because of their irregular shape, some centers of gravity fall outside their vortex. Vorticity threshold, $|\omega| = 0.9\omega_0'$.

choice for the particular experiment in the figure. Both quantities are also correlated, especially for badly fitting neighborhoods, whose kinetic energy is also low. As with the perturbation magnitude, the significant cells are just the low-approximation-error end of the joint p.d.f.

19.3 Vortices and Dipoles

Although a detailed discussion of the dynamics of two-dimensional turbulence is beyond the scope of the present notes, which are mostly concerned with the methodological aspects of how best to collect the information required for scientific discovery, it is still useful to review the most interesting physical aspect of the previous results, which is the classification of significant structures into vortices and dipoles.

Individual vortices are widely recognized to be the dominant coherent structures of two-dimensional turbulent flows (McWilliams 1990b). But vortex dipoles ("modons") have also been investigated as important components of rotating quasi-two-dimensional flows, particularly in geophysics (Flierl et al. 1980, McWilliams 1980), where they retain their individuality for long times in the atmosphere and in the ocean. However, although their presence in nonrotating two-dimensional turbulence is also well known, they have tended to be considered "transient" structures, whose importance is usually not emphasized (McWilliams 1990b). The present results suggest that they deserve a second look. Simulations of a point vortex gas, which can be viewed as a non-dissipative model for two-dimensional turbulence, suggest that the relevant property of dipoles in this context is that they carry no net circulation, so that their interaction with the rest of the flow is relatively weak. They are therefore able to move for long distances before they are destroyed by collisions with other objects, and they stir the flow in the process.

To gain a better sense of the prevalence of dipoles in two-dimensional turbulence, Figure 19.6 shows a segmentation of a sample flow field into individual positive and negative vortices. Vortices are defined as connected regions in which $|\omega| \geq H\omega_0'$, with $H = 0.9$, chosen to maximize the number of individual vortices (Moisy & Jiménez 2004). The vortices in Figure 19.6 have been grouped, whenever possible, into co and counterrotating pairs. Two vortices are considered a potential pair if their areas differ by less than a factor of m^2, which is an adjustable parameter. The figure uses $m = 2$, but statistics compiled with $m = 1.5$ and $m = 3$ show no substantial differences. Vortices are paired to the closest unpaired neighbor within their area class, and no vortex can have more than one partner. Some vortices find no suitable partner, and are left unpaired. Statistics compiled over approximately 8,500 independent flow fields show that, out of approximately 5×10^5 vortices, 48% are paired to form dipoles, 24% are in corotating pairs, and 28% are isolated. Most vortices in the flow are thus in the form of pairs, mostly dipoles. The difference between corotating and counterrotating pairs is interesting, but the reason is probably that corotating pairs tend to merge into single cores (Meunier et al. 2005), while modons are long-lasting (Flierl et al. 1980, McWilliams 1980).

19.4 Discussion and Conclusions

This chapter, together with Chapter 2, have explored two related but independent ideas in the context of scientific discovery in turbulence research.

The first idea is the well-known distinction between correlation and causation. We have noted that, because of the difficulty of doing experiments in turbulence (computational or otherwise), research has tended to center on correlations. Chapter 2 gave examples of what can be learned from this approach, and noted that the wealth of data generated by direct simulations in the 1990s probably contributed to that trend by creating the illusion that "we know everything," and that all questions can be answered.

There are two problems with this illusion. The first one is that, although large data sets contains many answers, the probability of finding them without active experimentation is very small. Consider how difficult it would be to study large asteroid strikes on Earth by observing the configuration of the solar system before spontaneous strikes happen. These are rare, and it is unclear whether the precursor to one of them applies to all the others. It is much more efficient to perturb the system (hopefully computationally) and observe the result. The same is true of many intermittent but significant processes in turbulence, especially if we are interested in controlling the flow by introducing local out-of-equilibrium perturbations. Studying them requires experiments that separate present causes from later effects. We noted in Chapter 2 that correlation is the tool of prediction, while causation is the tool of control.

The second problem is how to choose which experiments to perform. The classical search for scientific causation is hypothesis-driven. The entry point to the optimization cycle of scientific research is that the researcher thinks of a model, and designs experiments to test it. While this "hypothesis-driven search" has obvious theoretical appeal, it risks limiting creativity. A model has to be conceived before it is tested, and new ideas depend on the imagination of the researchers. However, we have noted at the beginning of this chapter that faster experimental and computational methods open the possibility of what we have called Monte Carlo searchers (perhaps describable as "search-driven hypotheses") in which experiments are performed randomly, and evaluated a posteriori in the hope that some of them may be interesting. This is more expensive than the classical procedure, but may be our best hope of avoiding ingrained prejudice.

We have illustrated these ideas by the application to two-dimensional turbulence in Sections 19.2 and 19.3, but the subject of these lectures is not turbulence itself but the method, and we now briefly discuss how far the hopes expressed earlier have been realized in the exercise. The first consideration is cost, which is always a pacing item for turbulence research. The whole program in this paper took about a month of computer time in a medium-sized department cluster, and was programmed by the author during his spare time over a year. Considerably more time was spent discussing with colleagues what should be done than in actually doing it. Since the effort was conceived as a proof of principle, the problem was chosen small on purpose, but more interesting problems are within reach. The basic simulation that had to be repeated many times for the experiments in Section 19.2 runs in about 10 core-seconds, but even a 256^3 simulation of three-dimensional isotropic turbulence can be run in two minutes in a modern GPU (Jiménez 2020a). A program to address causality in the three-dimensional turbulence cascade, about which considerably less is known than in two dimensions, could thus be performed in a modest GPU cluster in a few months. Again, the main roadblock would be to decide what to do.

The second question is whether something of physical interest has been learned. As mentioned at the beginning of Section 19.2, little was expected from a problem that is usually considered to be essentially understood, but the actual results were interesting. This chapter broadly follows the discussion in Jiménez (2020b), but two early versions (Jiménez 2018b, Jiménez 2020a) missed the dipoles completely, and concluded that the experiments confirmed the classical vortex-gas model of two-dimensional turbulence. The dipole template in Figure 19.4(b) was a mildly surprising result of further postprocessing and, if we admit that surprise is one of the defining ingredients of discovery (Schaffer 1994), it was a minor discovery. Unfortunately, we saw in Section 19.3 that dipoles were not completely unknown components of two-dimensional flows, and that finding them was rather an instance of recalling something that had been forgotten than of discovering something truly new. It is clear that the present chapter needs a lot of extra work before it becomes an article of independent interest in fluid mechanics, rather than in methodology, but the fact that something unexpected by the author was found without "prompting" is encouraging for the future of the Monte Carlo method in problems in which something is genuinely unknown.

The third question is what have we learned about the process of data exploitation. The scientific method can be seen as an optimization loop to search for the "best"

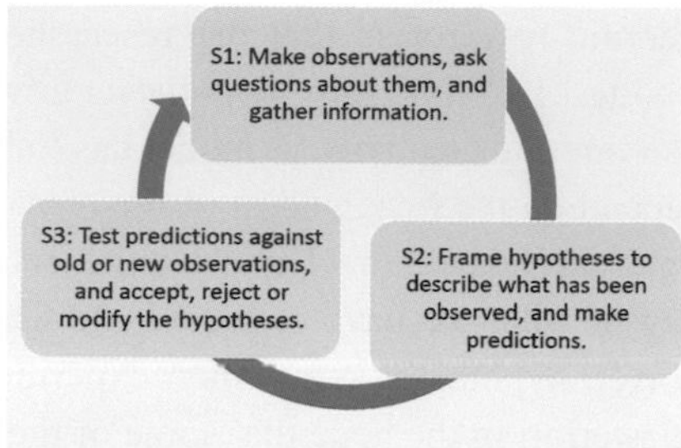

Figure 19.7 The scientific method. Reproduced from Chapter 2 for easy of reference.

hypothesis, and it is natural to ask which parts of it can be automated. Our case study in Sections 19.2 and 19.3 was conceived as an experimental test of how far the automation process could be pushed. Several things can be concluded. The first one is that step S1 in Figure 19.7 (observations) can be largely outsourced to the computer, including the reference to "asking questions." The experiments in Table 19.1 were decided (on purpose) with little thought about their significance, although it is difficult to say how much they were influenced by the previous experience of the author. The same can be said about the parts of step S3 that refer to testing predictions against observations. Simulations of two-dimensional turbulence are by now trivial, and the classification of the results was outsourced to library programs (Figures 19.2 and 19.3). An interesting, although not completely unexpected, outcome of the experience has been the importance of verification and validation, and a rereading of Section 19.2.2 and, up to a point, of Section 19.3, shows pitfalls that were avoided this way. Such problems are expected in any project involving data analysis, but they are especially dangerous in cases, like the present one in which the goal is to probe the unknown.

The last point to consider is the model generation and evaluation step (S2) in Figure 19.7, which is the core of the discovery process. Even here, something was automated. The templates in Figure 19.4 are flow models, and they were obtained automatically. Note that it is at this point that the transition from subsymbolic to symbolic AI took place in this example. Figures 19.2 and 19.3 are primarily useful for computer classification, but the templates in Figure 19.4 are intended for humans.

On the other hand, the interpretation in Section 19.3 of the templates was entirely manual, and it is difficult to see at the moment how it could be outsourced to a computer. This is not only because of the need for a level of intelligence somewhat above present computer software, but because what are we trying to do is not well defined. If the goal of a scientific project is to find a "good" model, we need to define precisely what we consider a good hypothesis. We mentioned in Chapter 2 that one of the goals of the scientific method was to produce "beautiful" theories, but we do not know very well how to define scientific beauty. The ultimate models for two-dimensional turbulence are the Navier–Stokes equations, and the ultimate causes of any observation are some notional initial conditions. We have restricted ourselves here to a time horizon shorter than the memory loss due to chaos, because of some generalized interest in flow control. This is a choice that influences the results and makes them less general, but it defines the metric that allows us to define templates.

Human supervision will probably still be required for some time to refine hypotheses, but Monte Carlo science can contribute something even now. Consider again the interpretation of the scientific method as an optimization loop to maximize "understanding." As such, it could conceivably be implemented as an automatic optimization procedure (e.g., a neural network). However, most classical optimizers, including humans, assume local convexity of the cost function, which is at the root of our misgivings about researcher originality and prejudice. Monte Carlo is a different way of looking for a maximum, which, in principle, bypasses the convexity constraint and escapes local maxima by injecting noise in the process (Deb & Gupta 2006). The best use of Monte Carlo science is probably as a partially randomized search and classification step, followed by repeated human fine-tuning. The encouraging news in fluid mechanics is that doing this is now becoming possible.

References

Abdi, H. (2003), 'Factor rotations in factor analyses', *Encyclopedia for Research Methods for the Social Sciences*, Sage, Thousand Oaks, CA, pp. 792–795.

Abraham, R., Marsden, J. E., and Marsden, J. E. (1978), *Foundations of Mechanics*, Benjamin/Cummings Publishing Company, Reading, MA.

Abraham, R., Marsden, J. E., and Ratiu, T. (1988), *Manifolds, Tensor Analysis, and Applications*, Vol. 75 of Applied Mathematical Sciences, Springer-Verlag, New York.

Abu-Mostafa, Y. S., Magndon-Ismail, M., and Lin, H.-T. (2012), *Learning from Data: A Short Course*, https://amlbook.com/.

Abu-Zurayk, M., Ilic, C., Schuster, A., and Liepelt, R. (2017), Effect of gradient approximations on aero-structural gradient-based wing optimization, in 'EUROGEN 2017', September 13–15, 2017, Madrid, Spain.

Adrian, R. J. (1975), On the role of conditional averages in turbulence theory, in 'Proceedings of the 4th Biennial Symposium on Turbulence in Liquids, University of Missouri-Rolla, 1975', Science Press, Princeton, NJ.

Adrian, R. J. (1991), 'Particle-imaging techniques for experimental fluid mechanics', *Annual Review of Fluid Mechanics* **23**(1), 261–304.

Adrian, R. J. (2007), 'Hairpin vortex organization in wall turbulence', *Physics of Fluids* **19**, 041301.

Adrian, R. J., Jones, B., Chung, M., Hassan, Y., Nithianandan, C., and Tung, A.-C. (1989), 'Approximation of turbulent conditional averages by stochastic estimation', *Physics of Fluids A: Fluid Dynamics* **1**(6), 992–998.

Ahuja, S. and Rowley, C. W. (2010), 'Feedback control of unstable steady states of flow past a flat plate using reduced-order estimators', *Journal of Fluid Mechanics* **645**, 447–478.

Albers, M., Meysonnat, P. S., Fernex, D., Semaan, R., Noack, B. R., and Schröder, W. (2020), 'Drag reduction and energy saving by spanwise traveling transversal surface waves for flat plate flow', *Flow, Turbulence and Combustion* **105**, 125–157.

Aleksic, K., Luchtenburg, D. M., King, R., Noack, B. R., and Pfeiffer, J. (2010), Robust nonlinear control versus linear model predictive control of a bluff body wake, in '5th AIAA Flow Control Conference', June 28–July 1, 2010, Chicago, AIAA-Paper 2010-4833, pp. 1–18.

Allaire, G. and Schoenauer, M. (2007), *Conception optimale de structures*, Vol. 58 of Mathématiques et Applications, Springer, Berlin.

Alsalman, M., Colvert, B., and Kanso, E. (2018), 'Training bioinspired sensors to classify flows', *Bioinspiration & Biomimetics* **14**(1), 016009.

Amsallem, D., Zahr, M. J., and Farhat, C. (2012), 'Nonlinear model order reduction based on local reduced-order bases', *International Journal for Numerical Methods in Engineering* **92**(10), 891–916.

Antoranz, A., Gonzalo, A., Flores, O., and Garcia-Villalba, M. (2015), 'Numerical simulation of heat transfer in a pipe with non-homogeneous thermal boundary conditions', *International Journal of Heat and Fluid Flow* **55**, 45–51.

Antoranz, A., Ianiro, A., Flores, O., and García-Villalba, M. (2018), 'Extended proper orthogonal decomposition of non-homogeneous thermal fields in a turbulent pipe flow', *International Journal of Heat and Mass Transfer* **118**, 1264–1275.

Antoulas, A. C. (2005), *Approximation of Large-Scale Dynamical Systems*, Society for Industrial and Applied Mathematics, Philadelphia.

Arbabi, H. and Mezić, I. (2017), 'Ergodic theory, dynamic mode decomposition, and computation of spectral properties of the Koopman operator', *SIAM Journal on Applied Dynamical Systems* **16**(4), 2096–2126.

Armstrong, E. and Sutherland, J. C. (2021), 'A technique for characterizing feature size and quality of manifolds', *Combustion Theory and Modelling*, **25**, 646–668.

Arthur, D. and Vassilvitskii, S. (2007), k-means++: The advantages of careful seeding, in 'Proceedings of the 18th Annual ACM-SIAM Symposium on Discrete Algorithms', Society for Industrial and Applied Mathematics, Philadelphia, pp. 1027–1035.

Ashurst, W. T. and Meiburg, E. (1988), 'Three-dimensional shear layers via vortex dynamics', *Journal of Fluid Mechanics* **189**, 87–116.

Aström, K. J. and Murray, R. M. (2010), *Feedback Systems: An Introduction for Scientists and Engineers*, Princeton University Press, Princeton, NJ.

Aubry, N., Holmes, P., Lumley, J. L., and Stone, E. (1988), 'The dynamics of coherent structures in the wall region of a turbulent boundary layer', *Journal of Fluid Mechanics* **192**(1), 115–173.

Baars, W. J. and Tinney, C. E. (2014), 'Proper orthogonal decomposition-based spectral higher-order stochastic estimation', *Physics of Fluids* **26**(5), 055112.

Babaee, H. and Sapsis, T. P. (2016), 'A variational principle for the description of time-dependent modes associated with transient instabilities', *Philosophical Transactions of the Royal Society of London* **472**, 20150779.

Bagheri, S., Brandt, L., and Henningson, D. (2009), 'Input–output analysis, model reduction and control of the flat-plate boundary layer', *Journal of Fluid Mechanics* **620**, 263–298.

Bagheri, S., Hoepffner, J., Schmid, P. J., and Henningson, D. S. (2009), 'Input–output analysis and control design applied to a linear model of spatially developing flows', *Applied Mechanics Reviews* **62**(2), 020803-1-020803-27.

Balajewicz, M., Dowell, E. H., and Noack, B. R. (2013), 'Low-dimensional modelling of high-Reynolds-number shear flows incorporating constraints from the Navier-Stokes equation', *Journal of Fluid Mechanics* **729**, 285–308.

Baldi, P. and Hornik, K. (1989), 'Neural networks and principal component analysis: Learning from examples without local minima', *Neural Networks* **2**(1), 53–58.

Banerjee, I. and Ierapetritou, M. (2006), 'An adaptive reduction scheme to model reactive flow', *Combustion and Flame* **144**(3), 619–633.

Baraniuk, R. G. (2007), 'Compressive sensing', *IEEE Signal Processing Magazine* **24**(4), 118–120.

Barbagallo, A., Sipp, D., and Schmid, P. J. (2009), 'Closed-loop control of an open cavity flow using reduced-order models', *Journal of Fluid Mechanics* **641**, 1–50.

Barkley, D. and Henderson, R. (1996), 'Three-dimensional Floquet stability analysis of the wake of a circular cylinder', *Journal of Fluid Mechanics* **322**, 215–241.

Barlow, R. S. and Frank, J. H. (1998), 'Effects of turbulence on species mass fractions in methane/air jet flames', *Symposium (International) on Combustion* **27**, 1087–1095.

Batchelor, G. K. (1969), 'Computation of the energy spectrum in homogeneous two dimensional turbulence', *Physics of Fluids* **12** (Suppl. II), 233–239.

Battaglia, P. W., Hamrick, J. B., Bapst, V., Sanchez-Gonzalez, A., Zambaldi, V., Malinowski, M., Tacchetti, A., Raposo, D., Santoro, A., Faulkner, R. Gulcehre, C., Song, F., Ballard, A., Gilmer, J., Dahl, G., Vaswani, A., Allen, K., Nash, C., Langston, V., Dyer, C., Heess, N., Wierstra, D., Kohli, P., Botvinick, M., Vinyals, O., Li, Y., and Pascanu, R. (2018), 'Relational inductive biases, deep learning, and graph networks', *arXiv preprint arXiv:1806.01261*.

Beerends, R. J., ter Morsche, H. G., van den Berg, J. C., and van de Vrie, E. M. (2003), *Fourier and Laplace Transforms*, Cambridge University Press, New York.

Beetham, S. and Capecelatro, J. (2020), 'Formulating turbulence closures using sparse regression with embedded form invariance', *arXiv preprint arXiv:2003.12884* .

Beintema, G., Corbetta, A., Biferale, L., and Toschi, F. (2020), 'Controlling Rayleigh–Bénard convection via reinforcement learning', *Journal of Turbulence* **21**(9–10), 585–605.

Bekemeyer, P., Ripepi, M., Heinrich, R., and Görtz, S. (2019), 'Nonlinear unsteady reduced-order modeling for gust-load predictions', *AIAA Journal* **57**(5), 1839–1850.

Bekemeyer, P., Wunderlich, T., Görtz, S., and Dähne, D. (2019), Effect of gradient approximations on aero-structural gradient-based wing optimization, in 'International Forum on Aeroelasticity and Structural Dynamics (IFASD-2019)', June 9–13, 2019, Savannah, GA.

Bekemeyer, P., Bertram, A., Chaves, D. A. H., Ribeiro, M. D., Garbo, A., Kiener, A., Sabater, C., Stradtner, M., Wassing, S., Widhalm, M., Goertz, S., Jaeckel, F., Hoppe, R. & Hoffmann, N. (2022), 'Data-driven aerodynamic modeling using the DLR SMARTy toolbox'.

Bellemans, A., Aversano, G., Coussement, A., and Parente, A. (2018), 'Feature extraction and reduced-order modelling of nitrogen plasma models using principal component analysis', *Computers & Chemical Engineering* **115**, 504–514.

Bellemare, M. G., Dabney, W., and Munos, R. (2017), A distributional perspective on reinforcement learning, in 'Proceedings of the 34th International Conference on Machine Learning – Volume 70', ICML 17, JMLR.org, pp. 449–458.

Bellman, R. (1957), 'A Markovian decision process', *Journal of Mathematics and Mechanics* **6**(5), 679–684.

Belson, B. A., Semeraro, O., Rowley, C. W., and Henningson, D. S. (2013), 'Feedback control of instabilities in the two-dimensional Blasius boundary layer: The role of sensors and actuators', *Physics of Fluids (1994–present)* **25**(5), 054106.

Belson, B. A., Tu, J. H., and Rowley, C. W. (2014), 'Algorithm 945: Modred – A parallelized model reduction library', *ACM Transactions on Mathematical Software (TOMS)* **40**(4), 30.

Belus, V., Rabault, J., Viquerat, J., Che, Z., Hachem, E., and Reglade, U. (2019), 'Exploiting locality and translational invariance to design effective deep reinforcement learning control of the 1-dimensional unstable falling liquid film', *AIP Advances* **9**(12), 125014.

Bence, J. R. (1995), 'Analysis of short time series: Correcting for autocorrelation', *Ecology* **76**(2), 628–639.

Benner, P., Gugercin, S., and Willcox, K. (2015), 'A survey of projection-based model reduction methods for parametric dynamical systems', *SIAM Review* **57**(4), 483–531.

Bergmann, M., Cordier, L., and Brancher, J.-P. (2005), 'Optimal rotary control of the cylinder wake using proper orthogonal decomposition reduced-order model', *Physics of Fluids (1994–present)* **17**(9), 097101.

Berkooz, G., Holmes, P., and Lumley, J. L. (1993), 'The proper orthogonal decomposition in the analysis of turbulent flows', *Annual Review of Fluid Mechanics* **23**, 539–575.

Bernal, L. P. (1981), The coherent structure of turbulent mixing layers, PhD thesis, California Institute of Technology, Pasadena.

Berry, M. W., Browne, M., Langville, A. N., Pauca, V. P., and Plemmons, R. J. (2007), 'Algorithms and applications for approximate nonnegative matrix factorization', *Computational Statistics & Data Analysis* **52**(1), 155–173.

Bertolotti, F., Herbert, T., and Spalart, P. (1992), 'Linear and nonlinear stability of the Blasius boundary layer', *Journal of Fluid Mechanics* **242**, 441–474.

Betchov, R. (1956), 'An inequality concerning the production of vorticity in isotropic turbulence', *Journal of Fluid Mechanics* **1**, 497–504.

Bieker, K., Peitz, S., Brunton, S. L., Kutz, J. N., and Dellnitz, M. (2019), 'Deep model predictive control with online learning for complex physical systems', *arXiv preprint arXiv:1905.10094*.

Biglari, A. and Sutherland, J. C. (2012), 'A filter-independent model identification technique for turbulent combustion modeling', *Combustion and Flame* **159**(5), 1960–1970.

Biglari, A. and Sutherland, J. C. (2015), 'An a-posteriori evaluation of principal component analysis-based models for turbulent combustion simulations', *Combustion and Flame* **162**(10), 4025–4035.

Bilger, R., Starner, S., and Kee, R. (1990), 'On reduced mechanisms for methane-air combustion in nonpremixed flames', *Combustion and Flame* **80**, 135–149.

Billings, S. A. (2013), *Nonlinear System Identification: NARMAX Methods in the Time, Frequency, and Spatio-Temporal Domains*, John Wiley & Sons, Hoboken, NJ.

Bishop, C. M. (2016), *Pattern Recognition and Machine Learning*, Springer, New York.

Bistrian, D. and Navon, I. (2016), 'Randomized dynamic mode decomposition for non-intrusive reduced order modelling', *International Journal for Numerical Methods in Engineering* **112**, 3–25.

Black, F., Schulze, P., and Unger, B. (2019), 'Nonlinear Galerkin model reduction for systems with multiple transport velocities', *arXiv preprint arXiv:1912.11138*.

Boffetta, G. and Ecke, R. E. (2012), 'Two-dimensional turbulence', *The Annual Review of Fluid Mechanics* **44**, 427–451.

Bohn, E., Coates, E. M., Moe, S., and Johansen, T. A. (2019), Deep reinforcement learning attitude control of fixed-wing UAVs using proximal policy optimization, in '2019 International Conference on Unmanned Aircraft Systems (ICUAS)', June 11–14, 2019, Atlanta, GA, IEEE, pp. 523–533.

Bongard, J. and Lipson, H. (2007), 'Automated reverse engineering of nonlinear dynamical systems', *Proceedings of the National Academy of Sciences* **104**(24), 9943–9948.

Borée, J. (2003), 'Extended proper orthogonal decomposition: A tool to analyse correlated events in turbulent flows', *Experiments in Fluids* **35**(2), 188–192.

Bourgeois, J. A., Martinuzzi, R. J., and Noack, B. R. (2013), 'Generalised phase average with applications to sensor-based flow estimation of the wall-mounted square cylinder wake', *Journal of Fluid Mechanics* **736**, 316–350.

Box, G. E., Jenkins, G. M., Reinsel, G. C., and Ljung, G. M. (2015), *Time Series Analysis: Forecasting and Control*, John Wiley & Sons, Hoboken, NJ.

Boyd, S., Parikh, N., Chu, E., Peleato, B., and Eckstein, J. (2011), 'Distributed optimization and statistical learning via the alternating direction method of multipliers', *Foundations and Trends in Machine Learning* **3**, 1–122.

Brackston, R., De La Cruz, J. G., Wynn, A., Rigas, G., and Morrison, J. (2016), 'Stochastic modelling and feedback control of bistability in a turbulent bluff body wake', *Journal of Fluid Mechanics* **802**, 726–749.

Breiman, L. (2001), 'Random forests', *Machine Learning* **45**(1), 5–32.

Brenner, M., Eldredge, J., and Freund, J. (2019a), 'Perspective on machine learning for advancing fluid mechanics', *Physical Review Fluids* **4**(10), 100501.

Bright, I., Lin, G., and Kutz, J. N. (2013), 'Compressive sensing and machine learning strategies for characterizing the flow around a cylinder with limited pressure measurements', *Physics of Fluids* **25**(127102), 1–15.

Brockwell, P. J. and Davis, R. A. (2010), *Introduction to Time Series and Forecasting* (Springer Texts in Statistics), Springer, New York.

Brooks, R. (1990), 'Elephants don't play chess', *Robotics and Autonomous Systems* **6**, 3–15.

Brown, G. L. and Roshko, A. (1974), 'On the density effects and large structure in turbulent mixing layers', *Journal of Fluid Mechanics* **64**, 775–816.

Brunton, B. W., Brunton, S. L., Proctor, J. L., and Kutz, J. N. (2016a), 'Sparse sensor placement optimization for classification', *SIAM Journal on Applied Mathematics* **76**(5), 2099–2122.

Brunton, S., Brunton, B., Proctor, J., Kaiser, E., and Kutz, J. (2017), 'Chaos as an intermittently forced linear system', *Nature Communication* **8**(19), 1–9.

Brunton, S. L., Brunton, B. W., Proctor, J. L., and Kutz, J. N. (2016b), 'Koopman invariant subspaces and finite linear representations of nonlinear dynamical systems for control', *PLoS ONE* **11**(2), e0150171.

Brunton, S. L., Dawson, S. T. M., and Rowley, C. W. (2014), 'State-space model identification and feedback control of unsteady aerodynamic forces', *Journal of Fluids and Structures* **50**, 253–270.

Brunton, S. L., Hemati, M. S., and Taira, K. (2020), 'Special issue on machine learning and data-driven methods in fluid dynamics', *Theoretical and Computational Fluid Dynamics* **34**, 333–337.

Brunton, S. L. and Kutz, J. N. (2019), *Data-Driven Science and Engineering: Machine Learning, Dynamical Systems, and Control*, Cambridge University Press, Cambridge.

Brunton, S. L. and Noack, B. R. (2015), 'Closed-loop turbulence control: Progress and challenges', *Applied Mechanics Reviews* **67**(5), 050801:01–48.

Brunton, S. L., Noack, B. R., and Koumoutsakos, P. (2020), 'Machine learning for fluid mechanics', *Annual Review of Fluid Mechanics* **52**, 477–508.

Brunton, S. L., Proctor, J. L., and Kutz, J. N. (2016a), 'Discovering governing equations from data by sparse identification of nonlinear dynamical systems', *Proceedings of the National Academy of Sciences* **113**(15), 3932–3937.

Brunton, S. L., Proctor, J. L., and Kutz, J. N. (2016b), 'Sparse identification of nonlinear dynamics with control (SINDYc)', *IFAC NOLCOS* **49**(18), 710–715.

Bucci, M. A., Semeraro, O., Allauzen, A., Wisniewski, G., Cordier, L., and Mathelin, L. (2019), 'Control of chaotic systems by deep reinforcement learning', *Proceedings of the Royal Society A: Mathematical, Physical and Engineering Sciences* **475**(2231), 20190351.

Budišić, M. and Mezić, I. (2009), An approximate parametrization of the ergodic partition using time averaged observables, in 'Proceedings of the 48th IEEE Conference on Decision and Control, 2009 held jointly with the 2009 28th Chinese Control Conference. CDC/CCC 2009. December 15–18, 2009, Shanghai, China, IEEE, Piscataway, NJ, pp. 3162–3168.

Budišić, M. and Mezić, I. (2012), 'Geometry of the ergodic quotient reveals coherent structures in flows', *Physica D: Nonlinear Phenomena* **241**(15), 1255–1269.

Budišić, M., Mohr, R. and Mezić, I. (2012), 'Applied Koopmanism', *Chaos: An Interdisciplinary Journal of Nonlinear Science* **22**(4), 047510.

Burda, Y., Edwards, H., Storkey, A. J., and Klimov, O. (2019), Exploration by random network distillation, in '7th International Conference on Learning Representations, ICLR 2019, New Orleans, LA, USA, May 6–9, 2019', OpenReview.net.

Burkardt, J., Gunzburger, M., and Lee, H.-C. (2006), 'POD and CVT-based reduced-order modeling of Navier–Stokes flows', *Computer Methods in Applied Mechanics and Engineering* **196**(1–3), 337–355.

Burkov, A. (2019), *The Hundred-Page Machine Learning Book*, Andriy Burkov Quebec City.

Busse, F. (1991), 'Numerical analysis of secondary and tertiary states of fluid flow and their stability properties', *Applied Scientific Research* **48**(3), 341–351.

Butler, K. and Farrell, B. (1992), 'Three-dimensional optimal perturbations in viscous shear flow', *Physics of Fluids* **4**, 1637–1650.

Butler, K. M. and Farrell, B. F. (1993), 'Optimal perturbations and streak spacing in wall-bounded shear flow', *Physics of Fluids A* **5**, 774–777.

Callaham, J., Maeda, K., and Brunton, S. L. (2019), 'Robust reconstruction of flow fields from limited measurements', *Physical Review Fluids* **4**, 103907.

Cammilleri, A., Gueniat, F., Carlier, J., Pastur, L., Memin, E., Lusseyran, F., and Artana, G. (2013), 'POD-spectral decomposition for fluid flow analysis and model reduction', *Theoretical and Computational Fluid Dynamics* **27**(6), 787–815.

Candès, E. J. (2006), Compressive sensing, in 'Proceedings of the International Congress of Mathematics, August 22–30, 2006', European Mathematical Society, Zurich.

Candès, E. J., Romberg, J., and Tao, T. (2006a), 'Robust uncertainty principles: Exact signal reconstruction from highly incomplete frequency information', *IEEE Transactions on Information Theory* **52**(2), 489–509.

Candès, E. J., Romberg, J., and Tao, T. (2006b), 'Stable signal recovery from incomplete and inaccurate measurements', *Communications in Pure and Applied Mathematics* **8**, 1207–1223.

Candès, E. J. and Tao, T. (2006), 'Near optimal signal recovery from random projections: Universal encoding strategies?', *IEEE Transactions on Information Theory* **52**(12), 5406–5425.

Cardesa, J. I., Vela-Martín, A., and Jiménez, J. (2017), 'The turbulent cascade in five dimensions', *Science* **357**, 782–784.

Carlberg, K., Barone, M., and Antil, H. (2017), 'Galerkin v. least-squares Petrov–Galerkin projection in nonlinear model reduction', *Journal of Computational Physics* **330**, 693–734.

Carlberg, K., Bou-Mosleh, C., and Farhat, C. (2011), 'Efficient non-linear model reduction via a least-squares Petrov–Galerkin projection and compressive tensor approximations', *International Journal for Numerical Methods in Engineering* **86**(2), 155–181.

Carlberg, K., Tuminaro, R., and Boggs, P. (2015), 'Preserving Lagrangian structure in nonlinear model reduction with application to structural dynamics', *SIAM Journal on Scientific Computing* **37**(2), B153–B184.

Carleman, T. (1932), 'Application de la théorie des équations intégrales linéaires aux systémes d'équations différentielles non linéaires', *Acta Mathematica* **59**(1), 63–87.

Carnevale, G. F., McWilliams, J. C., Pomeau, Y., Weiss, J. B., and Young, W. R. (1991), 'Evolution of vortex statistics in two-dimensional turbulence', *Physical Review Letters* **66**, 2735–2737.

Cassel, K. (2013), *Variational Methods with Applications in Science and Engineering*, Cambridge University Press, Cambridge.

Cattafesta III, L. N. and Sheplak, M. (2011), 'Actuators for active flow control', *Annual Review of Fluid Mechanics* **43**, 247–272.

Cattell, R. B. (1966), 'The scree test for the number of factors', *Multivariate Behavioral Research* **1**(2), 245–276.

Cavaliere, A. and de Joannon, M. (2004), 'Mild combustion', *Progress in Energy and Combustion Science* **30**, 329–366.

Champion, K., Lusch, B., Kutz, J. N., and Brunton, S. L. (2019), 'Data-driven discovery of coordinates and governing equations', *Proceedings of the National Academy of Sciences* **116**(45), 22445–22451.

Champion, K., Zheng, P., Aravkin, A. Y., Brunton, S. L., and Kutz, J. N. (2020), 'A unified sparse optimization framework to learn parsimonious physics-informed models from data', *IEEE Access* **8**, 169259–169271.

Chang, H., Yeung, D.-Y., and Xiong, Y. (2004), Super-resolution through neighbor embedding, in 'Proceedings of the 2004 IEEE Computer Society Conference on Computer Vision and Pattern Recognition, 2004. CVPR 2004.', Vol. 1, IEEE, pp. I–I.

Chartrand, R. (2011), 'Numerical differentiation of noisy, nonsmooth data', *ISRN Applied Mathematics* **2011**, article ID 164564.

Chen, K. K. and Rowley, C. W. (2011), 'H_2 optimal actuator and sensor placement in the linearised complex Ginzburg-Landau system', *Journal of Fluid Mechanics* **681**, 241–260.

Chen, K. K., Tu, J. H., and Rowley, C. W. (2012), 'Variants of dynamic mode decomposition: Boundary condition, Koopman, and Fourier analyses', *Journal of Nonlinear Science* **22**(6), 887–915.

Choi, H., Moin, P., and Kim, J. (1994), 'Active turbulence control for drag reduction in wall-bounded flows', *Journal of Fluid Mechanics* **262**, 75–110.

Chui, C. K. (1992), *An Introduction to Wavelets*, Academic Press, San Diego, CA.

Cimbala, J. M., Nagib, H. M., and Roshko, A. (1988), 'Large structure in the far wakes of two-dimensional bluff bodies', *Journal of Fluid Mechanics* **190**, 265–298.

Citriniti, J. H. and George, W. K. (2000), 'Reconstruction of the global velocity field in the axisymmetric mixing layer utilizing the proper orthogonal decomposition', *Journal of Fluid Mechanics* **418**, 137–166.

Cohen, L. (1995), *Time-Frequency Analysis*, Prentice Hall, Englewood Cliffs, NJ.

Colabrese, S., Gustavsson, K., Celani, A., and Biferale, L. (2017), 'Flow navigation by smart microswimmers via reinforcement learning', *Physical Review Letters* **118**(15), 158004.

Colvert, B., Alsalman, M., and Kanso, E. (2018), 'Classifying vortex wakes using neural networks', *Bioinspiration & Biomimetics* **13**(2), 025003.

Cooley, J. W. and Tukey, J. W. (1965), 'An algorithm for the machine calculation of complex Fourier series', *Mathematics of Computation* **19**(90), 297–301.

Cordier, L., El Majd, B. A., and Favier, J. (2010), 'Calibration of POD reduced-order models using Tikhonov regularization', *International Journal for Numerical Methods in Fluids* **63**(2), 269–296.

Cordier, L., Noack, B. R., Daviller, G., Delvile, J., Lehnasch, G., Tissot, G., Balajewicz, M., and Niven, R. (2013), 'Control-oriented model identification strategy', *Experiments in Fluids* **54**, Article 1580.

Corino, E. R. and Brodkey, R. S. (1969), 'A visual investigation of the wall region in turbulent flow', *Journal of Fluid Mechanics* **37**, 1.

Cornejo Maceda, G. Y., Li, Y., Lusseyran, F., Morzyński, M., and Noack, B. R. (2021), 'Stabilization of the fluidic pinball with gradient-based machine learning control', *Journal of Fluid Mechanics* **917**, A42, 1–43.

Cornejo Maceda, G., Bernd R. Noack, Lusseyran, F., Deng, N., Pastur, L., and Morzyński, M. (2019), 'Artificial intelligence control applied to drag reduction of the fluidic pinball', *Proceedings in Applied Mathematics and Mechanics* **19**(1), e201900268:1–2.

Corrsin, S. (1958), Local isotropy in turbulent shear flow, Res. Memo 58B11, National Advisory Committee for Aeronautics, Washington, DC.

Coussement, A., Gicquel, O., and Parente, A. (2013), 'MG-local-PCA method for reduced order combustion modeling', *Proceedings of the Combustion Institute* **34**(1), 1117–1123.

Coussement, A., Isaac, B. J., Gicquel, O., and Parente, A. (2016), 'Assessment of different chemistry reduction methods based on principal component analysis: Comparison of the MG-PCA and score-PCA approaches', *Combustion and Flame* **168**, 83–97.

Coveney, P. V., Dougherty, E. R., and Highfield, R. R. (2016), 'Big data need big theory too', *Philosophical Transactions of the Royal Society A* **374**, 20160153.

Cox, T. F. and Cox, M. A. A. (2000), *Multidimensional Scaling*, Vol. 88 of Monographs on Statistics and Applied Probability, 2nd ed., Chapman & Hall, London.

Cranmer, M. D., Xu, R., Battaglia, P., and Ho, S. (2019), 'Learning symbolic physics with graph networks', *arXiv preprint arXiv:1909.05862*.

Cranmer, M., Greydanus, S., Hoyer, S., Battaglia, P., Spergel, D., and Ho, S. (2020), 'Lagrangian neural networks', *arXiv preprint arXiv:2003.04630*.

Cristianini, N. and Shawe–Taylor, J. (2000), *An Introduction to Support Vector Machines and Other Kernel-based Learning Methods*, Cambridge University Press, Cambridge.

Cuoci, A., Frassoldati, A., Faravelli, T., and Ranzi, E. (2013), 'Numerical modeling of laminar flames with detailed kinetics based on the operator-splitting method', *Energy & Fuels* **27**(12), 7730–7753.

Cuoci, A., Frassoldati, A., Faravelli, T., and Ranzi, E. (2015), 'OpenSMOKE++: An object-oriented frame-work for the numerical modeling of reactive systems with detailed kinetic mechanisms', *Computer Physics Communications* **192**, 237–264.

Dalakoti, D. K., Wehrfritz, A., Savard, B., Day, M. S., Bell, J. B., and Hawkes, E. R. (2020), 'An a priori evaluation of a principal component and artificial neural network based combustion model in diesel engine conditions', *Proceedings of the Combustion Institute* **38**, 2701–2709.

D'Alessio, G., Attili, A., Cuoci, A., Pitsch, H., and Parente, A. (2020a), Analysis of turbulent reacting jets via principal component analysis, in 'Data Analysis for Direct Numerical Simulations of Turbulent Combustion', Springer, Cham, pp. 233–251.

D'Alessio, G., Attili, A., Cuoci, A., Pitsch, H., and Parente, A. (2020b), Unsupervised data analysis of direct numerical simulation of a turbulent flame via local principal component analysis and procustes analysis, in '15th International Workshop on Soft Computing Models in Industrial and Environmental Applications', September 16–18, 2020, Burgos, Spain, Springer Nature, Switzerland, pp. 460–469.

D'Alessio, G., Cuoci, A., Aversano, G., Bracconi, M., Stagni, A., and Parente, A. (2020), 'Impact of the partitioning method on multidimensional adaptive-chemistry simulations', *Energies* **13**(10), 2567.

D'Alessio, G., Cuoci, A., and Parente, A. (2020), 'OpenMORe: A Python framework for reduction, clustering and analysis of reacting flow data', *SoftwareX* **12**, 100630.

D'Alessio, G., Parente, A., Stagni, A., and Cuoci, A. (2020), 'Adaptive chemistry via pre-partitioning of composition space and mechanism reduction', *Combustion and Flame* **211**, 68–82.

Dam, M., Brøns, M., Juul Rasmussen, J., Naulin, V., and Hesthaven, J. S. (2017), 'Sparse identification of a predator-prey system from simulation data of a convection model', *Physics of Plasmas* **24**(2), 022310.

Daubechies, I. (1990), 'The wavelet transform, time-frequency localization and signal analysis', *IEEE Transactions on Information Theory* **36**(5), 961–1005.

Daubechies, I. (1992), *Ten Lectures on Wavelets*, Vol. 61, Society for Industrial and Applied Mathematics, Philadelphia.

Daubechies, I., Devore, R., Fornasier, M., and Güntürk, C. (2010), 'Iteratively reweighted least squares minimization for sparse recovery', *Communications on Pure and Applied Mathematics* **63**, 1–38.

de Silva, B., Higdon, D. M., Brunton, S. L., and Kutz, J. N. (2019), 'Discovery of physics from data: Universal laws and discrepancy models', *arXiv preprint arXiv:1906.07906*.

Deane, A. E., Kevrekidis, I. G., Karniadakis, G. E., and Orszag, S. A. (1991), 'Low-dimensional models for complex geometry flows: Application to grooved channels and circular cylinders', *Physics of Fluids A* **3**, 2337–2354.

Deb, K. and Gupta, H. (2006), 'Introducing robustness in multi-objective optimization', *Evolutionary Computation* **14**, 463–494.

Debien, A., von Krbek, K. A., Mazellier, N., Duriez, T., Cordier, L., Noack, B. R., Abel, M. W., and Kourta, A. (2016), 'Closed-loop separation control over a sharp edge ramp using genetic programming', *Experiments in Fluids* **57**(3), 40.

del Álamo, J. C., Jiménez, J., Zandonade, P., and Moser, R. D. (2006), 'Self-similar vortex clusters in the logarithmic region', *Journal of Fluid Mechanics* **561**, 329–358.

Deng, J., Dong, W., Socher, R., Li, L.-J., Li, K., and Fei-Fei, L. (2009), Imagenet: A large-scale hierarchical image database, in '2009 IEEE Conference on Computer Vision and Pattern Recognition', June 20–25, 2009, Miami, FL, IEEE, pp. 248–255.

Deng, N., Noack, B. R., Morzyński, M., and Pastur, L. R. (2020), 'Low-order model for successive bifurcations of the fluidic pinball', *Journal of Fluid Mechanics* **884**, A37.

Discetti, S., Bellani, G., Örlü, R., Serpieri, J., Vila, C. S., Raiola, M., Zheng, X., Mascotelli, L., Talamelli, A., and Ianiro, A. (2019), 'Characterization of very-large-scale motions in high-Re pipe flows', *Experimental Thermal and Fluid Science* **104**, 1–8.

Discetti, S. and Ianiro, A. (2017), *Experimental Aerodynamics*, CRC Press, Boca Raton, FL.

Discetti, S., Raiola, M., and Ianiro, A. (2018), 'Estimation of time-resolved turbulent fields through correlation of non-time-resolved field measurements and time-resolved point measurements', *Experimental Thermal and Fluid Science* **93**, 119–130.

Distefano, J. (2013), *Schaum's Outline of Feedback and Control Systems*, McGraw-Hill Education, New York.

Dong, C., Loy, C. C., He, K., and Tang, X. (2014), Learning a deep convolutional network for image super-resolution, in 'European Conference on Computer Vision', September 6–12, 2014, Zurich, Switzerland, Springer, Cham, pp. 184–199.

Dong, S., Lozano-Durán, A., Sekimoto, A., and Jiménez, J. (2017), 'Coherent structures in statistically stationary homogeneous shear turbulence', *Journal of Fluid Mechanics* **816**, 167–208.

Donoho, D. L. (1995), 'De-noising by soft-thresholding', *IEEE Transactions on Information Theory* **41**(3), 613–627.

Donoho, D. L. (2006), 'Compressed sensing', *IEEE Transactions on Information Theory* **52**(4), 1289–1306.

Donoho, D. L. and Johnstone, J. M. (1994), 'Ideal spatial adaptation by wavelet shrinkage', *Biometrika* **81**(3), 425–455.

Dracopoulos, D. C. (1997), *Evolutionary Learning Algorithms for Neural Adaptive Control*, Perspectives in Neural Computing, Springer-Verlag, London.

Drazin, P. and Reid, W. (1981), *Hydrodynamic Stability*, Cambridge University Press, Cambridge, UK.

Dullerud, G. E. and Paganini, F. (2000), *A Course in Robust Control Theory: A Convex Approach*, Texts in Applied Mathematics, Springer, Berlin, Heidelberg.

Duraisamy, K., Iaccarino, G., and Xiao, H. (2019), 'Turbulence modeling in the age of data', *Annual Reviews of Fluid Mechanics* **51**, 357–377.

Duriez, T., Brunton, S. L., and Noack, B. R. (2017), *Machine Learning Control: Taming Nonlinear Dynamics and Turbulence*, Springer, Cham.

Duwig, C. and Iudiciani, P. (2010), 'Extended proper orthogonal decomposition for analysis of unsteady flames', *Flow, Turbulence and Combustion* **84**(1), 25.

Echekki, T., Kerstein, A. R., and Sutherland, J. C. (2011), The one-dimensional-turbulence model, in T. Echekki & E. Mastorakos, eds., 'Turbulent Combustion Modeling', Springer, Dordrecht, pp. 249–276.

Echekki, T. and Mirgolbabaei, H. (2015), 'Principal component transport in turbulent combustion: A posteriori analysis', *Combustion and Flame* **162**(5), 1919–1933.

Eckart, C. and Young, G. (1936), 'The approximation of one matrix by another of lower rank', *Psychometrika* **1**(3), 211–218.

Ehlert, A., Nayeri, C. N., Morzyński, M., and Noack, B. R. (2020), 'Locally linear embedding for transient cylinder wakes'. *arXiv* manuscript **1906.07822** [physics.flu-dyn].

El Sayed M, Y., Semaan, R., and Radespiel, R. (2018), Sparse modeling of the lift gains of a high-lift configuration with periodic coanda blowing, in '2018 AIAA Aerospace Sciences Meeting', January 8–12, 2018, Kissimmee, FL, p. 1054.

Elman, J. L. (1990), 'Finding structure in time', *Cognitive Science* **14**(2), 179–211.

Encinar, M. P. and Jiménez, J. (2020), 'Momentum transfer by linearised eddies in turbulent channel flows', *arXiv e-prints* p. 1911.06096.

Engstrom, L., Ilyas, A., Santurkar, S., Tsipras, D., Janoos, F., Rudolph, L., and Madry, A. (2020), Implementation matters in deep RL: A case study on PPO and TRPO, in 'International Conference on Learning Representations' April 26–30, 2020, Addis Ababa, Ethiopia.

Epps, B. P. and Krivitzky, E. M. (2019), 'Singular value decomposition of noisy data: Noise filtering', *Experiments in Fluids* **60**(8), 126.

Erichson, N. B., Mathelin, L., Kutz, J. N., and Brunton, S. L. (2019), 'Randomized dynamic mode decomposition', *SIAM Journal on Applied Dynamical Systems* **18**(4), 1867–1891.

Erichson, N. B., Mathelin, L., Yao, Z., Brunton, S. L., Mahoney, M. W., and Kutz, J. N. (2020), 'Shallow neural networks for fluid flow reconstruction with limited sensors', *Proceedings of the Royal Society A* **476**(2238), 20200097.

Espeholt, L., Soyer, H., Munos, R., Simonyan, K., Mnih, V., Ward, T., Doron, Y., Firoiu, V., Harley, T., Dunning, I., Legg, S., and Kavukcuoglu, K. (2018), IMPALA: Scalable distributed deep-RL with importance weighted actor-learner architectures, in J. Dy and A. Krause, eds., 'Proceedings of the 35th International Conference on Machine Learning', Vol. 80 of Proceedings of Machine Learning Research, PMLR, Stockholmsmassan, Stockholm, Sweden, pp. 1407–1416.

Esposito, C., Mendez, M., Gouriet, J., Steelant, J., and Vetrano, M. (2021), 'Spectral and modal analysis of a cavitating flow through an orifice', *Experimental Thermal and Fluid Science* **121**, 110251.

Evans, W. R. (1948), 'Graphical analysis of control systems', *Transactions of the American Institute of Electrical Engineers* **67**(1), 547–551.

Everitt, B. and Skrondal, A. (2002), *The Cambridge Dictionary of Statistics*, Vol. 106, Cambridge University Press, Cambridge.

Ewing, D. and Citriniti, J. H. (1999), Examination of a LSE/POD complementary technique using single and multi-time information in the axisymmetric shear layer, in 'IUTAM

Symposium on Simulation and Identification of Organized Structures in Flows', May 25–29, 1997, Lyngby, Denmark, Springer, Dordrecht, pp. 375–384.

Fabbiane, N., Semeraro, O., Bagheri, S., and Henningson, D. S. (2014), 'Adaptive and model-based control theory applied to convectively unstable flows', *Applied Mechanics Reviews* **66**(6), 060801-1–060801-20.

Farge, M. (1992), 'Wavelet transforms and their applications to turbulence', *Annual Review of Fluid Mechanics* **24**(1), 395–458.

Fernex, D., Noack, B. R., and Semaan, R. (2021), 'Cluster-based network models – From snapshots to complex dynamical systems', *Science Advances* **7**(25).

Fernex, D., Semann, R., Albers, M., Meysonnat, P. S., Schröder, W., Ishar, R., Kaiser, E., and Noack, B. R. (2019), 'Cluster-based network model for drag reduction mechanisms of an actuated turbulent boundary layer', *Proceedings in Applied Mathematics and Mechanics* **19**(1), 1–2.

Ferziger, J. H., Perić, M., and Street, R. L. (2002), *Computational Methods for Fluid Dynamics*, Vol. 3, Springer, Berlin.

Feynman, R. P., Leighton, R. B., and Sands, M. (2013), *The Feynman Lectures on Physics*, Vol. 2, Basic Books, New York.

Finzi, M., Wang, K. A., and Wilson, A. G. (2020), 'Simplifying Hamiltonian and Lagrangian neural networks via explicit constraints', *Advances in Neural Information Processing Systems* **33**, 17605–17616.

Fleming, P. J. and Purshouse, R. C. (2002), 'Evolutionary algorithms in control systems engineering: A survey', *Control Engineering Practice* **10**, 1223–1241.

Fletcher, C. A. J. (1984), *Computational Galerkin Methods*, 1st ed., Springer, New York.

Flierl, G. R., Larichev, V. D., McWilliams, J. C., and Reznik, G. M. (1980), 'The dynamics of baroclinic and barotropic solitary eddies', *Dynamics of Atmospheres and Oceans* **5**, 1–41.

Flinois, T. L. and Morgans, A. S. (2016), 'Feedback control of unstable flows: A direct modelling approach using the eigensystem realisation algorithm', *Journal of Fluid Mechanics* **793**, 41–78.

Flores, O. and Jiménez, J. (2010), 'Hierarchy of minimal flow units in the logarithmic layer', *Physics of Fluids* **22**, 071704.

Floryan, D. and Graham, M. D. (2021), 'Discovering multiscale and self-similar structure with data-driven wavelets', *Proceedings of the National Academy of Sciences* **118**(1), e2021299118.

Föppl, L. (1913), 'Wirbelbewegung hinter einem kreiszylinder (Vortex movement behind a circular cylinder)', Bayerische Akademie der Wissenschaften, Munich.

Fornberg, B. (1977), 'A numerical study of 2-D turbulence', *Journal of Computational Physics* **25**, 1–31.

Fosas de Pando, M., Schmid, P., and Sipp, D. (2014), 'A global analysis of tonal noise in flows around aerofoils', *Journal of Fluid Mechanics* **754**, 5–38.

Fosas de Pando, M., Schmid, P., and Sipp, D. (2017), 'On the receptivity of aerofoil tonal noise: An adjoint analysis', *Journal of Fluid Mechanics* **812**, 771–791.

Fosas de Pando, M., Sipp, D., and Schmid, P. (2012), 'Efficient evaluation of the direct and adjoint linearized dynamics from compressible flow solvers', *Journal of Computational Physics* **231**(23), 7739–7755.

Franklin, G. F., Powell, J. D., and Emami-Naeini, A. (1994), *Feedback Control of Dynamic Systems*, Vol. 3, Addison-Wesley, Reading, MA.

Franz, T. and Held, M. (2017), Data fusion of CFD solutions and experimental aerodynamic data, in 'Proceedings Onera-DLR Aerospace Symposium (ODAS)' June 7–9, 2017, Aussois, France.

Franz, T., Zimmermann, R., and Görtz, S. (2017), Adaptive sampling for nonlinear dimensionality reduction based on manifold learning, in P. Benner, M. Ohlberger, A. Patera, G. Rozza, and K. Urban, eds., Model Reduction of Parametrized Systems, Springer, Cham, pp. 255–269.

Franz, T., Zimmermann, R., Görtz, S., and Karcher, N. (2014), 'Interpolation-based reduced-order modelling for steady transonic flows via manifold learning', *International Journal of Computational Fluid Dynamics* **28**(3–4), 106–121.

Frassoldati, A., Faravelli, T., and Ranzi, E. (2003), 'Kinetic modeling of the interactions between NO and hydrocarbons at high temperature', *Combustion and Flame* **135**, 97–112.

Freeman, W. T., Jones, T. R., and Pasztor, E. C. (2002), 'Example-based super-resolution', *IEEE Computer Graphics and Applications* **22**(2), 56–65.

Freymuth, P. (1966), 'On transition in a separated laminar boundary layer', *Journal of Fluid Mechanics* **25**, 683–704.

Friedman, J. (1991), 'Multivariate adaptive regression splines', *The Annals of Statistics* **19**(1), 1–67.

Fu, J., Luo, K., and Levine, S. (2018), Learning robust rewards with adverserial inverse reinforcement learning, in '6th International Conference on Learning Representations, ICLR 2018, Vancouver, BC, Canada, April 30–May 3, 2018, Conference Track Proceedings', OpenReview.net.

Fukami, K., Fukagata, K., and Taira, K. (2018), 'Super-resolution reconstruction of turbulent flows with machine learning', *arXiv preprint arXiv:1811.11328*.

Gabor, D. (1946), 'Theory of communication. Part 1: The analysis of information', *Journal of the Institution of Electrical Engineers-Part III: Radio and Communication Engineering* **93**(26), 429–441.

Gad-el Hak, M. (2007), *Flow Control: Passive, Active, and Reactive Flow Management*, Cambridge University Press, Cambridge.

Galerkin, B. G. (1915), 'Rods and plates: Series occuring in various questions regarding the elastic equilibrium of rods and plates (translated)', *Vestn. Inzhen.* **19**, 897—908.

Galletti, G., Bruneau, C. H., Zannetti, L., and Iollo, A. (2004), 'Low-order modelling of laminar flow regimes past a confined square cylinder', *Journal of Fluid Mechanics* **503**, 161–170.

Gardiner, W. C., Lissianski, V. V., Qin, Z., Smith, G. P., Golden, D. M., Frenklach, M., Moriarty, N. W., Eiteneer, B., Goldenberg, M., Bowman, C. T., Hanson, R. K., and Song Jr., S. (2012), 'Gri-mech 3.0'.

Gardner, E. (1988), 'The space of interactions in neural network models', *Journal of Physics A: Mathematical and General* **21**(1), 257.

Garnier, P., Viquerat, J., Rabault, J., Larcher, A., Kuhnle, A., and Hachem, E. (2019), 'A review on deep reinforcement learning for fluid mechanics', *arXiv preprint arXiv:1908.04127*.

Gaster, M., Kit, E., and Wygnanski, I. (1985), 'Large-scale structures in a forced turbulent mixing layer', *Journal of Fluid Mechanics* **150**, 23–39.

Gautier, N., Aider, J.-L., Duriez, T., Noack, B., Segond, M., and Abel, M. (2015), 'Closed-loop separation control using machine learning', *Journal of Fluid Mechanics* **770**, 442–457.

Gazzola, M., Hejazialhosseini, B., and Koumoutsakos, P. (2014), 'Reinforcement learning and wavelet adapted vortex methods for simulations of self-propelled swimmers', *SIAM Journal on Scientific Computing* **36**(3), B622–B639.

Gazzola, M., Tchieu, A. A., Alexeev, D., de Brauer, A., and Koumoutsakos, P. (2016), 'Learning to school in the presence of hydrodynamic interactions', *Journal of Fluid Mechanics* **789**, 726–749.

Gerhard, J., Pastoor, M., King, R., Noack, B. R., Dillmann, A., Morzyński, M., and Tadmor, G. (2003), Model-based control of vortex shedding using low-dimensional Galerkin models, in '33rd AIAA Fluids Conference and Exhibit', Orlando, FL, June 23–26, 2003. Paper 2003-4262.

Germano, M., Piomelli, U., Moin, P., and Cabot, W. H. (1991), 'A dynamic subgrid-scale eddy viscosity model', *Physics of Fluids A: Fluid Dynamics* **3**(7), 1760–1765.

Giannetti, F. and Luchini, P. (2007), 'Structural sensitivity of the first instability of the cylinder wake', *Journal of Fluid Mechanics* **581**, 167–197.

Glaz, B., Liu, L., and Friedmann, P. P. (2010), 'Reduced-order nonlinear unsteady aerodynamic modeling using a surrogate-based recurrence framework', *AIAA Journal* **48**(10), 2418–2429.

Goldstein, T. and Osher, S. (2009), 'The split Bregman method for L_1-regularized problems', *SIAM Journal on Imaging Sciences* **2**(2), 323–343.

Gonzalez-Garcia, R., Rico-Martinez, R., and Kevrekidis, I. (1998), 'Identification of distributed parameter systems: A neural net based approach', *Computers & Chemical Engineering* **22**, S965–S968.

Gonzalez, R. C. and Woods, R. E. (2017), *Digital Image Processing*, 4th ed., Pearson, New York.

Goodfellow, I., Bengio, Y., and Courville, A. (2016), *Deep Learning*, MIT Press, Cambridge, MA.

Goodfellow, I., Pouget-Abadie, J., Mirza, M., Xu, B., Warde-Farley, D., Ozair, S., Courville, A., and Bengio, Y. (2014), Generative adversarial nets, in 'Advances in Neural Information Processing Systems', December 8–13, 2014, Montreal, Canada, pp. 2672–2680.

Görtz, S., Ilic, C., Abu-Zurayk, M., Liepelt, R., Jepsen, J., Führer, T., Becker, R., Scherer, J., Kier, T., and Siggel, M. (2016), Collaborative multi-level MDO process development and application to long-range transport aircraft, in 'Proceedings of the 30th Congress of the International Council of the Aeronautical Sciences (ICAS)', September 25–30, 2016, Daejeon, Korea.

Gray, R. M. (2005), 'Toeplitz and circulant matrices: A review', *Foundations and Trends® in Communications and Information Theory* **2**(3), 155–239.

Greydanus, S., Dzamba, M., and Yosinski, J. (2019), 'Hamiltonian neural networks', *Advances in Neural Information Processing Systems* **32**, 15379–15389.

Griewank, A. and Walther, A. (2000), 'Algorithm 799: Revolve: An implementation of checkpointing for the reverse or adjoint mode of computational differentiation', *ACM Transactions on Mathematical Software* **26(1)**, 19–45.

Grossmann, A. and Morlet, J. (1984), 'Decomposition of hardy functions into square integrable wavelets of constant shape', *SIAM Journal on Mathematical Analysis* **15**(4), 723–736.

Gu, S., Lillicrap, T., Sutskever, I., and Levine, S. (2016), *Continuous deep Q-learning with model-based acceleration*, Vol. 48 of Proceedings of Machine Learning Research, PMLR, New York, pp. 2829–2838.

Guan, Y., Brunton, S. L., and Novosselov, I. (2020), 'Sparse nonlinear models of chaotic electroconvection', *arXiv preprint arXiv:2009.11862*.

Guastoni, L., Güemes, A., Ianiro, A., Discetti, S., Schlatter, P., Azizpour, H., and Vinuesa, R. (2020), 'Convolutional-network models to predict wall-bounded turbulence from wall quantities', *arXiv preprint arXiv:2006.12483*.

Guckenheimer, J. and Holmes, P. (1983), *Nonlinear Oscillations, Dynamical Systems, and Bifurcations of Vector Fields*, Springer Verlag, New York.

Güemes, A., Discetti, S., and Ianiro, A. (2019), 'Sensing the turbulent large-scale motions with their wall signature', *Physics of Fluids* **31**(12), 125112.

Guidorzi, R. (2003), *Multivariable System Identification. From Observations to Models*, Bononia University Press, Bologna.

Gunzburger, M. (2003), *Perspectives in Flow Control and Optimization*, Society for Industrial and Applied Mathematics, Philadelphia.

Ha, D. and Schmidhuber, J. (2018), 'World models', *CoRR, arXiv:1803.10122.*

Haar, A. (1911), 'Zur theorie der orthogonalen funktionensysteme', *Mathematische Annalen* **71**(1), 38–53.

Haarnoja, T., Zhou, A., Abbeel, P., and Levine, S. (2018), Soft actor-critic: Off-policy maximum entropy deep reinforcement learning with a stochastic actor, in J. Dy and A. Krause, eds, 'Proceedings of the 35th International Conference on Machine Learning', Vol. 80 of Proceedings of Machine Learning Research, PMLR, Stockholmsmassan, Stockholm, Sweden, pp. 1861–1870.

Halko, N., Martinsson, P.-G., Shkolnisky, Y., and Tygert, M. (2011), 'An algorithm for the principal component analysis of large data sets', *SIAM Journal on Scientific Computing* **33**, 2580–2594.

Halko, N., Martinsson, P.-G., and Tropp, J. A. (2011), 'Finding structure with randomness: Probabilistic algorithms for constructing approximate matrix decompositions', *SIAM Review* **53**(2), 217–288.

Hama, F. R. (1962), 'Streaklines in a perturbed shear flow', *The Physics of Fluids* **5**(6), 644–650.

Hamilton, J. M., Kim, J., and Waleffe, F. (1995), 'Regeneration mechanisms of near-wall turbulence structures', *Journal of Fluid Mechanics* **287**, 317–348.

Han, Z.-H., Abu-Zurayk, M., Görtz, S., and Ilic, C. (2018), Surrogate-based aerodynamic shape optimization of a wing-body transport aircraft configuration, in 'Notes on Numerical Fluid Mechanics and Multidisciplinary Design', October 13–14, 2015, Braunschweig, Germany, Springer, Berlin, pp. 257–282.

Han, Z.-H. and Görtz, S. (2012a), 'Alternative cokriging method for variable-fidelity surrogate modeling', *AIAA Journal* **50**(5), 1205–1210.

Han, Z.-H. and Görtz, S. (2012b), 'Hierarchical kriging model for variable-fidelity surrogate modeling', *AIAA Journal* **50**(9), 1885–1896.

Han, Z.-H., Görtz, S., and Zimmermann, R. (2013), 'Improving variable-fidelity surrogate modeling via gradient-enhanced kriging and a generalized hybrid bridge function', *Aerospace Science and Technology* **25**(1), 177–189.

Hansen, N., Niederberger, A. S., Guzzella, L., and Koumoutsakos, P. (2009), 'A method for handling uncertainty in evolutionary optimization with an application to feedback control of combustion', *IEEE Transactions on Evolutionary Computation* **13**(1), 180–197.

Harris, F. (1978), 'On the use of windows for harmonic analysis with the discrete Fourier transform', *Proceedings of the IEEE* **66**(1), 51–83.

Hasselmann, K. (1988), 'PIPs and POPs: The reduction of complex dynamical systems using principal interaction and oscillation patterns', *Journal of Geophysical Research* **93**(D9), 11015.

Hasselt, H. v., Guez, A., and Silver, D. (2016), Deep reinforcement learning with double Q-learning, in 'Proceedings of the Thirtieth AAAI Conference on Artificial Intelligence', AAAI 16, February 12–17, 2016, Phoenix, AZ, AAAI Press, Palo Alto, CA, pp. 2094–2100.

Hastie, T., Tibshirani, R., and Friedman, J. (2009), *The Elements of Statistical Learning*, Vol. 2, Springer, New York.

Hayes, M. (2011), *Schaums Outline of Digital Signal Processing*, McGraw-Hill Education, New York.

Hemati, M., Rowley, C., Deem, E., and Cattafesta, L. (2017), 'De-biasing the dynamic mode decomposition for applied Koopman spectral analysis', *Theoretical and Computational Fluid Dynamics* **31**(4), 349–368.

Henderson, P., Islam, R., Bachman, P., Pineau, J., Precup, D., and Meger, D. (2018), Deep reinforcement learning that matters, in S. A. McIlraith and K. Q. Weinberger, eds., 'Proceedings of the Thirty-Second AAAI Conference on Artificial Intelligence, (AAAI-18), the 30th Innovative Applications of Artificial Intelligence (IAAI-18), and the 8th AAAI Symposium on Educational Advances in Artificial Intelligence (EAAI-18), New Orleans, Louisiana, USA, February 2–7, 2018', AAAI Press, Palo Alto, CA, pp. 3207–3214.

Henningson, D. (2010), 'Description of complex flow behaviour using global and dynamic modes', *Journal of Fluid Mechanics* **656**, 1–4.

Herbert, T. (1988), 'Secondary instability of boundary layers', *Annual Review of Fluid Mechanics* **20**, 487–526.

Hernán, M. A. and Jiménez, J. (1982), 'Computer analysis of a high-speed film of the plane turbulent mixing layer', *Journal of Fluid Mechanics* **119**, 323–345.

Hervé, A., Sipp, D., Schmid, P. J. and Samuelides, M. (2012), 'A physics-based approach to flow control using system identification', *Journal of Fluid Mechanics* **702**, 26–58.

Hill, D. (1995), 'Adjoint systems and their role in the receptivity problem for boundary layers', *Journal of Fluid Mechanics* **292**, 183–204.

Hinton, G. E. (2006), 'Reducing the dimensionality of data with neural networks', *Science* **313**(5786), 504–507.

Hinton, G. E. and Sejnowski, T. J. (1986), 'Learning and releaming in Boltzmann machines', *Parallel Distributed Processing: Explorations in the Microstructure of Cognition* **1**(282–317), 2.

Hinze, M. and Kunisch, K. (2000), 'Three control methods for time-dependent fluid flow', *Flow, Turbulence and Combustion* **65**, 273–298.

Ho, B. L. and Kalman, R. E. (1965), Effective construction of linear state-variable models from input/output data, in 'Proceedings of the 3rd Annual Allerton Conference on Circuit and System Theory', October 20–22, 1965, Monticello, IL, pp. 449–459.

Ho, J. and Ermon, S. (2016), Generative adversarial imitation learning, in D. D. Lee, M. Sugiyama, U. V. Luxburg, I. Guyon, and R. Garnett, eds., 'Advances in Neural Information Processing Systems 29', December 5–10, 2016, Barcelona, Spain, Curran Associates, Red Hook, NY, pp. 4565–4573.

Hochbruck, M. and Ostermann, A. (2010), 'Exponential integrators', *Acta Numerica* **19**, 209–286.

Hochreiter, S. and Schmidhuber, J. (1997), 'Long short-term memory', *Neural Computation* **9**(8), 1735–1780.

Hof, B., Van Doorne, C. W., Westerweel, J., Nieuwstadt, F. T., Faisst, H., Eckhardt, B., Wedin, H., Kerswell, R. R., and Waleffe, F. (2004), 'Experimental observation of nonlinear traveling waves in turbulent pipe flow', *Science* **305**(5690), 1594–1598.

Hoffmann, M., Fröhner, C., and Noé, F. (2019), 'Reactive SINDy: Discovering governing reactions from concentration data', *Journal of Chemical Physics* **150**, 025101.

Holland, J. H. (1975), *Adaptation in Natural and Artificial Systems: An Introductory Analysis with Applications to Biology, Control, and Artificial Intelligence*, University of Michigan Press, Ann Arbor.

Holmes, P., Lumley, J. L., Berkooz, G., and Rowley, C. W. (2012), *Turbulence, Coherent Structures, Dynamical Systems and Symmetry*, 2nd paperback ed., Cambridge University Press, Cambridge.

Hopf, E. (1948), 'A mathematical example displaying features of turbulence', *Communications on Pure and Applied Mathematics* **1**(4), 303–322.

Hopfield, J. J. (1982), 'Neural networks and physical systems with emergent collective computational abilities', *Proceedings of the National Academy of Sciences* **79**(8), 2554–2558.

Horn, R. A. and Johnson, C. R. (2012), *Matrix Analysis*, Cambridge University Press, Cambridge.

Hornik, K., Stinchcombe, M., and White, H. (1989), 'Multilayer feedforward networks are universal approximators', *Neural Networks* **2**(5), 359–366.

Hosseini, Z., Martinuzzi, R. J., and Noack, B. R. (2015), 'Sensor-based estimation of the velocity in the wake of a low-aspect-ratio pyramid', *Experiments in Fluids* **56**(1), 13.

Hou, W., Darakananda, D., and Eldredge, J. (2019), Machine learning based detection of flow disturbances using surface pressure measurements, in 'AIAA Scitech 2019 Forum', January 7–11, 2019, San Diego, CA, p. 1148.

Hoyas, S. and Jiménez, J. (2006), 'Scaling of the velocity fluctuations in turbulent channels up to $Re_\tau = 2003$', *Physics of Fluids* **18**, 011702.

Hsu, H. (2013), *Schaum's Outline of Signals and Systems*, 3rd ed. (Schaum's Outlines), McGraw-Hill Education, New York.

Huang, Z., Du, W., and Chen, B. (2005), Deriving private information from randomized data, in 'Proceedings of the 2005 ACM SIGMOD International Conference on Management of Data', June 14–16, 2005, Baltimore, MD, pp. 37–48.

Hwangbo, J., Lee, J., Dosovitskiy, A., Bellicoso, D., Tsounis, V., Koltun, V., and Hutter, M. (2019), 'Learning agile and dynamic motor skills for legged robots', *arXiv preprint arXiv:1901.08652*.

Illingworth, S. J. (2016), 'Model-based control of vortex shedding at low Reynolds numbers', *Theoretical and Computational Fluid Dynamics* **30**(5), 1–20.

Illingworth, S. J., Morgans, A. S., and Rowley, C. W. (2010), 'Feedback control of flow resonances using balanced reduced-order models', *Journal of Sound and Vibration* **330**(8), 1567–1581.

Illingworth, S. J., Morgans, A. S., and Rowley, C. W. (2012), 'Feedback control of cavity flow oscillations using simple linear models', *Journal of Fluid Mechanics* **709**, 223–248.

Illingworth, S. J., Naito, H., and Fukagata, K. (2014), 'Active control of vortex shedding: An explanation of the gain window', *Physical Review E* **90**(4), 043014.

Ingle, V. K. and Proakis, J. G. (2011), *Digital Signal Processing Using MATLAB*, Cengage Learning, Boston.

Isaac, B. J., Coussement, A., Gicquel, O., Smith, P. J., and Parente, A. (2014), 'Reduced-order PCA models for chemical reacting flows', *Combustion and Flame* **161**(11), 2785–2800.

Isaac, B. J., Thornock, J. N., Sutherland, J., Smith, P. J., and Parente, A. (2015), 'Advanced regression methods for combustion modelling using principal components', *Combustion and Flame* **162**(6), 2592–2601.

Jackson, C. P. (1987), 'A finite-element study of the onset of vortex shedding in flow past variously shaped bodies', *Journal of Fluid Mechanics* **182**, 23–45.

Jaderberg, M., Mnih, V., Czarnecki, W. M., Schaul, T., Leibo, J. Z., Silver, D., and Kavukcuoglu, K. (2017), Reinforcement learning with unsupervised auxiliary tasks, in '5th International Conference on Learning Representations, ICLR 2017, Toulon, France, April 24–26, 2017, Conference Track Proceedings', OpenReview.net.

James, G., Witten, D., Hastie, T., and Tibshirani, R. (2013), *An Introduction to Statistical Learning*, Springer, New York.

Jho, H. (2018), 'Beautiful physics: Re-vision of aesthetic features of science through the literature review', *The Journal of the Korean Physical Society* **73**, 401–413.

Jiménez, J. (2002), Turbulence in 'Perspectives in Fluid Dynamics', Cambridge University Press, Cambridge.

Jiménez, J. (2013), 'How linear is wall-bounded turbulence?', *Physics of Fluids* **25**, 110814.

Jiménez, J. (2018a), 'Coherent structures in wall-bounded turbulence', *Journal of Fluid Mechanics* **842**, P1.

Jiménez, J. (2018b), 'Machine-aided turbulence theory', *Journal of Fluid Mechanics* **854**, R1.

Jiménez, J. (2020a), 'Computers and turbulence', *The European Journal of Mechanics - B/Fluids* **79**, 1–11.

Jiménez, J. (2020b), 'Monte Carlo science', *Journal of Turbulence* **21**(9–10), 544–566.

Jiménez, J., Kawahara, G., Simens, M. P., Nagata, M., and Shiba, M. (2005), 'Characterization of near-wall turbulence in terms of equilibrium and "bursting" solutions', *Physics of Fluids* **17**, 015105.

Jiménez, J. and Moin, P. (1991), 'The minimal flow unit in near-wall turbulence', *Journal of Fluid Mechanics* **225**, 213–240.

Jiménez, J. and Pinelli, A. (1999), 'The autonomous cycle of near-wall turbulence', *Journal of Fluid Mechanics* **389**, 335–359.

Jiménez, J. and Simens, M. P. (2001), 'Low dimensional dynamics of a turbulent wall flow', *Journal of Fluid Mechanics* **435**, 81–91.

Jiménez, J., Wray, A. A., Saffman, P. G., and Rogallo, R. S. (1993), 'The structure of intense vorticity in isotropic turbulence', *Journal of Fluid Mechanics* **255**, 65–90.

Johannink, T., Bahl, S., Nair, A., Luo, J., Kumar, A., Loskyll, M., Ojea, J. A., Solowjow, E., and Levine, S. (2019), Residual reinforcement learning for robot control, in '2019 International Conference on Robotics and Automation (ICRA)', May 20–24, 2019, Montreal, Canada, IEEE, pp. 6023–6029.

Jolliffe, I. (2002), *Principal Component Analysis*, Springer-Verlag, New York.

Jordan, D. and Smith, P. (1988), *Nonlinear Ordinary Differential Equations*, Clarendon Press, Oxford.

Joseph, D. (1976), *Stability of Fluid Motions I*, Springer Verlag, New York.

Joseph, D. and Carmi, S. (1969), 'Stability of Poiseuille flow in pipes, annuli and channels', *Quarterly of Applied Mathematics* **26**, 575–599.

Jovanović, M. R., Schmid, P. J., and Nichols, J. W. (2014), 'Sparsity-promoting dynamic mode decomposition', *Physics of Fluids* **26**(2), 024103.

Juang, J. N. (1994), *Applied System Identification*, Prentice Hall PTR, Upper Saddle River.

Juang, J. N. and Pappa, R. S. (1985), 'An eigensystem realization algorithm for modal parameter identification and model reduction', *Journal of Guidance, Control, and Dynamics* **8**(5), 620–627.

Juang, J. N., Phan, M., Horta, L. G., and Longman, R. W. (1991), Identification of observer/Kalman filter Markov parameters: Theory and experiments, Technical Memorandum 104069, NASA.

Kaiser, E., Kutz, J. N., and Brunton, S. L. (2018), 'Sparse identification of nonlinear dynamics for model predictive control in the low-data limit', *Proceedings of the Royal Society of London A* **474**(2219), 20180335.

Kaiser, E., Li, R., and Noack, B. R. (2017), On the control landscape topology, in 'The 20th World Congress of the International Federation of Automatic Control (IFAC)', Toulouse, France, pp. 1–4.

Kaiser, E., Noack, B. R., Cordier, L., Spohn, A., Segond, M., Abel, M., Daviller, G., Osth, J., Krajnovic, S., and Niven, R. K. (2014), 'Cluster-based reduced-order modelling of a mixing layer', *Journal of Fluid Mechanics* **754**, 365–414.

Kaiser, E., Noack, B. R., Spohn, A., Cattafesta, L. N., and Morzyński, M. (2017a), 'Cluster-based control of a separating flow over a smoothly contoured ramp', *Theoretical and Computational Fluid Dynamics* **31**(5–6), 579–593.

Kaiser, E., Noack, B. R., Spohn, A., Cattafesta, L. N., and Morzyński, M. (2017b), 'Cluster-based control of nonlinear dynamics', *Theoretical and Computational Fluid Dynamics* **31**(5–6), 1579–593.

Kaiser, G. (2010), *A Friendly Guide to Wavelets*, Springer Science & Business Media, Boston.

Kaiser, H. F. (1958), 'The varimax criterion for analytic rotation in factor analysis', *Psychometrika* **23**(3), 187–200.

Kambhatla, N. and Leen, T. K. (1997), 'Dimension reduction by local principal component analysis', *Neural Computation* **9**(7), 1493–1516.

Kaptanoglu, A. A., Morgan, K. D., Hansen, C. J., and Brunton, S. L. (2020), 'Physics-constrained, low-dimensional models for MHD: First-principles and data-driven approaches', *arXiv preprint arXiv:2004.10389* .

Kawahara, G. (2005), 'Laminarization of minimal plane Couette flow: Going beyond the basin of attraction of turbulence', *Physics of Fluids* **17**, 041702.

Kawahara, G., Uhlmann, M., and van Veen, L. (2012), 'The significance of simple invariant solutions in turbulent flows', *Annual Review of Fluid Mechanics* **44**, 203–225.

Kee, R. J., Coltrin, M. E., and Glarborg, P. (2005), *Chemically Reacting Flow: Theory and Practice*, John Wiley & Sons, Hoboken, NJ.

Keefe, L., Moin, P., and Kim, J. (1992), 'The dimension of attractors underlying periodic turbulent Poiseuille flow', *Journal of Fluid Mechanics* **242**, 1–29.

Kenney, J. and Keeping, E. (1951), *Mathematics of Statistics*, Vol. II, D. Van Nostrand Co., New York.

Kerhervé, F., Roux, S., and Mathis, R. (2017), 'Combining time-resolved multi-point and spatially-resolved measurements for the recovering of very-large-scale motions in high Reynolds number turbulent boundary layer', *Experimental Thermal and Fluid Science* **82**, 102–115.

Kerstein, A. R. (1999), 'One-dimensional turbulence: Model formulation and application to homogeneous turbulence, shear flows, and buoyant stratified flows', *Journal of Fluid Mechanics* **392**, 277–334.

Kerstens, W., Pfeiffer, J., Williams, D., King, R., and Colonius, T. (2011), 'Closed-loop control of lift for longitudinal gust suppression at low Reynolds numbers', *AIAA Journal* **49**(8), 1721–1728.

Khalil (2002), *Nonlinear Systems*, 3rd ed., Dover, I, New York.

Khalil, H. K. and Grizzle, J. W. (2002), *Nonlinear Systems*, Vol. 3, Prentice Hall, Upper Saddle River, NJ.

Kim, H. J., Jordan, M. I., Sastry, S., and Ng, A. Y. (2004), Autonomous helicopter flight via reinforcement learning, in 'Advances in Neural Information Processing Systems', December 8–13, 2003, Vancouver and Whistler, British Columbia, Canada, pp. 799–806.

Kim, H. T., Kline, S. J., and Reynolds, W. C. (1971), 'The production of turbulence near a smooth wall in a turbulent boundary layer', *Journal of Fluid Mechanics* **50**, 133–160.

Kim, J. (2011), 'Physics and control of wall turbulence for drag reduction', *Philosophical Transactions of the Royal Society A: Mathematical, Physical and Engineering Sciences* **369**(1940), 1396–1411.

Kim, J. and Bewley, T. (2007), 'A linear systems approach to flow control', *Annual Review of Fluid Mechanics* **39**, 383–417.

King, P., Rowland, J., Aubrey, W., Liakata, M., Markham, M., Soldatova, L. N., Whelan, K. E., Clare, A., Young, M., Sparkes, A., Oliver, S. G., and Pir, P. (2009), 'The robot scientist Adam', *Computer* **42**(7), 46–54.

Kingma, D. P. and Ba, J. (2014), 'Adam: A method for stochastic optimization', *arXiv preprint arXiv:1412.6980*.

Kline, S. J., Reynolds, W. C., Schraub, F. A., and Runstadler, P. W. (1967), 'Structure of turbulent boundary layers', *Journal of Fluid Mechanics* **30**, 741–773.

Knaak, M., Rothlubbers, C. and Orglmeister, R. (1997), A Hopfield neural network for flow field computation based on particle image velocimetry/particle tracking velocimetry image sequences, in 'International Conference on Neural Networks, 1997', Vol. 1, October 8–10, 1997, Lausanne, Switzerland, IEEE, pp. 48–52.

Knight, W. (2018), 'Google just gave control over data center cooling to an AI', `www.technologyreview.com/s/611902/google-just-gave-control-over-data-center-cooling-to-an-ai/`.

Koch, W., Bertolotti, F., Stolte, A., and Hein, S. (2000), 'Nonlinear equilibrium solutions in a three-dimensional boundary layer and their secondary instability', *Journal of Fluid Mechanics* **406**, 131–174.

Kochenderfer, M. and Wheeler, T. (2019), *Algorithms for Optimization*, MIT Press, Cambridge, MA.

Kolmogorov, A. N. (1941), 'The local structure of turbulence in incompressible viscous fluid for very large Reynolds numbers', *Doklady Akademii Nauk SSSR* **30**, 209–303.

Kot, M. (2015), *A First Course in the Calculus of Variations*, American Mathematical Society, Providence, RI.

Koza, J. R. (1992), *Genetic Programming: On the Programming of Computers by Means of Natural Selection*, The MIT Press, Boston.

Kraichnan, R. H. (1967), 'Inertial ranges in two-dimensional turbulence', *Physics of Fluids* **10**, 1417–1423.

Kraichnan, R. H. (1971), 'Inertial range transfer in two- and three-dimensional turbulence', *Journal of Fluid Mechanics* **47**, 525–535.

Kraichnan, R. H. and Montgomery, D. (1980), 'Two-dimensional turbulence', *Reports on Progress in Physics* **43**, 547–619.

Kramer, B., Grover, P., Boufounos, P., Benosman, M., and Nabi, S. (2015), 'Sparse sensing and DMD based identification of flow regimes and bifurcations in complex flows', *arXiv preprint arXiv:1510.02831*.

Krizhevsky, A., Sutskever, I., and Hinton, G. E. (2012), Imagenet classification with deep convolutional neural networks, in 'Advances in Neural Information Processing Systems', December 3–6, 2012, Lake Tahoe, NV, pp. 1097–1105.

Kroll, N., Abu-Zurayk, M., Dimitrov, D., Franz, T., Führer, T., Gerhold, T., Görtz, S., Heinrich, R., Ilic, C., Jepsen, J., Jägersküpper, J., Kruse, M., Krumbein, A., Langer, S., Liu, D., Liepelt, R., Reimer, L., Ritter, M., Schwöppe, A., Scherer, J., Spiering, F., Thormann, R., Togiti, V., Vollmer, D., and Wendisch, J.-H. (2015), 'DLR project digital-x: Towards virtual aircraft design and flight testing based on high-fidelity methods', *CEAS Aeronautical Journal* **7**(1), 3–27.

Kroll, N., Langer, S., and Schwöppe, A. (2014), The DLR flow solver TAU – Status and recent algorithmic developments, in '52nd Aerospace Sciences Meeting', January 3–17, 2014, National Harbor, MD, American Institute of Aeronautics and Astronautics, Reston, VA.

Kuhn, T. S. (1970), *The Structure of Scientific Revolutions*, 2nd ed., Chicago University Press, Chicago.

Kuhnle, A., Schaarschmidt, M., and Fricke, K. (2017), 'Tensorforce: a TensorFlow library for applied reinforcement learning', https://tensorforce.readthedocs.io/en/latest/#.

Kuipers, L. and Niederreiter, H. (2005), *Uniform Distribution of Sequences*, Dover, Mineola, NY, p. 129.

Kuo, K. and Acharya, R. (2012), *Fundamentals of Turbulent and Multi-Phase Combustion*, Wiley, New York.

Kutz, J., Brunton, S., Brunton, B., and Proctor, J. (2016), *Dynamic Mode Decomposition: Data-Driven Modeling of Complex Systems*, Society for Industrial and Applied Mathematics, Philadelphia.

Kutz, J. N. (2017), 'Deep learning in fluid dynamics', *Journal of Fluid Mechanics* **814**, 1–4.

Kutz, J. N., Fu, X., and Brunton, S. L. (2016), 'Multiresolution dynamic mode decomposition', *SIAM Journal on Applied Dynamical Systems* **15**(2), 713–735.

Labonté, G. (1999), 'A new neural network for particle-tracking velocimetry', *Experiments in Fluids* **26**(4), 340–346.

Lagaris, I. E., Likas, A., and Fotiadis, D. I. (1998), 'Artificial neural networks for solving ordinary and partial differential equations', *IEEE Transactions on Neural Networks* **9**(5), 987–1000.

Lai, Z. and Nagarajaiah, S. (2019), 'Sparse structural system identification method for nonlinear dynamic systems with hysteresis/inelastic behavior', *Mechanical Systems and Signal Processing* **117**, 813–842.

Lall, S., Marsden, J. E., and Glavaški, S. (1999), 'Empirical model reduction of controlled nonlinear systems,' *International Federation of Automatic Control* **32**(2), 2598–2603.

Lall, S., Marsden, J. E., and Glavaški, S. (2002), 'A subspace approach to balanced truncation for model reduction of nonlinear control systems', *International Journal of Robust and Nonlinear Control* **12**(6), 519–535.

Lan, Y. and Mezić, I. (2013), 'Linearization in the large of nonlinear systems and Koopman operator spectrum', *Physica D: Nonlinear Phenomena* **242**(1), 42–53.

Landau, L. D. (1944), 'On the problem of turbulence,' *C.R. Acad. Sci. USSR* **44**, 311–314.

Landau, L. D. and Lifshitz, E. M. (1987), *Fluid Mechanics*, Vol. 6 in Course of Theoretical Physics, 2nd English ed., Pergamon Press, Oxford.

Lasota, A. and Mackey, M. (1994), *Chaos, Fractals, and Noise: Stochastic Aspects of Dynamics*, Springer Verlag, New York.

Law, C. K. (2010), *Combustion Physics*, Cambridge University Press, Cambridge.

Le Clainche, S. and Vega, J. M. (2017), 'Higher order dynamic mode decomposition to identify and extrapolate flow patterns', *Physics of Fluids* **29**(8), 084102.

Leclercq, C., Demourant, F., Poussot-Vassal, C., and Sipp, D. (2019), 'Linear iterative method for closed-loop control of quasiperiodic flows', *Journal of Fluid Mechanics* **868**, 26–65.

LeCun, Y., Bengio, Y., and Hinton, G. (2015), 'Deep learning', *Nature* **521**(7553), 436–444.

Lee, C., Kim, J., Babcock, D., and Goodman, R. (1997), 'Application of neural networks to turbulence control for drag reduction', *Physics of Fluids* **9**(6), 1740–1747.

Lee, J.-H. and Sung, H. J. (2013), 'Comparison of very-large-scale motions of turbulent pipe and boundary layer simulations', *Physics of Fluids* **25**, 045103.

Lee, Y., Yang, H., and Yin, Z. (2017), 'PIV-DCNN: Cascaded deep convolutional neural networks for particle image velocimetry', *Experiments in Fluids* **58**(12), 171.

Leonard, A. (1980), 'Vortex methods for flow simulation', *Journal of Computational Physics* **37**(3), 289–335.

Li, H., Fernex, D., Semaan, R., Tan, J., Morzyński, M., and Noack, B. R. (2021), 'Cluster-based network model', *Journal of Fluid Mechanics* **906**, A21:1–41.

Li, Q., Dietrich, F., Bollt, E. M., and Kevrekidis, I. G. (2017), 'Extended dynamic mode decomposition with dictionary learning: A data-driven adaptive spectral decomposition of the Koopman operator', *Chaos: An Interdisciplinary Journal of Nonlinear Science* **27**(10), 103111.

Li, R., Noack, B. R., Cordier, L., Borée, J., Kaiser, E., and Harambat, F. (2018), 'Linear genetic programming control for strongly nonlinear dynamics with frequency crosstalk', *Archives of Mechanics* **70**(6), 505–534.

Li, Y., Perlman, E., Wan, M., Yang, Y., Meneveau, C., Burns, R., Chen, S., Szalay, A., and Eyink, G. (2008), 'A public turbulence database cluster and applications to study Lagrangian evolution of velocity increments in turbulence', *Journal of Turbulence* **9**, N31.

Liberty, E., Woolfe, F., Martinsson, P.-G., Rokhlin, V., and Tygert, M. (2007), 'Randomized algorithms for the low-rank approximation of matrices', *Proceedings of the National Academy of Sciences* **104**(51), 20167–20172.

Lin, C.-J. (2007), 'Projected gradient methods for nonnegative matrix factorization', *Neural Computation* **19**(10), 2756–2779.

Ling, J., Jones, R., and Templeton, J. (2016), 'Machine learning strategies for systems with invariance properties', *Journal of Computational Physics* **318**, 22–35.

Ling, J., Kurzawski, A., and Templeton, J. (2016), 'Reynolds averaged turbulence modelling using deep neural networks with embedded invariance', *Journal of Fluid Mechanics* **807**, 155–166.

Ling, J. and Templeton, J. (2015), 'Evaluation of machine learning algorithms for prediction of regions of high Reynolds averaged Navier Stokes uncertainty', *Physics of Fluids* **27**(8), 085103.

Liu, J. T. C. (1989), 'Coherent structures in transitional and turbulent free shear flows', *Annual Review of Fluid Mechanics* **21**(1), 285–315.

Ljung, L. (1999), *System Identification: Theory for the User*, Prentice Hall, Upper Saddle River, NJ.

Ljung, L. (2008), 'Perspectives on system identification', *IFAC Proceedings Volumes* **41**(2), 7172–7184.

Ljung, L. and Glad, T. (1994), *Modeling of Dynamic Systems*, Prentice Hall, Englewood Cliffs, NJ.

Loan, C. V. (1992), *Computational Frameworks for the Fast Fourier Transform*, Society for Industrial and Applied Mathematics (SIAM), Philadelphia.

Loiseau, J.-C. (2019), 'Data-driven modeling of the chaotic thermal convection in an annular thermosyphon', *arXiv preprint arXiv:1911.07920*.

Loiseau, J.-C. and Brunton, S. L. (2018), 'Constrained sparse Galerkin regression', *Journal of Fluid Mechanics* **838**, 42–67.

Loiseau, J.-C., Brunton, S. L., and Noack, B. R. (2021), From the POD-Galerkin method to sparse manifold models, in P. Benner, S. Grivet-Talocaia, A. Quarteroni, R. G., W. Schilders, and L. M. Silveria, eds., *Handbook of Model-Order Reduction. Volume 3: Applications*, De Gruyter, Berlin, pp. 279–320.

Loiseau, J.-C., Noack, B. R., and Brunton, S. L. (2018), 'Sparse reduced-order modeling: Sensor-based dynamics to full-state estimation', *Journal of Fluid Mechanics* **844**, 459–490.

Lorenz, E. (1956), Empirical orthogonal functions and statistical weather prediction, Statistical forecasting project, Scientific Report No. 1, MIT, Department of Meteorology, Cambridge, MA.

Lorenz, E. N. (1963), 'Deterministic nonperiodic flow', *Journal of the Atmospheric Sciences* **20**(2), 130–141.

Lozano-Durán, A., Flores, O., and Jiménez, J. (2012), 'The three-dimensional structure of momentum transfer in turbulent channels', *Journal of Fluid Mechanics* **694**, 100–130.

Lozano-Durán, A. and Jiménez, J. (2014), 'Time-resolved evolution of coherent structures in turbulent channels: Characterization of eddies and cascades', *Journal of Fluid Mechanics* **759**, 432–471.

Lu, L., Meng, X., Mao, Z., and Karniadakis, G. E. (2019), 'Deepxde: A deep learning library for solving differential equations', *arXiv preprint arXiv:1907.04502*.

Lu, S. S. and Willmarth, W. W. (1973), 'Measurements of the structure of the Reynolds stress in a turbulent boundary layer', *Journal of Fluid Mechanics* **60**, 481–511.

Lu, T. and Law, C. K. (2009), 'Toward accommodating realistic fuel chemistry in large-scale computations', *Progress in Energy and Combustion Science* **35**(2), 192–215.

Luchtenburg, D. M., Günter, B., Noack, B. R., King, R., and Tadmor, G. (2009), 'A generalized mean-field model of the natural and actuated flows around a high-lift configuration', *Journal of Fluid Mechanics* **623**, 283–316.

Luchtenburg, D. M. and Rowley, C. W. (2011), 'Model reduction using snapshot-based realizations', *Bulletin of the American Physical Society* **56**(18), 37pp.

Luchtenburg, D. M., Schlegel, M., Noack, B. R., Aleksić, K., King, R., Tadmor, G., and Günther, B. (2010), Turbulence control based on reduced-order models and nonlinear control design, in R. King, ed., Active Flow Control II, Vol. 108 of Notes on Numerical Fluid Mechanics and Multidisciplinary Design, May 26–28, 2010, Springer-Verlag, Berlin, pp. 341–356.

Luhar, M., Sharma, A. S., and McKeon, B. J. (2014), 'Opposition control within the resolvent analysis framework', *Journal of Fluid Mechanics* **749**, 597–626.

Lumley, J. (1970), *Stochastic Tools in Turbulence*, Academic Press, New York.

Lumley, J. L. (1967), The structure of inhomogeneous turbulent flows, in A. M. Yaglam and V. I. Tatarsky, eds., 'Proceedings of the International Colloquium on the Fine Scale Structure of the Atmosphere and Its Influence on Radio Wave Propagation', Doklady Akademii Nauk SSSR, Moscow, Nauka.

Lumley, J. L. and Poje, A. (1997), 'Low-dimensional models for flows with density fluctuations', *Physics of Fluids* **9**(7), 2023–2031.

Lusch, B., Kutz, J. N., and Brunton, S. L. (2018), 'Deep learning for universal linear embeddings of nonlinear dynamics', *Nature Communications* **9**(1), 4950.

Lyapunov, A. (1892), *The General Problem of the Stability of Motion*, The Kharkov Mathematical Society, Kharkov, 251pp.

Ma, Z., Ahuja, S., and Rowley, C. W. (2011), 'Reduced order models for control of fluids using the eigensystem realization algorithm', *Theoretical and Computational Fluid Dynamics* **25**(1), 233–247.

Mackie, J. (1974), *The Cement of the Universe: A Study of Causation*, Oxford University Press, New York.

MacQueen, J. (1967), Some methods for classification and analysis of multivariate observations, in 'Proceedings of the Fifth Berkeley Symposium on Mathematical Statistics and Probability', Vol. 1, Oakland, CA, pp. 281–297.

Magill, J., Bachmann, M., and Rixon, G. (2003), 'Dynamic stall control using a model-based observer.', *Journal of Aircraft* **40**(2), 355–362.

Magnussen, B. F. (1981), On the structure of turbulence and a generalized eddy dissipation concept for chemical reaction in turbulent flow, in '19th AIAA Aerospace Science Meeting', May 21–24, 2007, Williamsburg, VA.

Magri, L. (2019), 'Adjoint methods as design tools in thermoacoustics', *Applied Mechanics Reviews* **71(2)**, 020801.

Mahoney, M. W. (2011), 'Randomized algorithms for matrices and data', *Foundations and Trends in Machine Learning* **3**, 123–224.

Majda, A. J. and Harlim, J. (2012), 'Physics constrained nonlinear regression models for time series', *Nonlinearity* **26**(1), 201.

Malik, M. R., Isaac, B. J., Coussement, A., Smith, P. J., and Parente, A. (2018), 'Principal component analysis coupled with nonlinear regression for chemistry reduction', *Combustion and Flame* **187**, 30–41.

Malik, M. R., Obando Vega, P., Coussement, A., and Parente, A. (2020), 'Combustion modeling using Principal Component Analysis: A posteriori validation on Sandia flames D, E and F', *Proceedings of the Combustion Institute* **38**(2), 2635–2643.

Mallat, S. (2009), *A Wavelet Tour of Signal Processing*, Elsevier, Oxford.

Mallat, S. G. (1989), 'Multiresolution approximations and wavelet orthonormal bases of $\mathcal{L}^2(\mathbb{R})$', *Transactions of the American Mathematical Society* **315**(1), 69–87.

Mallor, F., Raiola, M., Vila, C. S., Örlü, R., Discetti, S., and Ianiro, A. (2019), 'Modal decomposition of flow fields and convective heat transfer maps: An application to wall-proximity square ribs', *Experimental Thermal and Fluid Science* **102**, 517–527.

Maltrud, M. E. and Vallis, G. K. (1991), 'Energy spectra and coherent structures in forced two-dimensional and beta-plane turbulence', *Journal of Fluid Mechanics* **228**, 321–342.

Mangan, N. M., Brunton, S. L., Proctor, J. L., and Kutz, J. N. (2016), 'Inferring biological networks by sparse identification of nonlinear dynamics', *IEEE Transactions on Molecular, Biological, and Multi-Scale Communications* **2**(1), 52–63.

Mangan, N. M., Kutz, J. N., Brunton, S. L., and Proctor, J. L. (2017), 'Model selection for dynamical systems via sparse regression and information criteria', *Proceedings of the Royal Society A* **473**(2204), 1–16.

Manohar, K., Brunton, B. W., Kutz, J. N., and Brunton, S. L. (2018), 'Data-driven sparse sensor placement: Demonstrating the benefits of exploiting known patterns', *IEEE Control Systems Magazine* **38**(3), 63–86.

Marčenko, V. A. and Pastur, L. A. (1967), 'Distribution of eigenvalues for some sets of random matrices', *Mathematics of the USSR-Sbornik* **1**(4), 457.

Mardt, A., Pasquali, L., Wu, H., and Noé, F. (2018), 'VAMPnets: Deep learning of molecular kinetics', *Nature Communications* **9**, 5.

Marsden, J. E. and Ratiu, T. S. (1999), *Introduction to Mechanics and Symmetry*, 2nd ed., Springer-Verlag, New York.

Maulik, R., San, O., Rasheed, A., and Vedula, P. (2019), 'Subgrid modelling for two-dimensional turbulence using neural networks', *Journal of Fluid Mechanics* **858**, 122–144.

Maurel, S., Borée, J., and Lumley, J. (2001), 'Extended proper orthogonal decomposition: Application to jet/vortex interaction', *Flow, Turbulence and Combustion* **67**(2), 125–136.

McKeon, B. and Sharma, A. (2010), 'A critical-layer framework for turbulent pipe flow', *Journal of Fluid Mechanics* **658**, 336–382.

McWilliams, J. C. (1980), 'An application of equivalent modons to atmospheric blocking', *Dynamics of Atmospheres and Oceans* **5**, 43–66.

McWilliams, J. C. (1984), 'The emergence of isolated coherent vortices in turbulent flow', *Journal of Fluid Mechanics* **146**, 21–43.

McWilliams, J. C. (1990a), 'A demonstration of the suppression of turbulent cascades by coherent vortices in two-dimensional turbulence', *Physics of Fluids A* **2**, 547–552.

McWilliams, J. C. (1990b), 'The vortices of two-dimensional turbulence', *Journal of Fluid Mechanics* **219**, 361–385.

Meena, M. G., Nair, A. G., and Taira, K. (2018), 'Network community-based model reduction for vortical flows', *Physical Review E* **97**(6), 063103.

Mendez, M. A., Hess, D., Watz, B. B., and Buchlin, J.-M. (2020), 'Multiscale proper orthogonal decomposition (MPOD) of TR-PIV data- A case study on stationary and transient cylinder wake flows', *Measurement Science and Technology* **80**, 981–1002.

Mendez, M. A. and Buchlin, J.-M. (2016), 'Notes on 2D pulsatile Poiseuille flows: An introduction to eigenfunctions and complex variables using MATLAB', Technical Report TN 215, von Karman Institute for Fluid Dynamics, Sint-Genesius-Rode, Belgium.

Mendez, M. A., Scelzo, M., and Buchlin, J.-M. (2018), 'Multiscale modal analysis of an oscillating impinging gas jet', *Experimental Thermal and Fluid Science* **91**, 256–276.

Mendez, M., Balabane, M., and Buchlin, J.-M. (2019), 'Multi-scale proper orthogonal decomposition of complex fluid flows', *Journal of Fluid Mechanics* **870**, 988–1036.

Mendez, M., Gosset, A., and Buchlin, J.-M. (2019), 'Experimental analysis of the stability of the jet wiping process, part II: Multiscale modal analysis of the gas jet-liquid film interaction', *Experimental Thermal and Fluid Science* **106**, 48–67.

Mendez, M., Raiola, M., Masullo, A., Discetti, S., Ianiro, A., Theunissen, R., and Buchlin, J.-M. (2017), 'POD-based background removal for particle image velocimetry', *Experimental Thermal and Fluid Science* **80**, 181–192.

Mendible, A., Brunton, S. L., Aravkin, A. Y., Lowrie, W., and Kutz, J. N. (2020), 'Dimensionality reduction and reduced-order modeling for traveling wave physics', *Theoretical and Computational Fluid Dynamics* **34**(4), 385–400.

Meneveau, C. (1991), 'Analysis of turbulence in the orthonormal wavelet representation', *Journal of Fluid Mechanics* **232**, 469–520.

Meneveau, C. and Katz, J. (2000), 'Scale-invariance and turbulence models for large-eddy simulation', *Annual Review of Fluid Mechanics* **32**(1), 1–32.

Meunier, P., Le Dizès, S., and Leweke, T. (2005), 'Physics of vortex merging', *C. R. Physique* **6**, 431–450.

Meyers, S. D., Kelly, B. G., and O'Brien, J. J. (1993), 'An introduction to wavelet analysis in oceanography and meteorology: With application to the dispersion of Yanai waves', *Monthly Weather Review* **121**(10), 2858–2866.

Mezić, I. (2005), 'Spectral properties of dynamical systems, model reduction and decompositions', *Nonlinear Dynamics* **41**(1–3), 309–325.

Mezić, I. (2013), 'Analysis of fluid flows via spectral properties of the Koopman operator', *Annual Review of Fluid Mechanics* **45**, 357–378.

Mezić, I. and Banaszuk, A. (2004), 'Comparison of systems with complex behavior', *Physica D: Nonlinear Phenomena* **197**(1), 101–133.

Mifsud, M., Vendl, A., Hansen, L.-U., and Görtz, S. (2019), 'Fusing wind-tunnel measurements and CFD data using constrained gappy proper orthogonal decomposition', *Aerospace Science and Technology* **86**, 312–326.

Mifsud, M., Zimmermann, R., and Görtz, S. (2014), 'Speeding-up the computation of high-lift aerodynamics using a residual-based reduced-order model', *CEAS Aeronautical Journal* **6**(1), 3–16.

Milano, M. and Koumoutsakos, P. (2002), 'Neural network modeling for near wall turbulent flow', *Journal of Computational Physics* **182**(1), 1–26.

Min, C. and Choi, H. (1999), 'Suboptimal feedback control of vortex shedding at low reynolds numbers', *Journal of Fluid Mechanics* **401**, 123–156.

Minelli, G., Dong, T., Noack, B. R., and Krajnović, S. (2020), 'Upstream actuation for bluff-body wake control driven by a genetically inspired optimization', *Journal of Fluid Mechanics* **893**, A1.

Mizuno, Y. and Jiménez, J. (2013), 'Wall turbulence without walls', *Journal of Fluid Mechanics* **723**, 429–455.

Mnih, V., Badia, A. P., Mirza, M., Graves, A., Lillicrap, T., Harley, T., Silver, D., and Kavukcuoglu, K. (2016), Asynchronous methods for deep reinforcement learning, in M. F. Balcan and K. Q. Weinberger, eds., 'Proceedings of the 33rd International Conference on Machine Learning', Vol. 48 of Proceedings of Machine Learning Research, PMLR, New York, pp. 1928–1937.

Mnih, V., Kavukcuoglu, K., Silver, D., Rusu, A. A., Veness, J., Bellemare, M. G., Graves, A., Riedmiller, M., Fidjeland, A. K., Ostrovski, G., Petersen, S., Beattie, C., Sadik, A., Antonoglou, I., King, H., Kumaran, D., Wierstra, D., Legg, S., and Hassabis, D. (2015), 'Human-level control through deep reinforcement learning', *Nature* **518**(7540), 529–533.

Moarref, R., Jovanovic, M., Tropp, J., Sharma, A., and McKeon, B. (2014), 'A low-order decomposition of turbulent channel flow via resolvent analysis and convex optimization', *Physics Fluids* **26**, 051701.

Mohammed, R., Tanoff, M., Smooke, M., Schaffer, A., and Long, M. (1998), 'Computational and experimental study of a forced, time-varying, axisymmetric, laminar diffusion flame', *Symposium (International) on Combustion* **27**(1), 693–702.

Moisy, F. and Jiménez, J. (2004), 'Geometry and clustering of intense structures in isotropic turbulence', *Journal of Fluid Mechanics* **513**, 111–133.

Mojgani, R. and Balajewicz, M. (2017), 'Lagrangian basis method for dimensionality reduction of convection dominated nonlinear flows', *arXiv preprint arXiv:1701.04343*.

Moore, B. C. (1981), 'Principal component analysis in linear systems: Controllability, observability, and model reduction', *IEEE Transactions on Automatic Control* **AC-26**(1), 17–32.

Morzyński, M., Afanasiev, K., and Thiele, F. (1999), 'Solution of the eigenvalue problems resulting from global non-parallel flow stability analysis', *Computer Methods in Applied Mechanics and Engineering* **169**, 161–176.

Morzyński, M., Noack, B. R., and Tadmor, G. (2007), 'Global stability analysis and reduced order modeling for bluff-body flow control', *Journal of Theoretical and Applied Mechanics* **45**, 621–642.

Morzyński, M., Stankiewicz, W., Noack, B. R., Thiele, F., and Tadmor, G. (2006), Generalized mean-field model for flow control using continuous mode interpolation, in '3rd AIAA Flow Control Conference', June 5–8, 2006, San Francisco, CA, Invited AIAA-Paper 2006-3488.

Mowlavi, S. and Sapsis, T. P. (2018), 'Model order reduction for stochastic dynamical systems with continuous symmetries', *SIAM Journal on Scientific Computing* **40**(3), A1669–A1695.

Murata, T., Fukami, K., and Fukagata, K. (2019), CNN-SINDY based reduced order modeling of unsteady flow fields, in 'Fluids Engineering Division Summer Meeting', July–August 2019, San Francisco, CA, Vol. 59032, American Society of Mechanical Engineers, New York, p. V002T02A074.

Murphy, K. P. (2012), *Machine Learning*, MIT Press, Cambridge.

Nagata, M. (1990), 'Three-dimensional finite-amplitude solutions in plane Couette flow: Bifurcation from infinity', *Journal of Fluid Mechanics* **217**, 519–527.

Nair, A. G., Brunton, S. L., and Taira, K. (2018), 'Networked-oscillator-based modeling and control of unsteady wake flows', *Physical Review E* **97**(6), 063107.

Nair, A. G. and Taira, K. (2015), 'Network-theoretic approach to sparsified discrete vortex dynamics', *Journal of Fluid Mechanics* **768**, 549–571.

Nair, A. G., Yeh, C.-A., Kaiser, E., Noack, B. R., Brunton, S. L., and Taira, K. (2019), 'Cluster-based feedback control of turbulent post-stall separated flows', *Journal of Fluid Mechanics* **875**, 345–375.

Narasingam, A. and Kwon, J. S.-I. (2018), 'Data-driven identification of interpretable reduced-order models using sparse regression', *Computers & Chemical Engineering* **119**, 101–111.

Neely, T. W., Samson, E. C., Bradley, A. S., Davis, M. J., and Anderson, B. P. (2010), 'Observation of vortex dipoles in an oblate Bose–Einstein condensate', *Physical Review Letters* **104**, 160401.

Newell, A. and Simon, H. A. (1976), 'Computer science as empirical inquiry: Symbols and search', *Communication of ACM* **19**(3), 113–126.

Newton, I. S. (1687), *Philosophiae naturalis principia mathematica*, Vol. 1, London.

Ng, A. Y., Harada, D., and Russell, S. J. (1999), Policy invariance under reward transformations: Theory and application to reward shaping, in 'Proceedings of the Sixteenth International Conference on Machine Learning', ICML 99, Morgan Kaufmann, San Francisco, CA, pp. 278–287.

Ng, A. Y., Jordan, M. I., and Weiss, Y. (2002), On spectral clustering: Analysis and an algorithm, in 'Advances in Neural Information Processing Systems', December 3–8, 2001, Vancouver, British Columbia, Canada, pp. 849–856.

Nielsen, A. (2019), *Practical Time Series Analysis*, O'Reilly Media, Sebastopol, CA.

Nilsson, N. (1998), *Artificial Intelligence: A New Synthesis*, Morgan Kaufmann, San Francisco, CA.

Ninni, D. and Mendez, M. A. (2021), 'Modulo: A software for multiscale proper orthogonal decomposition of data', *SoftwareX* **12**, 100622.

Noack, B. R. (2019), Closed-loop turbulence control – From human to machine learning (and retour), in Y. Zhou, M. Kimura, G. Peng, A. D. Lucey, and L. Hung, eds., 'Fluid-Structure-Sound Interactions and Control. Proceedings of the 4th Symposium on Fluid-Structure-Sound Interactions and Control', Springer, Berlin, pp. 23–32.

Noack, B. R., Afanasiev, K., Morzyński, M., Tadmor, G., and Thiele, F. (2003), 'A hierarchy of low-dimensional models for the transient and post-transient cylinder wake', *Journal of Fluid Mechanics* **497**, 335–363.

Noack, B. R. and Eckelmann, H. (1994), 'A global stability analysis of the steady and periodic cylinder wake', *Journal of Fluid Mechanics* **270**, 297–330.

Noack, B. R., Mezić, I., Tadmor, G., and Banaszuk, A. (2004), 'Optimal mixing in recirculation zones', *Physics of Fluids* **16**(4), 867–888.

Noack, B. R., Morzynski, M., and Tadmor, G. (2011), *Reduced-Order Modelling for Flow Control*, Vol. 528 of International Centre for Mechanical Sciences: Courses and Lectures, Springer Science & Business Media, Dordrecht.

Noack, B. R., Papas, P., and Monkewitz, P. A. (2005), 'The need for a pressure-term representation in empirical Galerkin models of incompressible shear flows', *Journal of Fluid Mechanics* **523**, 339.

Noack, B. R., Pelivan, I., Tadmor, G., Morzyński, M., and Comte, P. (2004), Robust low-dimensional Galerkin models of natural and actuated flows, in 'Fourth Aeroacoustics Workshop', RWTH, Aachen, February 26–27, 2004, pp. 0001–0012.

Noack, B. R., Schlegel, M., Ahlborn, B., Mutschke, G., Morzyński, M., Comte, P., and Tadmor, G. (2008), 'A finite-time thermodynamics of unsteady fluid flows', *Journal of Non-Equilibrium Thermodynamics* **33**(2), 103–148.

Noack, B. R., Stankiewicz, W., Morzynski, M., and Schmid, P. J. (2016), 'Recursive dynamic mode decomposition of a transient cylinder wake', *Journal of Fluid Mechanics* **809**, 843–872.

Nocedal, J. and Wright, S. (2006), *Numerical Optimization*, Springer Verlag, New York.

Noé, F. and Nuske, F. (2013), 'A variational approach to modeling slow processes in stochastic dynamical systems', *Multiscale Modeling Simulation* **11**(2), 635–655.

Noé, F., Olsson, S., Köhler, J., and Wu, H. (2019), 'Boltzmann generators: Sampling equilibrium states of many-body systems with deep learning', *Science* **365**(6457), eaaw1147.

Noether, E. (1918), 'Invariante Variationsprobleme', *Nachrichten von der Gesellschaft der Wissenschaften zu Göttingen, Mathematisch-Physikalische Klasse* **1918**, 235–257. English Reprint: physics/0503066, http://dx. doi. org/10.1080/00411457108231446, p. 57.

Novati, G., Mahadevan, L., and Koumoutsakos, P. (2019), 'Controlled gliding and perching through deep-reinforcement-learning', *Physical Review Fluids* **4**(9), 093902.

Novati, G., Verma, S., Alexeev, D., Rossinelli, D., Van Rees, W. M., and Koumoutsakos, P. (2017), 'Synchronisation through learning for two self-propelled swimmers', *Bioinspiration & Biomimetics* **12**(3), aa6311.

Nüske, F., Schneider, R., Vitalini, F., and Noé, F. (2016), 'Variational tensor approach for approximating the rare-event kinetics of macromolecular systems', *The Journal of Chemical Physics* **144**(5), 054105.

Obukhov, A. M. (1941), 'On the distribution of energy in the spectrum of turbulent flow', *Doklady Akademii Nauk SSSR* **32**, 22–24.

Ogata, K. (2009), *Modern Control Engineering*, 5th ed., Pearson, Upper Saddle River, NJ.

Onsager, L. (1949), 'Statistical hydrodynamics', *Nuovo Cimento Suppl.* **6**, 279–286.

Oppenheim, A. V. (2015), *Signals, Systems and Inference*, Pearson Education, Harlow.

Oppenheim, A. V. and Schafer, R. W. (2009), *Discrete-Time Signal Processing*, 3rd ed. (Prentice-Hall Signal Processing Series), Pearson, Harlow.

Oppenheim, A. V., Willsky, A. S., and with S. Hamid (1996), *Signals and Systems,* 2nd ed., Prentice-Hall, Upper Saddle River, NJ.

Orr, W. (1907a), 'The stability or instability of the steady motions of a perfect liquid and of a viscous liquid', *Proceedings of the Royal Irish Academy* **27**, 9–138.

Orr, W. M. (1907b), 'The stability or instability of the steady motions of a perfect liquid, and of a viscous liquid. Part I: A perfect liquid', *Proceedings of the Royal Irish Academy* **27**, 9–68.

Orszag, S. (1971), 'Accurate solution of the Orr-Sommerfeld stability equation', *Journal of Fluid Mechanics* **50**, 689–703.

Orszag, S. A. and Patera, A. T. (1983), 'Secondary instability of wall-bounded shear flows', *Journal of Fluid Mechanics* **128**, 347–385.

Östh, J., Noack, B. R., and Krajnović, S. (2014), 'On the need for a nonlinear subscale turbulence term in POD models as exemplified for a high Reynolds number flow over an Ahmed body', *Journal of Fluid Mechanics* **747**, 518–544.

Otto, S. E. and Rowley, C. W. (2019), 'Linearly-recurrent autoencoder networks for learning dynamics', *SIAM Journal on Applied Dynamical Systems* **18**(1), 558–593.

Ouellette, N. T., Xu, H., and Bodenschatz, E. (2006), 'A quantitative study of three-dimensional Lagrangian particle tracking algorithms', *Experiments in Fluids* **40**(2), 301–313.

Paatero, P. and Tapper, U. (1994), 'Positive matrix factorization: A non-negative factor model with optimal utilization of error estimates of data values', *Environmetrics* **5**(2), 111–126.

Pan, S., Arnold-Medabalimi, N., and Duraisamy, K. (2020), 'Sparsity-promoting algorithms for the discovery of informative Koopman invariant subspaces', *arXiv preprint arXiv:2002.10637*.

Pao, H.-T. (2008), 'A comparison of neural network and multiple regression analysis in modeling capital structure', *Expert Systems with Applications* **35**, 720–727.

Parente, A. and Sutherland, J. C. (2013), 'Principal component analysis of turbulent combustion data: Data pre-processing and manifold sensitivity', *Combustion and Flame* **160**(2), 340–350.

Parente, A., Sutherland, J., Dally, B., Tognotti, L., and Smith, P. (2011), 'Investigation of the MILD combustion regime via Principal Component Analysis', *Proceedings of the Combustion Institute* **33**(2), 3333–3341.

Parente, A., Sutherland, J., Tognotti, L., and Smith, P. (2009), 'Identification of low-dimensional manifolds in turbulent flames', *Proceedings of the Combustion Institute* **32**(1), 1579–1586.

Parezanovic, V., Cordier, L., Spohn, A., Duriez, T., Noack, B. R., Bonnet, J.-P., Segond, M., Abel, M., and Brunton, S. L. (2016), 'Frequency selection by feedback control in a turbulent shear flow', *Journal of Fluid Mechanics* **797**, 247–283.

Parish, E. J. and Duraisamy, K. (2016), 'A paradigm for data-driven predictive modeling using field inversion and machine learning', *Journal of Computational Physics* **305**, 758–774.

Pastoor, M. (2008), Niederdimensionale Wirbelmodelle zur Kontrolle von Scher- und Nachlaufströmungen, PhD thesis, Berlin Institute of Technology, Germany.

Pastoor, M., Henning, L., Noack, B. R., King, R., and Tadmor, G. (2008), 'Feedback shear layer control for bluff body drag reduction', *Journal of Fluid Mechanics* **608**, 161–196.

Pathak, D., Agrawal, P., Efros, A. A. and Darrell, T. (2017), Curiosity-driven exploration by self-supervised prediction, in D. Precup and Y. W. Teh, eds., 'Proceedings of the 34th International Conference on Machine Learning, ICML 2017, Sydney, NSW, Australia, 6–11

August 2017', Vol. 70 of Proceedings of Machine Learning Research, PMLR, pp. 2778–2787.

Peerenboom, K., Parente, A., Kozák, T., Bogaerts, A., and Degrez, G. (2015), 'Dimension reduction of non-equilibrium plasma kinetic models using principal component analysis', *Plasma Sources Science and Technology* **24**(2), 025004.

Penland, C. (1996), 'A stochastic model of IndoPacific sea surface temperature anomalies', *Physica D: Nonlinear Phenomena* **98**(2–4), 534–558.

Penland, C. and Magorian, T. (1993), 'Prediction of Niño 3 sea surface temperatures using linear inverse modeling', *Journal of Climate* **6**(6), 1067–1076.

Pepiot-Desjardins, P. and Pitsch, H. (2008), 'An efficient error-propagation-based reduction method for large chemical kinetic mechanisms', *Combustion and Flame* **154**(1–2), 67–81.

Perko, L. (2013), *Differential Equations and Dynamical Systems*, Vol. 7 of Texts in Applied Mathematics, Springer Science & Business Media, New York.

Perlman, E., Burns, R., Li, Y., and Meneveau, C. (2007), Data exploration of turbulence simulations using a database cluster, in 'Proceedings of the SC07', ACM, New York, pp. 23.1–23.11.

Perrin, R., Braza, M., Cid, E., Cazin, S., Barthet, A., Sevrain, A., Mockett, C., and Thiele, F. (2007), 'Obtaining phase averaged turbulence properties in the near wake of a circular cylinder at high Reynolds number using POD', *Experiments in Fluids* **43**(2–3), 341–355.

Peters, N. (1984), 'Laminar diffusion flamelet models in non-premixed turbulent combustion', *Progress in Energy and Combustion Science* **10**(3), 319–339.

Phan, M. Q., Juang, J.-N., and Hyland, D. C. (1995), 'On neural networks in identification and control of dynamic systems,' in Ardéshir Guran and Daniel J Inman, eds., 'Wave Motion, Intelligent Structures and Nonlinear Mechanics', World Scientific, Singapore, pp. 194–225.

Picard, C. and Delville, J. (2000), 'Pressure velocity coupling in a subsonic round jet', *International Journal of Heat and Fluid Flow* **21**(3), 359–364.

Poincaré, H. (1920), *Science et méthode*, Flammarion, Paris. English translation in Dover books, 1952.

Pope, S. B. (2000), *Turbulent Flows*, Cambridge University Press, Cambridge.

Pope, S. B. (2013), 'Small scales, many species and the manifold challenges of turbulent combustion', *Proceedings of the Combustion Institute* **34**(1), 1–31.

Proctor, J. L., Brunton, S. L., and Kutz, J. N. (2016), 'Dynamic mode decomposition with control', *SIAM Journal on Applied Dynamical Systems* **15**(1), 142–161.

Protas, B. (2004), 'Linear feedback stabilization of laminar vortex shedding based on a point vortex model', *Physics of Fluids* **16**(12), 4473–4488.

Quarteroni, A. and Rozza, G. (2013), *Reduced Order Methods for Modeling and Computational Reduction*, Vol. 9 of MS&A – Modeling, Simulation & Appplications, Springer, New York.

Rabault, J. (2019), 'Deep reinforcement learning applied to fluid mechanics: Materials from the 2019 Flow/Interface School on Machine Learning and Data Driven Methods', DOI:10.13140/RG.2.2.11533.90086.

Rabault, J., Belus, V., Viquerat, J., Che, Z., Hachem, E., Reglade, U., and Jensen, A. (2019), Exploiting locality and physical invariants to design effective deep reinforcement learning control of the unstable falling liquid film, in 'The 1st Graduate Forum of CSAA and the 7th International Academic Conference for Graduates', NUAA, November 21–22, 2019, Nanjing, China.

Rabault, J., Kuchta, M., Jensen, A., Reglade, U., and Cerardi, N. (2019), 'Artificial neural networks trained through deep reinforcement learning discover control strategies for active flow control', *Journal of Fluid Mechanics* **865**, 281–302.

Rabault, J. and Kuhnle, A. (2019), 'Accelerating deep reinforcement learning strategies of flow control through a multi-environment approach', *Physics of Fluids* **31**(9), 094105.

Rabault, J. and Kuhnle, A. (2020), 'Deep reinforcement learning applied to active flow control', ResearchGate Preprint https://doi.org/10.13140/RG 2.

Rabault, J., Ren, F., Zhang, W., Tang, H., and Xu, H. (2020), 'Deep reinforcement learning in fluid mechanics: A promising method for both active flow control and shape optimization', *Journal of Hydrodynamics* **32**(2), 234–246.

Raibaudo, C., Zhong, P., Noack, B. R., and Martinuzzi, R. J. (2020), 'Machine learning strategies applied to the control of a fluidic pinball', *Physics of Fluids* **32**(1), 015108.

Raiola, M., Discetti, S., and Ianiro, A. (2015), 'On PIV random error minimization with optimal POD-based low-order reconstruction', *Experiments in Fluids* **56**(4), 75.

Raiola, M., Ianiro, A., and Discetti, S. (2016), 'Wake of tandem cylinders near a wall', *Experimental Thermal and Fluid Science* **78**, 354–369.

Raissi, M. and Karniadakis, G. E. (2018), 'Hidden physics models: Machine learning of nonlinear partial differential equations', *Journal of Computational Physics* **357**, 125–141.

Raissi, M., Perdikaris, P., and Karniadakis, G. (2019), 'Physics-informed neural networks: A deep learning framework for solving forward and inverse problems involving nonlinear partial differential equations', *Journal of Computational Physics* **378**, 686–707.

Raissi, M., Yazdani, A., and Karniadakis, G. E. (2020), 'Hidden fluid mechanics: Learning velocity and pressure fields from flow visualizations', *Science* **367**(6481), 1026–1030.

Ranzi, E., Frassoldati, A., Stagni, A., Pelucchi, M., Cuoci, A., and Faravelli, T. (2014), 'Reduced kinetic schemes of complex reaction systems: fossil and biomass-derived transportation fuels', *International Journal of Chemical Kinetics* **46**(9), 512–542.

Rasmussen, C. (2006), *Gaussian Processes for Machine Learning*, MIT Press, Cambridge, MA.

Rayleigh, L. (1887), 'On the stability of certain fluid motions', *The Proceedings of the London Mathematical Society* **19**, 67–74.

Recht, B. (2019), 'A tour of reinforcement learning: The view from continuous control', *Annual Review of Control, Robotics, and Autonomous Systems* **2**, 253–279.

Reddy, G., Celani, A., Sejnowski, T. J., and Vergassola, M. (2016), 'Learning to soar in turbulent environments', *Proceedings of the National Academy of Sciences of the United States of America* **113**(33), E4877–E4884.

Reddy, G., Wong-Ng, J., Celani, A., Sejnowski, T. J., and Vergassola, M. (2018), 'Glider soaring via reinforcement learning in the field', *Nature* **562**(7726), 236–239.

Reddy, S. and Henningson, D. (1993), 'Energy growth in viscous channel flows', *Journal of Fluid Mechanics* **252**, 209–238.

Reimer, L., Heinrich, R., and Ritter, M. (2019), Towards higher-precision maneuver and gust loads computations of aircraft: Status of related features in the CFD-based multidisciplinary simulation environment FlowSimulator, in 'Notes on Numerical Fluid Mechanics and Multidisciplinary Design', Springer, Cham, pp. 597–607.

Reiss, J., Schulze, P., Sesterhenn, J., and Mehrmann, V. (2018), 'The shifted proper orthogonal decomposition: A mode decomposition for multiple transport phenomena', *SIAM Journal on Scientific Computing* **40**(3), A1322–A1344.

Rempfer, D. (1995), *Empirische Eigenfunktionen und Galerkin-Projektionen zur Beschreibung des laminar-turbulenten Grenzschichtumschlags (transl.: Empirical eigenfunctions and*

Galerkin projection for the description of the laminar-turbulent boundary-layer transition), Habilitationsschrift, Fakultät für Luft- und Raumfahrttechnik, Universität Stuttgart.

Rempfer, D. (2000), 'On low-dimensional Galerkin models for fluid flow', *Theoretical and Computational Fluid Dynamics* **14**(2), 75–88.

Rempfer, D. and Fasel, F. H. (1994a), 'Dynamics of three-dimensional coherent structures in a flat-plate boundary-layer', *Journal of Fluid Mechanics* **275**, 257–283.

Rempfer, D. and Fasel, F. H. (1994b), 'Evolution of three-dimensional coherent structures in a flat-plate boundary-layer', *Journal of Fluid Mechanics* **260**, 351–375.

Ren, F., Hu, H.-b., and Tang, H. (2020), 'Active flow control using machine learning: A brief review', *Journal of Hydrodynamics* **32**(2), 247–253.

Ren, F., Rabault, J., and Tang, H. (2020), 'Applying deep reinforcement learning to active flow control in turbulent conditions', *arxiv preprint arXiv:2006.10683*.

Ren, Z. and Pope, S. (2008), 'Second-order splitting schemes for a class of reactive systems', *Journal of Computational Physics* **227**(17), 8165–8176.

Reynolds, O. (1883), 'An experimental investigation of the circumstances which determine whether the motion of water shall be direct or sinuous, and of the law of resistance in parallel channels', *Proceedings of the Royal Society of London* **35**, 84–99.

Reynolds, O. (1894), 'On the dynamical theory of incompressible viscous fluids and the determination of the criterion', *Proceedings of the Royal Society of London* **56**, 40–45.

Reynolds, W. C. and Tiederman, W. G. (1967), 'Stability of turbulent channel flow, with application to Malkus' theory', *Journal of Fluid Mechanics* **27**, 253–272.

Richards, J. I. and Youn, H. K. (1990), *The Theory of Distributions*, Cambridge University Press, Cambridge.

Richardson, L. F. (1920), 'The supply of energy from and to atmospheric eddies', *Proceedings of the Royal Society A* **97**, 354–373.

Ripepi, M., Verveld, M. J., Karcher, N. W., Franz, T., Abu-Zurayk, M., Görtz, S., and Kier, T. M. (2018), 'Reduced-order models for aerodynamic applications, loads and MDO', *CEAS Aeronautical Journal* **9**(1), 171–193.

Robinson, S. K. (1991), 'Coherent motions in the turbulent boundary layer', *Annual Review Fluid Mechanics* **23**, 601–639.

Rokhlin, V., Szlam, A., and Tygert, M. (2009), 'A randomized algorithm for principal component analysis', *SIAM Journal on Matrix Analysis and Applications* **31**, 1100–1124.

Rosenblatt, F. (1958), 'The perceptron: A probabilistic model for information storage and organization in the brain', *Psychological Review* **65**(6), 386.

Roussopoulos, K. (1993), 'Feedback control of vortex shedding at low Reynolds numbers', *Journal of Fluid Mechanics* **248**, 267–296.

Roweis, S. and Lawrence, S. (2000), 'Nonlinear dimensionality reduction by locally linear embedding', *Science* **290**(5500), 2323–2326.

Rowley, C. W. (2005), 'Model reduction for fluids using balanced proper orthogonal decomposition.', *International Journal of Bifurcation and Chaos* **15**(3), 997–1013.

Rowley, C. W., Colonius, T., and Murray, R. M. (2004), 'Model reduction for compressible flows using POD and Galerkin projection', *Physica D: Nonlinear Phenomena* **189**(1–2), 115–129.

Rowley, C. W. and Dawson, S. T. M. (2017), 'Model reduction for flow analysis and control', *Annual Review of Fluid Mechanics* **49**(1), 387–417.

Rowley, C. W., Kevrekidis, I. G., Marsden, J. E., and Lust, K. (2003), 'Reduction and reconstruction for self-similar dynamical systems', *Nonlinearity* **16**(4), 1257.

Rowley, C. W. and Marsden, J. E. (2000), 'Reconstruction equations and the Karhunen–Loève expansion for systems with symmetry', *Physica D: Nonlinear Phenomena* **142**(1), 1–19.

Rowley, C. W., Mezić, I., Bagheri, S., Schlatter, P., and Henningson, D. (2009), 'Spectral analysis of nonlinear flows', *Journal of Fluid Mechanics* **645**, 115–127.

Rowley, C. W. and Williams, D. R. (2006), 'Dynamics and control of high-Reynolds-number flow over open cavities', *Annual Review of Fluid Mechanics* **38**, 251–276.

Rudy, S. H., Brunton, S. L., Proctor, J. L., and Kutz, J. N. (2017), 'Data-driven discovery of partial differential equations', *Science Advances* **3**, e1602614.

Rumelhart, D. E., Hinton, G. E., Williams, R. J. (1986), 'Learning representations by back-propagating errors', *Nature* **323**(6088), 533–536.

Russell, B. (1912), 'On the notion of cause', *The Proceedings of the Aristotelian Society* **13**, 1–26.

Sabater, C., Bekemeyer, P., and Görtz, S. (2020), 'Efficient bilevel surrogate approach for optimization under uncertainty of shock control bumps', *AIAA Journal* **58**(12), 5228–5242.

Salimans, T. and Chen, R. (2018), 'Learning Montezuma's Revenge from a single demonstration', *arXiv preprint arXiv:1812.03381*.

Salimans, T., Ho, J., Chen, X., and Sutskever, I. (2017), 'Evolution strategies as a scalable alternative to reinforcement learning', *CoRR, arXiv:1703.03864*.

Saltzman, B. (1962), 'Finite amplitude free convection as an initial value problem – I', *Journal of Atmospheric Sciences* **19**(4), 329–341.

Sarlos, T. (2006), Improved approximation algorithms for large matrices via random projections, in '47th Annual IEEE Symposium on Foundations of Computer Science', October 21–24, 2006, Washington, DC, pp. 143–152.

Savinov, N., Raichuk, A., Marinier, R., Vincent, D., Pollefeys, M., Lillicrap, T., and Gelly, S. (2018), 'Episodic curiosity through reachability', *arXiv preprint 1810.02274*.

Schaeffer, H. (2017), Learning partial differential equations via data discovery and sparse optimization, 'Proceedings of the Royal Society A', **473**, 20160446.

Schaeffer, H. and McCalla, S. G. (2017), 'Sparse model selection via integral terms', *Physical Review E* **96**(2), 023302.

Schäfer, M., Turek, S., Durst, F., Krause, E., and Rannacher, R. (1996), Benchmark computations of laminar flow around a cylinder, in 'Flow Simulation with High-Performance Computers II', Springer, Berlin, pp. 547–566.

Schaffer, S. (1994), Making up discovery, in M. A. Boden, ed., 'Dimensions of creativity', MIT Press, Cambridge, pp. 13–51.

Scherl, I., Strom, B., Shang, J. K., Williams, O., Polagye, B. L., and Brunton, S. L. (2020), 'Robust principal component analysis for particle image velocimetry', *Physical Review Fluids* **5**, 054401.

Schlegel, M. and Noack, B. R. (2015), 'On long-term boundedness of Galerkin models', *Journal of Fluid Mechanics* **765**, 325–352.

Schlegel, M., Noack, B. R., and Tadmor, G. (2004), Low-dimensional Galerkin models and control of transitional channel flow, Technical Report 01/2004, Hermann-Föttinger-Institut für Strömungsmechanik, Technische Universität Berlin, Germany.

Schmelzer, M., Dwight, R. P., and Cinnella, P. (2020), 'Discovery of algebraic Reynolds-stress models using sparse symbolic regression', *Flow, Turbulence and Combustion* **104**(2), 579–603.

Schmid, P. (2007), 'Nonmodal stability theory', *Annual Review of Fluid Mechanics* **39**, 129–162.

Schmid, P. and Henningson, D. (2001), *Stability and Transition in Shear Flows*, Springer Verlag, New York.

Schmid, P. J. (2010), 'Dynamic mode decomposition of numerical and experimental data', *Journal of Fluid Mechanics* **656**, 5–28.

Schmid, P., Li, L., Juniper, M., and Pust, O. (2011), 'Applications of the dynamic mode decomposition', *Theoretical and Computational Fluid Dynamics* **25**(1–4), 249–259.

Schmid, P., Violato, D., and Scarano, F. (2012), 'Decomposition of time-resolved tomographic PIV', *Experiments in Fluids* **52**(6), 1567–1579.

Schmidt, M. and Lipson, H. (2009), 'Distilling free-form natural laws from experimental data', *Science* **324**(5923), 81–85.

Schmidt, O. T. and Towne, A. (2019), 'An efficient streaming algorithm for spectral proper orthogonal decomposition', *Computer Physics Communications* **237**, 98–109.

Schneider, K. and Vasilyev, O. V. (2010), 'Wavelet methods in computational fluid dynamics', *Annual Review of Fluid Mechanics* **42**, 473–503.

Schölkopf, B. and Smola, A. J. (2002), *Learning With Kernels: Support Vector Machines, Regularization, Optimization, and Beyond*, MIT Press, Cambridge.

Schoppa, W. and Hussain, F. (2002), 'Coherent structure generation in near-wall turbulence', *Journal of Fluid Mechanics* **453**, 57–108.

Schulman, J., Levine, S., Abbeel, P., Jordan, M., and Moritz, P. (2015), Trust region policy optimization, in F. Bach and D. Blei, eds., 'Proceedings of the 32nd International Conference on Machine Learning', Vol. 37 of Proceedings of Machine Learning Research, PMLR, Lille, France, pp. 1889–1897.

Schulman, J., Wolski, F., Dhariwal, P., Radford, A., and Klimov, O. (2017), 'Proximal policy optimization algorithms', *CoRR, arXiv:1707.06347.*

Schulze, J., Schmid, P., and Sesterhenn, J. (2009), 'Exponential time integration using Krylov subspaces', *The International Journal for Numerical Methods in Fluids* **60(6)**, 591–609.

Schumm, M., Berger, E., and Monkewitz, P. A. (1994), 'Self-excited oscillations in the wake of two-dimensional bluff bodies and their control', *Journal of Fluid Mechanics* **271**, 17–53.

Schwer, D., Lu, P. and Green Jr., W. (2003), 'An adaptive chemistry approach to modeling complex kinetics in reacting flows', *Combustion and Flame* **133**(4), 451–465.

Seber, G. A. (2009), *Multivariate Observations*, John Wiley & Sons, Hoboken, NJ.

Semaan, R., Fernex, D., Weiner, A., and Noack, B. R. (2020), *xROM – A Toolkit for Reduced-Order Modeling of Fluid Flows*, Vol. 1 of Machine Learning Tools for Fluid Mechanics, Technische Universität Braunschweig, Braunschweig.

Semaan, R., Kumar, P., Burnazzi, M., Tissot, G., Cordier, L. and Noack, B. R. (2015), 'Reduced-order modelling of the flow around a high-lift configuration with unsteady Coanda blowing', *Journal of Fluid Mechanics* **800**, 72–110.

Semeraro, O., Bagheri, S., Brandt, L., and Henningson, D. S. (2011), 'Feedback control of three-dimensional optimal disturbances using reduced-order models', *Journal of Fluid Mechanics* **677**, 63–102.

Semeraro, O., Bagheri, S., Brandt, L., and Henningson, D. S. (2013), 'Transition delay in a boundary layer flow using active control', *Journal of Fluid Mechanics* **731**, 288–311.

Semeraro, O., Lusseyran, F., Pastur, L. and Jordan, P. (2017), 'Qualitative dynamics of wave packets in turbulent jets', *Physical Review Fluids* **2**(9), 094605.

Semeraro, O. and Mathelin, L. (2016), 'An open-source toolbox for data-driven linear system identification', Technical report, LIMSI-CNRS, Orsay, France.

Sharma, A. S., Mezić, I., and McKeon, B. J. (2016), 'Correspondence between Koopman mode decomposition, resolvent mode decomposition, and invariant solutions of the Navier-Stokes equations', *Physical Review Fluids* **1**(3), 032402.

Sieber, M., Paschereit, C. O., and Oberleithner, K. (2016), 'Spectral proper orthogonal decomposition', *Journal of Fluid Mechanics* **792**, 798–828.

Siegel, S. G., Seidel, J., Fagley, C., Luchtenburg, D. M., Cohen, K., and McLaughlin, T. (2008), 'Low dimensional modelling of a transient cylinder wake using double proper orthogonal decomposition', *Journal of Fluid Mechanics* **610**, 1–42.

Sigurd Skogestad, I. P. (2005), *Multivariable Feedback Control*, John Wiley & Sons, Hoboken, NJ.

Sillero, J. A. (2014), High Reynolds numbers turbulent boundary layers, PhD thesis, Aeronautics, Universidad Politécnica, Madrid. https://tesis.biblioteca.upm.es/tesis/7538.

Silver, D., Huang, A., Maddison, C. J., Guez, A., Sifre, L., Van Den Driessche, G., Schrittwieser, J., Antonoglou, I., Panneershelvam, V., Lanctot, M. Sander, D., Dominik, G., John, N., Nal, K., Ilya, S., Timothy, L., Madeleine, L., Koray, K., Thore, G., and Demis H. (2016), 'Mastering the game of Go with deep neural networks and tree search', *Nature* **529**(7587), 484–489.

Silver, D., Hubert, T., Schrittwieser, J., Antonoglou, I., Lai, M., Guez, A., Lanctot, M., Sifre, L., Kumaran, D., Graepel, T., Lillicrap, T., Simonyan, K., and Hassabis, D. (2018), 'A general reinforcement learning algorithm that masters Chess, Shogi, and Go through self-play', *Science* **362**(6419), 1140–1144.

Silver, D., Lever, G., Heess, N., Degris, T., Wierstra, D., and Riedmiller, M. (2014), Deterministic policy gradient algorithms, in E. P. Xing and T. Jebara, eds., 'Proceedings of the 31st International Conference on Machine Learning', Vol. 32 of Proceedings of Machine Learning Research, PMLR, Bejing, China, pp. 387–395.

Silver, D., Schrittwieser, J., Simonyan, K., Antonoglou, I., Huang, A., Guez, A., Hubert, T., Baker, L., Lai, M., and Bolton, A. (2017), 'Mastering the game of Go without human knowledge', *Nature* **550**(7676), 354.

Singh, A. P., Medida, S., and Duraisamy, K. (2017), 'Machine-learning-augmented predictive modeling of turbulent separated flows over airfoils', *AIAA Journal* **55**, 2215–2227.

Sirovich, L. (1987), 'Turbulence and the dynamics of coherent structures. I–Coherent structures. II–Symmetries and transformations. III–Dynamics and scaling', *Quarterly of Applied Mathematics* **45**, 561–571.

Sirovich, L. and Kirby, M. (1987), 'A low-dimensional procedure for the characterization of human faces', *Journal of the Optical Society of America A* **4**(3), 519–524.

Skene, C., Eggl, M., and Schmid, P. (2020), 'A parallel-in-time approach for accelerating direct-adjoint studies', *Journal of Computational Physics* **429**(2), 110033.

Skogestad, S. and Postlethwaite, I. (2005), *Multivariable Feedback Control: Analysis and Design*, 2 ed., John Wiley & Sons, Hoboken, NJ.

Skogestad, S. and Postlethwaite, I. (2007), *Multivariable Feedback Control: Analysis and Design*, Vol. 2, Wiley, New York.

Smith, J. O. (2007a), *Introduction to Digital Filters: With Audio Applications*, W3K Publishing, http://books.w3k.org/.

Smith, J. O. (2007b), *Mathematics of the Discrete Fourier Transform (DFT): With Audio Applications*, W3K Publishing, http://books.w3k.org/.

Smith, S. W. (1997), *The Scientist & Engineer's Guide to Digital Signal Processing*, California Technical Publishing, San Diego, www.dspguide.com/.

Sommerfeld, A. (1908), 'Ein Beitrag zur hydrodynamischen Erklärung der turbulenten Flüssigkeitsbewegungen', *Atti. del 4 Congr. Int. dei Mat. III*, pp. 116–124.

Sorokina, M., Sygletos, S., and Turitsyn, S. (2016), 'Sparse identification for nonlinear optical communication systems: SINO method', *Optics Express* **24**(26), 30433–30443.

Spalart, P. R. (1988), 'Direct simulation of a turbulent boundary layer up to $Re_\theta = 1410$', *Journal of Fluid Mechanics* **187**, 61–98.

Stengel, R. F. (1994), *Optimal Control and Estimation*, Dover, New York.

Strang, G. (1968), 'On the construction and comparison of difference schemes', *SIAM Journal on Numerical Analysis* **5**(3), 506–517.

Strang, G. (1988), *Linear Algebra and Its Applications*, Harcourt Brace Jovanovich College Publishers, San Diego, CA.

Strang, G. (1989), 'Wavelets and dilation equations: A brief introduction', *SIAM Review* **31**(4), 614–627.

Strang, G. (1996), *Wavelets and Filter Banks*, Wellesley-Cambridge Press, Wellesley, MA.

Strang, G. (2007), *Computational Science and Engineering*, Wellesley-Cambridge Press, Wellesley, MA.

Strang, G. and Nguyen, T. (1996), *Wavelets and Filter Banks*, SIAM, Philadelphia.

Strom, B., Brunton, S. L., and Polagye, B. (2017), 'Intracycle angular velocity control of cross-flow turbines', *Nature Energy* **2**(17103), 1–9.

Stuart, J. (1958), 'On the non-linear mechanics of hydrodynamic stability', *Journal of Fluid Mechanics* **4**, 1–21.

Stuart, J. T. (1971), 'Nonlinear stability theory', *Annual Review of Fluid Mechanics* **3**, 347–370.

Succi, S. and Coveney, P. V. (2018), 'Big data: the end of the scientific method', *Philosophical Transactions of the Royal Society A* **377**, 20180145.

Suh, Y. K. (1993), 'Periodic motion of a point vortex in a corner subject to a potential flow', *Journal of the Physical Society of Japan* **62**(10), 3441–3445.

Sundqvist, B. (1991), 'Thermal diffusivity measurements by Ångström method in a fluid environment', *International Journal of Thermophysics* **12**(1), 191–206.

Sutherland, J. C. and Parente, A. (2009), 'Combustion modeling using principal component analysis', *Proceedings of the Combustion Institute* **32**(1), 1563–1570.

Sutherland, J. C., Smith, P. J., and Chen, J. H. (2007), 'A quantitative method for a priori evaluation of combustion reaction models', *Combustion Theory and Modelling* **11**(2), 287–303.

Sutton, R. S. and Barto, A. G. (2018), *Reinforcement Learning: An Introduction*, 2nd ed., MIT Press, Cambridge, MA.

Szegedy, C., Ioffe, S., Vanhoucke, V., and Alemi, A. A. (2017), Inception-v4, inception-resnet and the impact of residual connections on learning, in 'Thirty-First AAAI Conference on Artificial Intelligence', February 4–9, 2017, San Francisco, CA.

Tadmor, G., Lehmann, O., Noack, B. R., Cordier, L., Delville, J., Bonnet, J.-P., and Morzyński, M. (2011), 'Reduced order models for closed-loop wake control', *Philosophical Transactions of the Royal Society A* **369**(1940), 1513–1524.

Tadmor, G. and Noack, B. R. (2004), Dynamic estimation for reduced Galerkin models of fluid flows, in 'The 2004 American Control Conference', Boston, MA, June 30–July 2, 2004, pp. 0001–0006. Paper **WeM18.1**.

Taira, K., Brunton, S., Dawson, S., Rowley, C., Colonius, T., McKeon, B., Schmidt, O., Gordeyev, S., Theofilis, V., and Ukeiley, S. (2017), 'Modal analysis of fluid flows: An overview', *AIAA Journal* **55**(12), 4013–4041.

Taira, K., Hemati, M. S., Brunton, S. L., Sun, Y., Duraisamy, K., Bagheri, S., Dawson, S., and Yeh, C.-A. (2019), 'Modal analysis of fluid flows: Applications and outlook', *arXiv preprint arXiv:1903.05750* .

Takeishi, N., Kawahara, Y., and Yairi, T. (2017), Learning Koopman invariant subspaces for dynamic mode decomposition, in 'Advances in Neural Information Processing Systems', December 4–9, 2017, Long Beach, CA, pp. 1130–1140.

Takens, F. (1981), Detecting strange attractors in turbulence, in 'Dynamical systems and turbulence, Warwick 1980', Vol. 898 in Lecture Notes in Math., Springer, Heidelberg and New York, pp. 366–381.

Talamelli, A., Persiani, F., Fransson, J. H., Alfredsson, P. H., Johansson, A. V., Nagib, H. M., Rüedi, J.-D., Sreenivasan, K. R., and Monkewitz, P. A. (2009), 'CICLoPE —- A response to the need for high Reynolds number experiments', *Fluid Dynamics Research* **41**(2), 021407.

Tanahashi, M., Kang, S., Miyamoto, T., and Shiokawa, S. (2004), 'Scaling law of fine scale eddies in turbulent channel flows up to $Re_\tau = 800$', *International Journal of Heat and Fluid Flow* **25**, 331–341.

Tang, H., Rabault, J., Kuhnle, A., Wang, Y., and Wang, T. (2020), 'Robust active flow control over a range of reynolds numbers using an artificial neural network trained through deep reinforcement learning', *Physics of Fluids* **32**(5), 053605.

Tedrake, R., Jackowski, Z., Cory, R., Roberts, J. W., and Hoburg, W. (2009), Learning to fly like a bird, in '14th International Symposium on Robotics Research', Lucerne, Switzerland.

Tenenbaum, J. B. (2000), 'A global geometric framework for nonlinear dimensionality reduction', *Science* **290**(5500), 2319–2323.

Tennekes, H. and Lumley, J. L. (1972), *A First Course in Turbulence*, MIT Press, Cambridge, MA.

Thaler, S., Paehler, L., and Adams, N. A. (2019), 'Sparse identification of truncation errors', *Journal of Computational Physics* **397**, 108851.

Theofilis, V. (2011), 'Global linear instability', *Annual Review of Fluid Mechanics* **43**, 319–352.

Thiria, B., Goujon-Durand, S., and Wesfreid, J. E. (2006), 'The wake of a cylinder performing rotary oscillations', *Journal of Fluid Mechanics* **560**, 123–147.

Thormann, R. and Widhalm, M. (2013), 'Linear-frequency-domain predictions of dynamic-response data for viscous transonic flows', *AIAA Journal* **51**(11), 2540–2557.

Tibshirani, R. (1996), 'Regression shrinkage and selection via the lasso', *Journal of the Royal Statistical Society. Series B (Methodological)* **58**(1), 267–288.

Tinney, C. E., Ukeiley, L., and Glauser, M. N. (2008), 'Low-dimensional characteristics of a transonic jet. Part 2. Estimate and far-field prediction', *Journal of Fluid Mechanics* **615**, 53–92.

Toedtli, S. S., Luhar, M., and McKeon, B. J. (2019), 'Predicting the response of turbulent channel flow to varying-phase opposition control: Resolvent analysis as a tool for flow control design', *Physical Review Fluids* **4**(7), 073905.

Torrence, C. and Compo, G. P. (1998), 'A practical guide to wavelet analysis', *Bulletin of the American Meteorological Society* **79**(1), 61–78.

Towne, A., Schmidt, O. T., and Colonius, T. (2018), 'Spectral proper orthogonal decomposition and its relationship to dynamic mode decomposition and resolvent analysis', *Journal of Fluid Mechanics* **847**, 821–867.

Townsend, A. A. (1976), *The Structure of Turbulent Shear Flow*, Cambridge University Press, Cambridge.

Tran, G. and Ward, R. (2016), 'Exact recovery of chaotic systems from highly corrupted data', *arXiv preprint arXiv:1607.01067.*

Trefethen, L. and Bau, D. (1997), *Numerical Linear Algebra*, Society for Industrial and Applied Mathematics, Philadelphia.

Trefethen, L., Trefethen, A., Reddy, S., and Driscoll, T. (1993), 'Hydrodynamic stability without eigenvalues', *Science* **261**, 578–584.

Tropp, J. A. and Gilbert, A. C. (2007), 'Signal recovery from random measurements via orthogonal matching pursuit', *IEEE Transactions on Information Theory* **53**(12), 4655–4666.

Tu, J. H. and Rowley, C. W. (2012), 'An improved algorithm for balanced POD through an analytic treatment of impulse response tails', *Journal of Computational Physics* **231**(16), 5317–5333.

Tu, J. H., Rowley, C. W., Kutz, J. N., and Shang, J. K. (2014), 'Spectral analysis of fluid flows using sub-Nyquist-rate PIV data', *Experiments in Fluids* **55**(9), 1–13.

Tu, J., Rowley, C., Luchtenburg, D., Brunton, S., and Kutz, J. (2014), 'On dynamic mode decomposition: theory and applications', *Journal of Computational Dynamics* **1**(2), 391–421.

Turns, S. R. (1996), *Introduction to Combustion*, Vol. 287, McGraw-Hill Companies, New York.

Uruba, V. (2012), 'Decomposition methods in turbulence research', *EPJ Web of Conferences* **25**, 01095.

Van den Berg, J. (2004), *Wavelets in Physics*, Cambridge University Press, Cambridge.

Van Loan, C. F. and Golub, G. H. (1983), *Matrix Computations*, Johns Hopkins University Press, Baltimore.

Vassberg, J., Dehaan, M., Rivers, M., and Wahls, R. (2008), Development of a common research model for applied CFD validation studies, in '26th AIAA Applied Aerodynamics Conference', August 18-21, 2008, Honolulu, HI, American Institute of Aeronautics and Astronautics, Reston, VA.

Venturi, D. (2006), 'On proper orthogonal decomposition of randomly perturbed fields with applications to flow past a cylinder and natural convection over a horizontal plate', *Journal of Fluid Mechanics* **559**, 215–254.

Verma, S., Novati, G., and Koumoutsakos, P. (2018), 'Efficient collective swimming by harnessing vortices through deep reinforcement learning', *Proceedings of the National Academy of Sciences* **115**(23), 5849–5854.

Vincent, A. and Meneguzzi, M. (1991), 'The spatial structure and statistical properties of homogeneous turbulence', *Journal of Fluid Mechanics* **225**, 1–20.

Viquerat, J., Rabault, J., Kuhnle, A., Ghraieb, H., and Hachem, E. (2019), 'Direct shape optimization through deep reinforcement learning', *arXiv preprint arXiv:1908.09885.*

Vlachas, P. R., Byeon, W., Wan, Z. Y., Sapsis, T. P., and Koumoutsakos, P. (2018), 'Data-driven forecasting of high-dimensional chaotic systems with long short-term memory networks', *Proceedings of the Royal Society A* **474**(2213), 20170844.

Vladimir Cherkassky, F. M. M. (2008), *Learning from Data*, John Wiley & Sons, New York.

Voit, E. O. (2019), 'Perspective: dimensions of the scientific method', *PLoS: Computational Biology* **15**(9), e1007279.

Voltaire, F. (1764), *Dictionnaire Philosophique: Atomes*, Oxford University Press, 1994.

von Helmholtz, H. (1858), 'Über integrale der hydrodynamischen gleichungen, welche den wirbelbewegungen entsprechen', *J. für die reine und angewandte Mathematik* **55**, 25–55.

von Kármán, T. (1911), 'über den Mechanismus des Widerstandes, den ein bewegter Körper in einer Flüssigkeit erfährt', *Nachrichten der Kaiserlichen Gesellschaft der Wissenschaften zu Göttingen*, pp. 509–517.

von Storch, H. and Xu, J. (1990), 'Principal oscillation pattern analysis of the 30- to 60-day oscillation in the tropical troposphere', *Climate Dynamics* **4**(3), 175–190.

Voropayev, S. I., Afanasyev, Y. D., and Filippov, I. A. (1991), 'Horizontal jets and vortex dipoles in a stratified fluid', *Journal of Fluid Mechanics* **227**, 543–566.

Voss, R., Tichy, L., and R.Thormann (2011), A ROM based flutter prediction process and its validation with a new reference model, in 'International Forum on Aeroelasticity and Structural Dynamics (IFASD-2011)', June 26–30, 2011, Paris, France.

Vukasonivic, B., Rusak, Z., and Glezer, A. (2010), 'Dissipative small-scale actuation of a turbulent shear layer', *Journal of Fluid Mechanics* **656**, 51–81.

Wahde, M. (2008), *Biologically Inspired Optimization Methods: An Introduction*, WIT Press, Southampton.

Wan, Z. Y., Vlachas, P., Koumoutsakos, P., and Sapsis, T. (2018), 'Data-assisted reduced-order modeling of extreme events in complex dynamical systems', *PloS One* **13**(5), e0197704.

Wang, J.-X., Wu, J.-L., and Xiao, H. (2017), 'Physics-informed machine learning approach for reconstructing reynolds stress modeling discrepancies based on DNS data', *Physical Review Fluids* **2**(3), 034603.

Wang, M. and Hemati, M. S. (2017), 'Detecting exotic wakes with hydrodynamic sensors', *arXiv preprint arXiv:1711.10576* .

Wang, Q., Moin, P., and Iaccarino, G. (2009), 'Minimal repetition dynamic checkpointing algorithm for unsteady adjoint calculation', *SIAM Journal of Scientific Computing* **31**(4), 2549–2567.

Wang, R. (2009), *Introduction to Orthogonal Transforms*, Cambridge University Press, New York.

Wang, W. X., Yang, R., Lai, Y. C., Kovanis, V., and Grebogi, C. (2011), 'Predicting catastrophes in nonlinear dynamical systems by compressive sensing', *Physical Review Letters* **106**, 154101-1–154101-4.

Wang, Y., Yao, H., and Zhao, S. (2016), 'Auto-encoder based dimensionality reduction', *Neuro-Computing* **184**, 232–242.

Wang, Z., Akhtar, I., Borggaard, J., and Iliescu, T. (2012), 'Proper orthogonal decomposition closure models for turbulent flows: a numerical comparison', *Computer Methods in Applied Mechanics and Engineering* **237**, 10–26.

Wang, Z., Bapst, V., Heess, N., Mnih, V., Munos, R., Kavukcuoglu, K., and de Freitas, N. (2017), Sample efficient actor-critic with experience replay, in '5th International Conference on Learning Representations, ICLR 2017, Toulon, France, April 24–26, 2017, Conference Track Proceedings', OpenReview.net.

Wang, Z., Schaul, T., Hessel, M., Van Hasselt, H., Lanctot, M., and De Freitas, N. (2016), Dueling network architectures for deep reinforcement learning, in 'Proceedings of the 33rd International Conference on International Conference on Machine Learning – Volume 48', ICML 16, JMLR.org, pp. 1995–2003.

Watkins, C. J. C. H. (1989), Learning from Delayed Rewards, PhD thesis, King's College, Cambridge, UK.

Wehmeyer, C. and Noé, F. (2018), 'Time-lagged autoencoders: Deep learning of slow collective variables for molecular kinetics', *The Journal of Chemical Physics* **148**(241703), 1–9.

Welch, P. (1967), 'The use of fast Fourier transform for the estimation of power spectra: A method based on time averaging over short, modified periodograms', *IEEE Transactions on Audio and Electroacoustics* **15**(2), 70–73.

Weller, J., Camarri, S., and Iollo, A. (2009), 'Feedback control by low-order modelling of the laminar flow past a bluff body', *Journal of Fluid Mechanics* **634**, 405.

White, A. P., Zhu, G., and Choi, J. (2013), *Linear Parameter-Varying Control for Engineering Applications*, Springer, London.

Whittle, P. (1951), *Hypothesis Testing in Time Series Analysis*, Almqvist & Wiksells, Uppsala.

Wiener, N. (1948), *Cybernetics: Control and Communication in the Animal and the Machine*, Wiley, New York.

Wigner, E. P. (1960), 'The unreasonable effectiveness of mathematics in the natural sciences', *Communications on Pure and Applied Mathematics* **13**, 1–14.

Willcox, K. and Peraire, J. (2002), 'Balanced model reduction via the proper orthogonal decomposition', *AIAA Journal* **40**(11), 2323–2330.

Willert, C. E. and Gharib, M. (1991), 'Digital particle image velocimetry', *Experiments in Fluids* **10**(4), 181–193.

Williams, M., Kevrekidis, I., and Rowley, C. (2015), 'A data-driven approximation of the Koopman operator: extending dynamic mode decomposition', *Journal of Nonlinear Science* **6**, 1307–1346.

Williams, M., Rowley, C., and Kevrekidis, I. (2015), 'A kernel approach to data-driven Koopman spectral analysis', *Journal of Computational Dynamics* **2**(2), 247–265.

Williams, R. J. (1992), 'Simple statistical gradient-following algorithms for connectionist reinforcement learning', *Machine Learning* **8**(3–4), 229–256.

Williamson, D. (1999), *Discrete-Time Signal Processing*, Springer, London.

Wright, J., Yang, A., Ganesh, A., Sastry, S., and Ma, Y. (2009), 'Robust face recognition via sparse representation', *IEEE Transactions on Pattern Analysis and Machine Intelligence (PAMI)* **31**(2), 210–227.

Wünning, J. and Wünning, J. (1997), 'Flameless oxidation to reduce thermal no-formation', *Progress in Energy and Combustion Science* **23**(1), 81–94.

Xiao, H., Wu, J.-L., Wang, J.-X., Sun, R., and Roy, C. (2016), 'Quantifying and reducing model-form uncertainties in Reynolds-averaged Navier–Stokes simulations: A data-driven, physics-informed Bayesian approach', *Journal of Computational Physics* **324**, 115–136.

Xu, H., Zhang, W., Deng, J., and Rabault, J. (2020), 'Active flow control with rotating cylinders by an artificial neural network trained by deep reinforcement learning', *Journal of Hydrodynamics* **32**(2), 254–258.

Yan, X., Zhu, J., Kuang, M., and Wang, X. (2019), 'Aerodynamic shape optimization using a novel optimizer based on machine learning techniques', *Aerospace Science and Technology* **86**, 826–835.

Yang, J., Wright, J., Huang, T. S., and Ma, Y. (2010), 'Image super-resolution via sparse representation', *IEEE Transactions on Image Processing* **19**(11), 2861–2873.

Yang, Y., Pope, S. B., and Chen, J. H. (2013), 'Empirical low-dimensional manifolds in composition space', *Combustion and Flame* **160**(10), 1967–1980.

Yeh, C.-A. and Taira, K. (2019), 'Resolvent-analysis-based design of airfoil separation control', *Journal of Fluid Mechanics* **867**, 572–610.

Yetter, R. A., Dryer, F., and Rabitz, H. (1991), 'A comprehensive reaction mechanism for carbon monoxide/hydrogen/oxygen kinetics', *Combustion Science and Technology* **79**(1–3), 97–128.

Yeung, E., Kundu, S. and Hodas, N. (2017), 'Learning deep neural network representations for Koopman operators of nonlinear dynamical systems', *arXiv preprint arXiv:1708.06850*.

Zdravkovich, M. (1987), 'The effects of interference between circular cylinders in cross flow', *Journal of Fluids and Structures* **1**(2), 239–261.

Zdybał, K., Armstrong, E., Parente, A., and Sutherland, J. C. (2020), 'PCAfold: Python software to generate, analyze and improve PCA-derived low-dimensional manifolds', *SoftwareX* **12**, 100630.

Zdybał, K., Sutherland, J. C., Armstrong, E., and Parente, A. (2021), 'State-space informed data sampling on combustion manifolds', *Combustion and Flame (manuscript in preparation)*.

Zdybał, K., Sutherland, J. C. & Parente, A. (2022), 'Manifold-informed state vector subset for reduced-order modeling', *Proceedings of the Combustion Institute* .

Zdybał, K., Armstrong, E., Sutherland, J. C. & Parente, A. (2022), 'Cost function for low-dimensional manifold topology assessment', *Scientific Reports* **12**(1), 1–19.

Zebib, A. (1987), 'Stability of viscous flow past a circular cylinder', *Journal of Engineering Mathematics* **21**, 155–165.

Zhang, H.-Q., Fey, U., Noack, B. R., König, M., and Eckelmann, H. (1995), 'On the transition of the cylinder wake', *Physics of Fluids* **7**(4), 779–795.

Zhang, W., Wang, B., Ye, Z., and Quan, J. (2012), 'Efficient method for limit cycle flutter analysis based on nonlinear aerodynamic reduced-order models', *AIAA Journal* **50**(5), 1019–1028.

Zheng, P., Askham, T., Brunton, S. L., Kutz, J. N., and Aravkin, A. Y. (2019), 'Sparse relaxed regularized regression: SR3', *IEEE Access* **7**(1), 1404–1423.

Zhong, Y. D. and Leonard, N. (2020), 'Unsupervised learning of lagrangian dynamics from images for prediction and control', *Advances in Neural Information Processing Systems* **33**.

Zhou, K. and Doyle, J. C. (1998), *Essentials of Robust Control*, Prentice Hall, Upper Saddle River, NJ.

Zhou, Y., Fan, D., Zhang, B., Li, R., and Noack, B. R. (2020), 'Artificial intelligence control of a turbulent jet', *Journal of Fluid Mechanics* **897**, 1–46.

Zimmermann, R. and Görtz, S. (2010), 'Non-linear reduced order models for steady aerodynamics', *Procedia Computer Science* **1**(1), 165–174.

Zimmermann, R. and Görtz, S. (2012), 'Improved extrapolation of steady turbulent aerodynamics using a non-linear POD-based reduced order model', *The Aeronautical Journal* **116**(1184), 1079–1100.

Zimmermann, R., Vendl, A., and Görtz, S. (2014), 'Reduced-order modeling of steady flows subject to aerodynamic constraints', *AIAA Journal* **52**(2), 255–266.

Zou, H. and Hastie, T. (2005), 'Regularization and variable selection via the elastic net', *Journal of the Royal Statistical Society: Series B (Statistical Methodology)* **67**(2), 301–320.